Synthesis Series in Computational Electromagnetics

Synthesis Series in Computational Electromagnetics

Other volumes in this series include:

Volume 1

Mapped Vector Basis Function for Electromagnetic Integral Equations
Andrew F. Peterson
2006

MRTD (Multi Resolution Time Domain) Method in Electromagnetics
Nathan A. Bushyager, Manos M. Tentzeris
2005

Higher Order FDTD Schemes for Waveguide and Antenna Structures
Nikolaos V. Kantartzis, Theodoros D. Tsiboukis
2006

Volume 2

Computational Electronics
Dragica Vasileska and Stephen M. Goodnick
2006

Support Vector Machines for Antenna Array Processing and Electromagnetics
Manel Martínez-Ramón, Christos Christodoulou
2006

Introduction to the Finite Element Method in Electromagnetics
Anastasis C. Polycarpou
2006

Volume 3

Frequency Domain Hybrid Finite Element Methods for Electromagnetics
John L. Volakis, Kubilay Sertel, Brian C. Usner
2006

Exploitation of A Ship's Magnetic Field Signatures
John J. Holmes
2006

Perfectly Matched Layer (PML) for Computational Electromagnetics
Jean-Pierre Bérenger
2007

The Transmission-Line Modeling (TLM) Method in Electromagnetics
Christos Christopoulos
2006

Volume 4

Mellin-Transform Method for Integral Evaluation
Introduction and Applications to Electromagnetics
George Fikioris
2007

Integral Equation Methods for Electromagnetic and Elastic Waves
Weng Cho Chew, Mei Song Tong, Bin Hu
2009

Adaptive Mesh Refinement for Time-DomainNumerical Electromagnetics
Costas D. Sarris
2007

Volume 5

Modeling a Ship's Ferromagnetic Signatures
John J. Holmes
2007

Fundamentals of Electromagnetics 1: Internal Behavior of Lumped Elements
David Voltmer
2007

Fundamentals of Electromagnetics 2: Quasistatics and Waves
David Voltmer
2007

Volume 6

Particle Swarm Optimization: A Physics-Based Approach
Said M. Mikki, Ahmed A. Kishk
2008

Electromagnetic Scattering Using the Iterative Multiregion Technique
Mohamed H. Al Sharkawy, Veysel Demir, Atef Z. Elsherbeni
2007

Electromagnetics and Antenna Optimization Using Taguchi's Method
Wei-Chung Weng, Fan Yang, Atef Elsherbeni
2007

Three-Dimensional Integration and Modeling: A Revolution in RF and Wireless Packaging
Jong-Hoon Lee, Manos M. Tentzeris
2008

Volume 7

Modern EMC Analysis Techniques
Volume I: Time-Domain Computational Schemes
Nikolaos V. Kantartzis, Theodoros D. Tsiboukis
2008

Modern EMC Analysis Techniques
Volume II: Models and Applications
Nikolaos V. Kantartzis, Theodoros D. Tsiboukis
2008

Reduction of a Ship's Magnetic Field Signatures
John J. Holmes
2008

Volume 8

Introduction to Smart Antennas
Constantine A. Balanis, Panayiotis I. Ioannides
2007

Antennas withNon-Foster MatchingNetworks
James T. Aberle, Robert Loepsinger-Romak
2007

Implanted Antennas in Medical Wireless Communications
Yahya Rahmat-Samii, Jaehoon Kim
2006

Transient Signals on Transmission Lines
An Introduction to Non-Ideal Effects and Signal Integrity Issues in Electrical Systems
Andrew F. Peterson, Gregory D. Durgin
2009

Volume 9

Multiantenna Systems for MIMO Communications
Franco De Flaviis, Lluis Jofre, Jordi Romeu, AlfredGrau
2008

Reconfigurable Antennas
Jennifer T. Bernhard
2007

Electronically Scanned Arrays
Robert J. Mailloux
2007

Volume 10

Analysis and Design of Substrate Integrated Waveguide Using Efficient 2D Hybrid Method
Xuan HuiWu, Ahmed A. Kishk
2010

An Introduction to the Locally-Corrected Nyström Method
Andrew F. Peterson, Malcolm M. Bibby
2010

Generalized Transmission Line Method to Study the Far-zone Radiation of Antennas under a Multilayer Structure
Xuan HuiWu, Ahmed A. Kishk, AllenW. Glisson
2008

Narrowband Direction of Arrival Estimation for Antenna Arrays
Jeffrey Foutz, Andreas Spanias, Mahesh K. Banavar
2008

Synthesis Series in Computational Electromagnetics and Antennas Volume 1

Mapped Vector Basis Function for Electromagnetic Integral Equations
Andrew F. Peterson

MRTD (Multi Resolution Time Domain) Method in Electromagnetics
Nathan A. Bushyager, Manos M. Tentzeris

Higher Order FDTD Schemes for Waveguide and Antenna Structures
Nikolaos V. Kantartzis, Theodoros D. Tsiboukis

ISBN: 9781608453030

www.morganclaypool.com

A Publication in the Morgan & Claypool Publishers

Synthesis Series in Computational Electromagnetics and Antennas

Series Editor: Constantine A. Balanis

Series ISSN: 2152-2685

Synthesis Series in Computational Electromagnetics Volume 1

This volume includes:

Mapped Vector Basis Function for Electromagnetic Integral Equations
Andrew F. Peterson

MRTD (Multi Resolution Time Domain) Method in Electromagnetics
Nathan A. Bushyager, Manos M. Tentzeris

Higher Order FDTD Schemes for Waveguide and Antenna Structures
Nikolaos V. Kantartzis, Theodoros D. Tsiboukis

MAPPED VECTOR BASIS FUNCTIONS FOR ELECTROMAGNETIC INTEGRAL EQUATIONS

Andrew F. Peterson
School of Electrical and Computer Engineering
Georgia Institute of Technology

MORGAN & CLAYPOOL PUBLISHERS

ABSTRACT

The method-of-moments solution of the electric field and magnetic field integral equations (EFIE and MFIE) is extended to conducting objects modeled with curved cells. These techniques are important for electromagnetic scattering, antenna, radar signature, and wireless communication applications. Vector basis functions of the divergence-conforming and curl-conforming types are explained, and specific interpolatory and hierarchical basis functions are reviewed. Procedures for mapping these basis functions from a reference domain to a curved cell, while preserving the desired continuity properties on curved cells, are discussed in detail. For illustration, results are presented for examples that employ divergence-conforming basis functions with the EFIE and curl-conforming basis functions with the MFIE. The intended audience includes electromagnetic engineers with some previous familiarity with numerical techniques.

KEYWORDS

boundary element method, computational electromagnetics, integral equations, method of moments, parametric mapping

Contents

Preface

The discipline of computational electromagnetics has matured to the extent that there are now a number of commercial products in widespread use, and these tools are perceived to be useful to the community. In recent years, research has shifted from the basic foundations to specialized techniques, such as the various "fast methods" that accelerate the matrix solution process. Another research topic, which forms the focus of this monograph, is the use of higher order basis functions and curved cells to enhance the accuracy of the results. These techniques have not significantly penetrated the commercial marketplace.

My understanding of this subject has been largely influenced by two publications. The 1980 paper *Mixed finite elements in R3* by Jean-Claude Nedelec provided a comprehensive introduction to the two classes of vector basis functions considered in Chapters 3 and 4. The 1988 Ph.D. dissertation *Mixed-order Covariant Projection Finite Elements for Vector Fields* of Christopher W. Crowley contained a fairly complete discussion of the mapping procedures (for both divergence-conforming and curl-conforming functions) that I have expanded into Chapter 5. I am grateful for the efforts of these authors. Their papers arose from work directed toward the numerical solution of differential equations; it is hoped that the present text helps disseminate these ideas into the integral equation community. I am also grateful for the suggestions of John Shaeffer and Malcolm Bibby, who read an early draft of this manuscript.

A question remains to be answered: Are higher order techniques "better" than the status quo? At present, it is clear that high-accuracy results cannot be realized with low-order methods. It is also apparent that enhanced accuracy permits error estimation, improved confidence in numerical solutions, and the potential for "dialable" accuracy in the future. On the other hand, high accuracy will also require better representations for currents at all edges, and it will not prove meaningful without a commensurate improvement in the resolution of surface geometries and material parameters. As to which approach will prove best in the long run, the answer will be left to the user community.

CHAPTER 1

Introduction

Electromagnetics problems have been formulated in terms of integral equations since Pocklington analyzed wave propagation along wires in 1897, although the numerical solution of these equations (at least in the modern sense) is a relatively recent development. Rather than review the formulation and solution of these equations in detail, we refer the reader to a number of texts encompassing the scope of this discipline [1–6]. It is interesting that relatively sophisticated numerical techniques employing curved subsectional cells appeared more than three decades ago [7], but are not widely used even today. (Nor are they discussed in the standard texts [1–5] on the subject, except for a brief summary in [6].) One reason for this is that the majority of approaches in use rely on low-order polynomial basis functions, negating some of the accuracy advantages of curved cells. A second reason is that the implementation of curved cells in the full three-dimensional vector case involves a number of subtleties.

The present monograph attempts to walk the reader through the details of a curved-cell based discretization process. Our intended audience includes researchers and practitioners having some familiarity with flat-cell techniques. The scope of the text is restricted to perfectly conducting objects in three-dimensional space, with the surface of the object represented by a collection of contiguous patches. Wire structures are not considered, although the readers should be able to specialize these ideas to wires (or extend them to penetrable objects) without difficulty.

The remainder of Chapter 1 briefly reviews the formulation of two types of integral equation in widespread use for conducting bodies: the electric-field integral equation (EFIE) and the magnetic field integral equation (MFIE). The numerical discretization of these equations by the method of moments is also reviewed. Chapter 2 discusses several ways of representing curved-cell surface models. Chapters 3 and 4 provide motivation for the two types of vector basis functions in use for representing the surface current density within the integral equations, and reviews several families of these functions. Chapter 5 describes the manner in which those basis functions may be mapped to curved cells, while maintaining the critical continuity properties of the basis functions. Chapters 6 and 7 illustrate the complete numerical solution process, by applying the techniques to the EFIE and MFIE, respectively. A few numerical results are incorporated. A sample computer program that implements the EFIE approach is available as a companion to this text.

1.1 INTEGRAL EQUATIONS

Consider a perfect electric conducting body in an infinite homogeneous environment of permittivity ε and permeability μ. The body is illuminated by a sinusoidal steady-state source of electromagnetic radiation, having radian frequency ω, and all the time-varying quantities are represented by phasors with suppressed time dependence $e^{j\omega t}$. In the absence of the body, the source would produce an electric field $\bar{E}^{\text{inc}}(x, y, z)$ and a magnetic field $\bar{H}^{\text{inc}}(x, y, z)$ throughout the surrounding space. (These are denoted the *incident* fields.) In the presence of the conducting body, the fields are perturbed from these to the *total* fields $\bar{E}^{\text{tot}}(x, y, z)$ and $\bar{H}^{\text{tot}}(x, y, z)$. The perturbation can be accounted for by the presence of equivalent induced currents on the body.

The induced surface current density can be expressed as $\bar{J}(s, t)$, where s and t are parametric variables on the surface. [Henceforth, all references to the variable t are intended to denote position, not time.] The current density, if treated as a source function that exists in the absence of the conducting body, produces the *scattered* fields $\bar{E}^s(x, y, z)$ and $\bar{H}^s(x, y, z)$. The various fields are related by

$$\bar{E}^{\text{inc}} + \bar{E}^s = \bar{E}^{\text{tot}} \tag{1.1}$$

$$\bar{H}^{\text{inc}} + \bar{H}^s = \bar{H}^{\text{tot}} \tag{1.2}$$

Since the scattered fields are produced in infinite homogeneous space by $\bar{J}$, they are readily determined using any of the standard source-field relations,[1] such as

$$\bar{E}^s = \frac{\nabla(\nabla \bullet \bar{A}) + k^2\bar{A}}{j\omega\varepsilon} \tag{1.3}$$

$$\bar{H}^s = \nabla \times \bar{A} \tag{1.4}$$

where the wavenumber of the medium is given by $k = \omega\sqrt{\mu\varepsilon}$, and the magnetic vector potential function is

$$\bar{A}(x, y, z) = \iint_{\text{surface}} \bar{J}(s', t')\frac{e^{-jkR}}{4\pi R}ds'\,dt' \tag{1.5}$$

In (1.5), R is the distance from a point (s', t') on the surface to the point (x, y, z) where the field is evaluated. Throughout the text, primed coordinates will be used to describe the "source" of the electromagnetic field (the current density), while unprimed coordinates denote the "observer" location where that field is evaluated.

The total fields in (1.1) and (1.2) must satisfy the electromagnetic boundary conditions on the surface of the perfect electric body:

$$\hat{n} \times \bar{E}^{\text{tot}}|_{\text{surface}} = 0 \tag{1.6}$$

$$\hat{n} \times \bar{H}^{\text{tot}}|_{\text{surface}} = \bar{J} \tag{1.7}$$

where $\hat{n}$ is the outward normal unit vector, and in (1.7) the surface is approached from the exterior. By combining (1.1), (1.3), and (1.6), we obtain

$$\hat{n} \times \bar{E}^{\text{inc}}|_{\text{surface}} = -\hat{n} \times \left.\frac{\nabla(\nabla \bullet \bar{A}) + k^2\bar{A}}{j\omega\varepsilon}\right|_{\text{surface}} \tag{1.8}$$

[1]As in [6], (1.4) is employed as an alternative to the more traditional $\bar{B} = \nabla \times \bar{A}$.

This equation is one form of the *electric field integral equation*. (While it is in fact an integro-differential equation, the term "integral equation" is commonly used for simplicity.) The equation can be solved in principle for the surface current density $\bar{J}$ appearing within the magnetic vector potential $\bar{A}$.

By combining (1.2), (1.4), and (1.7), we obtain the *magnetic field integral equation*

$$\hat{n} \times \bar{H}^{\text{inc}}|_{\text{surface}} = \bar{J} - \hat{n} \times (\nabla \times \bar{A})|_{\text{surface}} \tag{1.9}$$

Equation (1.9) is enforced in the limiting case from the exterior of the surface of the body. This equation may also be solved for $\bar{J}$. Once $\bar{J}$ is determined by a solution of (1.8) or (1.9), the fields and other observable quantities associated with the electromagnetic scattering problem can be determined by direct calculation.

The MFIE in (1.9) is only applicable to closed conducting bodies where (1.7) is a valid boundary condition, but the EFIE can be used for thin shells as well as solid structures where (1.6) holds. This restriction on the MFIE is due to the fact that (1.7) has been specialized to the situation where the magnetic field is zero on the inner side of the surface. The form of the MFIE in (1.9) is not valid if the surface is open and the field is nonzero on both sides of the surface.

Both equations can fail for closed bodies whose surfaces coincide with resonant cavities. In that situation, neither (1.6) nor (1.7) alone is sufficient to guarantee that the surface current density is the proper exterior solution. However, the two equations can be combined together to form the *combined field integral equation*. In this manner, the two boundary conditions are both invoked, and are sufficient to produce the proper solution. Reference [6] contains a discussion of this issue, known as the *interior resonance* problem, as well as the combined field equation and alternative remedies.

There are other integral equations that describe electromagnetic fields [8]; the scope here is limited to the EFIE and MFIE. Surface integral equations for application to dielectric bodies involve integral operators similar in form to (1.8) and (1.9). The mapping procedures used with Eqs. (1.8) and (1.9) can easily be extended to other equations.

1.2 THE METHOD OF MOMENTS

The vector integral equations in (1.8) and (1.9) have the form

$$L\{\bar{J}\} = \bar{V} \tag{1.10}$$

where L denotes a linear vector operator, $\bar{V}$ the given excitation function, and the equality holds for tangential vector components over the scatterer surface. To solve (1.10) numerically, it must be projected from the continuous infinite-dimensional space to a finite dimensional subspace (or in less mathematical terms, converted into a matrix equation). This process is known as *discretization*. One procedure for accomplishing this discretization is known as the *method of moments* (MoM) [1].

Our implementation of the MoM procedure involves approximating the quantity to be determined, $\bar{J}$, by an expansion in linearly independent vector basis functions

$$\bar{J}(s,t) \cong \sum_{n=1}^{N} I_n \bar{B}_n(s,t) \tag{1.11}$$

The N coefficients $\{I_n\}$ in (1.11) become the unknowns to be determined. To obtain N linearly independent equations, both sides of (1.10) are multiplied (using a scalar or dot product) with suitable vector testing functions $\{\bar{T}_m(s,t)\}$, and integrated over the surface of the body, to obtain

$$\iint_{\text{surface}} \bar{T}_m \bullet L\{\bar{J}\} ds\, dt = \iint_{\text{surface}} \bar{T}_m \bullet \bar{V}\, ds\, dt, \quad m = 1, 2, \ldots, N \tag{1.12}$$

By enforcing the equation in this manner, the boundary condition imbedded in the integral equation (1.10) is imposed throughout the domain of the testing function.

The equations in (1.12) can be organized into a matrix of the form $\mathbf{AI} = \mathbf{V}$, where the $N \times N$ matrix $\mathbf{A}$ has entries

$$A_{mn} = \iint_{\text{surface}} \bar{T}_m \bullet L\{\bar{B}_n\} ds\, dt \tag{1.13}$$

and the $N \times 1$ column vector $\mathbf{V}$ has entries

$$V_m = \iint_{\text{surface}} \bar{T}_m \bullet \bar{V}\, ds\, dt \tag{1.14}$$

The numerical solution of the matrix equation yields the coefficients $\{I_n\}$.

The method of moments is one of several general-purpose discretization procedures; it is closely related to the *weighted residual* method, the *Rayleigh Ritz* method, and the *finite element* or *boundary element* methods. Readers may consult [1–6] for specific examples of its use with a variety of scalar and vector electromagnetics equations, and discussions of the accuracy of the approximate results. The procedure can obviously be generalized in many ways from the specific implementation outlined above—for instance, more testing functions than basis functions can be employed to yield an overdetermined system [9]. Since such aspects are beyond the scope of the present text, we limit our consideration to the above.

As expressed in (1.11)–(1.14), the basis and testing functions may be spread over the entire surface, or defined with a more limited domain of support. The use of entire-domain basis functions, which are spread over the entire surface, is usually limited to codes developed for specific geometries. Because they allow flexibility in the geometry, codes based on subsectional functions are applicable to a broad class of problems. Henceforth, we restrict our attention to surfaces that are divided into a relatively large number of cells, and basis and testing functions with their domain limited to one or two adjacent cells each. Chapter 2 describes subsectional surface representations.

The choice of suitable basis and testing functions is the primary task before the user of the MoM procedure. These functions must be selected with a specific operator $L\{\bar{J}\}$ in mind. For instance, constraints on the continuity of the representation in (1.11) are dictated by the operator and usually play a large role in the choice of $\{\bar{B}_n\}$. Chapters 3 and 4 summarize two classes of vector basis functions known respectively as *divergence-conforming* functions and *curl-conforming* functions. Divergence-conforming functions maintain normal-vector continuity from cell to cell on the surface, while curl-conforming functions maintain tangential-vector continuity between adjacent cells. These basis functions are easily defined in square or triangular cells. To implement more general cell shapes, the basis function definition must be translated to curved cells, while maintaining the appropriate type of continuity between adjacent cells. This procedure is what we call *mapping*. Because the type of cell-to-cell continuity is different for divergence-conforming and curl-conforming functions, the type of mapping (Chapter 5) is also different. Since the EFIE involves a divergence operator, it is natural to employ divergence-conforming basis functions for its discretization. In contrast, the MFIE incorporates a curl operation, which suggests the use of curl-conforming basis functions. Chapters 6 and 7 provide specific details for curved-cell discretizations of the EFIE and MFIE.

REFERENCES

[1] R. F. Harrington, *Field Computation by Moment Methods.* New York: IEEE Press, 1993. (Reprint of the original 1968 edition.)

[2] R. Mittra, Ed., *Computer Techniques for Electromagnetics.* New York: Hemisphere, 1987. (Reprint of the original 1973 edition.)

[3] N. Morita, N. Kumagai, and J. R. Mautz, *Integral Equation Methods for Electromagnetics.* Boston: Artech House, 1990. (Translation of the original 1987 edition.)

[4] J. J. H. Wang, *Generalized Moment Methods in Electromagnetics—Formulation and Solution of Integral Equations.* New York: Wiley, 1991.

[5] K. Umashankar and A. Taflove, *Computational Electromagnetics.* Boston: Artech House, 1993.

[6] A. F. Peterson, S. L. Ray, and R. Mittra, *Computational Methods for Electromagnetics*. New York: IEEE Press, 1998.

[7] D. L. Knepp and J. Goldhirsh, "Numerical analysis of electromagnetic radiation properties of smooth conducting bodies of arbitrary shape," *IEEE Trans. Antennas Propagat.*, vol. AP-20, pp. 383–388, 1972.

[8] H. Contopanagos, B. Dembart, M. Epton, J. J. Ottusch, V. Rokhlin, J. L. Visher, and S. M. Wandzura, "Well-conditioned boundary integral equations for three-dimensional electromagnetic scattering," *IEEE Trans. Antennas Propagat.*, vol. 50, pp. 1824–1830, 2002.

[9] M. M. Bibby and A. F. Peterson, "On the use of over-determined systems in the adaptive numerical solution of integral equations," *IEEE Trans. Antennas Propagat.*, vol. 53, pp. 2267–2273, July 2005.

CHAPTER 2

The Surface Model

The numerical representation of an object's surface by an idealized subsectional model determines how well that surface can be approximated—how smooth it is, its precise curvature, its actual location, etc. Any subsequent analysis is limited by the accuracy of the surface representation. Furthermore, a computer code used for electromagnetic analysis is usually tightly coupled to the specific type of representation: the number of nodes and degree of polynomial, for instance, used to describe each "patch" or subsection of the surface. In this chapter, the basic framework for modeling surfaces on a cell-by-cell basis is reviewed.

Scalar mappings to define curved cell shapes are widely used in connection with the finite element solution of differential equations [1–4]. A cell-by-cell parametric mapping was used in connection with the numerical solution of electromagnetic integral equations as early as 1972 [5]. The benefits of employing curved-cell models include the fact that often the design and manufacturing process employs such models; furthermore the improvement in accuracy can be substantial [6].

2.1 DIFFERENTIAL GEOMETRY

In all the examples under consideration, the surface of the structure being analyzed is represented by a collection of curved (if necessary) cells or patches. Each patch in x–y–z space is defined by a mapping from a square or triangular reference cell in u–v space. The mapping, and the resulting cell shape, is therefore determined by three functions $x(u, v)$, $y(u, v)$, and $z(u, v)$. These functions are different for each cell of the model. Various ways of systematically obtaining these functions from the surface coordinates of the desired cell are described in the sections of this chapter that follow. One way of describing the curved patch that results from this mapping is by the position vector

$$\bar{r}(u, v) = x(u, v)\hat{x} + y(u, v)\hat{y} + z(u, v)\hat{z} \tag{2.1}$$

The vector in (2.1) is directed from the origin (0, 0, 0) to various points (x, y, z) on the curved cell. These points are defined in turn by the parameters u and v. Tangent vectors along the patch may be expressed as a function of u and v as

$$\frac{\partial \bar{r}}{\partial u} = \frac{\partial x}{\partial u}\hat{x} + \frac{\partial y}{\partial u}\hat{y} + \frac{\partial z}{\partial u}\hat{z} \tag{2.2}$$

$$\frac{\partial \bar{r}}{\partial v} = \frac{\partial x}{\partial v}\hat{x} + \frac{\partial y}{\partial v}\hat{y} + \frac{\partial z}{\partial v}\hat{z} \tag{2.3}$$

A normal vector to the patch is obtained from

$$\frac{\partial \bar{r}}{\partial u} \times \frac{\partial \bar{r}}{\partial v} = \left(\frac{\partial y}{\partial u}\frac{\partial z}{\partial v} - \frac{\partial z}{\partial u}\frac{\partial y}{\partial v}\right)\hat{x} + \left(\frac{\partial z}{\partial u}\frac{\partial x}{\partial v} - \frac{\partial x}{\partial u}\frac{\partial z}{\partial v}\right)\hat{y} + \left(\frac{\partial x}{\partial u}\frac{\partial y}{\partial v} - \frac{\partial y}{\partial u}\frac{\partial x}{\partial v}\right)\hat{z} \tag{2.4}$$

If u or v is held constant, the preceding equations describe tangent curves on the surface. Appropriate differential length tangent vectors along these curves are given in x–y–z space by

$$d\bar{s} = \frac{\partial \bar{r}}{\partial u} du \tag{2.5}$$

$$d\bar{t} = \frac{\partial \bar{r}}{\partial v} dv \tag{2.6}$$

The differential surface area is

$$dS = |d\bar{s} \times d\bar{t}\,| = \left|\frac{\partial \bar{r}}{\partial u} \times \frac{\partial \bar{r}}{\partial v}\right| du\, dv \tag{2.7}$$

The magnitude of the cross product in (2.7) plays the same role as the determinant of a Jacobian matrix, to be defined in Chapter 5. We will denote this quantity by $D(u, v)$. Using (2.4), we obtain

$$D(u,v) = \left| \frac{\partial \bar{r}}{\partial u} \times \frac{\partial \bar{r}}{\partial v} \right|$$
$$= \sqrt{\left(\frac{\partial y}{\partial u}\frac{\partial z}{\partial v} - \frac{\partial z}{\partial u}\frac{\partial y}{\partial v} \right)^2 + \left(\frac{\partial z}{\partial u}\frac{\partial x}{\partial v} - \frac{\partial x}{\partial u}\frac{\partial z}{\partial v} \right)^2 + \left(\frac{\partial x}{\partial u}\frac{\partial y}{\partial v} - \frac{\partial y}{\partial u}\frac{\partial x}{\partial v} \right)^2} \tag{2.8}$$

This quantity may be used to express an integral over a curvilinear patch surface in terms of an integral over the reference cell. Suppose that the reference cell occupies the domain $-1 < u < 1, -1 < v < 1$. The integral over the curved patch is given by

$$\iint f[x, y, z] dS = \int_{-1}^{1}\int_{-1}^{1} f[x(u,v),\ y(u,v),\ z(u,v)] D(u,v) du\, dv \tag{2.9}$$

Other uses of $D(u, v)$ in the mapping process will be explored in Chapter 5.

For the electromagnetic analysis of interest, we desire to preserve the first-order vector continuity of the basis functions used within the numerical solution process (either normal or tangential components of the vector functions). In order to accomplish this using the methods of Chapter 5, the surface-patch models must be *conforming*. Conforming cells have their edges and corners aligned so that they yield a continuous surface (although not necessarily a smooth one). If the cells are defined by interpolation polynomials, for instance, adjacent cells must share all the nodes used on the common boundary to define either cell's shape. The corner of one cell will be aligned with the corner of the adjacent cell, not with the center of an adjacent cell's edge, and so on. Nonconforming models might have gaps between adjacent cells, and be unsuitable for providing a continuous expansion.

2.2 MAPPING FROM SQUARE CELLS USING LAGRANGIAN INTERPOLATION POLYNOMIALS

Lagrangian interpolation polynomials find widespread use in computational analysis. A set of M polynomials, each of degree $M-1$, may be defined on an interval $u_1 < u < u_M$ by the expression

$$P_i^M(u) = \frac{(u-u_1)(u-u_2)\cdots(u-u_{i-1})(u-u_{i+1})\cdots(u-u_M)}{(u_i-u_1)(u_i-u_2)\cdots(u_i-u_{i-1})(u_i-u_{i+1})\cdots(u_i-u_M)} \tag{2.10}$$

where $\{u_i\}$ denote the locations of the interpolation points or *nodes*. These functions satisfy the conditions that

$$P_i^M(u_i) = 1 \tag{2.11}$$

and

$$P_i^M(u_j) = 0, \quad j \neq i \tag{2.12}$$

Since only one function is nonzero at each node, they provide an interpolation. The entire set of M polynomials must be used over each interval to provide a degree $M-1$ representation.

For example, the two Lagrangian polynomials of degree 1 are given by

$$P_1^2(u) = \frac{(u-u_2)}{(u_1-u_2)} \tag{2.13}$$

$$P_2^2(u) = \frac{(u-u_1)}{(u_2-u_1)} \tag{2.14}$$

while the three quadratic polynomials are given by

$$P_1^3(u) = \frac{(u-u_2)(u-u_3)}{(u_1-u_2)(u_1-u_3)} \tag{2.15}$$

$$P_2^3(u) = \frac{(u-u_1)(u-u_3)}{(u_2-u_1)(u_2-u_3)} \tag{2.16}$$

$$P_3^3(u) = \frac{(u-u_1)(u-u_2)}{(u_3-u_1)(u_3-u_2)} \tag{2.17}$$

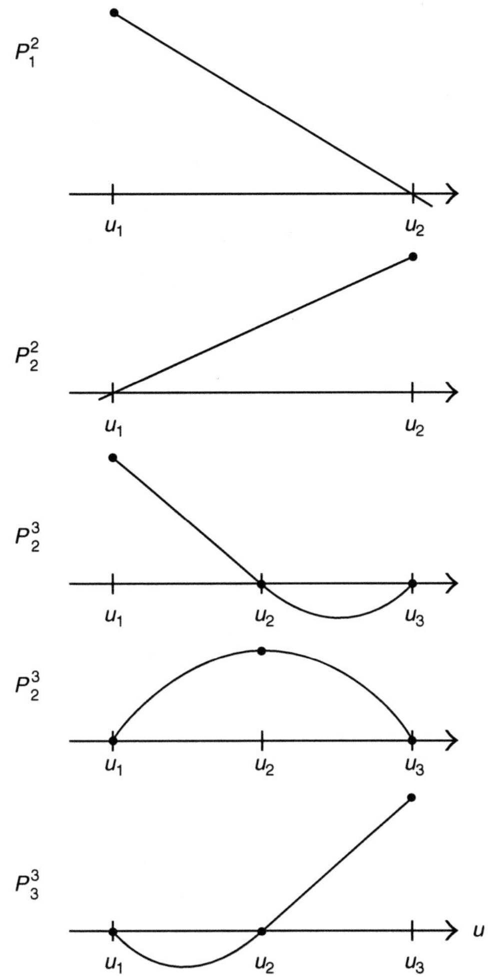

FIGURE 2.1: Lagrangian polynomials.

These are sketched in Fig. 2.1. Although the nodes may be located arbitrarily, in practice they are typically distributed uniformly throughout the interval $u_1 < u < u_M$.

Consider a square reference cell on the domain $-1 < u < 1$, $-1 < v < 1$, as shown in Fig. 2.2. [The domain $0 < u < 1$, $0 < v < 1$ would work equally well; our choice appears more frequently in the finite element literature and will be used throughout this text as the reference coordinates for square domains.] A set of

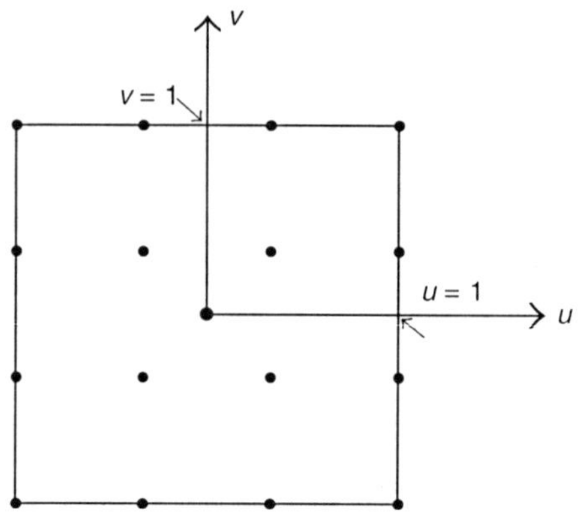

FIGURE 2.2: Interpolation points for $M = 4$.

two-dimensional Lagrangian interpolation polynomials of degree $M - 1$ may be defined by the product rule

$$S_{ij}(u, v) = P_i^M(u) P_j^M(v), \quad i, j = 1, 2, \ldots, M \tag{2.18}$$

Each of these two-dimensional functions involves M^2 nodes, and each interpolate at one of the nodes. The nodes for $M = 4$ are illustrated in Fig. 2.2 for the case when they are uniformly distributed throughout the domain.

To illustrate the use of Lagrangian interpolation polynomials for the definition of a curvilinear patch, the following section considers the specific case of quadratic polynomials in more detail.

2.3 A SPECIFIC EXAMPLE: QUADRATIC POLYNOMIALS MAPPED FROM A SQUARE REFERENCE CELL

The use of interpolation polynomials facilitates a representation of the desired object in terms of some number of coordinates on its surface. As an example, suppose a piecewise-quadratic model is to be used. It is convenient to define a quadratic cell that coincides with a portion of the actual surface at nine coordinate values. Using the set of Lagrangian interpolation polynomials for $M = 3$, defined by (2.18), a curved patch can be expressed by the coordinates

$$
\begin{aligned}
x &= \sum_{i=1}^{3}\sum_{j=1}^{3} x_{ij} S_{ij}(u, v) \\
y &= \sum_{i=1}^{3}\sum_{j=1}^{3} y_{ij} S_{ij}(u, v) \\
z &= \sum_{i=1}^{3}\sum_{j=1}^{3} z_{ij} S_{ij}(u, v)
\end{aligned}
\tag{2.19}
$$

where the functions $\{S_{ij}\}$ denote quadratic Lagrangian interpolation polynomials for $M = 3$. The nine specific interpolation polynomials, assuming that their nodes are distributed uniformly throughout the domain $-1 < u < 1$, $-1 < v < 1$, are

$$S_{11} = \frac{u(u-1)}{2}\frac{v(v-1)}{2} \tag{2.20}$$

$$S_{12} = \frac{u(u-1)}{2}(1+v)(1-v) \tag{2.21}$$

$$S_{13} = \frac{u(u-1)}{2}\frac{v(v+1)}{2} \tag{2.22}$$

$$S_{21} = (1+u)(1-u)\frac{v(v-1)}{2} \tag{2.23}$$

$$S_{22} = (1+u)(1-u)(1+v)(1-v) \tag{2.24}$$

$$S_{23} = (1+u)(1-u)\frac{v(v+1)}{2} \tag{2.25}$$

$$S_{31} = \frac{u(u+1)}{2}\frac{v(v-1)}{2} \tag{2.26}$$

$$S_{32} = \frac{u(u+1)}{2}(1+v)(1-v) \tag{2.27}$$

$$S_{33} = \frac{u(u+1)}{2}\frac{v(v+1)}{2} \tag{2.28}$$

In (2.19), each two-dimensional interpolation polynomial is scaled by a coordinate (x_{ij}, for instance) that usually is an actual value from the surface being modeled. Because of the interpolatory nature of the polynomials, the value of x returned at that node will be that same value (x_{ij}). Thus, the representation in (2.19) will provide an approximation of the surface that coincides with the true surface at the nine nodes. Eight of the nine nodes are located on the cell boundary, with the other node placed in the interior.

Equation (2.19) defines a single cell. Normally, a large number of cells are required to adequately represent a complicated surface. Cells must be smaller in regions where the surface is highly curved to provide an accurate representation.

Each variable within the representation may be explicitly written in terms of the interpolation polynomials. By expanding (2.19), $x(u, v)$ may be written as

$$\begin{aligned} x = x_{22} &+ \frac{u}{2}(x_{32} - x_{12}) + \frac{v}{2}(x_{23} - x_{21}) + \frac{uv}{4}(x_{11} - x_{13} - x_{31} + x_{33}) \\ &+ \frac{u^2}{2}(x_{12} - 2x_{22} + x_{32}) + \frac{v^2}{2}(x_{21} - 2x_{22} + x_{23}) \\ &+ \frac{uv^2}{4}(-x_{11} + 2x_{12} - x_{13} + x_{31} - 2x_{32} + x_{33}) \\ &+ \frac{u^2v}{4}(-x_{11} + 2x_{21} - x_{31} + x_{13} - 2x_{23} + x_{33}) \\ &+ \frac{u^2v^2}{4}(x_{11} - 2x_{12} + x_{13} - 2x_{21} + 4x_{22} - 2x_{23} + x_{31} - 2x_{32} + x_{33}) \end{aligned} \tag{2.29}$$

Derivatives with respect to u and v may be obtained as

$$\begin{aligned} \frac{\partial x}{\partial u} = &\frac{x_{32} - x_{12}}{2} + u(x_{12} - 2x_{22} + x_{32}) + \frac{v}{4}(x_{11} - x_{13} - x_{31} + x_{33}) \\ &+ \frac{uv}{2}(-x_{11} + 2x_{21} - x_{31} + x_{13} - 2x_{23} + x_{33}) \\ &+ \frac{v^2}{4}(-x_{11} + 2x_{12} - x_{13} + x_{31} - 2x_{32} + x_{33}) \\ &+ \frac{uv^2}{2}(x_{11} - 2x_{12} + x_{13} - 2x_{21} + 4x_{22} - 2x_{23} + x_{31} - 2x_{32} + x_{33}) \end{aligned} \tag{2.30}$$

and

$$\begin{aligned} \frac{\partial x}{\partial v} = &\frac{x_{23} - x_{21}}{2} + \frac{u}{4}(x_{11} - x_{13} - x_{31} + x_{33}) + v(x_{21} - 2x_{22} + x_{23}) \\ &+ \frac{uv}{2}(-x_{11} + 2x_{12} - x_{13} + x_{31} - 2x_{32} + x_{33}) \\ &+ \frac{u^2}{4}(-x_{11} + 2x_{21} - x_{31} + x_{13} - 2x_{23} + x_{33}) \\ &+ \frac{u^2v}{2}(x_{11} - 2x_{12} + x_{13} - 2x_{21} + 4x_{22} - 2x_{23} + x_{31} - 2x_{32} + x_{33}) \end{aligned} \tag{2.31}$$

Expressions for $y(u, v)$ and $z(u, v)$ and their derivatives are identical to the above, with x replaced by y or z in all occurrences within (2.29)–(2.31), and the subscripts kept the same.

Equations (2.30), (2.31), and the analogous equations involving $y(u, v)$ and $z(u, v)$, provide explicit components of the tangent vectors in (2.2) and (2.3), and the function $D(u, v)$ used in integrals such as (2.9). These quantities are all determined by the mapping of (2.19).

2.4 MAPPING FROM TRIANGULAR CELLS VIA INTERPOLATION POLYNOMIALS

Triangular cells provide additional flexibility when modeling arbitrary surfaces, and also permit a mapping based on interpolation polynomials. Consider the unit right triangle depicted in Fig. 2.3, occupying the domain $0 \leq u \leq 1, 0 \leq v \leq 1$, with $u + v \leq 1$. For this domain, the usual coordinates (u, v) happen to coincide with two of the three so-called *simplex* coordinates (u, v, w) often used to represent quantities on triangles [1–4]. Each simplex coordinate is the relative distance from one side of a triangle to the opposing corner, with value 0 at the side and value 1 at the corner. The third coordinate can be obtained in this case as

$$w = 1 - u - v \tag{2.32}$$

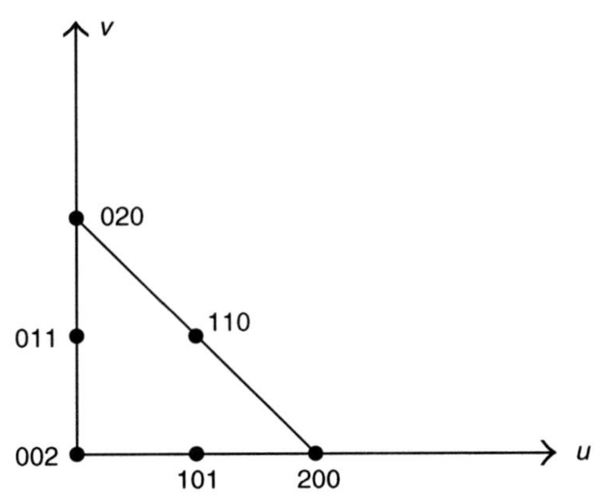

FIGURE 2.3: The indices ijk correspond to the interpolation points of S_{ijk} given in (2.39)–(2.44).

and is clearly linearly dependent on the other two. Simplex coordinates offer a symmetrical way of describing a point within the triangle, and are convenient variables for use in developing interpolation polynomials for triangular domains.

As in the previous sections, interpolation polynomials interpolate at specific nodes within the triangle. For linear polynomials, the obvious interpolation points are the three corners. For a polynomial of degree M, uniformly spaced node locations can be succinctly expressed in terms of simplex coordinates as

$$\left(\frac{i}{M}, \frac{j}{M}, \frac{k}{M}\right), \quad i, j, k = 0, 1, \ldots, M \tag{2.33}$$

with the restriction that $i + j + k = M$. There are a total of

$$\frac{(M+1)(M+2)}{2} \tag{2.34}$$

nodes involved.[1]

Two-dimensional interpolation polynomials are defined in simplex coordinates by the product rule

$$S_{ijk}(u, v, w) = R_i^{(M)}(u) R_j^{(M)}(v) R_k^{(M)}(w) \tag{2.35}$$

where $R_m^{(M)}$ is the Silvester polynomial [3]

$$R_m^{(M)}(u) = \begin{cases} 1 & m = 0 \\ \dfrac{1}{m!} \displaystyle\prod_{i=0}^{m-1} (Mu - i) & m > 0 \end{cases} \tag{2.36}$$

The Silvester polynomial exhibits m equally spaced zeros located at $u = 0, 1/M, \ldots, (m-1)/M$ and has unity value at $u = m/M$. The interpolation polynomial S_{ijk} has unity value at node ijk and vanishes at all the other nodes within the triangle.

Equations (2.35) and (2.36) provide a general framework for interpolation polynomials of any degree. The following section illustrates a specific example.

[1] Readers may observe that these nodes are the locations of sample points for Newton–Cotes quadrature rules for triangular domains.

2.5 EXAMPLE: QUADRATIC POLYNOMIALS MAPPED FROM A TRIANGULAR REFERENCE CELL

Consider a curvilinear patch defined in terms of six quadratic interpolation polynomials on a triangular domain, distributed around the reference cell as illustrated in Fig. 2.3. The patch coordinates may be expressed as

$$
\begin{aligned}
x &= \sum_{i=0}^{2} \sum_{j=0}^{2} x_{ijk} S_{ijk}(u, v, w) \\
y &= \sum_{i=0}^{2} \sum_{j=0}^{2} y_{ijk} S_{ijk}(u, v, w) \\
z &= \sum_{i=0}^{2} \sum_{j=0}^{2} z_{ijk} S_{ijk}(u, v, w)
\end{aligned}
\tag{2.37}
$$

where the index k is defined by

$$k = 2 - i - j \tag{2.38}$$

As in section 2.3, the coefficients x_{ijk}, y_{ijk}, and z_{ijk} are usually points on the desired surface. From Fig. 2.3, it should be apparent that these points define the boundary of the patch. (For polynomials of greater degree than quadratic, there are also interior points involved.) The six interpolation polynomials in (2.37) are defined

$$S_{200} = (2u - 1)u \tag{2.39}$$

$$S_{020} = (2v - 1)v \tag{2.40}$$

$$S_{002} = (2w - 1)w \tag{2.41}$$

$$S_{110} = 4uv \tag{2.42}$$

$$S_{101} = 4uw \tag{2.43}$$

$$S_{011} = 4vw \tag{2.44}$$

where (u, v, w) denote simplex coordinates. Since the dummy variable w is dependent on the other two, (2.32) may be used to redefine these functions in terms of u and v.

The explicit representation for S_{ijk} may be used to obtain

$$\begin{aligned} x = {} & x_{002} + u(4x_{101} - 3x_{002} - x_{200}) + v(4x_{011} - 3x_{002} - x_{020}) \\ & + u^2(2x_{200} + 2x_{002} - 4x_{101}) + uv(4x_{002} + 4x_{110} - 4x_{101} - 4x_{011}) \\ & + v^2(2x_{020} + 2x_{002} - 4x_{011}) \end{aligned} \tag{2.45}$$

where $y(u, v)$ and $z(u, v)$ have an identical form with x replaced by y or z throughout. Derivatives are obtained as

$$\begin{aligned} \frac{\partial x}{\partial u} = {} & 4x_{101} - 3x_{002} - x_{200} + u(4x_{200} + 4x_{002} - 8x_{101}) \\ & + v(4x_{002} + 4x_{110} - 4x_{101} - 4x_{011}) \end{aligned} \tag{2.46}$$

$$\begin{aligned} \frac{\partial x}{\partial v} = {} & 4x_{011} - 3x_{002} - x_{020} + u(4x_{002} + 4x_{110} - 4x_{101} - 4x_{011}) \\ & + v(4x_{020} + 4x_{002} - 8x_{011}) \end{aligned} \tag{2.47}$$

Derivatives of y and z are similar, with x replaced by y or z on the right-hand sides of (2.46) and (2.47).

Equations (2.46) and (2.47), and the corresponding equations obtained from $y(u, v)$ and $z(u, v)$, provide components of the tangent vectors in (2.2) and (2.3), and the function $D(u, v)$ used in the integral (2.9).

It should be noted that the quadratic interpolation functions defined on quadrilateral cells (Section 2.3) and triangular cells (Section 2.5) exhibit the same behavior along the cell boundaries and thus can be mixed into the same surface model. This combination will be employed with the examples used for illustration in Chapters 6 and 7.

2.6 CONSTRAINTS ON NODE DISTRIBUTION

There are obvious limitations associated with the type of mappings defined in (2.19) and (2.37). For polynomial degrees greater than one, the nodes must be fairly uniformly spaced on the patch in x–y–z space, or the mapping may not be one-to-one: cells may be folded back over themselves. (References [1–4] address this issue.) In the following chapters, we will assume that the node distribution is adequate to ensure a one-to-one mapping. This is straightforward to ensure for simple surface shapes, but may require additional effort for complex bodies.

2.7 HERMITIAN MAPPING FROM SQUARE CELLS

A second limitation associated with the patch-by-patch Lagrangian mappings is that they provide a smooth (highly differentiable) representation within each patch, but do nothing to ensure smoothness beyond first-order continuity across patch boundaries. Electromagnetic fields are sensitive to the presence of corners or edges, and even slight curvature discontinuities. Therefore it may be necessary to consider a means for ensuring patch-to-patch derivative continuity. One possible way of achieving this is to employ a mapping based on Hermite polynomials. We briefly review the procedure for quadrilateral cell shapes.

The lowest degree Hermitian polynomials on the interval $-1 < u < 1$ consist of the four functions

$$H_1^0(u) = \frac{(1-u)^2(2+u)}{4} \tag{2.48}$$

$$H_2^0(u) = \frac{(1+u)^2(2-u)}{4} \tag{2.49}$$

$$H_1^1(u) = \frac{(1-u)^2(1-u)}{4} \tag{2.50}$$

$$H_2^1(u) = \frac{(-1+u)^2(1+u)}{4} \tag{2.51}$$

The functions in (2.48) and (2.49) interpolate at $u = -1$ and $u = +1$, respectively, while those in (2.50) and (2.51) interpolate to the first derivative at those points. (In other words, they equal zero at those points but their derivative equals unity there, while the first derivatives of (2.48) and (2.49) are zero there.) If extended to a square domain $-1 < u < 1$, $-1 < v < 1$, the corresponding two-dimensional functions can be written as

$$S_{ij}^{pq}(u, v) = H_i^p(u) H_j^q(v) \tag{2.52}$$

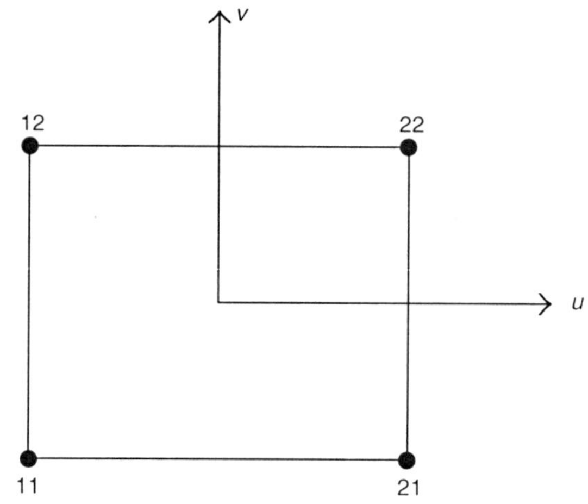

FIGURE 2.4: The indices ij correspond to (2.52).

By allowing indices i and j to equal 1 or 2, while p and q assume values 0 or 1, Eq. (2.52) defines 16 cubic polynomials. Figure 2.4 illustrates the interpolation points in the reference cell.

Using the 16 functions in (2.52) a mapping can be constructed that ensures the continuity of the surface from patch to patch, and its first derivative from patch to patch. Such a mapping has an x-coordinate of the form

$$\begin{aligned} x &= x_{11} S_{11}^{00}(u, v) + x_{12} S_{12}^{00}(u, v) + x_{21} S_{21}^{00}(u, v) + x_{22} S_{22}^{00}(u, v) \\ &+ \left.\frac{\partial x}{\partial s}\right|_{11} S_{11}^{10}(u, v) + \left.\frac{\partial x}{\partial s}\right|_{12} S_{12}^{10}(u, v) + \left.\frac{\partial x}{\partial s}\right|_{21} S_{21}^{10}(u, v) + \left.\frac{\partial x}{\partial s}\right|_{22} S_{22}^{10}(u, v) \\ &+ \left.\frac{\partial x}{\partial t}\right|_{11} S_{11}^{01}(u, v) + \left.\frac{\partial x}{\partial t}\right|_{12} S_{12}^{01}(u, v) + \left.\frac{\partial x}{\partial t}\right|_{21} S_{21}^{01}(u, v) + \left.\frac{\partial x}{\partial t}\right|_{22} S_{22}^{01}(u, v) \\ &+ \left.\frac{\partial^2 x}{\partial s\, \partial t}\right|_{11} S_{11}^{11}(u, v) + \left.\frac{\partial^2 x}{\partial s\, \partial t}\right|_{12} S_{12}^{11}(u, v) + \left.\frac{\partial^2 x}{\partial s\, \partial t}\right|_{21} S_{21}^{11}(u, v) + \left.\frac{\partial^2 x}{\partial s\, \partial t}\right|_{22} S_{22}^{11}(u, v) \end{aligned} \tag{2.53}$$

where s and t are tangential variables (with units of length) along the directions of parameters u and v. (These may be replaced by u and v, or any other convenient parameters, as long as the interpretation of the coefficients of the derivative terms is not an issue.) Similar expressions can be constructed for y and z. Explicit derivatives must be obtained from (2.53), following an approach similar to that used in Sections

2.3 and 2.5, to obtain $D(u, v)$ in (2.9) and other parameters to be identified in Chapter 5.

2.8 CONNECTIVITY

The representation of a surface in terms of cells involves more than just a list of the coordinates (nodes). The model also requires *connectivity*. In its simplest form, connectivity is provided by a pointer array that identifies the specific nodes belonging to each cell (Fig. 2.5). Each node and each cell in the model are assigned an integer; the connectivity array assigns a group of nodes to each cell. In practice, the connectivity array simultaneously provides an ordering of the nodes within each cell to eliminate ambiguity as to the cell orientation, and to clearly indicate which

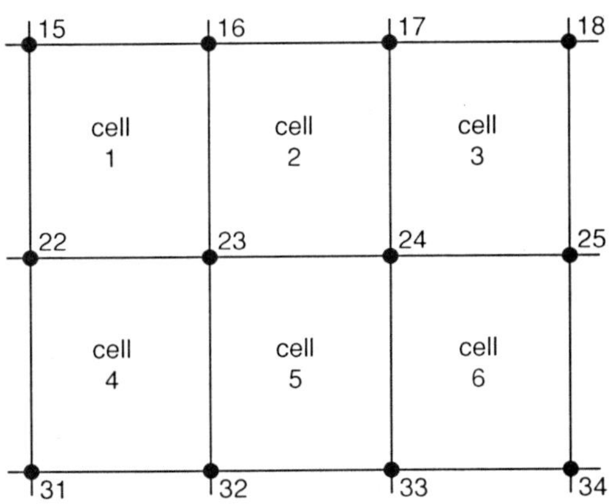

cell index	node 1	node 2	node 3	node 4
1	22	23	15	16
2	23	24	16	17
3	24	25	17	18
4	31	32	22	23
5	32	33	23	24
6	33	34	24	25
.				
.				
.				

FIGURE 2.5: Portion of mesh and pointer indicating the cell-to-node connectivity.

face of the cell is inside or outside. For cells with relatively few nodes, a clockwise or counter-clockwise orientation of nodes around the cell center may be adequate for this purpose. For scalar problems (and especially in connection with scalar finite element solutions of differential equations), the cell-to-node connectivity is sufficient to describe the geometry of the structure. It is often convenient to incorporate a separate list that identifies nodes residing on the outer boundary of the domain of interest.

For problems involving vector quantities, it is usually necessary to extend the connectivity to the cell edges. Each edge in the model is assigned an integer index. Additional pointers may be generated that identify edges associated with cells, nodes associated with edges, or cells associated with edges. For simple domains, these pointers can be generated automatically from the basic cell-to-node pointer described above. Their availability simplifies the computer program that performs the analysis.

REFERENCES

[1] D. S. Burnett, *Finite Element Analysis.* Reading, MA: Addison-Wesley, 1987.

[2] O. C. Zienkiewicz and R. L. Taylor, *The Finite Element Method.* London: McGraw-Hill, 1988.

[3] P. P. Silvester and R. L. Ferrari, *Finite Elements for Electrical Engineers.* Cambridge: Cambridge University Press, 1990.

[4] J. E. Akin, *Finite Elements for Analysis and Design.* San Diego, CA: Academic Press, 1994.

[5] D. L. Knepp and J. Goldhirsh, "Numerical analysis of electromagnetic radiation properties of smooth conducting bodies of arbitrary shape," *IEEE Trans. Antennas Propagat.*, vol. AP-20, pp. 383–388, 1972.

[6] M. I. Sancer, R. L. McClary, and K. J. Glover, "Electromagnetic computation using parametric geometry" *Electromagnetics*, vol. 10, pp. 85–103, 1990.

CHAPTER 3

Divergence-Conforming Basis Functions

The vector basis functions used in electromagnetic modeling usually belong in one of two classes: divergence-conforming and curl-conforming. This chapter considers the former, as defined for square and triangular reference cells. Curl-conforming functions will be the topic of Chapter 4. As will be explained in Chapter 4, these two types of functions are closely related, and therefore developments and extensions of basis functions of one type motivate a simultaneous development for the other type. Similarly, the literature on these two types of basis functions includes many interrelated articles.

Most integral equation numerical techniques reported during the past two decades have employed $p = 0$ divergence conforming basis functions. The $p = 0$ functions for square and triangular cells are presented below, and used for illustration in the EFIE implementation of Chapter 6. However, the advantages associated with curvilinear cell shapes are not likely to be fully realized without the use of higher order bases. Therefore, methods for constructing basis functions of higher polynomial degrees are also described and tables of these functions are included.

3.1 CHARACTERISTICS OF VECTOR FIELDS AND VECTOR BASIS FUNCTIONS

The term *field* is used in engineering to denote any function of position (which may also be a function of time). A *scalar field* is a function associating a parameter or number (some scalar quantity) with each point. One example of a scalar field is the temperature throughout some region. Electromagnetic fields are *vector* quantities, meaning that a direction in x−y−z space is also associated with their value at each point. The velocity of water throughout a river is another example of a vector field. In the present context the vector fields are phasor quantities representing a sinusoidal time dependence, and are consequently complex valued functions of position, with a direction at each point.

Vector fields exhibit two properties: divergence and curl. *Divergence* refers to the tendency of the field to spread apart at any point, and is obtained for a general 3D vector $\bar{B}$ by the operation

$$\nabla \bullet \bar{B} = \frac{\partial B_x}{\partial x} + \frac{\partial B_y}{\partial y} + \frac{\partial B_z}{\partial z} \tag{3.1}$$

It is also of interest to consider the surface divergence of a vector function $\bar{B}$ that is tangential to some surface, which can be expressed as

$$\nabla_s \bullet \bar{B} = \frac{\partial B_s}{\partial s} + \frac{\partial B_t}{\partial t} \tag{3.2}$$

where s and t are local orthogonal variables with units of length along the surface. The divergence is a scalar quantity.

The *curl* of a vector field is related to the tendency of the field to rotate or twist about a point, and will be defined mathematically in Chapter 4. Since there are three planes in which the vector field can twist, the curl of a general 3D vector is itself expressed as a vector quantity. It is also possible to define a surface curl operation, which can be thought of as a scalar property associated with a vector function tangential to the surface.

The importance of the divergence and curl of electromagnetic fields should be apparent from a study of the key equations describing electromagnetic fields—those named for James Clerk Maxwell. In their modern form for phasor fields, they relate the curl of each field to the other field and to a current density source, and the divergence of each field to a charge density source.

Within numerical techniques for electromagnetics, basis functions are used to represent the fields directly or the surface currents that produce the fields. Historically, numerical procedures of this type were initially developed for scalar quantities, such as temperature in heat transfer problems or the voltage field in electrostatics. It was natural to develop families of scalar basis functions for the purpose of representing scalar fields. In principle, each component of a vector field can also be represented by scalar basis functions. However, there are difficulties with representing vector fields this way. For general cell shapes, a component-by-component representation is not convenient for imposing continuity and boundary conditions. Nor does such a representation allow the user to easily control the divergence or curl of the quantities being represented.

In recent years, a wide variety of vector basis functions have been developed to address some of the apparent drawbacks of scalar representations of vector fields. Each basis function provides a local vector direction typically oriented along or across cell edges. The simplest functions exhibit either a constant divergence or a constant curl within a cell. The basis functions fall into two classes: divergence-conforming and curl-conforming functions.

3.2 WHAT DOES DIVERGENCE-CONFORMING MEAN?

The term *conforming* was defined in Chapter 2, where it was used to describe a subsectional model of a surface that maintained first order continuity (no gaps between cells). The term *divergence-conforming* is used to denote a function $\bar{B}$ that maintains the first-order continuity needed by the divergence operator in (3.1) or (3.2). If this condition is met, it will result in the complete absence of Dirac delta functions in $\nabla \cdot \bar{B}$. In practice, discontinuities associated with a basis function for subsectional cells occur at the cell edges. The divergence operator involves a differentiation along the vector direction of $\bar{B}$. Since the normal-vector component of the function is differentiated in the normal direction (across the cell edge), the important continuity is that of the normal component. It does not matter if the function maintains cell-to-cell tangential-vector continuity, since the divergence operation does not differentiate the tangential component across the edge. Consequently, *a divergence-conforming basis function is one that maintains first-order normal-vector continuity across cell edges*, but not necessarily any tangential-vector continuity.

The complementary behavior is provided by *curl-conforming* basis functions. As described in Chapter 4, curl-conforming functions maintain tangential-vector continuity across cell edges.

3.3 HISTORY OF THE USE OF DIVERGENCE-CONFORMING BASIS FUNCTIONS

Divergence-conforming basis functions were first investigated for finite element solutions by Raviart and Thomas [1]. The lowest order ($p = 0$) functions were developed independently by Glisson for numerical solutions of the EFIE [2], and they have been widely used within the electromagnetics community [3, 4]. These low-order functions were generalized to arbitrary polynomial degree p by Nedelec [5], initially for use with time domain finite element solutions of Maxwell's equations. Nedelec proposed a family of divergence-conforming spaces that could be adapted to basis functions of any polynomial degree, for cells of triangular, tetrahedral, quadrilateral, or hexahedral shape. The $p = 0$ functions for rectangular and triangular cells are known as "rooftop" (or "triangular rooftop") functions, because of their obvious similarity to the roof of an A-frame house (Fig. 3.1). The triangular cell $p = 0$ functions are also known as RWG basis functions after the authors of [4].

Most types of higher order basis functions are either *interpolatory* or *hierarchical.* The coefficient of each interpolatory function represents one vector component of the unknown quantity at a specific point within the domain. The basis function "interpolates" at that point. Interpolatory representations usually involve a number of basis functions of the same polynomial degree p. To improve the order of the representation, all the interpolatory functions are exchanged for functions of greater

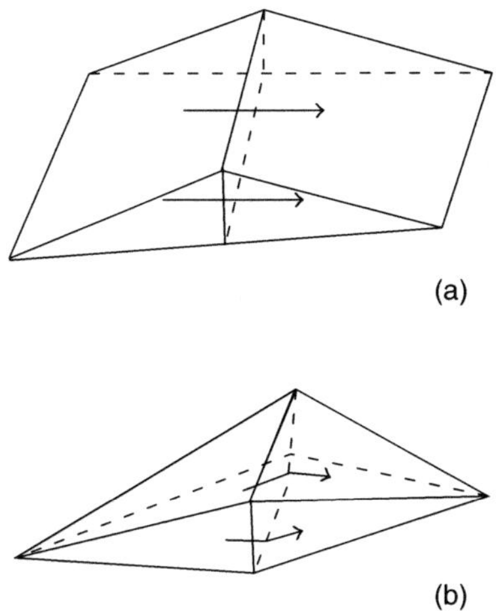

FIGURE 3.1: Rooftop function straddling two square cells (a) or two triangular cells (b). The vector direction is roughly indicated by the arrows.

degree. Interpolatory functions offer relatively good linear independence, and their coefficients have the simple physical interpretation as being the unknown current density or field at various locations.

Hierarchical functions, in contrast, are designed to build systematically upon basis functions of lower order. If the lowest-order members provide a constant representation, the next order will provide the linear degrees of freedom, followed by the quadratic, and so on. Each order is added to the previous orders, so the computations associated with the lower order functions may not need to be repeated as the representation is refined. Hierarchical functions make it more efficient to refine the representation to higher order, but usually provide weaker linear independence than interpolatory functions. In addition, the coefficients of hierarchical functions have no physical interpretation (the full set of functions must be superimposed to obtain the field or current density at some location). It is more difficult in some cases to ensure the satisfaction of boundary conditions or continuity conditions with hierarchical representations [6].

A number of different interpolatory and hierarchical families of higher order basis functions have been proposed for quadrilateral and triangular cell shapes. Interpolatory functions of arbitrary order, which have the rooftop type of basis as the $p = 0$ members, will be summarized in the sections following immediately. Hierarchical functions are considered in Sections 3.9 and 3.10.

3.4 BASIS FUNCTIONS OF ORDER $p = 0$ FOR A SQUARE REFERENCE CELL

Figure 3.2 shows a square reference cell occupying the region $-1 < u < 1, -1 < v < 1$. Four divergence-conforming basis functions of the lowest order can be defined within this cell as

$$\bar{R}_1^{\text{div}} = \frac{u-1}{2}\hat{u} \tag{3.3}$$

$$\bar{R}_2^{\text{div}} = \frac{u+1}{2}\hat{u} \tag{3.4}$$

$$\bar{R}_3^{\text{div}} = \frac{v-1}{2}\hat{v} \tag{3.5}$$

$$\bar{R}_4^{\text{div}} = \frac{v+1}{2}\hat{v} \tag{3.6}$$

Each of these functions has a nonzero normal-vector component along one cell edge, and contributes no normal component along any other edge. The normal components are constant, and it is convenient to think of each function as interpolating at the center of the appropriate edge. As an example, the function $\bar{R}_1$ is zero

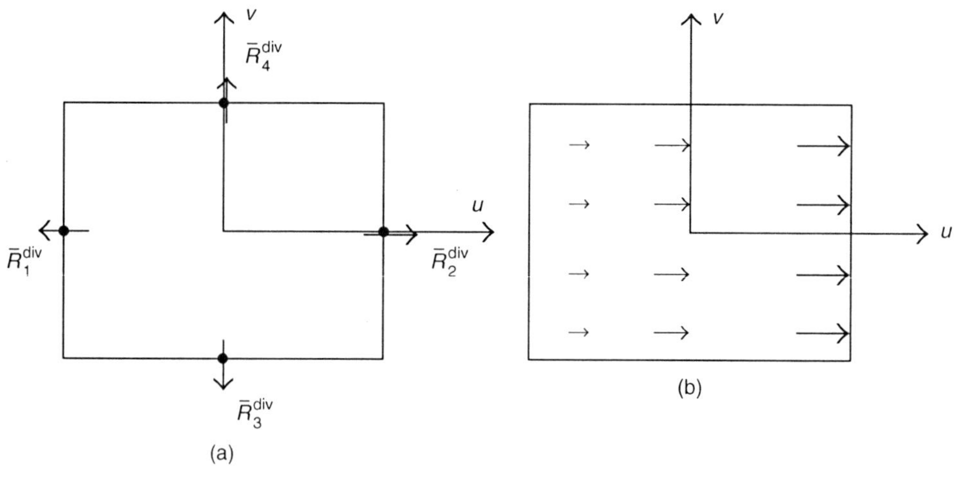

FIGURE 3.2: (a) The four $p = 0$ divergence-conforming basis functions on a square cell. (b) The function $\bar{R}_2^{\text{div}}$.

at $u = +1$, and is entirely tangential at $v = \pm 1$. Its normal component at $u = -1$ is directed out of the cell (in the $-\hat{u}$ direction).

For these functions to be divergence-conforming, they must be matched with basis functions in adjacent cells that maintain the continuity of the nonzero normal component. Conceptually, functions similar to (3.3)–(3.6) are assigned to the adjacent cells for this purpose, and attached to these in order to construct the rooftop functions of Fig. 3.1. The sign of some of the basis functions must be adjusted so that all maintain the same normal-vector direction across cell boundaries. Divergence-conforming functions usually do not attempt to maintain the continuity of the tangential component of the vector unknown, and a rooftop representation will generally exhibit jump discontinuities in the tangential components at cell edges.

The coefficients of the preceding basis functions have an obvious physical interpretation: they are the field or current density component at the center of the cell edge, perpendicular to that edge. For this interpretation to survive the process of mapping these functions to curvilinear cells, their normalization in the x–y–z space must be maintained so that they have a unit normal component at the appropriate interpolation point on the cell edge. The proper normalization will be determined in connection with the mapping process discussed in Chapter 5.

The divergence of the preceding functions in the reference cell coordinates may be calculated using

$$\nabla \bullet \bar{R} = \frac{\partial R_u}{\partial u} + \frac{\partial R_v}{\partial v} \tag{3.7}$$

For the functions in (3.3)–(3.6), the reference cell divergence is constant:

$$\frac{\partial R_{1u}^{\text{div}}}{\partial u} + \frac{\partial R_{1v}^{\text{div}}}{\partial v} = \frac{1}{2} \tag{3.8}$$

$$\frac{\partial R_{2u}^{\text{div}}}{\partial u} + \frac{\partial R_{2v}^{\text{div}}}{\partial v} = \frac{1}{2} \tag{3.9}$$

$$\frac{\partial R_{3u}^{\text{div}}}{\partial u} + \frac{\partial R_{3v}^{\text{div}}}{\partial v} = \frac{1}{2} \tag{3.10}$$

$$\frac{\partial R_{4u}^{\text{div}}}{\partial u} + \frac{\partial R_{4v}^{\text{div}}}{\partial v} = \frac{1}{2} \tag{3.11}$$

After being mapped to a curvilinear cell in x–y–z space, the divergence of the basis functions also incorporates a (nonconstant) scale factor related to the Jacobian of the mapping, a possible adjustment in sign, and a normalization factor. The appropriate divergence calculation is considered in Chapter 5.

3.5 BASIS FUNCTIONS OF ORDER $p = 0$ FOR A TRIANGULAR REFERENCE CELL

Figure 3.3 shows a triangular reference cell occupying the region $0 < u < 1, 0 < v < 1, u + v < 1$. Three divergence conforming basis functions of the lowest order can be defined within this cell as

$$\bar{R}_1^{\text{div}} = (u - 1)\hat{u} + v\hat{v} \tag{3.12}$$

$$\bar{R}_2^{\text{div}} = u\hat{u} + (v - 1)\hat{v} \tag{3.13}$$

$$\bar{R}_3^{\text{div}} = \sqrt{2}(u\hat{u} + v\hat{v}) \tag{3.14}$$

Each of these functions interpolates to the vector component normal to one edge, and contributes no normal component along any other edge. For example, the function $\bar{R}_1$ contributes a constant normal component along $u = 0$ axis, and is

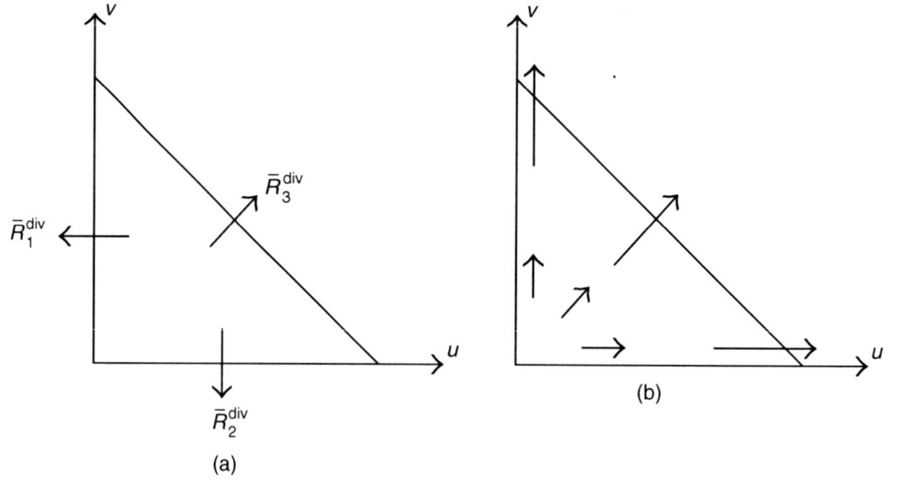

FIGURE 3.3: (a) The three $p = 0$ divergence-conforming basis functions on a triangular cell. (b) The function $\bar{R}_3^{\text{div}}$.

entirely tangential or zero along the other two edges. Its normal component at $u = 0$ is directed out of the cell in the $-\hat{u}$ direction. Function $\bar{R}_2$ provides an outward normal component along the $v = 0$ edge, while $\bar{R}_3$ contributes a constant normal component along the edge at $u + v = 1$.

For these functions to be divergence-conforming, they must be matched with basis functions in adjacent cells to create the triangular rooftop functions of Fig. 3.1. In the process, the sign of the local basis function in each cell must be adjusted to maintain a consistent normal-vector direction across cell boundaries. The tangential components of (3.12)–(3.14) are linear functions along cell boundaries; cell-to-cell tangential continuity is not maintained by the triangular rooftop functions.

As in the square cell case, the coefficients of the preceding basis functions have the physical interpretation that they are the normal component of the field or current density at the center of each cell edge. The proper normalization is considered in Chapter 5.

In the local coordinates of the reference cell, these functions have divergence

$$\frac{\partial R_{1u}^{\text{div}}}{\partial u} + \frac{\partial R_{1v}^{\text{div}}}{\partial v} = 2 \tag{3.15}$$

$$\frac{\partial R_{2u}^{\text{div}}}{\partial u} + \frac{\partial R_{2v}^{\text{div}}}{\partial v} = 2 \tag{3.16}$$

$$\frac{\partial R_{3u}^{\text{div}}}{\partial u} + \frac{\partial R_{3v}^{\text{div}}}{\partial v} = 2\sqrt{2} \tag{3.17}$$

The surface divergence on curvilinear cells in x–y–z space is different due to a nonconstant scale factor, a normalization factor, and a possible change in sign (Chapter 5).

3.6 NEDELEC'S MIXED-ORDER SPACES AND THE EFIE

The $p = 0$ basis functions of (3.3)–(3.6) and (3.12)–(3.14) correspond to the lowest order members of Nedelec's mixed-order divergence-conforming spaces [5]. Unlike most representations that are mathematically *complete* to the same polynomial degree in each variable, these spaces deliberately provide a representation with one additional polynomial degree along the primary vector direction of the basis function, compared with the orthogonal direction. [For instance, the $p = 0$ functions provide a linear dependence for the tangential components along cell edges, but only a constant dependence in the normal components.] As a consequence, the divergence of each function is complete to the same degree as the basis function itself. From another point of view, Nedelec's mixed-order spaces discard some of the degrees of freedom from a polynomial-complete expansion. When used to discretize an equation whose leading-order derivative is a divergence operator, the discarded degrees of freedom are essentially those that do not contribute to the balance of terms in the discretized equation. The EFIE is one such equation.

The EFIE of Eq. (1.8) is usually manipulated into a more suitable form for discretization (see Chapter 6). The resulting equation involves two terms: one proportional to the surface current density and the other proportional to the divergence of the surface current density. The latter quantity is essentially the electric charge density, related to the current density by

$$\rho_s = \frac{-1}{j\omega}\nabla_s \bullet \bar{J} \tag{3.18}$$

Because of the nature of the EFIE operator, the charge density provides the dominant contribution to the electric field near the source, while the current density is the dominant contributor to the electric field far from the source. The accuracy of the overall numerical solution is limited by the near-field interactions, or equivalently by the representation for charge density. Although the current is the quantity that is explicitly represented by the basis functions, the behavior of the divergence of those basis functions (the charge density) is critical. In fact, the charge representation appears to be more important than the current representation.

When used to expand the current density $\bar{J}$, in conjunction with (3.18) for ρ_s, Nedelec's mixed-order spaces provide a mathematically complete representation of the surface charge density. With mixed-order basis functions used for $\bar{J}$, the charge associated with the $\hat{s}$-component of $\bar{J}$ and the charge associated with the $\hat{t}$-component of $\bar{J}$ are both complete to the same degree in each variable. This interpretation may explain the longstanding observation that mixed-order solutions of the EFIE are more accurate than those obtained using polynomial-complete expansions for $\bar{J}$ [7, 8].

Nedelec's mixed-order divergence-conforming spaces appear ideally suited for use with the EFIE. Consequently, we restrict our consideration to basis functions of this type.

3.7 HIGHER-ORDER INTERPOLATORY FUNCTIONS FOR SQUARE CELLS

For constructing higher-order basis functions, it will be convenient to introduce the *shifted Silvester polynomial*

$$S_m^{(M)}(\xi) = \begin{cases} 1 & m = 1 \\ \dfrac{1}{(m-1)!}\displaystyle\prod_{i=1}^{m-1}(M\xi - i) & 2 \le m \le M+1 \end{cases} \tag{3.19}$$

for use on the domain $0 \le \xi \le 1$ [9]. This polynomial is related to the Silvester polynomial of (2.36) through

$$S_m^{(M)}(\xi) = R_{m-1}^{(M)}\left(\xi - \frac{1}{M}\right) \tag{3.20}$$

The zeros of several shifted Silvester polynomials are plotted in Fig. 3.4. The polynomial in (3.19) is of degree $m - 1$, and divides the domain $0 \le \xi \le 1$ into M intervals. Unlike the Silvester polynomial of (2.36), however, (3.19) has no zero at $\xi = 0$.

Nedelec's prescription for quadrilateral spaces requires $2(p+1)(p+2)$ degrees of freedom for a representation of minimum degree p [5]. The $p = 0$ representation of (3.3)–(3.6) involves four functions, conveniently distributed so that each one provides a perpendicular vector component at one edge. Higher order functions associate some of the degrees of freedom with each edge and some with

$S_1^{(p)} = 1$

$S_2^{(1)} = \xi - 1$

$S_2^{(2)} = 2\xi - 1$

$S_2^{(3)} = 3\xi - 1$

$S_2^{(4)} = 4\xi - 1$

$S_3^{(2)} = \frac{1}{2}(2\xi - 1)(2\xi - 2)$

$S_3^{(3)} = \frac{1}{2}(3\xi - 1)(3\xi - 2)$

$S_3^{(4)} = \frac{1}{2}(4\xi - 1)(4\xi - 2)$

$S_4^{(3)} = \frac{1}{6}(3\xi - 1)(3\xi - 2)(3\xi - 3)$

$S_4^{(4)} = \frac{1}{6}(4\xi - 1)(4\xi - 2)(4\xi - 3)$

$S_4^{(5)} = \frac{1}{6}(5\xi - 1)(5\xi - 2)(5\xi - 3)$

$S_4^{(6)} = \frac{1}{6}(6\xi - 1)(6\xi - 2)(6\xi - 3)$

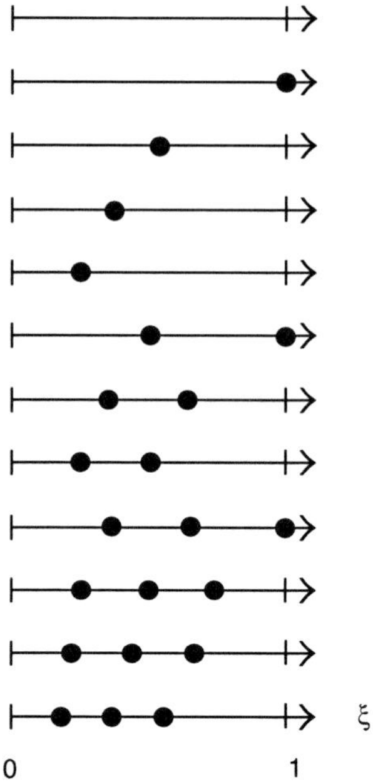

FIGURE 3.4: Examples of shifted Silvester polynomials defined on the interval $0 \leq \xi \leq 1$. The figure at right shows the location of the zeros of the polynomial at left.

the cell interior, as summarized in Table 3.1. There are a variety of forms that the corresponding basis functions can assume. For the $p = 1$ case, there are eight functions associated with cell edges, where they provide a linear normal behavior, and four functions that contribute no normal component to the edges. (The $p = 1$ functions collectively provide a quadratic tangential dependence along the cell edges.) Figure 3.5 shows one approach for assigning the interpolation points for the 12 $p = 1$ basis functions, after [9]. Specific functions can be obtained by combining the shifted Silvester polynomials from Fig. 3.4 to produce zeros at all the interpolation points except one.

As an example, the $p = 1$ function $\bar{R}_3$ that interpolates at the point $u = 1, v = 1/3$ in Fig. 3.5 can be constructed from the combination

$$\bar{R}_3(u, v) = \hat{u}\, S_3^{(2)}\left(\frac{1-u}{2}\right) S_2^{(3)}\left(\frac{1+v}{2}\right) \tag{3.21}$$

TABLE 3.1: Number of Degrees of Freedom and Polynomial Behavior of Normal and Tangential Vector Components Associated with Nedelec's Spaces of Minimal Degree p for Quadrilateral Cells.

p	$2(p+1)(p+2)$	NORMAL	TANGENTIAL	# ON EDGES	# IN CELL
0	4	Constant	Linear	4	0
1	12	Linear	Quadratic	8	4
2	24	Quadratic	Cubic	12	12
3	40	Cubic	Degree 4	16	24
4	60	Degree 4	Degree 5	20	40
5	84	Degree 5	Degree 6	24	60

Note. The column labeled "# on edges" indicates the number of basis functions associated with the normal component at the cell edges, which must be made normally continuous with analogous functions in adjacent cells. The remaining degrees of freedom ("# in cell") are entirely local and are not tied to functions in adjacent cells.

where the form of the arguments adapts the domain $0 \le \xi \le 1$ to either $-1 < u < 1$ or $-1 < v < 1$. The shifted Silvester polynomial $S_3^{(2)}$ equals zero when its argument is 1/2 or 1, meaning that (3.21) is zero at $u = -1$ and $u = 0$. The polynomial $S_2^{(3)}$ assigns a zero when its argument is 1/3, or when $v = -1/3$. Thus, the combination in (3.21) places zeros at the five other interpolation points indicated in Fig. 3.5. At the interpolation point $u = 1$, $v = 1/3$, (3.21) has unit magnitude. The basis function is of degree 2 in u and degree 1 in v.

As a second example, the $p = 1$ function $\bar{R}_5$, interpolating at $u = 0$, $v = -1/3$ in Fig. 3.5 can be constructed as

$$\bar{R}_5(u, v) = \hat{u} S_2^{(1)}\left(\frac{1-u}{2}\right) S_2^{(1)}\left(\frac{1+u}{2}\right) S_2^{(3)}\left(\frac{1-v}{2}\right) \tag{3.22}$$

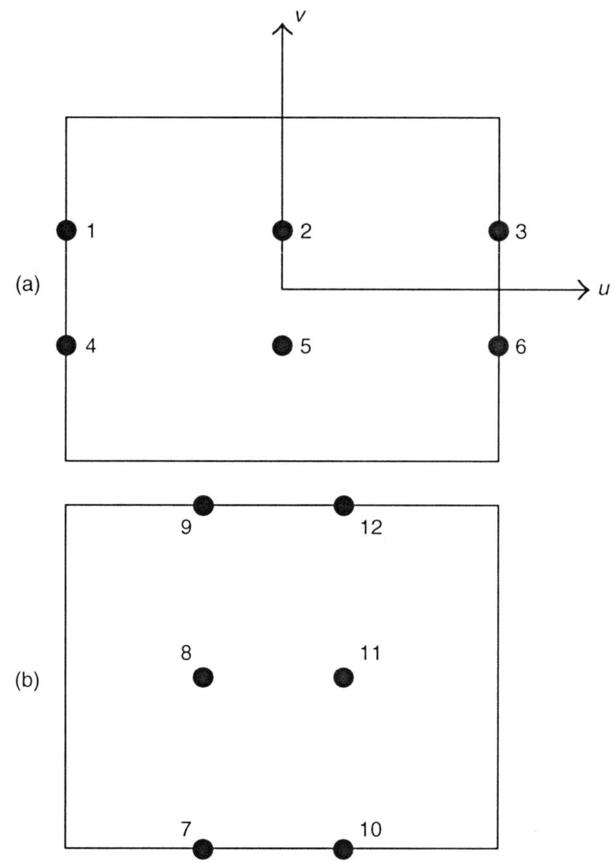

FIGURE 3.5: Interpolation points for the 12 $p = 1$ basis functions. (a) Points for $\hat{u}$ components, (b) points for $\hat{v}$ components.

The polynomial $S_2^{(1)}$ is zero when its argument equals 1, therefore the function

$$S_2^{(1)}\left(\frac{1-u}{2}\right) S_2^{(1)}\left(\frac{1+u}{2}\right) \tag{3.23}$$

places zeros at $u = -1$ and $u = +1$. The polynomial $S_2^{(3)}$ assigns a zero when its argument is 1/3, or when $v = 1/3$ in (3.22). Thus, the construction in (3.22) places zeros at all the interpolation points except the point labeled "5" in Fig. 3.5. The function $\bar{R}_5$ is of degree 2 in u and degree 1 in v, and has a value of 1/4 at the interpolation point. Therefore, it must be scaled by a factor of 4 in order to exhibit a unit value there.

Using the preceding approach, the set of $p = 1$ basis functions consists of

$$\bar{R}_1 = (-\hat{u})\frac{u(u-1)(3v+1)}{4} \tag{3.24}$$

$$\bar{R}_2 = \hat{u}\frac{(1-u)(1+u)(3v+1)}{2} \tag{3.25}$$

$$\bar{R}_3 = \hat{u}\frac{u(u+1)(3v+1)}{4} \tag{3.26}$$

$$\bar{R}_4 = (-\hat{u})\frac{u(u-1)(1-3v)}{4} \tag{3.27}$$

$$\bar{R}_5 = \hat{u}\frac{(1-u)(1+u)(1-3v)}{2} \tag{3.28}$$

$$\bar{R}_6 = \hat{u}\frac{u(u+1)(1-3v)}{4} \tag{3.29}$$

$$\bar{R}_7 = (-\hat{v})\frac{(1-3u)v(v-1)}{4} \tag{3.30}$$

$$\bar{R}_8 = \hat{v}\frac{(1-3u)(1-v)(1+v)}{2} \tag{3.31}$$

$$\bar{R}_9 = \hat{v}\frac{(1-3u)v(v+1)}{4} \tag{3.32}$$

$$\bar{R}_{10} = (-\hat{v})\frac{(3u+1)v(v-1)}{4} \tag{3.33}$$

$$\bar{R}_{11} = \hat{v}\frac{(3u+1)(1-v)(1+v)}{2} \tag{3.34}$$

$$\bar{R}_{12} = \hat{v}\frac{(3u+1)v(v+1)}{4} \tag{3.35}$$

These functions have been normalized to unity at the interpolation points, and are adjusted so that those with interpolation points on the edges point out of the cell. The edge-based functions must be paired with corresponding functions in the adjacent cells to maintain normal-vector continuity between cells.

The $p = 2$ case involves 24 basis functions per cell, with 12 that are edge-based (and interpolate to normal components on edges) and 12 that contribute no

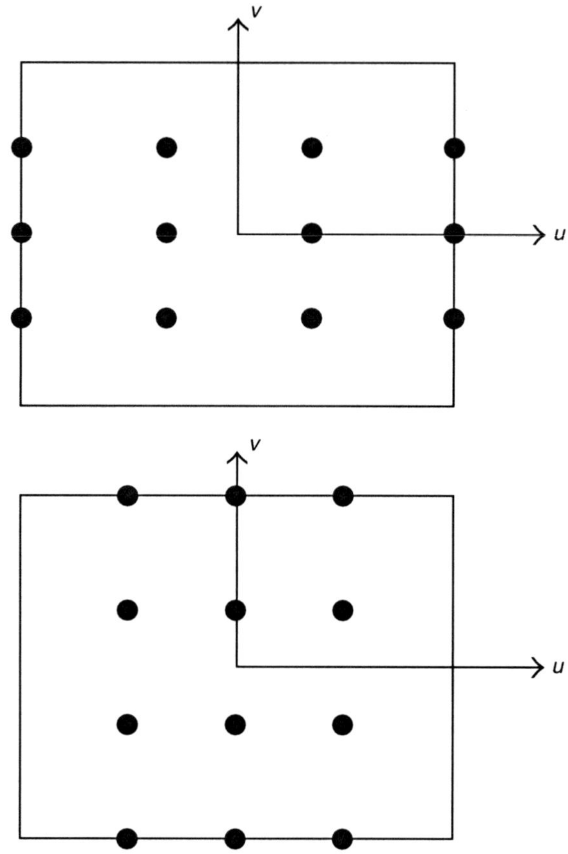

FIGURE 3.6: Interpolation points for the 24 $p = 2$ basis functions. (a) $\hat{u}$ components, (b) $\hat{v}$ components.

normal components along cell edges. (Since the edge-based functions are shared with neighboring cells, the global number of unknowns is less than 24 times the number of cells.) The overall representation is quadratic for the normal vector components, and cubic for the tangential components. Figure 3.6 illustrates one possibility for the 24 interpolation points, based on [9]. The construction of specific basis functions follows in the same manner as the $p = 1$ functions discussed above. Additional information and alternative expressions may be found in [9]. We note that there are other ways of constructing interpolatory vector bases that may offer advantages such as improved matrix condition numbers [10].

3.8 HIGHER-ORDER INTERPOLATORY FUNCTIONS FOR TRIANGULAR CELLS

Interpolatory basis functions of the divergence-conforming type may be developed for triangular cells in much the same manner as those for square cells in Section 3.6. Since there are only three edges per cell, it is possible to eliminate additional degrees of freedom compared to the basis functions for square cells, while ensuring that the divergence of the function is mathematically complete to the same degree as the basis function itself. Nedelec's spaces require a total of $(p+1)(p+3)$ degrees of freedom for a representation of minimum degree p [5], with $3(p+1)$ of those degrees of freedom contributing a nonzero normal component at the cell edges.

The $p = 0$ basis functions are those in (3.12)–(3.14). Each function exhibits a constant normal-vector component on one edge of the triangle, and can be thought of as interpolating to the normal-vector function at the center of the appropriate edge.

Reference [9] proposed a systematic approach for higher-degree representations, as illustrated for the $p = 1$ and $p = 2$ cases in Fig. 3.7. Interpolation points are defined at the cell edges and at regularly spaced intervals within the cell. One basis function is assigned to each of the edge points, while two are assigned at each interior point. These basis functions are constructed from products of the shifted Silvester polynomials with the $p = 0$ functions of (3.12)–(3.14). The $p = 1$ representation involves two basis functions per edge, and two functions that can be made to interpolate at the interior point, for a total of eight degrees of freedom.

For example, consider the $p = 1$ basis function for node 201 in Fig. 3.7(a). This basis function builds on the $p = 0$ function $\bar{R}_2^{\text{div}}$ in (3.13), which provides the underlying vector direction. However, the $p = 1$ function must also vanish at the points labeled 102 and 111 in Figure 3.7(a), meaning that it must be zero at $u = 1/3$. (It is not necessary to zero the function at points 012, 021, 120, or 210 since $\bar{R}_2^{\text{div}}$ has no normal vector component at those locations.) The desired behavior may be obtained from the shifted Silvester polynomial $\hat{S}_2^{(3)}$. Thus, the

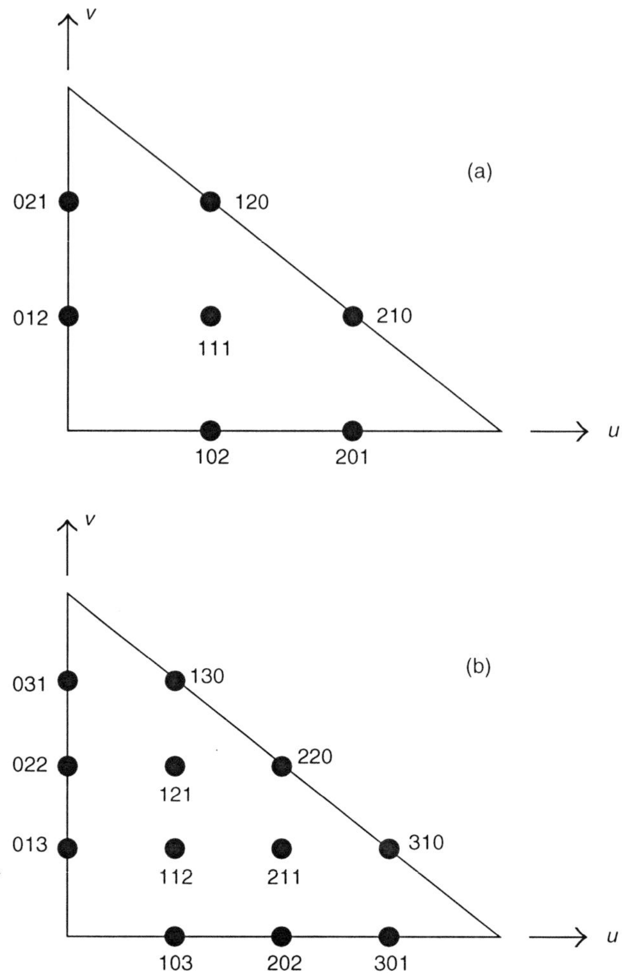

FIGURE 3.7: Interpolation points for (a) $p = 1$ basis functions and (b) $p = 2$ basis functions. Two functions interpolate at each of the interior nodes.

basis function may be constructed as

$$\bar{R}_{201}(u, v) = \bar{R}_2^{\mathrm{div}} S_2^{(3)}(u) \tag{3.36}$$

where the arguments range over $0 < u < 1$ and $0 < v < 1$. It is easily verified that (3.36) has a linear normal component along the edge $v = 0$, and no normal component along the other two edges of the cell. The tangential-vector component is quadratic. This function must be matched with a basis function in the adjacent cell that also has a linear normal component in the same global direction.

Two functions must be assigned to interpolate at the interior node labeled 111 in Fig. 3.7(a). These functions may utilize any two of the three $p = 0$ basis functions in (3.12)–(3.14), multiplied by shifted Silvester polynomials to zero the result at the other interpolation points. There are three suitable constructions possible:

$$\bar{R}_{111a} = \bar{R}_1^{\text{div}} S_2^{(1)}(1-u) \tag{3.37}$$

$$\bar{R}_{111b} = \bar{R}_2^{\text{div}} S_2^{(1)}(1-v) \tag{3.38}$$

$$\bar{R}_{111c} = \bar{R}_3^{\text{div}} S_2^{(1)}(u+v) \tag{3.39}$$

However, only two of these are linearly independent functions, so one must be discarded. The functions in (3.37)–(3.39) each have an identically zero normal component at all three of the cell edges.

The set of eight $p = 1$ basis functions obtained from the preceding construction is as follows:

$$\begin{aligned}\bar{R}_{012} &= \{(u-1)\hat{u} + v\hat{v}\}(3w-1) \\ &= \{(u-1)\hat{u} + v\hat{v}\}(2-3u-3v)\end{aligned} \tag{3.40}$$

$$\bar{R}_{021} = \{(u-1)\hat{u} + v\hat{v}\}(3v-1) \tag{3.41}$$

$$\begin{aligned}\bar{R}_{102} &= \{u\hat{u} + (v-1)\hat{v}\}(3w-1) \\ &= \{u\hat{u} + (v-1)\hat{v}\}(2-3u-3v)\end{aligned} \tag{3.42}$$

$$\bar{R}_{201} = \{u\hat{u} + (v-1)\hat{v}\}(3u-1) \tag{3.43}$$

$$\bar{R}_{120} = \left\{\sqrt{2}(u\hat{u} + v\hat{v})\right\}(3v-1) \tag{3.44}$$

$$\bar{R}_{210} = \left\{\sqrt{2}(u\hat{u} + v\hat{v})\right\}(3u-1) \tag{3.45}$$

$$\bar{R}_{111a} = \{(u-1)\hat{u} + v\hat{v}\}(-u) \tag{3.46}$$

$$\bar{R}_{111b} = \{u\hat{u} + (v-1)\hat{v}\}(-v) \tag{3.47}$$

The edge-based functions in (3.40)–(3.45) are normalized so that the outward component is unity at the interpolation point. These functions must be paired with analogous functions in adjacent cells to ensure normal-vector continuity at cell boundaries. The cell-based functions in (3.46) and (3.47) are not normalized, and are not paired with functions outside the cell.

The preceding idea may be extended to develop basis functions of any degree p. Additional information and alternative expressions may be found in [9].

3.9 HIGHER-ORDER HIERARCHICAL FUNCTIONS FOR SQUARE CELLS

Because interpolatory basis functions of one degree are not re-used in representations of greater polynomial degree, hierarchical functions are more computationally efficient when p-refinement is used to obtain a more accurate solution. Hierarchical vector functions can be constructed to satisfy the requirements of Nedelec's mixed-order spaces. It is also desirable to construct functions that provide reasonable linear independence (to avoid ill-conditioned matrices). Ideally, the higher-order functions should not duplicate the principal degrees of freedom provided by the lower-order functions, in order to provide a representation where the coefficients of lower-order functions only change slightly as the higher-order functions are added. (This situation suggests some type of orthogonality between lower-order and higher-order functions.) To date, these constraints are only partially realized in the available basis sets.

Hierarchical vector basis functions for quadrilateral or hexahedral cell shapes have been proposed by Wang [11], Ilic and Notaros [12], and Jorgensen *et al.* [13]. For illustration, we discuss those of [12] in the following. (The bases in [12] are actually curl-conforming functions; for square cell shapes we have converted these to divergence-conforming bases using the simple relation between the two types as discussed in Chapter 4.) Hierarchical vector bases can be organized by their maximum polynomial degree p, and also by their designation as "edge-based" or "cell-based" functions: edge-based functions have a nonzero normal component that must be made continuous with analogous functions in adjacent cells, while cell-based functions are entirely local.

The divergence-conforming basis functions obtained from [12] have the form

$$\hat{u}\,P_i(u)v^j, \quad i = 1, 2, \ldots, N; j = 1, 2, \ldots, N-1 \tag{3.48}$$

$$\hat{v}u^i\,P_j(v), \quad i = 1, 2, \ldots, N-1; j = 1, 2, \ldots, N \tag{3.49}$$

where

$$P_i(u) = \begin{cases} 1-u & i=0 \\ 1+u & i=1 \\ u^i-1 & i=2,4,6,\ldots \\ u^i-u & i=3,5,7,\ldots \end{cases} \tag{3.50}$$

Table 3.2 presents the first 24 divergence-conforming basis functions, following [12]. The basis functions are not normalized. The first 4 functions are essentially the $p = 0$ functions of (3.3)–(3.6), which are edge-based. The second 4 functions are also edge-based, and serve to elevate the polynomial degree of the representation to a consistently linear behavior in the normal and tangential vector components. The next 4 functions are cell-based, meaning that they provide no normal component on any of the cell edges. The cell-based functions collectively serve to provide a quadratic tangential-vector component along the cell edges. Use of the first 12 basis functions in Table 3.2 provides a representation equivalent to Nedelec's mixed-order $p = 1$ space.

The philosophy underlying the Nedelec spaces suggests that, for a general problem where the leading order derivative in an equation is a divergence operator, the best accuracy will be obtained by consistently using the first 4 degrees of freedom (Nedelec $p = 0$), or the first 12 (Nedelec $p = 1$), or the first 24 (Nedelec $p = 2$). Little improvement is expected by adding the 4 $p = 2$ edge-based basis functions to the set of 12 $p = 1$ functions, since those degrees of freedom will not be balanced

TABLE 3.2: Hierarchical Divergence-Conforming Vector Bases for Quadrilateral Cells

$p = 0$	$\hat{u}(1 - u)$	Four edge-based functions
	$\hat{u}\,(1 + u)$	
	$\hat{v}\,(1 - v)$	
	$\hat{v}\,(1 + v)$	Total degree of freedom = 4
$p = 1$, edge	$\hat{u}\,(1 - u)v$	Four edge-based functions
	$\hat{u}\,(1 + u)v$	
	$\hat{v}\,u\,(1 - v)$	
	$\hat{v}\,u\,(1 + v)$	Total degrees of freedom = 8
$p = 1$, cell	$\hat{u}\,(u^2 - 1)$	Four cell-based functions
	$\hat{u}\,(u^2 - 1)v$	
	$\hat{v}\,(v^2 - 1)$	
	$\hat{v}\,u\,(v^2 - 1)$	Total degrees of freedom = 12
$p = 2$, edge	$\hat{u}\,(1 - u)v^2$	Four edge-based functions
	$\hat{u}\,(1 + u)v^2$	
	$\hat{v}\,u^2\,(1 - v)$	
	$\hat{v}\,u^2\,(1 + v)$	Total degrees of freedom = 16
$p = 2$, cell	$\hat{u}\,(u^2 - 1)v^2$	Eight cell-based functions
	$\hat{u}\,(u^3 - u)$	
	$\hat{u}\,(u^3 - u)v$	
	$\hat{u}\,(u^3 - u)v^2$	
	$\hat{v}u^2\,(v^2 - 1)$	
	$\hat{v}\,(v^3 - v)$	
	$\hat{v}u\,(v^3 - v)$	
	$\hat{v}u^2\,(v^3 - v)$	Total degrees of freedom = 24

Note. $-1 \leq u, v \leq 1$; basis functions are not normalized; edge-based functions must be made normally continuous with neighboring cells. Adapted from [12].

in the equation. However, while a user of interpolatory basis functions is usually forced to work with an expansion that is either complete to a given polynomial degree or a given Nedelec space, hierarchical functions provide more freedom to adapt the representation to a particular situation. This adaption may be carried out dynamically within a p-refinement algorithm [14].

As previously mentioned, once the coefficients are determined the current density at specific locations is obtained by superimposing the set of hierarchical basis functions (scaled by their coefficients). Thus, the normalization of these basis functions is arbitrary. However, there are times when hierarchical testing functions are used within error estimators to drive adaptive refinement procedures [14]. In that situation, some normalization is usually necessary for the error estimate to be meaningful.

3.10 HIGHER-ORDER HIERARCHICAL FUNCTIONS FOR TRIANGULAR CELLS

Hierarchical basis functions for triangular or tetrahedral cell shapes have been proposed by Andersen and Volakis [15], Webb [16], Preissig and Peterson [17], and others. Most of the published bases are curl-conforming, but these can easily be converted into divergence-conforming functions for triangles using the relation described in Chapter 4. For illustration, Table 3.3 shows the first 24 divergence-conforming basis functions from [17], grouped in order of polynomial degree and separated by classification into "edge-based" and "cell-based" functions. The first three functions are essentially the $p = 0$ functions from (3.12)–(3.14). The next three functions are edge-based and elevate the polynomial degree of the representation to linear. The seventh and eighth entries in the table are cell-based functions that bring the representation up to the Nedelec $p = 1$ space [5]. The edge-based functions must be paired with similar functions in the adjacent cells to provide normal-vector continuity.

TABLE 3.3: Hierarchical Divergence-Conforming Vector Bases for Triangular Cells

Mixed order 0/1	$(u-1)\hat{u}+v\,\hat{v}$	Three edge-based functions
	$u\hat{u}+(v-1)\,\hat{v}$	
	$u\,\hat{u}+v\,\hat{v}$	Total degrees of freedom = 3
Complete order 1	$u\,\hat{u}-v\,\hat{v}$	Three edge-based functions
	$-u\,\hat{u}+(2u+v-1)\,\hat{v}$	
	$(1-u-2v)\hat{u}+v\,\hat{v}$	Total degrees of freedom = 6
Mixed order 1/2	$u[(u-1)\hat{u}+v\,\hat{v}]$	Two cell-based functions
	$v[u\,\hat{u}+(v-1)\,\hat{v}]$	Total degrees of freedom = 8
Complete order 2	$(u^2-2uv)\,\hat{u}-(2uv-v^2)\,\hat{v}$	Three edge-based functions
	$(2uw-u^2)\,\hat{u}-(2uw-[w-u]^2)\,\hat{v}$	One cell-based function
	$(2vw-[w-v]^2)\,\hat{u}-(2vw-v^2)\,\hat{v}$	
	$u(w-v)\,\hat{u}-v(w-u)\,\hat{v}$	Total degrees of freedom = 12
Mixed-order 2/3	$uv[(u-1)\,\hat{u}+v\,\hat{v}]$	Three cell-based functions
	$uw[(u-1)\,\hat{u}+v\,\hat{v}]$	
	$vw[u\hat{u}+(v-1)\,\hat{v}]$	Total degrees of freedom = 15

Complete order 3	$2u(u^2 - 5uv + 3v^2)\,\hat{u} - 2v(3u^2 - 5uv + v^2)\,\hat{v}$	Three edge-based functions
	$-2u(u^2 - 5uw + 3w^2)\,\hat{u} - 2(w - u)(u^2 - 7uw + w^2)\,\hat{v}$	Two cell-based functions
	$2(w - v)(v^2 - 7vw + w^2)\,\hat{u} + 2v(v^2 - 5vw + 3w^2)\,\hat{v}$	
	$u[v(v - 2w) - u(v - w)]\,\hat{u} - v[u(v - u) + w(v - 2u)]\,\hat{v}$	
	$u[u(w - v) - w(w - 2v)]\,\hat{u} - v[w(w - 4u) - u^2)]\,\hat{v}$	Total degrees of freedom = 20
Mixed order 3/4	$uv(u - v)[(u - 1)\,\hat{u} + v\,\hat{v}]$	Four cell-based functions
	$uw(u - w)[(u - 1)\,\hat{u} + v\,\hat{v}]$	
	$vw(v - w)[u\hat{u} + (v - 1)\,\hat{v}]$	
	$uvw[u\hat{u} + (v - 1)\hat{v}]$	Total degrees of freedom = 24

Note. Simplex coordinates $u, v, w = 1 - u - v; 0 \leq u, v, w \leq 1$; basis functions not normalized; edge-based functions must have normal components continuous with neighboring cells. Adapted from [17].

REFERENCES

[1] P. A. Raviart and J. M. Thomas, "Primal hybrid finite element methods for 2nd order elliptic equations," *Math. Comput.*, vol. 31, pp. 391–413, April 1977.

[2] A. W. Glisson, "On the development of numerical techniques for treating arbitrarily shaped surfaces," Ph.D. Dissertation, University of Mississippi, 1978.

[3] A. W. Glisson and D. R. Wilton, "Simple and efficient numerical methods for problems of electromagnetic radiation and scattering from surfaces," *IEEE Trans. Antennas Propagat.*, vol. AP-28, pp. 593–603, Sept. 1980.

[4] S. M. Rao, D. R. Wilton, and A. W. Glisson, "Electromagnetic scattering by surfaces of arbitrary shape," *IEEE Trans. Antennas Propagat.*, vol. AP-30, pp. 409–418, May 1982.

[5] J. C. Nedelec, "Mixed finite elements in R3," *Numer. Math.*, vol. 35, pp. 315–341, 1980.

[6] J. P. Webb, "Matching a given field using hierarchal vector basis functions," *Electromagnetics*, vol. 24, pp. 113–122, 2004.

[7] L. Gurel, K. Sertel, and I. K. Sendur, "On the choice of basis functions to model surface electric current densities in computational electromagnetics," *Radio Sci.*, vol. 34, pp. 1373–1387, Nov./Dec. 1999.

[8] A. W. Glisson, S. M. Rao, and D. R. Wilton, "Physically-based approximation of electromagnetic field quantities," in *IEEE Antenn. Propagat. Int. Symp. Digest*, San Antonio, TX, pp. 78–81, July 2002.

[9] R. D. Graglia, D. R. Wilton, and A. F. Peterson, "Higher order interpolatory vector bases for computational electromagnetics," *IEEE Trans. Antenn. Propagat.*, vol. 45, no. 3, pp. 329–342, March 1997.

[10] R. N. Rieben, D. A. White, and G. H. Rodrigue, "Improved conditioning of finite element matrices using new high-order interpolatory bases," *IEEE Trans. Antenn. Propagat.*, vol. 52, pp. 2675–2683, Oct. 2004.

[11] J.-S. Wang, "Hierarchic "edge" elements for high-frequency problems," *IEEE Trans. Magnetics*, vol. 33, pp. 1536–1539, March 1997.

[12] M. M. Ilic and B. M. Notaros, "Higher order hierarchical curved hexahedral vector finite elements for electromagnetic modeling," *IEEE Trans. Microwave Theory Tech.*, vol. 51, pp. 1026–1033, March 2003.

[13] E. Jorgensen, J. L. Volakis, P. Meincke, and O. Breinbjerg, "Higher order hierarchical Legendre basis functions for electromagnetic modeling," *IEEE Trans. Antenn. Propagat.*, vol. 52, pp. 2985–2995, Nov. 2004.

[14] M. Salazar-Palma, *et al.*, *Iterative and Self-Adaptive Finite Elements in Electromagnetic Modeling*. Boston: Artech House, 1998.

[15] L. S. Andersen and J. L. Volakis, "Hierarchical tangential vector finite elements for tetrahedra," *IEEE Microwave Guided Wave Lett.*, vol. 8, pp. 127–129, March 1998.

[16] J. P. Webb, "Hierarchical vector basis functions of arbitrary order for triangular and tetrahedral finite elements," *IEEE Trans. Antenn. Propagat.*, vol. 47, pp. 1244–1253, August 1999.

[17] R. S. Preissig and A. F. Peterson, "A rationale for p-refinement with the vector Helmholtz equation and two-dimensional vector finite elements," *ACES J.*, vol. 19, pp. 65–75, July 2004.

CHAPTER 4

Curl-Conforming Basis Functions

Curl-conforming basis functions are complementary to the divergence-conforming functions discussed in Chapter 3. Here, curl-conforming functions are considered for square and triangular reference cells. The lowest order functions will be used to discretize the MFIE in Chapter 7. Tables of higher-order functions are included for completeness. The mapping of these functions to curvilinear cells will be the focus of Chapter 5.

4.1 WHAT DOES CURL-CONFORMING MEAN?

The curl of a vector function is

$$\nabla \times \bar{B} = \hat{x}\left\{\frac{\partial B_z}{\partial y} - \frac{\partial B_y}{\partial z}\right\} + \hat{y}\left\{\frac{\partial B_x}{\partial z} - \frac{\partial B_z}{\partial x}\right\} + \hat{z}\left\{\frac{\partial B_y}{\partial x} - \frac{\partial B_x}{\partial y}\right\} \tag{4.1}$$

In a local coordinate system $(\hat{s}, \hat{t}, \hat{n})$, where $\hat{n}$ is normal to the surface, $\hat{n} = \hat{s} \times \hat{t}$, and $\bar{B}$ is a tangential vector function, we may also refer to the surface curl operation

$$\nabla_s \times \bar{B} = \hat{n}\left\{\frac{\partial B_t}{\partial s} - \frac{\partial B_s}{\partial t}\right\} \tag{4.2}$$

Each component of the curl is proportional to the "twist" of the vector function about a point in the plane perpendicular to that component.

A *curl-conforming* basis function is one that maintains enough continuity to allow it to be differentiated via the curl, while yielding a bounded, well-defined result without the appearance of Dirac delta functions in $\nabla \times \bar{B}$. To achieve this for subsectional representations, the tangential-vector components of $\bar{B}$ must be continuous across cell edges. The normal-vector components are not differentiated across cell edges, and thus cell-to-cell normal-vector continuity is not required. Therefore, *a curl-conforming basis function is one that maintains first-order tangential-vector continuity across cell edges.*

4.2 HISTORY OF THE USE OF CURL-CONFORMING BASIS FUNCTIONS

Curl-conforming basis functions were first proposed for use with finite element solutions by Nedelec [1]. They have been used and extended primarily in connection with solutions of the vector Helmholtz equation (the so-called "curl-curl" form of that equation) in electromagnetic field problems. Early research using lower-order functions were reported by Bossavit and Verite [2], Mur and de Hoop [3], and Barton and Cendes [4], all who used triangular or tetrahedral cells. Crowley employed curl-conforming functions for hexahedral cells [5]. Rao and Wilton appear to have been the first to use curl-conforming basis functions for discretizing integral equations [6].

Interpolatory and hierarchical functions of arbitrary order for square and triangular cell shape have subsequently been proposed. These functions correspond to the Nedelec mixed-order spaces of minimum degree p [1].

4.3 RELATION BETWEEN THE DIVERGENCE-CONFORMING AND CURL-CONFORMING FUNCTIONS

Unlike representations that are mathematically complete to the same polynomial degree in all variables, Nedelec's two-dimensional mixed-order curl-conforming spaces deliberately provide an additional polynomial degree in the variable orthogonal to the vector direction [1]. Because of this behavior, the curl of the basis function is mathematically complete to the same degree as the basis function itself. Equivalently, these spaces discard some of the possible degrees of freedom. When discretizing an equation whose leading-order term is a curl operator, the discarded degrees of freedom are essentially those that do not contribute to a balance of terms in the equation. Consequently, the mixed-order representations are believed to provide a more efficient discretization (fewer unknowns for comparable accuracy).

A similar philosophy was discussed in Chapter 3 in the context of Nedelec's divergence-conforming spaces. In fact, in two dimensions the curl-conforming and divergence-conforming functions are closely related. Given a divergence-conforming function, one can obtain a curl-conforming function of the same Nedelec degree p from the simple operation

$$\hat{n} \times \bar{R}^{\text{div}} = \bar{R}^{\text{curl}} \tag{4.3}$$

where $\hat{n}$ is a unit vector perpendicular to the two-dimensional domain in which the basis functions are defined. It follows that

$$\hat{n} \times \bar{R}^{\text{curl}} = -\bar{R}^{\text{div}} \tag{4.4}$$

Because of this simple relation, curl-conforming functions can immediately be obtained from the divergence-conforming functions already developed in Chapter 3. In the following sections, we employ this relationship as a short-cut to obtain the curl-conforming functions in the reference cells.

4.4 BASIS FUNCTIONS OF ORDER $p = 0$ FOR A SQUARE REFERENCE CELL

Figure 4.1 shows a square reference cell occupying the region $-1 < u < 1$, $-1 < v < 1$. Four curl-conforming basis functions of the lowest order can be defined within this cell as

$$\bar{R}_1^{\text{curl}} = \frac{u-1}{2}\hat{v} \tag{4.5}$$

$$\bar{R}_2^{\text{curl}} = \frac{u+1}{2}\hat{v} \tag{4.6}$$

$$\bar{R}_3^{\text{curl}} = -\frac{v-1}{2}\hat{u} \tag{4.7}$$

$$\bar{R}_4^{\text{curl}} = -\frac{v+1}{2}\hat{u} \tag{4.8}$$

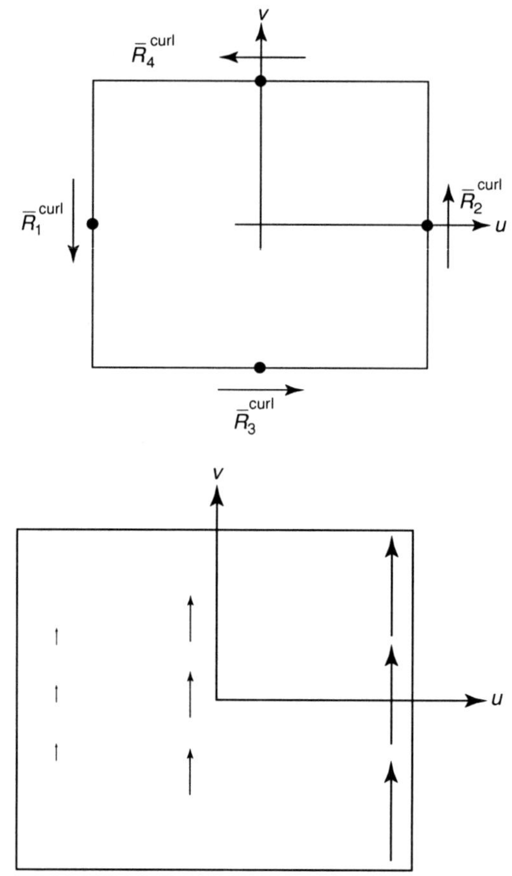

FIGURE 4.1: (a) The interpolation points and vector directions for the $p = 0$ curl-conforming basis functions on a square cell. (b) The function $\bar{R}_2^{\text{curl}}$.

Each of these functions has a nonzero vector component tangential along one edge, and contributes no tangential component along any other edge. For example, the function $\bar{R}_1^{\text{curl}}$ is zero at $u = +1$, and is entirely normal at $v = \pm 1$. Its only nonzero tangential component is along the edge at $u = -1$, and is in the $-\hat{v}$ direction. Each function provides a nonzero tangential component at a different cell edge. It is convenient to think of each as interpolating to the tangential-vector component at the center of that edge.

For these functions to be curl-conforming, they must be matched with basis functions in adjacent cells that maintain the continuity of the nonzero tangential component. Conceptually, similar basis functions are located in cells around those occupied by the preceding functions, for this purpose. The vector direction of the basis functions must be adjusted to maintain a consistent tangential direction across cell boundaries. In addition, the tangential-vector continuity must be maintained by the mapping to curvilinear cells. Curl-conforming functions do not usually maintain the continuity of the normal component of the vector unknown at cell boundaries, and a representation therefore exhibits jump discontinuities in the normal components at cell edges.

The coefficients of the preceding basis functions have an obvious physical interpretation: they are the field or current density component at the cell edge, tangential to that edge. For this interpretation to survive the process of mapping these functions to curvilinear cells, their normalization in the x–y–z space must be controlled. The proper normalization will be considered in connection with the mapping process in Chapter 5.

The reader should compare these curl-conforming basis functions to the divergence-conforming functions given in (3.3)–(3.6), and note that the only difference is the direction of the vector has been rotated by 90° in the u–v plane.

The curl of the preceding functions, carried out in the reference cell coordinates, is given by

$$\nabla \times \bar{R}^{\text{curl}} = (\hat{u} \times \hat{v}) \left\{ \frac{\partial R_v^{\text{curl}}}{\partial u} - \frac{\partial R_u^{\text{curl}}}{\partial v} \right\} \tag{4.9}$$

For the functions in (4.5)–(4.8), the curl within the cell is given by the simple results

$$\frac{\partial R_{1v}^{\text{curl}}}{\partial u} - \frac{\partial R_{1u}^{\text{curl}}}{\partial v} = \frac{1}{2} \tag{4.10}$$

$$\frac{\partial R_{2v}^{\text{curl}}}{\partial u} - \frac{\partial R_{2u}^{\text{curl}}}{\partial v} = \frac{1}{2} \tag{4.11}$$

$$\frac{\partial R_{3v}^{\text{curl}}}{\partial u} - \frac{\partial R_{3u}^{\text{curl}}}{\partial v} = \frac{1}{2} \tag{4.12}$$

$$\frac{\partial R_{4v}^{\text{curl}}}{\partial u} - \frac{\partial R_{4u}^{\text{curl}}}{\partial v} = \frac{1}{2} \tag{4.13}$$

After being mapped to a curvilinear cell in x–y–z space, the curl of the basis functions includes an additional nonconstant scale factor related to the Jacobian of the mapping (Section 5.7). The sign of these results must be adjusted in accordance with any sign change in direction as mentioned above, and any other normalization factor that may be included to achieve a unit component at a desired location within the curvilinear cell.

4.5 BASIS FUNCTIONS OF ORDER $p = 0$ FOR A TRIANGULAR REFERENCE CELL

Figure 4.2 shows a triangular reference cell occupying the region $0 < u < 1, 0 < v < 1, u + v < 1$. Three curl-conforming basis functions of the lowest order can be defined within this cell as

$$\bar{R}_1^{\text{curl}} = v\nabla w - w\nabla v = (u-1)\hat{v} - v\hat{u} \tag{4.14}$$

$$\bar{R}_2^{\text{curl}} = w\nabla u - u\nabla w = u\hat{v} - (v-1)\hat{u} \tag{4.15}$$

$$\bar{R}_3^{\text{curl}} = \sqrt{2}(u\nabla v - v\nabla u) = \sqrt{2}(u\hat{v} - v\hat{u}) \tag{4.16}$$

where w is the third simplex coordinate, satisfying $w = 1 - u - v$. Each of these functions interpolates to the vector component tangential to one edge, and contributes no tangential component along any other edge. For example, the function $\bar{R}_1^{\text{curl}}$ contributes a tangential component along the $u = 0$ axis, and is entirely normal or zero along the other two edges. Its tangential component at $u = 0$ is in the $-\hat{v}$ direction. Function $\bar{R}_2^{\text{curl}}$ provides a tangential component along the $v = 0$ axis, in the $+\hat{u}$ direction, while $\bar{R}_3^{\text{curl}}$ contributes a tangential component only along the edge at $u + v = 1$. These functions are the same as the three divergence-conforming basis functions of (3.12)–(3.14), with their vector directions rotated by 90° in the u–v plane.

For these functions to be curl-conforming, they must be matched with basis functions in adjacent cells that maintain the continuity of the tangential component. In addition, the vector direction of the basis function in cells sharing an edge must be adjusted to maintain a continuous tangential component across that edge.

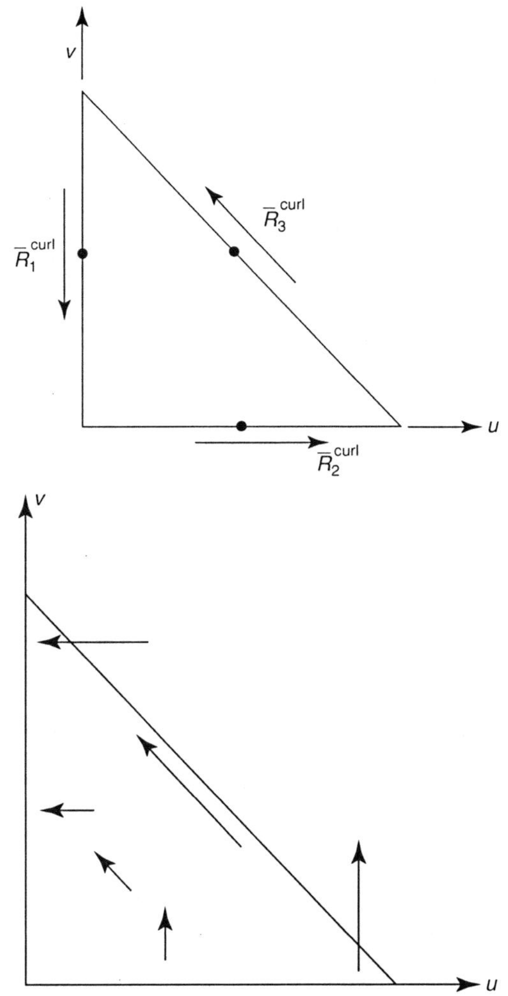

FIGURE 4.2: (a) The interpolation points and vector directions for the $p = 0$ curl-conforming basis functions on the unit triangle. (b) The function $\bar{R}_3^{\text{curl}}$.

The continuity of the normal-vector components is not maintained across cell boundaries.

As in the square cell case, the coefficients of the preceding basis functions have the physical interpretation that they are the tangential component of the field or current density at each cell edge. The normalization necessary to achieve a unit component after mapping is considered in Chapter 5.

The curl of these functions in the local coordinates of the reference cell is obtained as

$$\frac{\partial R^{\text{curl}}_{1v}}{\partial u} - \frac{\partial R^{\text{curl}}_{1u}}{\partial v} = 2 \tag{4.17}$$

$$\frac{\partial R^{\text{curl}}_{2v}}{\partial u} - \frac{\partial R^{\text{curl}}_{2u}}{\partial v} = 2 \tag{4.18}$$

$$\frac{\partial R^{\text{curl}}_{3v}}{\partial u} - \frac{\partial R^{\text{curl}}_{3u}}{\partial v} = 2\sqrt{2} \tag{4.19}$$

As explained in Chapter 5, the curl of the basis functions in the x–y–z space is different due to the presence of a nonconstant scale factor, possible sign change, and normalization constant.

4.6 HIGHER-ORDER INTERPOLATORY FUNCTIONS FOR SQUARE CELLS

Curl-conforming functions may be constructed for square cells using the same arrangement of interpolation points depicted in Fig. 3.5 for the 12 Nedelec $p = 1$ functions, or Fig. 3.6 for the 24 $p = 2$ functions, based on [7]. It is convenient to employ shifted Silvester polynomials to generate the higher-order functions, as explained in Section 3.6 for divergence conforming bases. Table 4.1 presents the functions for degrees $p = 0$, $p = 1$, and $p = 2$. The functions labeled "edge-based" must be paired with analogous functions in adjacent cells in order to maintain tangential-vector continuity. The "cell-based" functions are entirely local and contribute no tangential component on any of the cell edges. As described in [7], the procedure can be continued to obtain bases of any degree.

4.7 HIGHER-ORDER INTERPOLATORY FUNCTIONS FOR TRIANGULAR CELLS

Figure 3.7 presented triangular-cell interpolation points that can be used to construct Nedelec $p = 1$ and $p = 2$ functions, also following the approach of [7]. The construction of curl-conforming basis functions is similar to that carried out

TABLE 4.1: Interpolatory Curl-Conforming Vector Bases for Square Cells

$p = 0$	$\frac{u-1}{2}\hat{v}$	Four edge-based functions
	$\frac{u+1}{2}\hat{v}$	
	$-\frac{v-1}{2}\hat{u}$	Total degrees of freedom = 4
	$-\frac{v+1}{2}\hat{u}$	
$p = 1$, edge	$-\hat{v}\frac{u(u-1)(3v+1)}{4}$	Eight edge-based functions
	$\hat{v}\frac{u(u+1)(3v+1)}{4}$	
	$-\hat{v}\frac{u(u-1)(1-3v)}{4}$	
	$\hat{v}\frac{u(u+1)(1-3v)}{4}$	
	$\hat{u}\frac{(1-3u)v(v-1)}{4}$	
	$-\hat{u}\frac{(1-3u)v(v+1)}{4}$	
	$\hat{u}\frac{(3u+1)v(v-1)}{4}$	
	$-\hat{u}\frac{(3u+1)v(v+1)}{4}$	
$p = 1$, cell	$\hat{v}\frac{(1-u)(1+u)(3v+1)}{2}$	Four cell-based functions
	$\hat{v}\frac{(1-u)(1+u)(1-3v)}{2}$	
	$-\hat{u}\frac{(1-3u)(1-v)(1+v)}{2}$	
	$-\hat{u}\frac{(3u+1)(1-v)(1+v)}{2}$	Total degrees of freedom = 12

(*Continued*)

TABLE 4.1: (*Continued*)

$p = 2$, edge	$\hat{u}\,u(1-2u)(1+3v)(1-3v)(1-v)/16$	Twelve edge-based functions
	$-\hat{u}\,(1-2u)(1+2u)(1+3v)(1-3v)(1-v)/16$	
	$-\hat{u}\,u(1+2u)(1+3v)(1-3v)(1-v)/16$	
	$\hat{v}\,(1+u)(1+3u)(1-3u)v(1-2v)/16$	
	$-\hat{v}\,(1+u)(1+3u)(1-3u)(1-2v)(1+2v)/16$	
	$-\hat{v}\,(1+u)(1+3u)(1-3u)v(1+2v)/16$	
	$-\hat{u}\,u(1-2u)(1+v)(1+3v)(1-3v)/16$	
	$\hat{u}\,(1-2u)(1+2u)(1+v)(1+3v)(1-3v)/16$	
	$\hat{u}u(1+2u)(1+v)(1+3v)(1-3v)/16$	
	$\hat{v}\,(1+3u)(1-3u)(1-u)v(1+2v)/16$	
	$\hat{v}\,(1+3u)(1-3u)(1-u)(1-2v)(1+2v)/16$	
	$-\hat{v}\,(1+3u)(1-3u)(1-u)v(1-2v)/16$	
$p = 2$, cell	$\hat{u}u(1-2u)(1-3v)(1-v^2)$	Twelve cell-based functions
	$\hat{u}\,(1-2u)(1+2u)(1-3v)(1-v^2)$	
	$\hat{u}\,u(1+2u)(1-3v)(1-v^2)$	
	$\hat{u}\,u(1-2u)(1+3v)(1-v^2)$	
	$\hat{u}\,(1-2u)(1+2u)(1+3v)(1-v^2)$	
	$\hat{u}\,u(1+2u)(1+3v)(1-v^2)$	
	$\hat{v}\,(1-3u)(1-u^2)v(1-2v)$	
	$\hat{v}\,(1-3u)(1-u^2)(1-2v)(1+2v)$	
	$\hat{v}\,(1-3u)(1-u^2)v(1+2v)$	
	$\hat{v}\,(1+3u)(1-u^2)v(1-2v)$	
	$\hat{v}\,(1+3u)(1-u^2)(1-2v)(1+2v)$	
	$\hat{v}\,(1+3u)(1-u^2)v(1+2v)$	Total degrees of freedom = 24

Note. $-1 \le u, v \le 1$; basis functions normalized to unity; edge-based functions must be made normally continuous with neighboring cells. After [7].

in Chapter 3 for divergence-conforming functions, by forming products of the $p = 0$ functions with appropriate shifted Silvester polynomials. For example, the functions that interpolate to nodes labeled 103, 202, and 301 in Fig. 3.7(b) are constructed from

$$\bar{R}_{103} = \bar{R}_2^{\text{curl}} S_3^{(4)}(w) \tag{4.20}$$
$$\bar{R}_{202} = \bar{R}_2^{\text{curl}} S_2^{(4)}(u) S_2^{(4)}(w) \tag{4.21}$$
$$\bar{R}_{301} = \bar{R}_2^{\text{curl}} S_3^{(4)}(u) \tag{4.22}$$

where w is the third simplex coordinate, and the shifted Silvester polynomial is defined in (3.19). Table 4.2 presents the $p = 0$, $p = 1$, and $p = 2$ functions. Tangential-vector continuity is maintained by pairing the functions labeled "edge-based" with analogous functions in adjacent cells.

4.8 HIGHER-ORDER HIERARCHICAL FUNCTIONS FOR SQUARE CELLS

Hierarchical functions for quadrilateral or hexahedral cells were summarized in Chapter 3, in the context of divergence-conforming bases. Most of the functions reported in the literature are actually curl-conforming, intended for the finite element solution of the curl-curl form of the vector Helmholtz equation. For illustration, we consider the curl-conforming bases of Ilic and Notaros [8], which have the form

$$\hat{u} u^i P_j(v), \quad i = 1, 2, \ldots, N-1; j = 1, 2, \ldots, N \tag{4.23}$$
$$\hat{v} P_i(u) v^j, \quad i = 1, 2, \ldots, N; j = 1, 2, \ldots, N-1 \tag{4.24}$$

where P_i is defined in Eq. (3.50). Table 4.3 presents the first 24 basis functions, organized by Nedelec degree and by the edge-based or cell-based nature of the functions. The cell-based functions are entirely local to a cell and have no tangential components along the cell boundaries. The edge-based functions must be paired with similar functions in adjacent cells to ensure tangential-vector continuity.

TABLE 4.2: Interpolatory Curl-Conforming Vector Bases for Triangular Cells

Mixed order 0/1: ($p = 0$)	$(u-1)\hat{v} - v\hat{u}$ $u\hat{v} - (v-1)\hat{u}$ $\sqrt{2}(u\hat{v} - v\hat{u})$	Three edge-based functions Total degrees of freedom = 3
Mixed order 1/2: ($p = 1$)	$\bar{R}_{012} = \{(u-1)\hat{v} - v\hat{u}\}(3w-1)$ $\bar{R}_{021} = \{(u-1)\hat{v} - v\hat{u}\}(3v-1)$ $\bar{R}_{102} = \{u\hat{v} - (v-1)\hat{u}\}(3w-1)$ $\bar{R}_{201} = \{u\hat{v} - (v-1)\hat{u}\}(3u-1)$ $\bar{R}_{120} = \{\sqrt{2}(u\hat{v} - v\hat{u}\}(3v-1)$ $\bar{R}_{210} = \{\sqrt{2}(u\hat{v} - v\hat{u})\}(3u-1)$	Six edge-based functions
	$\bar{R}_{111a} = \{(u-1)\hat{v} - v\hat{u}\}u$ $\bar{R}_{111b} = \{u\hat{v} - (v-1)\hat{u}\}v$	Two cell-based functions Total degrees of freedom = 8
Mixed-order 2/3: ($p = 2$)	$\bar{R}_{013} = \{(u-1)\hat{v} - v\hat{u}\}(4w-1)(4w-2)/2$ $\bar{R}_{022} = \{(u-1)\hat{v} - v\hat{u}\}(4v-1)(4w-1)$ $\bar{R}_{031} = \{(u-1)\hat{v} - v\hat{u}\}(4v-1)(4v-2)/2$ $\bar{R}_{103} = \{u\hat{v} - (v-1)\hat{u}\}(4w-1)(4w-2)/2$ $\bar{R}_{202} = \{u\hat{v} - (v-1)\hat{u}\}(4u-1)(4w-1)$ $\bar{R}_{301} = \{u\hat{v} - (v-1)\hat{u}\}(4u-1)(4u-2)/2$ $\bar{R}_{130} = \{\sqrt{2}(u\hat{v} - v\hat{u})\}(4v-1)(4v-2)/2$ $\bar{R}_{220} = \{\sqrt{2}(u\hat{v} - v\hat{u})\}(4u-1)(4v-1)$ $\bar{R}_{310} = \{\sqrt{2}(u\hat{v} - v\hat{u})\}(4u-1)(4u-2)/2$	Nine edge-based functions
	$\bar{R}_{112a} = \{(u-1)\hat{v} - v\hat{u}\}u(4w-1)$	Six cell-based functions

TABLE 4.2: (*Continued*)

$\bar{R}_{112b} = \{u\hat{v} - (v-1)\hat{u}\}v(4w-1)$ $\bar{R}_{121a} = \{(u-1)\hat{v} - v\hat{u}\}u(4v-1)$ $\bar{R}_{121b} = \{u\hat{v} - (v-1)\hat{u}\}v(4v-1)$ $\bar{R}_{211a} = \{(u-1)\hat{v} - v\hat{u}\}u(4u-1)$ $\bar{R}_{211b} = \{u\hat{v} - (v-1)\hat{u}\}v(4u-1)$	Total degrees of freedom = 15

Note. Simplex coordinates u, v, $w = 1 - u - v$; $0 \leq u, v, w \leq 1$; edge basis functions normalized to unity; edge-based functions must have tangential components continuous with neighboring cells. After [7].

In contrast to the organization of Tables 4.1 and 4.2, the functions in Table 4.3 are intended to build upon the lower-order bases. Thus, the full $p = 2$ representation must include the $p = 0$ and $p = 1$ functions in the table.

4.9 HIGHER-ORDER HIERARCHICAL FUNCTIONS FOR TRIANGULAR CELLS

The literature contains a variety of hierarchical bases for triangular or tetrahedral cells. Table 4.4 summarizes the first 24 curl-conforming basis functions from [9], organized by Nedelec degree and by their edge-based or cell-based nature. The edge-based functions must be paired with similar functions in adjacent cells to provide tangential-vector continuity. A subset of the triangular basis functions of any mixed-order Nedelec degree p form a polynomial-complete set of degree p, and this subset is also noted in the table. The full representation of any degree includes all the lower-degree functions listed in Table 4.4.

TABLE 4.3: Hierarchical Curl-Conforming Vector Bases for Square Cells

$p = 0$	$\hat{u}(1 - v)$	Four edge-based functions
	$\hat{u}\,(1 + v)$	
	$\hat{v}\,(1 - u)$	
	$\hat{v}\,(1 + u)$	Total degrees of freedom = 4
$p = 1$, edge	$\hat{u}u(1 - v)$	Four edge-based functions
	$\hat{u}\,u(1 + v)$	
	$\hat{v}\,(1 - u)v$	
	$\hat{v}\,(1 + u)v$	Total degrees of freedom = 8
$p = 1$, cell	$\hat{u}(v^2 - 1)$	Four cell-based functions
	$\hat{u}\,u(v^2 - 1)$	
	$\hat{v}\,(u^2 - 1)$	
	$\hat{v}\,(u^2 - 1)v$	Total degrees of freedom = 12
$p = 2$, edge	$\hat{u}u^2(1 - v)$	Four edge-based functions
	$\hat{u}\,u^2(1 + v)$	
	$\hat{v}\,(1 - u)v^2$	
	$\hat{v}(1 + u)v^2$	Total degrees of freedom = 16
$p = 2$, cell	$\hat{u}u^2(v^2 - 1)$	Eight cell-based functions
	$\hat{u}\,(v^3 - v)$	
	$\hat{u}\,u(v^3 - v)$	
	$\hat{u}\,u^2(v^3 - v)$	
	$\hat{v}\,(u^2 - 1)v^2$	
	$\hat{v}\,(u^3 - u)$	
	$\hat{v}\,(u^3 - u)v$	
	$\hat{v}\,(u^3 - u)v^2$	Total degrees of freedom = 24

Note. $-1 \leq u, v \leq 1$; basis functions are not normalized; edge-based functions must be made tangentially continuous with neighboring cells. After [8].

TABLE 4.4: Hierarchical Curl-Conforming Vector Bases for Triangular Cells

Mixed order 0/1: ($p = 0$)	$v\nabla w - w\nabla v$	Three edge-based functions
	$w\nabla u - u\nabla w$	
	$u\nabla v - v\nabla u$	Total degrees of freedom = 3
Complete degree 1	$\nabla(uv)$	Three edge-based functions
	$\nabla(uw)$	
	$\nabla(vw)$	Total degrees of freedom = 6
Mixed order 1/2: ($p = 1$)	$u(v\nabla w - w\nabla v)$	Two cell-based functions
	$v(w\nabla u - u\nabla w)$	Total degrees of freedom = 8
Complete degree 2	$\nabla\{uv(u - v)\}$	Three edge-based functions
	$\nabla\{uw(u - w)\}$	One cell-based function
	$\nabla\{vw(v - w)\}$	
	$\nabla\{uvw\}$	Total degrees of freedom = 12
Mixed-order 2/3: ($p = 2$)	$uv(v\nabla w - w\nabla v)$	Three cell-based functions
	$uw(v\nabla w - w\nabla v)$	
	$vw(w\nabla u - u\nabla w)$	Total degrees of freedom = 15

(*Continued*)

TABLE 4.4: (*Continued*)

Complete degree 3	$\nabla\{uv(2u - v)(u - 2v)\}$	Three edge-based functions, two cell-based functions
	$\nabla\{uw(2u - w)(u - 2w)\}$	
	$\nabla\{vw(2v - w)(v - 2w)\}$	
	$\nabla\{uvw\,(u - v)\}$	
	$\nabla\{uvw(u - w)\}$	Total degrees of freedom = 20
Mixed-order 3/4: ($p = 3$)	$uv(u - v)(v\nabla w - w\nabla v)$	Four cell-based functions
	$uw(u - w)(v\nabla w - w\nabla v)$	
	$vw(v - w)(w\nabla u - u\nabla w)$	
	$uvw(w\nabla u - u\nabla w)$	Total degrees of freedom = 24

Note. Simplex coordinates $u, v, w = 1 - u - v; 0 \leq u, v, w \leq 1$; basis functions not normalized; edge-based functions must have tangential components continuous with neighboring cells. After [9].

REFERENCES

[1] J. C. Nedelec, "Mixed finite elements in R3," *Numer. Math.*, vol. 35, pp. 315–341, 1980.

[2] A. Bossavit and J. C. Verite, "A mixed FEM-BIEM method to solve 3D eddy current problems," *IEEE Trans. Magnetics*, vol. MAG-18, pp. 431–435, March 1982.

[3] G. Mur and A. T. de Hoop, "A finite element method for computing three-dimensional electromagnetic fields in inhomogeneous media," *IEEE Trans. Magnetics*, vol. MAG-21, pp. 2188–2191, Nov. 1985.

[4] M. L. Barton and Z. J. Cendes, "New vector finite elements for three-dimensional magnetic field computation," *J. Appl. Phys.*, vol. 61, pp. 3919–3921, April 1987.

[5] C. W. Crowley, "Mixed-order covariant projection finite elements for vector fields," Ph.D. Dissertation, McGill University, Montreal, Quebec, Feb. 1988.

[6] S. M. Rao and D. R. Wilton, "E-field, H-field, and combined-field solution for arbitrarily shaped three-dimensional dielectric bodies," *Electromagnetics*, vol. 10, pp. 407–421, 1990.

[7] R. D. Graglia, D. R. Wilton, and A. F. Peterson, "Higher order interpolatory vector bases for computational electromagnetics," *IEEE Trans. Antenn. Propagat.*, vol. 45, no. 3, pp. 329–342, March 1997.

[8] M. M. Ilic and B. M. Notaros, "Higher order hierarchical curved hexahedral vector finite elements for electromagnetic modeling," *IEEE Trans. Microwave Theory Tech.*, vol. 51, pp. 1026–1033, March 2003.

[9] R. S. Preissig and A. F. Peterson, "A rationale for p-refinement with the vector Helmholtz equation and two-dimensional vector finite elements," *ACES J.*, vol. 19, pp. 65–75, July 2004.

CHAPTER 5

Transforming Vector Basis Functions to Curved Cells

This chapter explains the steps required to map a vector basis function from a 2D reference cell to a curvilinear surface in x–y–z space. To begin in a logical manner, we first consider a mapping from 3-space to 3-space. The process of mapping vector functions is somewhat more complicated than the scalar mapping used in Chapter 2 to define the cell shape and location. For this reason, it is necessary to introduce base vectors, reciprocal base vectors, and a number of other vector relations. After the basis function transformation is discussed, we consider the conversion of vector derivatives on the curved cells to equivalent derivatives in the reference coordinates.

Vector parametric mappings were described by Stratton in his 1941 textbook [1], and will be reviewed below in a slightly different notation. Several curved-cell implementations that used the same type of vector mapping described here may be studied for reference [2–5]. These approaches will be used in Chapters 6 and 7 for the numerical solution of electromagnetic integral equations.

5.1 BASE VECTORS AND RECIPROCAL BASE VECTORS

Suppose we have the general three-dimensional situation involving a mapping from a 3D reference space described by coordinates (u, v, w)[1] to the x–y–z space. For

[1] In previous chapters, w was used as the third simplex coordinate in 2D space. Here, it will be used for the third dimension in the reference space.

our purposes, it is convenient to assume that the mapping functions $x(u, v, w)$, $y(u, v, w)$, and $z(u, v, w)$ are defined on a cell-by-cell basis using Lagrangian interpolation polynomials (as described in Chapter 2, but generalized to a 3D reference cell). A position vector from the origin (0, 0, 0) to a point (x, y, z) in the curved cell is expressed

$$\bar{r}(u, v, w) = x(u, v, w)\hat{x} + y(u, v, w)\hat{y} + z(u, v, w)\hat{z} \tag{5.1}$$

From a study of the differential displacement

$$d\bar{r} = \frac{\partial \bar{r}}{\partial u} du + \frac{\partial \bar{r}}{\partial v} dv + \frac{\partial \bar{r}}{\partial w} dw \tag{5.2}$$

it is apparent that three displacement vectors can be defined as

$$\bar{s} = \frac{\partial \bar{r}}{\partial u} = \frac{\partial x}{\partial u}\hat{x} + \frac{\partial y}{\partial u}\hat{y} + \frac{\partial z}{\partial u}\hat{z} \tag{5.3}$$

$$\bar{t} = \frac{\partial \bar{r}}{\partial v} = \frac{\partial x}{\partial v}\hat{x} + \frac{\partial y}{\partial v}\hat{y} + \frac{\partial z}{\partial v}\hat{z} \tag{5.4}$$

$$\bar{n} = \frac{\partial \bar{r}}{\partial w} = \frac{\partial x}{\partial w}\hat{x} + \frac{\partial y}{\partial w}\hat{y} + \frac{\partial z}{\partial w}\hat{z} \tag{5.5}$$

If parameters v and w are held constant, while u is varied, the mapping creates a curve. The vector $\bar{s}$ is tangential to that curve. Similarly, $\bar{t}$ is tangential to a curve defined by constant values of parameters u and w, and $\bar{n}$ is tangential to a curve defined by constant values of u and v. These three vectors are known as *base vectors*. If all three parameters are varied between constant limits to create a curvilinear cell in x–y–z space, two of the three base vectors are tangential to each face of that curvilinear cell. The base vectors are not necessarily mutually perpendicular at a point, nor are they unit vectors in general.

Alternatively, we can define three independent vectors in terms of the gradients

$$\bar{s}' = \nabla u = \frac{\partial u}{\partial x}\hat{x} + \frac{\partial u}{\partial y}\hat{y} + \frac{\partial u}{\partial z}\hat{z} \tag{5.6}$$

$$\bar{t}' = \nabla v = \frac{\partial v}{\partial x}\hat{x} + \frac{\partial v}{\partial y}\hat{y} + \frac{\partial v}{\partial z}\hat{z} \tag{5.7}$$

$$\bar{n}' = \nabla w = \frac{\partial w}{\partial x}\hat{x} + \frac{\partial w}{\partial y}\hat{y} + \frac{\partial w}{\partial z}\hat{z} \tag{5.8}$$

It should be apparent from the gradient operation that the vector $\bar{s}'$ is normal to a surface over which u is constant, while $\bar{t}'$ is normal to a surface on which v is constant. Similarly, the vector $\bar{n}'$ is normal to a surface of constant w. These vectors are known as *reciprocal base vectors.* The reciprocal base vectors are also not necessarily mutually perpendicular or of unit length. If parameters u, v, and w are varied between constant limits to create a curvilinear cell in x–y–z space, one reciprocal base vector is normal at every point on each face of the resulting curvilinear cell.

5.2 JACOBIAN RELATIONS

In the mapping between the reference cell and the curvilinear cell in x–y–z space, derivatives transform according to the relation

$$\begin{bmatrix} \frac{\partial}{\partial u} \\ \frac{\partial}{\partial v} \\ \frac{\partial}{\partial w} \end{bmatrix} = \begin{bmatrix} \frac{\partial x}{\partial u} & \frac{\partial y}{\partial u} & \frac{\partial z}{\partial u} \\ \frac{\partial x}{\partial v} & \frac{\partial y}{\partial v} & \frac{\partial z}{\partial v} \\ \frac{\partial x}{\partial w} & \frac{\partial y}{\partial w} & \frac{\partial z}{\partial w} \end{bmatrix} \begin{bmatrix} \frac{\partial}{\partial x} \\ \frac{\partial}{\partial y} \\ \frac{\partial}{\partial z} \end{bmatrix} \tag{5.9}$$

The three-by-three matrix in (5.9) is known as the Jacobian matrix

$$\mathbf{J} = \begin{bmatrix} \frac{\partial x}{\partial u} & \frac{\partial y}{\partial u} & \frac{\partial z}{\partial u} \\ \frac{\partial x}{\partial v} & \frac{\partial y}{\partial v} & \frac{\partial z}{\partial v} \\ \frac{\partial x}{\partial w} & \frac{\partial y}{\partial w} & \frac{\partial z}{\partial w} \end{bmatrix} \tag{5.10}$$

The differential volumes of the two spaces are related by the determinant of the Jacobian matrix

$$dx\,dy\,dz = (\det \mathbf{J})du\,dv\,dw \tag{5.11}$$

It is also useful to consider the inverse relation

$$\begin{bmatrix} \dfrac{\partial}{\partial x} \\ \dfrac{\partial}{\partial y} \\ \dfrac{\partial}{\partial z} \end{bmatrix} = \mathbf{J}^{-1} \begin{bmatrix} \dfrac{\partial}{\partial u} \\ \dfrac{\partial}{\partial v} \\ \dfrac{\partial}{\partial w} \end{bmatrix} \tag{5.12}$$

where the inverse of the Jacobian matrix is given directly in the form

$$\mathbf{J}^{-1} = \begin{bmatrix} \dfrac{\partial u}{\partial x} & \dfrac{\partial v}{\partial x} & \dfrac{\partial w}{\partial x} \\ \dfrac{\partial u}{\partial y} & \dfrac{\partial v}{\partial y} & \dfrac{\partial w}{\partial y} \\ \dfrac{\partial u}{\partial z} & \dfrac{\partial v}{\partial z} & \dfrac{\partial w}{\partial z} \end{bmatrix} \tag{5.13}$$

Observe that the rows of $\mathbf{J}$ are the components of the base vectors, while the columns of $\mathbf{J}^{-1}$ are those of the reciprocal base vectors. Of course, a matrix and its inverse are also related by

$$\mathbf{J}\,\mathbf{J}^{-1} = \mathbf{I} \tag{5.14}$$

where $\mathbf{I}$ is the identity matrix. Embodied in (5.14) are the nine equations

$$\frac{\partial x}{\partial u}\frac{\partial u}{\partial x} + \frac{\partial y}{\partial u}\frac{\partial u}{\partial y} + \frac{\partial z}{\partial u}\frac{\partial u}{\partial z} = 1 \tag{5.15}$$

$$\frac{\partial x}{\partial v}\frac{\partial v}{\partial x} + \frac{\partial y}{\partial v}\frac{\partial v}{\partial y} + \frac{\partial z}{\partial v}\frac{\partial v}{\partial z} = 1 \tag{5.16}$$

$$\frac{\partial x}{\partial w}\frac{\partial w}{\partial x} + \frac{\partial y}{\partial w}\frac{\partial w}{\partial y} + \frac{\partial z}{\partial w}\frac{\partial w}{\partial z} = 1 \tag{5.17}$$

$$\frac{\partial x}{\partial u}\frac{\partial v}{\partial x} + \frac{\partial y}{\partial u}\frac{\partial v}{\partial y} + \frac{\partial z}{\partial u}\frac{\partial v}{\partial z} = 0 \tag{5.18}$$

$$\frac{\partial x}{\partial u}\frac{\partial w}{\partial x} + \frac{\partial y}{\partial u}\frac{\partial w}{\partial y} + \frac{\partial z}{\partial u}\frac{\partial w}{\partial z} = 0 \tag{5.19}$$

$$\frac{\partial x}{\partial v}\frac{\partial u}{\partial x} + \frac{\partial y}{\partial v}\frac{\partial u}{\partial y} + \frac{\partial z}{\partial v}\frac{\partial u}{\partial z} = 0 \tag{5.20}$$

$$\frac{\partial x}{\partial v}\frac{\partial w}{\partial x} + \frac{\partial y}{\partial v}\frac{\partial w}{\partial y} + \frac{\partial z}{\partial v}\frac{\partial w}{\partial z} = 0 \tag{5.21}$$

$$\frac{\partial x}{\partial w}\frac{\partial u}{\partial x}+\frac{\partial y}{\partial w}\frac{\partial u}{\partial y}+\frac{\partial z}{\partial w}\frac{\partial u}{\partial z}=0 \tag{5.22}$$

$$\frac{\partial x}{\partial w}\frac{\partial v}{\partial x}+\frac{\partial y}{\partial w}\frac{\partial v}{\partial y}+\frac{\partial z}{\partial w}\frac{\partial v}{\partial z}=0 \tag{5.23}$$

From the definition of the base vectors and reciprocal base vectors, we observe that Eqs. (5.15)–(5.17) are equivalent to

$$\bar{s} \bullet \bar{s}' = 1 \tag{5.24}$$

$$\bar{t} \bullet \bar{t}' = 1 \tag{5.25}$$

and

$$\bar{n} \bullet \bar{n}' = 1 \tag{5.26}$$

Equations (5.18)–(5.23) are equivalent to

$$\bar{s} \bullet \bar{t}' = 0 \tag{5.27}$$

$$\bar{s} \bullet \bar{n}' = 0 \tag{5.28}$$

$$\bar{t} \bullet \bar{s}' = 0 \tag{5.29}$$

$$\bar{t} \bullet \bar{n}' = 0 \tag{5.30}$$

$$\bar{n} \bullet \bar{s}' = 0 \tag{5.31}$$

$$\bar{n} \bullet \bar{t}' = 0 \tag{5.32}$$

Thus, the base vectors and reciprocal base vectors satisfy these orthogonality relations.

To further explore the interrelation of the base and reciprocal base vectors, consider a constant-w surface, with base vectors $\bar{s}$ and $\bar{t}$ tangential to that surface. Their cross product can be expanded to obtain

$$\begin{aligned}
\bar{s} \times \bar{t} &= \frac{\partial \bar{r}}{\partial u} \times \frac{\partial \bar{r}}{\partial v} \\
&= \left\{\frac{\partial x}{\partial u}\hat{x}+\frac{\partial y}{\partial u}\hat{y}+\frac{\partial z}{\partial u}\hat{z}\right\} \times \left\{\frac{\partial x}{\partial v}\hat{x}+\frac{\partial y}{\partial v}\hat{y}+\frac{\partial z}{\partial v}\hat{z}\right\} \\
&= \hat{x}\left\{\frac{\partial y}{\partial u}\frac{\partial z}{\partial v}-\frac{\partial y}{\partial v}\frac{\partial z}{\partial u}\right\}+\hat{y}\left\{\frac{\partial z}{\partial u}\frac{\partial x}{\partial v}-\frac{\partial z}{\partial v}\frac{\partial x}{\partial u}\right\} \\
&\quad +\hat{z}\left\{\frac{\partial x}{\partial u}\frac{\partial y}{\partial v}-\frac{\partial x}{\partial v}\frac{\partial y}{\partial u}\right\}
\end{aligned} \tag{5.33}$$

This vector must point in a direction normal to the surface of constant w. However, the reciprocal base vector $\bar{n}'$ also points normal to a constant-w surface. That suggests that

$$K\bar{n}' = \bar{s} \times \bar{t} \tag{5.34}$$

for some scalar function $K(u, v, w)$. From a consideration of the determinant of the Jacobian matrix, it is apparent that

$$\begin{aligned}
\det \mathbf{J} &= \frac{\partial x}{\partial w}\left\{\frac{\partial y}{\partial u}\frac{\partial z}{\partial v} - \frac{\partial y}{\partial v}\frac{\partial z}{\partial u}\right\} + \frac{\partial y}{\partial w}\left\{\frac{\partial z}{\partial u}\frac{\partial x}{\partial v} - \frac{\partial z}{\partial v}\frac{\partial x}{\partial u}\right\} \\
&\quad + \frac{\partial z}{\partial w}\left\{\frac{\partial x}{\partial u}\frac{\partial y}{\partial v} - \frac{\partial x}{\partial v}\frac{\partial y}{\partial u}\right\} \\
&= \frac{\partial \bar{r}}{\partial w} \bullet \frac{\partial \bar{r}}{\partial u} \times \frac{\partial \bar{r}}{\partial v} \\
&= \bar{n} \bullet \bar{s} \times \bar{t}
\end{aligned} \tag{5.35}$$

As an aside, we note that the Jacobian determinant can be written in several equivalent forms:

$$\det \mathbf{J} = \bar{n} \bullet \bar{s} \times \bar{t} = \bar{s} \bullet \bar{t} \times \bar{n} = \bar{t} \bullet \bar{n} \times \bar{s} \tag{5.36}$$

It follows from (5.34) and (5.35) that

$$\bar{n} \bullet (K\bar{n}') = \bar{n} \bullet (\bar{s} \times \bar{t}) = \det \mathbf{J} \tag{5.37}$$

However, since $\bar{n} \bullet \bar{n}' = 1$, we obtain the result that the function K is given by

$$K = \det \mathbf{J} \tag{5.38}$$

and we conclude from (5.34) that, at all points within the curvilinear cell,

$$\bar{n}' = \frac{1}{\det \mathbf{J}} \bar{s} \times \bar{t} \tag{5.39}$$

Similarly, the other reciprocal base vectors can be written as

$$\bar{s}' = \frac{1}{\det \mathbf{J}} \bar{t} \times \bar{n} \tag{5.40}$$

$$\bar{t}' = \frac{1}{\det \mathbf{J}} \bar{n} \times \bar{s} \tag{5.41}$$

It is easy to show that

$$\det(\mathbf{J}^{-1}) = \frac{1}{\det \mathbf{J}} = \bar{s}' \bullet \bar{t}' \times \bar{n}' = \bar{t}' \bullet \bar{n}' \times \bar{s}' = \bar{n}' \bullet \bar{s}' \times \bar{t}' \tag{5.42}$$

and it follows by analogous reasoning that the base vectors can be expressed as

$$\bar{s} = \det \mathbf{J}\,(\bar{t}' \times \bar{n}') \tag{5.43}$$

$$\bar{t} = \det \mathbf{J}\,(\bar{n}' \times \bar{s}') \tag{5.44}$$

$$\bar{n} = \det \mathbf{J}\,(\bar{s}' \times \bar{t}') \tag{5.45}$$

5.3 REPRESENTATION OF VECTOR FIELDS

An understanding of the base and reciprocal base vectors is critical for the representation of vector quantities, such as the surface current density or the electric field within some region. If represented in terms of projections onto the base vectors, known as *covariant* components, one obtains the expression

$$\bar{E} = (\bar{E} \bullet \bar{s})\bar{s}' + (\bar{E} \bullet \bar{t})\bar{t}' + (\bar{E} \bullet \bar{n})\bar{n}' \tag{5.46}$$

Alternatively, a vector can be represented in terms of its projections onto the reciprocal base vectors, or in terms of its *contravariant* components, leading to

$$\bar{E} = (\bar{E} \bullet \bar{s}')\bar{s} + (\bar{E} \bullet \bar{t}')\bar{t} + (\bar{E} \bullet \bar{n}')\bar{n} \tag{5.47}$$

The covariant components (such as $\bar{E} \bullet \bar{s}$) are the tangential components along the various curves defined by holding two of the three parameters (u, v, w) constant, while the contravariant components (such as $\bar{E} \bullet \bar{s}'$) are the components perpendicular to the constant parameter surfaces. Of course, (5.46) and (5.47) are just different ways of expressing the same vector function $\bar{E}$.

When discretizing vector integral equations, we employ either divergence-conforming or curl-conforming basis functions to represent the unknown vector quantities. With divergence-conforming functions, our goal is to maintain the normal-vector continuity of the function across cell boundaries, and the interpolation properties (if any) associated with "normal" components at various locations within the cell. Since the quantities of interest are the normal vector components, it will be convenient to work with the contravariant components as expressed in (5.47).

In contrast, when employing curl-conforming basis functions, our goal is usually to maintain the tangential-vector continuity between cells and to maintain the interpolation properties, if any, of the "tangential" components of those basis functions. Thus, the natural approach is to work with the covariant components of the functions as in (5.46).

5.4 RESTRICTION TO SURFACES

Our immediate problem of interest involves the mapping from a square or triangular planar reference cell (2D) to a curved surface in a 3D space. In the reference cell, the third variable (w) is not involved. The mapping functions x, y, and z are only functions of u and v. In common with our previous approach, the position vector from the origin to a point (x, y, z) is given by

$$\bar{r}(u, v) = x(u, v)\hat{x} + y(u, v)\hat{y} + z(u, v)\hat{z} \tag{5.48}$$

The base vectors for this cell are

$$\bar{s} = \frac{\partial \bar{r}}{\partial u} = \frac{\partial x}{\partial u}\hat{x} + \frac{\partial y}{\partial u}\hat{y} + \frac{\partial z}{\partial u}\hat{z} \tag{5.49}$$

$$\bar{t} = \frac{\partial \bar{r}}{\partial v} = \frac{\partial x}{\partial v}\hat{x} + \frac{\partial y}{\partial v}\hat{y} + \frac{\partial z}{\partial v}\hat{z} \tag{5.50}$$

while the reciprocal base vectors are

$$\bar{s}' = \nabla u = \frac{\partial u}{\partial x}\hat{x} + \frac{\partial u}{\partial y}\hat{y} + \frac{\partial u}{\partial z}\hat{z} \tag{5.51}$$

$$\bar{t}' = \nabla v = \frac{\partial v}{\partial x}\hat{x} + \frac{\partial v}{\partial y}\hat{y} + \frac{\partial v}{\partial z}\hat{z} \tag{5.52}$$

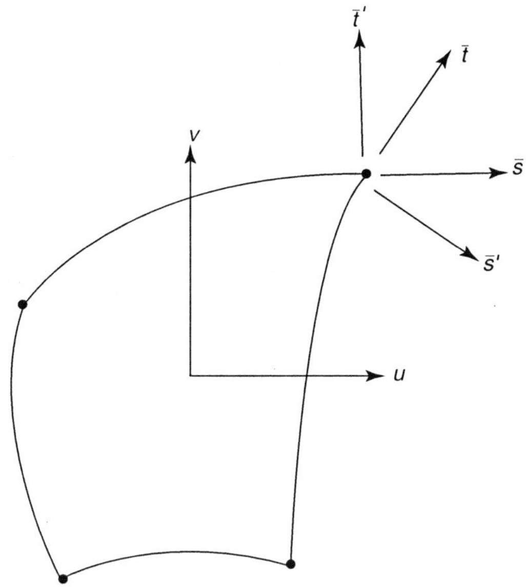

FIGURE 5.1: Base vectors $\bar{s}$, $\bar{t}$ and reciprocal base vectors $\bar{s}'$, $\bar{t}'$ on a curvilinear patch.

The base vectors are tangential to the surface defined by (5.48), while the reciprocal base vectors are normal to constant-u or constant-v surfaces, respectively. These vectors are illustrated in Fig. 5.1.

Derivatives transform according to

$$\begin{bmatrix} \dfrac{\partial}{\partial u} \\ \dfrac{\partial}{\partial v} \end{bmatrix} = \begin{bmatrix} \dfrac{\partial x}{\partial u} & \dfrac{\partial y}{\partial u} & \dfrac{\partial z}{\partial u} \\ \dfrac{\partial x}{\partial v} & \dfrac{\partial y}{\partial v} & \dfrac{\partial z}{\partial v} \end{bmatrix} \begin{bmatrix} \dfrac{\partial}{\partial x} \\ \dfrac{\partial}{\partial y} \\ \dfrac{\partial}{\partial z} \end{bmatrix} \tag{5.53}$$

and

$$\begin{bmatrix} \dfrac{\partial}{\partial x} \\ \dfrac{\partial}{\partial y} \\ \dfrac{\partial}{\partial z} \end{bmatrix} = \begin{bmatrix} \dfrac{\partial u}{\partial x} & \dfrac{\partial v}{\partial x} \\ \dfrac{\partial u}{\partial y} & \dfrac{\partial v}{\partial y} \\ \dfrac{\partial u}{\partial z} & \dfrac{\partial v}{\partial z} \end{bmatrix} \begin{bmatrix} \dfrac{\partial}{\partial u} \\ \dfrac{\partial}{\partial v} \end{bmatrix} \tag{5.54}$$

The Jacobian matrix in (5.53) has dimension 2×3 in accordance with the dimensionality of the mapping, while the inverse Jacobian matrix in (5.54) has dimension 3×2.

Since the functions x, y, and z are defined explicitly (Chapter 2), the entries of the Jacobian matrix in (5.53) are readily available. However, the matrix entries in (5.54) are not. In situations where the inverse Jacobian entries are required (to be encountered below), it is necessary to be able to invert the Jacobian matrix numerically, which is not possible if it is not a square matrix! Thus, it is desirable to introduce a dummy parameter w, to fill the equations in (5.53) and (5.54) out to 3×3 systems. Since the third variable is arbitrary, the dummy parameter may be constrained so that

$$\bar{n}' = \nabla w = \frac{\bar{s} \times \bar{t}}{|\bar{s} \times \bar{t}|} \tag{5.55}$$

which, by virtue of its definition, is a unit vector. (The only assumption being made here is that w can be chosen to make $\bar{n}'$ a unit vector.) From (5.33),

$$\begin{aligned} |\bar{s} \times \bar{t}| &= \sqrt{\left(\frac{\partial y}{\partial u}\frac{\partial z}{\partial v} - \frac{\partial z}{\partial u}\frac{\partial y}{\partial v}\right)^2 + \left(\frac{\partial z}{\partial u}\frac{\partial x}{\partial v} - \frac{\partial x}{\partial u}\frac{\partial z}{\partial v}\right)^2 + \left(\frac{\partial x}{\partial u}\frac{\partial y}{\partial v} - \frac{\partial y}{\partial u}\frac{\partial x}{\partial v}\right)^2} \\ &= D(u, v) \end{aligned} \tag{5.56}$$

However, Eq. (5.39) still holds in this situation, implying that the quantity equivalent to the determinant in this case is given by

$$\det \mathbf{J} = D \tag{5.57}$$

From (5.35), it appears that one way by which (5.57) can be true is if

$$\begin{aligned} \bar{n} = \frac{\partial \bar{r}}{\partial w} = \frac{1}{D}\left(\frac{\partial y}{\partial u}\frac{\partial z}{\partial v} - \frac{\partial z}{\partial u}\frac{\partial y}{\partial v}\right)\hat{x} + \frac{1}{D}\left(\frac{\partial z}{\partial u}\frac{\partial x}{\partial v} - \frac{\partial x}{\partial u}\frac{\partial z}{\partial v}\right)\hat{y} \\ + \frac{1}{D}\left(\frac{\partial x}{\partial u}\frac{\partial y}{\partial v} - \frac{\partial y}{\partial u}\frac{\partial x}{\partial v}\right)\hat{z} \end{aligned} \tag{5.58}$$

which is the same as

$$\bar{n} = \frac{1}{D}\bar{s} \times \bar{t} = \bar{n}' \tag{5.59}$$

The implications of this choice for w are summarized as follows:

1. $\bar{n} = \bar{n}'$, and both are unit vectors normal to the surface of constant w.

2. Vectors $\bar{s}$ and $\bar{t}$ are tangential to the surface of constant w, by definition. Since $\bar{n}$ is perpendicular to the surface (by our choice of w), the reciprocal base vectors $\bar{s}'$ and $\bar{t}'$ are also tangential to the surface of constant w. This conclusion follows from (5.40) and (5.41), which in this case simplify to

$$\bar{s}' = \frac{1}{D}\bar{t} \times \bar{n} \tag{5.60}$$

$$\bar{t}' = \frac{1}{D}\bar{n} \times \bar{s} \tag{5.61}$$

3. Since (a) $\bar{n}$ is a unit vector, (b) $\bar{n}$ is perpendicular to $\bar{t}$, and (c) $\bar{n}$ is perpendicular to $\bar{s}$, the magnitudes of the reciprocal base vectors are related to those of the base vectors by

$$|\bar{s}'| = \left|\frac{1}{D}\bar{t} \times \hat{n}\right| = \frac{|\bar{t}|}{D} \tag{5.62}$$

$$|\bar{t}'| = \left|\frac{1}{D}\hat{n} \times \bar{s}\right| = \frac{|\bar{s}|}{D} \tag{5.63}$$

4. The explicit form of the 3×3 Jacobian matrix is given by

$$\mathbf{J} = \begin{bmatrix} \frac{\partial x}{\partial u} & \frac{\partial y}{\partial u} & \frac{\partial z}{\partial u} \\ \frac{\partial x}{\partial v} & \frac{\partial y}{\partial v} & \frac{\partial z}{\partial v} \\ \frac{1}{D}\left(\frac{\partial y}{\partial u}\frac{\partial z}{\partial v} - \frac{\partial z}{\partial u}\frac{\partial y}{\partial v}\right) & \frac{1}{D}\left(\frac{\partial z}{\partial u}\frac{\partial x}{\partial v} - \frac{\partial x}{\partial u}\frac{\partial z}{\partial v}\right) & \frac{1}{D}\left(\frac{\partial x}{\partial u}\frac{\partial y}{\partial v} - \frac{\partial y}{\partial u}\frac{\partial x}{\partial v}\right) \end{bmatrix} \tag{5.64}$$

and can be used to compute the inverse Jacobian matrix entries in (5.54).

In summary, when working with the base and reciprocal base vectors on a 2D surface in 3D space, it will be convenient to invoke the assumption in (5.55).

This ensures that base vectors $\bar{s}$ and $\bar{t}$, and reciprocal base vectors $\bar{s}'$ and $\bar{t}'$, are tangential to the surface and permits the computation of all needed parameters.

5.5 CURL-CONFORMING BASIS FUNCTIONS ON CURVILINEAR CELLS

We now turn our attention to the mapping of the vector basis functions from the reference cell in (u, v) space to the curvilinear cell in (x, y, z) space. It is simpler to discuss the mapping of curl-conforming basis functions, so we consider those first. Curl-conforming functions maintain tangential-vector continuity across cell boundaries, and our mapping procedure must ensure that behavior. Since the principal quantities of interest are the tangential vector components at cell boundaries, it is natural to work with the covariant components of the basis functions as expressed in (5.46). We desire to match the tangential components of the basis functions (where "tangential" corresponds to the directions of the base vectors $\bar{s}$ and $\bar{t}$, even at locations away from the cell boundary) on the curvilinear cell with those on the reference cell:

$$\bar{B} \bullet \bar{s} = R_u^{\text{curl}} \tag{5.65}$$

$$\bar{B} \bullet \bar{t} = R_v^{\text{curl}} \tag{5.66}$$

In accordance with (5.46), the basis function can be written as

$$\bar{B} = R_u^{\text{curl}} \bar{s}' + R_v^{\text{curl}} \bar{t}' \tag{5.67}$$

On a component-by-component level, (5.67) is equivalent to the matrix relation

$$\begin{bmatrix} B_x \\ B_y \\ B_z \end{bmatrix} = \begin{bmatrix} \dfrac{\partial u}{\partial x} & \dfrac{\partial v}{\partial x} \\ \dfrac{\partial u}{\partial y} & \dfrac{\partial v}{\partial y} \\ \dfrac{\partial u}{\partial z} & \dfrac{\partial v}{\partial z} \end{bmatrix} \begin{bmatrix} R_u^{\text{curl}} \\ R_v^{\text{curl}} \end{bmatrix} = \mathbf{J}^{-1} \begin{bmatrix} R_u^{\text{curl}} \\ R_v^{\text{curl}} \end{bmatrix} \tag{5.68}$$

where $\mathbf{J}^{-1}$ is used symbolically to denote the 3×2 matrix in (5.54) and (5.68) and not necessarily the "inverse" of $\mathbf{J}$, which must be obtained from the 3×3 system in

(5.64). There are two aspects of (5.68) to consider, (1) the continuity of tangential fields across cell boundaries, and (2) the normalization of the basis functions in the curvilinear domain.

As a consequence of the way that the mapping of the cell coordinates is defined (Chapter 2), the vectors $\bar{s}$ and $\bar{t}$ are tangential to the boundaries of the curvilinear patch at the extreme values of the parameters v and u. Within two adjacent cells sharing a boundary that coincides with a constant-u surface, and common cell endpoints in v, the base vector tangential to that boundary is defined by the derivatives of x, y, and z with respect to v, and will therefore be the same in either cell. The imposition of (5.65) and (5.66) in adjacent cells, for basis functions that are adjusted to maintain tangential continuity in the reference space, will ensure that the resulting vector basis functions in (x, y, z) space also maintain tangential continuity.

The curl-conforming basis functions (Chapter 4) can be normalized so that their tangential components have unity value at appropriate locations along a cell boundary. Consider one such location v_i along a boundary where $\bar{t}$ is tangential and R_v^{curl} has a unit tangent. In the curvilinear space, it follows that

$$\bar{B} \bullet \bar{t}\Big|_{v=v_i} = 1 \tag{5.69}$$

Since the vector $\bar{t}$ is not a unit vector, (5.69) implies that the tangential component in the curvilinear space is

$$\bar{B} \bullet \hat{t}\Big|_{v=v_i} = \frac{1}{|\bar{t}|}\Big|_{v=v_i} \tag{5.70}$$

where $\hat{t}$ is a unit vector in the $\bar{t}$ direction. Consequently, an additional scaling is necessary before the curl-conforming basis functions in the curvilinear space exhibit unity tangential components. In a similar manner, at a location along a boundary where $\bar{s}$ is tangential and R_u^{curl} has a unity tangential component,

$$\bar{B} \bullet \hat{s}\Big|_{u=u_i} = \frac{1}{|\bar{s}|}\Big|_{u=u_i} \tag{5.71}$$

These scale factors are constant within a cell, since they are base vectors evaluated at specific locations, but they differ with the basis function and from cell to cell in general. If it is necessary to compute the value of $\bar{B}$ at a point within the cell, the 3×3 matrix in (5.64) may be evaluated at that point and inverted to provide numerical values for the reciprocal base vector components.

5.6 DIVERGENCE-CONFORMING BASIS FUNCTIONS ON CURVILINEAR CELLS

Divergence-conforming functions maintain normal-vector continuity across cell boundaries, and our mapping procedure must ensure that behavior in x–y–z space. Since the principal quantities of interest are the normal-vector components on cell boundaries, it will be necessary to work with the contravariant components of the basis functions as expressed in (5.47).

While the mapping of the cell coordinates as defined in Chapter 2 ensures that the tangential base vectors on either side of a common cell boundary are the same, it does not guarantee that property for the reciprocal base vectors normal to a common boundary. For a boundary defined by a constant value of parameter u, for instance, the base vector $\bar{t}$ (tangential to that boundary) is a sole function of v, which has the same value at any location along the boundary for both cells sharing that boundary. The reciprocal base vector $\bar{s}'$ (normal to that same boundary) is also a function of u, meaning that it generally will not be the same on either side of a constant-u boundary.

The magnitude of $\bar{s}'$ at any location within a cell was determined in (5.62) to be

$$|\bar{s}'| = \frac{|\bar{t}|}{D} \tag{5.72}$$

where D is defined in (5.56). Since this general expression is the same in either cell sharing a constant-u boundary, and $\bar{t}$ is the same on either side of that same cell boundary, it is sufficient to scale by the function D to obtain vectors of the same length along the boundary in either cell. Similarly, on a boundary defined by a constant value for v, the magnitude of $\bar{t}'$ is

$$|\bar{t}'| = \frac{|\bar{s}|}{D} \tag{5.73}$$

Since $\bar{s}$ is the same on either side of a constant-v boundary, it is again sufficient to scale by D. (Note that D is generally not a constant along the boundary or within the cell.)

Thus, to impose normal-vector continuity, we will need to equate the contravariant components of the basis functions in (x, y, z) space with the corresponding components of divergence-conforming functions on the reference cell, scaled by the function $D(u, v)$:

$$\bar{B} \bullet \bar{s}' = \frac{1}{D} R_u^{\text{div}} \tag{5.74}$$

$$\bar{B} \bullet \bar{t}' = \frac{1}{D} R_v^{\text{div}} \tag{5.75}$$

Equivalently, in accordance with (5.47), the basis function can be written as

$$\bar{B} = \frac{1}{D} R_u^{\text{div}} \bar{s} + \frac{1}{D} R_v^{\text{div}} \bar{t} \tag{5.76}$$

Equation (5.76) is the same as the matrix relation

$$\begin{bmatrix} B_x \\ B_y \\ B_z \end{bmatrix} = \frac{1}{D} \begin{bmatrix} \frac{\partial x}{\partial u} & \frac{\partial x}{\partial v} \\ \frac{\partial y}{\partial u} & \frac{\partial y}{\partial v} \\ \frac{\partial z}{\partial u} & \frac{\partial z}{\partial v} \end{bmatrix} \begin{bmatrix} R_u^{\text{div}} \\ R_v^{\text{div}} \end{bmatrix} = \frac{1}{D} \mathbf{J}^{\text{T}} \begin{bmatrix} R_u^{\text{div}} \\ R_v^{\text{div}} \end{bmatrix} \tag{5.77}$$

where $\mathbf{J}^{\text{T}}$ is the transpose of the matrix in (5.53).

The divergence-conforming basis functions presented in Chapter 3 are normalized so that their normal components have unity value at appropriate locations along a cell boundary. From the definition in (5.77), at some location v_i along a constant-u boundary where $\bar{s}'$ is normal and R_u^{div} has a unit normal component, the basis function in the curvilinear space satisfies

$$\bar{B} \bullet \bar{s}'\big|_{v=v_i} = \frac{1}{D} \tag{5.78}$$

Because $\bar{s}'$ is not a unit vector, (5.78) implies that the normal component of the basis function is

$$\bar{B} \bullet \hat{s}'\Big|_{v=v_i} = \frac{1}{D\,|\bar{s}'|}\Bigg|_{v=v_i} \tag{5.79}$$

Since the reciprocal base vectors are not explicitly defined by the mapping, it is convenient to employ (5.62) to obtain the equivalent result

$$\bar{B} \bullet \hat{s}'\Big|_{v=v_i} = \frac{1}{|\bar{t}|}\Bigg|_{v=v_i} \tag{5.80}$$

Therefore, this basis function must be scaled by the magnitude of $\bar{t}$ at the interpolation point v_i so that its normal component exhibits a unity value there. Similarly, at a location along a boundary where $\hat{t}'$ is normal and R_v^{div} has a unity normal component, analogous reasoning leads to the result

$$\bar{B} \bullet \hat{t}'\Big|_{u=u_i} = \frac{1}{|\bar{s}|}\Bigg|_{u=u_i} \tag{5.81}$$

These scale factors are constant within a cell, but may be different for each basis function and each cell.

Since the mapping (Chapter 2) provides an explicit expression for x, y, z and their derivatives, the basis functions are easily computed from (5.77) at points within the curvilinear cell.

5.7 THE IMPLEMENTATION OF VECTOR DERIVATIVES

An intrinsic feature of the type of analysis under consideration is that all operations involving the basis functions on the curvilinear cell can be transferred to the reference cell. First, we note that the surface gradient operator may be expressed as

$$\begin{aligned}\nabla f &= \frac{\partial f}{\partial u}\nabla u + \frac{\partial f}{\partial v}\nabla v \\ &= \frac{\partial f}{\partial u}\bar{s}' + \frac{\partial f}{\partial v}\bar{t}'\end{aligned} \tag{5.82}$$

Consider a curl-conforming basis function defined by

$$\begin{aligned}\bar{B} &= R_u^{\text{curl}}\bar{s}' + R_v^{\text{curl}}\bar{t}' \\ &= R_u^{\text{curl}}\nabla u + R_v^{\text{curl}}\nabla v\end{aligned} \tag{5.83}$$

Using the standard vector identity

$$\nabla \times (f\nabla g) = \nabla f \times \nabla g \tag{5.84}$$

the curl of $\bar{B}$ can be expanded to produce

$$\begin{aligned}\nabla \times \bar{B} &= \nabla \times \left(R_u^{\text{curl}}\nabla u\right) + \nabla \times \left(R_v^{\text{curl}}\nabla v\right) \\ &= \nabla\left(R_u^{\text{curl}}\right) \times \nabla u + \nabla\left(R_v^{\text{curl}}\right) \times \nabla v \\ &= \nabla\left(R_u^{\text{curl}}\right) \times \bar{s}' + \nabla\left(R_v^{\text{curl}}\right) \times \bar{t}'\end{aligned} \tag{5.85}$$

From (5.82)

$$\nabla\left(R_u^{\text{curl}}\right) = \frac{\partial R_u^{\text{curl}}}{\partial u}\bar{s}' + \frac{\partial R_u^{\text{curl}}}{\partial v}\bar{t}' \tag{5.86}$$

$$\nabla\left(R_v^{\text{curl}}\right) = \frac{\partial R_v^{\text{curl}}}{\partial u}\bar{s}' + \frac{\partial R_v^{\text{curl}}}{\partial v}\bar{t}' \tag{5.87}$$

Thus, Eq. (5.85) can be written as

$$\nabla \times \bar{B} = \frac{\partial R_u^{\text{curl}}}{\partial u}\bar{s}' \times \bar{s}' + \frac{\partial R_u^{\text{curl}}}{\partial v}\bar{t}' \times \bar{s}' + \frac{\partial R_v^{\text{curl}}}{\partial u}\bar{s}' \times \bar{t}' + \frac{\partial R_v^{\text{curl}}}{\partial v}\bar{t}' \times \bar{t}' \tag{5.88}$$

The first and last terms on the right side of (5.88) vanish, leaving

$$\begin{aligned}\nabla \times \bar{B} &= \frac{\partial R_u^{\text{curl}}}{\partial v}\bar{t}' \times \bar{s}' + \frac{\partial R_v^{\text{curl}}}{\partial u}\bar{s}' \times \bar{t}' \\ &= \left\{\frac{\partial R_v^{\text{curl}}}{\partial u} - \frac{\partial R_u^{\text{curl}}}{\partial v}\right\}(\bar{s}' \times \bar{t}')\end{aligned} \tag{5.89}$$

Finally, using (5.45), we obtain

$$\nabla \times \bar{B} = \frac{1}{D}\left\{\frac{\partial R_v^{\text{curl}}}{\partial u} - \frac{\partial R_u^{\text{curl}}}{\partial v}\right\}\bar{n} \tag{5.90}$$

where $\bar{n}$ is a unit vector in accordance with the development in Section 5.4. The curl of $\bar{B}$ in x–y–z space may be obtained directly in terms of the reference coordinates u and v, using an expression (5.90) that is in fact a local curl operation scaled by D.

Consider a divergence-conforming basis function defined by

$$\begin{aligned} \bar{B} &= \tfrac{1}{D} R_u^{\text{div}} \bar{s} + \tfrac{1}{D} R_v^{\text{div}} \bar{t} \\ &= R_u^{\text{div}} (\bar{t}' \times \bar{n}') + R_v^{\text{div}} (\bar{n}' \times \bar{s}') \end{aligned} \tag{5.91}$$

Since $\bar{t}' \times \bar{n}' = \nabla v \times \nabla w$, the vector identities

$$\nabla \bullet (\bar{A} \times \bar{B}) = \bar{B} \bullet \nabla \times \bar{A} - \bar{A} \bullet \nabla \times \bar{B} \tag{5.92}$$

and

$$\nabla \times \nabla f = 0 \tag{5.93}$$

can be used to show that

$$\nabla \bullet (\bar{t}' \times \bar{n}') = 0 \tag{5.94}$$

It follows from the additional vector identity

$$\nabla \bullet (f \bar{V}) = f \nabla \bullet \bar{V} + \nabla f \bullet \bar{V} \tag{5.95}$$

that the divergence of $\bar{B}$ is given by

$$\begin{aligned} \nabla \bullet \bar{B} &= \nabla \left(R_u^{\text{div}} \right) \bullet (\bar{t}' \times \bar{n}') + \nabla \left(R_v^{\text{div}} \right) \bullet (\bar{n}' \times \bar{s}') \\ &= \nabla \left(R_u^{\text{div}} \right) \bullet \frac{1}{D} \bar{s} + \nabla \left(R_v^{\text{div}} \right) \bullet \frac{1}{D} \bar{t} \\ &= \frac{1}{D} \left\{ \frac{\partial R_u^{\text{div}}}{\partial u} \bar{s}' + \frac{\partial R_u^{\text{div}}}{\partial v} \bar{t}' \right\} \bullet \bar{s} + \left\{ \frac{\partial R_v^{\text{div}}}{\partial u} \bar{s}' + \frac{\partial R_v^{\text{div}}}{\partial v} \bar{t}' \right\} \bullet \bar{t} \\ &= \frac{1}{D} \left\{ \frac{\partial R_u^{\text{div}}}{\partial u} + \frac{\partial R_v^{\text{div}}}{\partial v} \right\} \end{aligned} \tag{5.96}$$

Therefore, the divergence of (5.91) in x–y–z space can be calculated directly from the application of a divergence operator in the reference coordinates, scaled by the function D.

5.8 SUMMARY

This chapter has presented a procedure for mapping divergence-conforming and curl-conforming basis functions to a curvilinear surface in 3D space. This process ensures the continuity of normal components at cell boundaries (divergence-conforming) or tangential components at cell boundaries (curl-conforming) as required. The normalization of the basis functions is considered, along with a procedure for transferring vector derivatives from the curvilinear coordinates to the reference coordinates. These results will be used in Chapters 6 and 7.

REFERENCES

[1] J. A. Stratton, *Electromagnetic Theory*. New York: McGraw-Hill, 1941.

[2] C. W. Crowley, "Mixed-order Covariant Projection Finite Elements for Vector Fields," Ph.D. Dissertation, McGill University, Montreal, Quebec, 1988.

[3] S. Wandzura, "Electric current basis functions for curved surfaces," *Electromagnetics*, vol. 12, pp. 77–91, 1992.

[4] G. E. Antilla and N. G. Alexopoulos, "Scattering from complex three-dimensional geometries by a curvilinear hybrid finite element-integral equation approach," *J. Opt. Soc. Am. A*, vol. 11, pp. 1445–1457, 1994.

[5] A. F. Peterson and K. R. Aberegg, "Parametric mapping of vector basis functions for surface integral equation formulations," *ACES J.*, vol. 10, pp. 107–115, 1995.

CHAPTER 6

Use of Divergence-conforming Basis Functions with the Electric Field Integral Equation

To demonstrate the vector mapping procedures developed in Chapter 5, we consider the electric field integral equation (EFIE) for scattering from perfectly conducting objects with curved surfaces. The EFIE was derived in Chapter 1. Since the EFIE operator involves a divergence of the surface current density, the method of moments (MoM) procedure will be illustrated in conjunction with divergence-conforming basis and testing functions. Expressions for the MoM matrix entries incorporating the curved-cell mappings are developed. Several numerical results are presented for illustration.

6.1 TESTED FORM OF THE EFIE

The MoM procedure employs a testing function to convert the equation under consideration into a weak form, which reduces the number of derivatives acting on the unknown function. If the EFIE of Eq. (1.8) is multiplied (by scalar product) with a vector testing function $\bar{T}(s, t)$, tangential to the surface, one obtains the equation

$$\iint_{\text{surface}} \bar{T} \bullet \bar{E}^{\text{inc}} ds\, dt = -\iint_{\text{surface}} \bar{T} \bullet \bar{E}^{s} ds\, dt \tag{6.1}$$

where the integration is performed over the surface of the conductor. The incident electric field $\bar{E}^{\text{inc}}$ is the field in the absence of the structure (the excitation) and is assumed known. The function $\bar{E}^{\text{s}}$ is the "scattered" electric field produced by the surface current $\bar{J}$, also in the absence of the structure, and may be obtained from the expression

$$\bar{E}^{\text{s}} = -j\omega\mu \iint_{\text{surface}} \bar{J}(s', t')G\, ds'\, dt' + \frac{1}{j\omega\varepsilon}\nabla \iint_{\text{surface}} \nabla' \bullet \bar{J}(s', t')G\, ds'\, dt' \tag{6.2}$$

where G is the free space Green's function

$$G(R) = \frac{e^{-jkR}}{4\pi R} \tag{6.3}$$

and

$$R = \sqrt{[x(s, t) - x(s', t')]^2 + [y(s, t) - y(s', t')]^2 + [z(s, t) - z(s', t')]^2} \tag{6.4}$$

To obtain (6.2) from Eq. (1.8), we employed the well-known relation

$$\nabla \bullet \bar{A} = \iint_{\text{surface}} \nabla' \bullet \bar{J}(s', t')G(R)ds'\, dt' \tag{6.5}$$

The operator in (6.2) involves a divergence of the surface current density, and provides motivation for the use of divergence-conforming basis functions.

6.2 THE SUBSECTIONAL MODEL

The surface of the conducting object is to be represented by curved cells having either three sides or four sides. Such a mapping may be obtained using the quadratic Lagrangian interpolation polynomials of Sections 2.3 and 2.5, and reference cells that are triangular or square. The cell shapes are defined by explicit functions of local parameters u and v, including Eqs. (2.29) and (2.45) for x, and equivalent expressions for y and z. Since these equations are to be incorporated into the computer program that performs the analysis, the curved-cell model that must be provided as input to the analysis program consists of a list of the coordinates of the nodes that define each cell (nine nodes for square cells, six for triangular) and connectivity arrays that link the cells, edges, and nodes on the surface.

The surface current density in (6.2) is represented by vector basis functions $\{\bar{B}_n\}$ that are distributed over the curved-cell surface model. As discussed in Chapter 3, the divergence-conforming basis functions either straddle two adjacent cells or are entirely confined within a single cell. The connectivity arrays associated with the surface can be used to provide a systematic way of organizing the basis functions within a cell and linking edge-based basis functions that straddle cells to appropriate basis functions in adjacent cells. The current is represented by the summation

$$\bar{J}(s,\ t) = \sum_{n=1}^{N} I_n \bar{B}_n(s,\ t) \tag{6.6}$$

where $\{I_n\}$ denote N complex-valued coefficients that henceforth are the unknowns to be determined. The discretized EFIE exhibits symmetry between the basis functions and testing functions—look ahead to Eq. (6.13)—and it is therefore convenient to define the testing functions to be the same as the divergence-conforming basis functions

$$\bar{T}_m(s,\ t) = \bar{B}_m(s,\ t) \tag{6.7}$$

When used within (6.1), the testing functions provide a means for obtaining N linearly independent equations from the EFIE. The equations can be organized into the form of a matrix equation

$$\mathbf{E} = \mathbf{Z}\mathbf{I} \tag{6.8}$$

where **E** and **I** are $N \times 1$ column vectors and **Z** is an $N \times N$ matrix. The entries of **I** are the coefficients $\{I_n\}$, while those of **E** are given by

$$E_m = \iint\limits_{\text{surface}} \bar{T}_m \bullet \bar{E}^{\text{inc}} ds\, dt \tag{6.9}$$

The entries of **Z** have the general form

$$Z_{mn} = -\iint\limits_{\text{surface}} \bar{T}_m \bullet \bar{E}_n^{\text{s}} ds\, dt \tag{6.10}$$

where $\bar{E}_n^{\text{s}}$ denotes the electric field produced by the nth basis function. Although this quantity is defined by the expression

$$\bar{E}_n^{\text{s}} = -j\omega\mu \iint\limits_{\text{surface}} \bar{B}_n(s', t')G\, ds'dt' + \frac{1}{j\omega\varepsilon}\nabla \iint\limits_{\text{surface}} \nabla' \bullet \bar{B}_n(s', t')G\, ds'dt' \tag{6.11}$$

the form of (6.11) does not exploit the testing function to eliminate a derivative. Instead, using the vector identity

$$\nabla \bullet (f\,\bar{T}) = \bar{T} \bullet \nabla f + f\,\nabla \bullet \bar{T} \tag{6.12}$$

in conjunction with the divergence theorem, (6.10) may be recast as

$$\begin{aligned} Z_{mn} = j\omega\mu \iint\limits_{s,t} \iint\limits_{s',t'} \bar{T}_m(s, t) \bullet \bar{B}_n(s', t')G\, ds'\, dt'\, ds\, dt \\ + \frac{1}{j\omega\varepsilon} \iint\limits_{s,t} \nabla \bullet \bar{T}_m \iint\limits_{s',t'} \nabla' \bullet \bar{B}_n G\, ds'dt'ds\, dt \end{aligned} \tag{6.13}$$

The matrix equation in (6.8) may be solved to produce the coefficients $\{I_n\}$, after which any other quantity of interest may be obtained by integrating over the current density in (6.6).

If (6.13) is used with a basis function that is not divergence-conforming, the divergence operation would produce a Dirac delta function type of behavior at any locations where the function lacks normal-vector continuity. In special cases when it might be necessary to employ nondivergence-conforming basis functions (such as to incorporate lumped-element feeds or loads, junctions between surfaces and wires, etc.) the Dirac delta function must be included as part of the expression. The result is equivalent to including a line integral along with the surface integrals in (6.13).

6.3 MAPPED MoM MATRIX ENTRIES

Since each cell comprising the surface is defined by an independent mapping, the preceding integral expressions must be evaluated separately for each cell. The *observer* cells where the testing functions reside are generally different from the *source* cells where the basis functions reside. A single matrix entry may require integrals over as many as four cells, since each edge-based basis or testing function straddles one or two cells. In the following, an index (m or n) is used to denote specific testing and basis functions, and the discussion considers a single observer cell and a single source cell. The pointer arrays associated with the surface model are used to identify the appropriate cells where these functions reside.

The integrals required in (6.9) and (6.13) must be transformed into integrals in the local coordinates of each reference cell, as explained in Chapter 5. The divergence-conforming basis functions are defined by the transformation in Eq. (5.77):

$$\begin{bmatrix} B_x \\ B_y \\ B_z \end{bmatrix} = \frac{1}{D} \begin{bmatrix} \frac{\partial x}{\partial u} & \frac{\partial x}{\partial v} \\ \frac{\partial y}{\partial u} & \frac{\partial y}{\partial v} \\ \frac{\partial z}{\partial u} & \frac{\partial z}{\partial v} \end{bmatrix} \begin{bmatrix} R_u^{\text{div}} \\ R_v^{\text{div}} \end{bmatrix} = \frac{1}{D} \mathbf{J}^{\text{T}} \begin{bmatrix} R_u^{\text{div}} \\ R_v^{\text{div}} \end{bmatrix} \tag{6.14}$$

where $\mathbf{J}^{\text{T}}$ is used to denote the transpose of the Jacobian matrix in (5.53), and $\bar{R}^{\text{div}}$ is the basis function in the reference cell. (The basis functions may be normalized so that desired components have unity value at certain locations on the curved cell; for the present discussion we omit the normalization constant.) The testing functions are also to be divergence-conforming functions, defined by a mapping of the same form. (The normalization of the testing functions, as described in Section 5.6, is optional.)

The evaluation of the preceding integrals proceeds with the assistance of matrix relations such as

$$\begin{bmatrix} T_x^{\mathrm{div}} \\ T_y^{\mathrm{div}} \\ T_z^{\mathrm{div}} \end{bmatrix}^{\mathrm{T}} = \frac{1}{D(s,t)|_{\mathrm{observer}}} \begin{bmatrix} R_u^{\mathrm{div}} & R_v^{\mathrm{div}} \end{bmatrix} \mathbf{J}|_{\mathrm{observer}} \tag{6.15}$$

For instance, the entries of the excitation column vector $\mathbf{E}$ in (6.9) may be obtained as

$$\iint\limits_{\mathrm{surface}} \bar{T} \bullet \bar{E}^{\mathrm{inc}} \, ds\,dt = \iint\limits_{\mathrm{reference\ cell\ (observer)}} \left[R_u^{\mathrm{div}} \quad R_v^{\mathrm{div}} \right]_m \mathbf{J}|_{\mathrm{observer}} \begin{bmatrix} E_x^{\mathrm{inc}} \\ E_y^{\mathrm{inc}} \\ E_z^{\mathrm{inc}} \end{bmatrix} du\,dv \tag{6.16}$$

Note that the scale factor $D(s, t)$ in (6.15) is cancelled by the same factor within the differential surface area

$$ds\;dt = D(s,t)|_{\mathrm{observer}} du\;dv \tag{6.17}$$

Standard matrix multiplication is used to collect terms in the integrand of (6.16), reducing the chain of three matrices to a single scalar quantity.

A similar approach can be used to express the dot product in the first integral of (6.13) as

$$\begin{aligned} \bar{T}_m \bullet \bar{B}_n &= \begin{bmatrix} T_{mx}^{\mathrm{div}} & T_{my}^{\mathrm{div}} & T_{mz}^{\mathrm{div}} \end{bmatrix} \begin{bmatrix} B_{nx}^{\mathrm{div}} \\ B_{ny}^{\mathrm{div}} \\ B_{nz}^{\mathrm{div}} \end{bmatrix} \\ &= \frac{1}{D(s,t)\big|_{\mathrm{observer}}} \frac{1}{D(s',t')\big|_{\mathrm{source}}} \left[R_u^{\mathrm{div}} \; R_v^{\mathrm{div}} \right]_m \mathbf{J}|_{\mathrm{observer}} \mathbf{J}^{\mathrm{T}}\Big|_{\mathrm{source}} \begin{bmatrix} R_u^{\mathrm{div}} \\ R_v^{\mathrm{div}} \end{bmatrix}_n \end{aligned} \tag{6.18}$$

In (6.18), it is explicit that the testing function is located in the observer cell, where the Jacobian matrix is $\mathbf{J}\,|_{\mathrm{observer}}$, while the basis function resides in the source cell, where the matrix is $\mathbf{J}\,|_{\mathrm{source}}$ (and the quantities are defined in terms of primed coordinates). The differential surface area for the source cell is

$$ds'dt' = D(s',t')\big|_{\mathrm{source}} du'dv' \tag{6.19}$$

The index of the testing function, m, is also independent of the index of the basis function, n. (Depending on the context, these indices may refer to global numbering system throughout the entire surface or a local numbering system within the individual cell.)

Using (6.18), the first integral in (6.13) may be written in the reference cell coordinates as

$$\iint_{s,\,t}\iint_{s',\,t'} \bar{T}_m(s,\ t) \bullet \bar{B}_n(s',\ t')G\,ds'\,dt'\,ds\,dt$$

$$= \iint_{\substack{\text{reference cell}\\ \text{(observer)}}} \iint_{\substack{\text{reference cell}\\ \text{(source)}}} \left[R_u^{\text{div}}\ R_v^{\text{div}}\right]_m \mathbf{J}|_{\text{observer}}\, \mathbf{J}^{\mathbf{T}}|_{\text{source}} \begin{bmatrix} R_u^{\text{div}} \\ R_v^{\text{div}} \end{bmatrix}_n G\,du'\,dv'\,du\,dv \tag{6.20}$$

In the integrand of (6.20), the scale factors $D(s,t)|_{\text{observer}}$ and $D(s',t')|_{\text{source}}$ in (6.18) cancel with those in (6.17) and (6.19). This integral must be evaluated by numerical quadrature carried out in the reference coordinates. The basis and testing functions, and the mapping functions $x(u,v)$, $y(u,v)$, and $z(u,v)$, are computed at the quadrature points for both the source and observer reference cells. The entries of the two Jacobian matrices in (6.20) are also required at the quadrature points; these are easily obtained from (5.53) and the explicit expressions for the derivatives of x, y, and z.

The second integral in (6.13) involves the divergence of the testing function and basis function. These can be obtained using (5.96), which is equivalent to

$$\nabla \bullet \bar{T}_m = \frac{1}{D(s,t)|_{\text{observer}}} \left\{ \frac{\partial R_u^{\text{div}}}{\partial u} + \frac{\partial R_v^{\text{div}}}{\partial v} \right\}_m \tag{6.21}$$

$$\nabla' \bullet \bar{B}_n = \frac{1}{D(s',t')|_{\text{source}}} \left\{ \frac{\partial R_u^{\text{div}}}{\partial u'} + \frac{\partial R_v^{\text{div}}}{\partial v'} \right\}_n \tag{6.22}$$

Therefore, the second integral in (6.13) can be written as

$$\iint_{s,\,t} \nabla \bullet \bar{T}_m \iint_{s',\,t'} \nabla' \bullet \bar{B}_n G\,ds'\,dt'\,ds\,dt$$

$$= \iint_{\text{reference cell (observer)}} \left\{ \frac{\partial R_u^{\text{div}}}{\partial u} + \frac{\partial R_v^{\text{div}}}{\partial v} \right\}_m \iint_{\text{reference cell (source)}} \left\{ \frac{\partial R_u^{\text{div}}}{\partial u'} + \frac{\partial R_v^{\text{div}}}{\partial v'} \right\}_n G\,du'\,dv'\,du\,dv \tag{6.23}$$

As in (6.20), the scale factors $D(s,t)|_{\text{observer}}$ and $D(s',t')|_{\text{source}}$ are cancelled.

6.4 NORMALIZATION OF DIVERGENCE-CONFORMING BASIS FUNCTIONS

If it is desired to interpret the coefficients $\{I_n\}$ in Eq. (6.6) as values of the surface current density (interpolatory basis functions), the divergence-conforming basis functions used for $\bar{J}$ must be normalized so that the normal component of each $\bar{B}_{\mathrm{n}}$ has unity value at an appropriate location within each cell (in the curvilinear x–y–z space). This is accomplished by scaling each basis function by a constant equal to the magnitude of the appropriate base vector at that location, as explained in Section 5.6. For functions that straddle two cells, the normalization constant may also contain a sign that ensures that the basis function points in a consistent normal-vector direction from a global perspective. The normalization is incorporated into the matrix entries of the preceding section by multiplying the basis functions in the reference cell by the appropriate constants. These constants are cell-specific, and can be determined at the start of the analysis and stored in an array for convenient reference as required during the matrix construction phase of the MoM procedure, and any post-processing that involves the basis functions.

6.5 TREATMENT OF THE SINGULARITY OF THE GREEN'S FUNCTION

When the source and observer cells in (6.20) and (6.23) coincide, there will generally be points where the function R within G vanishes. The resulting $1/R$ singularity complicates the evaluation of the integrals by quadrature. There are several possible approaches to evaluating the singular integral; here we consider the use of a Duffy transformation [1].

When the source and observer regions coincide, for any quadrature point arising in the evaluation of the *outer* integral, the domain of the inner integral in either (6.20) or (6.23) can be divided into three triangular subcells (if the reference cell is a triangle) or four triangular subcells (if the reference cell is square) with the singularity at one corner of each subcell. This approach reduces the integration to one of the form

$$I = \int_{v=0}^{1} \int_{u=0}^{v-1} f(u,\ v) \frac{1}{\sqrt{u^2 + (1-v)^2}} du\, dv \tag{6.24}$$

where the function f incorporates the basis and testing functions, the numerator of the Green's function, the matrix products in (6.20), etc. The Duffy transformation involves a change of variable from u to w, where

$$u = (1 - v)w \tag{6.25}$$

It follows from (6.25) that

$$du = (1 - v)dw \tag{6.26}$$

The limits of integration are modified as well, with the lower limit $u = 0$ replaced by a new limit of $w = 0$, and the upper limit of $u = 1 - v$ replaced by $w = 1$. The transformation produces

$$\begin{aligned} I &= \int_{v=0}^{1} \int_{w=0}^{1} f(u,\ v) \frac{1-v}{\sqrt{u^2 + (1-v)^2}} dw\, dv \\ &= \int_{v=0}^{1} \int_{w=0}^{1} f(u,\ v) \frac{1}{\sqrt{w^2 + 1}} dw\, dv \end{aligned} \tag{6.27}$$

The integrand in (6.27) is bounded over the entire domain. The transformation converts the triangular domain of (6.24) into a square domain, and in essence spreads the singularity over the additional edge to eliminate it. The evaluation by quadrature is straightforward, with the primary difference from the nonsingular case being that each inner integral is evaluated as three to four separate integrals.

The Duffy transformation may lead to irregular integrands in certain situations, such as the case when the aspect ratio of the subdivided cells is extreme (the observation point is very close to the original cell boundary). The matrix entries may also be difficult to evaluate when the observation point is outside the source cell but close to it. For these reasons, it is generally prudent to employ adaptive quadrature algorithms that can estimate the error in the integrations, seek a prescribed error level, and inform the user when they fail! Alternative approaches have been proposed that may improve upon the Duffy transformation [2, 3].

6.6 QUADRATURE RULES

Algorithms for numerical quadrature continue to evolve. The baseline approach for multidimensional domains is to implement Gauss–Legendre rules in product form. An adaptive alternative may be obtained using the Gauss–Kronrod–Patterson rules, which provide complete sample point reuse [4]. However, to treat 2D domains, it may be more efficient to employ rules that are specifically designed for square or triangular cells. A website with links to a large number of available rules has been developed by Cools [5].

6.7 EXAMPLE: SCATTERING CROSS SECTION OF A SPHERE

A computer program was constructed that implements the discretization of the EFIE using the $p = 0$ divergence-conforming functions of Chapter 3 with the quadratic cell shapes described in Sections 2.3 and 2.5. Table 6.1 shows some results obtained from the program, for flat and curved triangular-cell models of a sphere of radius 0.5λ, where λ is the free-space wavelength. It is convenient to consider a sphere since exact solutions are available for comparison. It is also convenient to examine the bistatic scattering cross section, which is defined in [6], since that parameter is a composite that depends on the currents produced on the entire sphere in response to a plane wave excitation. For the purpose of the comparisons, several models were created that employed flat triangular cells, and compared to similar models where the cells were curved to conform to the spherical surface. These models were constructed by dividing the sphere uniformly along θ, and subdividing in ϕ so that along the equator the triangle sides have the same dimension as they do in θ. In all cases, the surface area of the models is scaled to the same surface area as the desired sphere.

Table 6.1 presents the scattering cross section as a function of θ, for ϕ fixed at 0, in response to a plane wave propagating in the $\theta = 0$, $\phi = 0$ direction with the electric field polarized in the $\hat{x}$ direction. The table suggests that the curved-cell results converge to the exact solution as the model is refined. Furthermore, these results indicate that the 300-edge flat-cell model yields approximately the same accuracy as the 108-edge curved-cell model. The 300-edge model exhibits an average density of 96 unknowns/λ^2, roughly in the range where reasonably good solutions are expected from this type of formulation (at least for geometries as simple

TABLE 6.1: Scattering Cross Section Results

	FLAT		CURVED				EXACT
EDGES	192	300	48	108	192	300	
θ							
0	9.38	9.50	8.32	9.49	9.59	9.62	9.66
30	6.59	6.69	5.49	6.61	6.75	6.79	6.83
60	4.21	4.19	4.04	4.19	4.16	4.16	4.15
90	−6.45	−6.52	−5.37	−6.67	−6.60	−6.60	−6.58
120	1.65	1.64	1.13	1.88	1.67	1.64	1.63
150	−1.95	−1.77	−2.03	−1.28	−1.33	−1.37	−1.42
180	−2.86	−2.62	0.12	−2.60	−2.35	−2.30	−2.26

Note. Results given in dBλ^2 for sphere radius of 0.5λ, for flat triangular-cell models and curved quadratic cells mapped from triangular cells. The results are organized by the number of edges in model (the number of unknowns) and by the observer angle θ, for an observer position with $\phi = 0$.

as a sphere). The 108-edge model only employs 34 unknowns/λ^2, but because of the curved cells is able to produce similar accuracy for this problem with only 36% of the unknowns. Results at other angles in ϕ exhibit a comparable accuracy.

REFERENCES

[1] M. G. Duffy, "Quadrature over a pyramid or cube of integrands with a singularity at a vertex," *SIAM J. Num. Anal.*, vol. 19, pp. 1260–1262, 1982.

[2] S. Järvenpää, M. Taskinen, and P. Ylä-Oijala, "Singularity extraction technique for integral equation methods with higher order basis functions on plane triangles and tetrahedra," *Int. J. Num. Meth. Eng.*, vol. 58, pp. 1149–1165, 2003.

[3] M. A. Khayat and D. R. Wilton, "Numerical evaluation of singular and near-singular potential integrals," *IEEE Trans. Antenn. Propagat.*, vol. 53, pp. 3180–3190. October 2005.

[4] T. N. L. Patterson, "Generation of interpolatory quadrature rules of the highest degree of precision with preassigned nodes for general weight functions," *ACM Trans. Math Software*, vol. 15, pp. 137–143, June 1989.

[5] R. Cools, "An encyclopaedia of cubature formulas," http://www.cs.kuleuven.ac.be/~nines/ecf/.

[6] A. F. Peterson, S. L. Ray, and R. Mittra, *Computational Methods for Electromagnetics*. New York: IEEE Press, 1998.

CHAPTER 7

Use of Curl-conforming Bases with the Magnetic Field Integral Equation

As an illustration of the use of curl-conforming bases, the vector mapping procedures developed in Chapter 5 are applied to the magnetic field integral equation for scattering from closed perfectly conducting bodies with curved surfaces. The MFIE was derived in Section 1.1. The following sections discuss the process of discretizing the equation using the method of moments (MoM) procedure. Results are presented for scattering from conducting spheres, where they are compared to the exact solution and to results of the EFIE approach of Chapter 6.

7.1 TESTED FORM OF THE MFIE

As part of the MoM discretization process, the MFIE from Eq. (1.9) can be enforced by multiplication (by scalar product) with a vector testing function. This yields the equation

$$\begin{aligned}&\iint\limits_{\text{surface}} \bar{T} \bullet \{\hat{n} \times \bar{H}^{\text{inc}}\} ds\, dt = \\&\iint\limits_{\text{surface}} \bar{T} \bullet \bar{J}\, ds\, dt - \iint\limits_{\text{surface}} \bar{T} \bullet \{\hat{n} \times \bar{H}^{s}\} ds\, dt\end{aligned} \tag{7.1}$$

where the integration is performed over the surface of the conductor, using a "testing" function $\bar{T}(s, t)$ that is tangential to that surface. The vector $\hat{n}$ is the outward normal unit vector. The excitation function $\bar{H}^{\text{inc}}$ represents the magnetic field incident on the structure (in the absence of the structure) and is assumed known. The function $\bar{H}^s$ is the "scattered" magnetic field produced by the surface current $\bar{J}$, also in the structure's absence, which may be obtained from the expression

$$\bar{H}^s = \nabla \times \bar{A} \tag{7.2}$$

where

$$\bar{A}(s, t) = \iint_{\text{surface}} \bar{J}(s', t') G(R) ds' dt' \tag{7.3}$$

In (7.3), G is the free space Green's function

$$G(R) = \frac{e^{-jkR}}{4\pi R} \tag{7.4}$$

and

$$R = \sqrt{[x(s, t) - x(s', t')]^2 + [y(s, t) - y(s', t')]^2 + [z(s, t) - z(s', t')]^2} \tag{7.5}$$

When the source and observation locations coincide on the surface, R vanishes and the expressions in (7.1) and (7.3) are interpreted as the limiting cases when the observer approaches the surface from the exterior.

To provide motivation for using curl-conforming basis functions with (7.1), observe that the scattered magnetic field in (7.2) can be obtained from the equivalent expression

$$\bar{H}^s = \iiint \{\nabla' \times \bar{J}\} G \, ds' dt' dn' \tag{7.6}$$

In (7.6), despite $\bar{J}$ being confined to the surface, $\nabla \times \bar{J}$ must be interpreted as a three-dimensional generalized function (by including Dirac delta functions and their derivatives that arise when the three-dimensional curl operator is applied to $\bar{J}$). While it is seldom convenient to use (7.6) to calculate the tangential field components on the surface, this expression *is* convenient for use if it is necessary to calculate the normal component of the magnetic field in the source region.

If curl-conforming basis functions are used for $\bar{J}$, (7.6) suggests that all components of the magnetic field will be well behaved (bounded, continuous) at cell boundaries on the surface. On the other hand, basis functions that do not impose cell-to-cell tangential-vector continuity result in an infinite normal component of the magnetic field at cell edges. We note that most numerical approaches for solving the MFIE have *not* used curl-conforming bases in the past, and it may not be essential to do so. However, curl-conforming basis functions appear to be appropriate for the MFIE operator.

7.2 ENTRIES OF THE MoM MATRIX

Suppose that the surface is subdivided into a mesh of curved cells based on a mapping of square or triangular reference cells. For instance, a mapping in terms of the quadratic Lagrangian interpolation polynomials of Sections 2.3 and 2.5 is convenient for illustration. The cell shapes are provided by explicit functions x, y, and z of local parameters u and v, as described in Chapter 2.

The MoM process requires the expansion of the surface current density in terms of basis functions. For illustration, curl-conforming vector bases $\{\bar{B}_\text{n}\}$ are distributed over the surface in accordance with the curved-cell model. The current is represented by the summation

$$\bar{J}(s,\ t) \cong \sum_{n=1}^{N} I_n \bar{B}_n(s,\ t) \tag{7.7}$$

where $\{I_n\}$ denote N complex-valued coefficients that henceforth are the unknowns to be determined. The curl-conforming functions are also employed as testing functions

$$\bar{T}_m(s,\ t) = \bar{B}_m(s,\ t) \tag{7.8}$$

to provide a means for obtaining N linearly independent equations from (7.1). The equations can be organized into matrix form to yield

$$\mathbf{H} = \mathbf{YI} \tag{7.9}$$

where $\mathbf{H}$ and $\mathbf{I}$ are $N \times 1$ column vectors and Y is an $N \times N$ matrix. The entries of $\mathbf{I}$ are the coefficients $\{I_n\}$, while those of $\mathbf{H}$ are given by

$$H_m = \iint_{\text{surface}} \bar{T}_m \bullet \{\hat{n} \times \bar{H}^{\text{inc}}\} ds\, dt \tag{7.10}$$

The entries of $\mathbf{Y}$ are

$$Y_{mn} = \iint_{\text{surface}} \bar{T}_m \bullet \bar{B}_n\, ds\, dt - \iint_{\text{surface}} \bar{T}_m \bullet \left\{\hat{n} \times \bar{H}_n^s\right\} ds\, dt \tag{7.11}$$

where $\bar{H}_n^s$ denotes the magnetic field produced by the nth basis function. If the source region does not coincide with the observer location, the magnetic field can be easily obtained from the expression

$$\bar{H}_n^s = -\iint_{\text{surface}} \bar{B}_n(s', t') \times \nabla G\, ds'\, dt' \tag{7.12}$$

where

$$\begin{aligned} \nabla G = -\frac{e^{-jkR}}{4\pi R^3}(1 + jkR)\big\{&[x(s, t) - x(s', t')]\hat{x} \\ + [y(s, t) - y(s', t')]\hat{y} + &[z(s, t) - z(s', t')]\hat{z}\big\} \end{aligned} \tag{7.13}$$

When it is necessary to calculate $\bar{H}_n^s$ at a location within the source region, (7.12) must be evaluated in the limiting case from the exterior of the surface.

7.3 MAPPED MoM MATRIX ENTRIES

Since the surface is divided into cells, and each cell is defined by an independent mapping, the preceding integral expressions must be evaluated separately for each cell. Some or all of the basis and testing functions can be viewed as straddling cell pairs, so generally each matrix entry may involve integrals over more than one cell. Below, we consider the part of the matrix entry arising from a single observer cell (where the testing function is located) and a single source cell (where the basis function is located). The index (m or n) is used to denote specific testing and basis functions; the pointer arrays associated with the surface model can be used to identify the appropriate cells where these functions reside.

All the integrals required in Section 7.2 can be transformed into the local coordinates of each reference cell, as explained in Chapter 5. Here, explicit expressions are provided.

The curl-conforming basis functions on the curvilinear surface are defined according to the mapping from (5.68):

$$\begin{bmatrix} B_x^{\text{curl}} \\ B_y^{\text{curl}} \\ B_z^{\text{curl}} \end{bmatrix} = \begin{bmatrix} \dfrac{\partial u}{\partial x} & \dfrac{\partial v}{\partial x} \\ \dfrac{\partial u}{\partial y} & \dfrac{\partial v}{\partial y} \\ \dfrac{\partial u}{\partial z} & \dfrac{\partial v}{\partial z} \end{bmatrix} \begin{bmatrix} R_u^{\text{curl}} \\ R_v^{\text{curl}} \end{bmatrix} = \mathbf{J}^{-1} \begin{bmatrix} R_u^{\text{curl}} \\ R_v^{\text{curl}} \end{bmatrix} \tag{7.14}$$

In (7.14), $\bar{R}^{\text{curl}}$ denotes the basis function in the reference coordinates. (These functions may be normalized so that desired components have unity value at certain locations; for the moment we omit the normalization constant.) The testing functions are also to be curl-conforming functions, defined by a mapping of the same form as (7.14). As explained in Section 4.5, these curl-conforming functions are related to analogous divergence conforming functions by

$$\hat{n} \times \bar{B}^{\text{div}} = \bar{B}^{\text{curl}} \tag{7.15}$$

$$\hat{n} \times \bar{B}^{\text{curl}} = -\bar{B}^{\text{div}} \tag{7.16}$$

Because of the form of the MFIE in (7.1), it is convenient to exploit (7.15) and (7.16) as follows: Using the vector identity

$$\bar{A} \bullet \bar{B} \times \bar{C} = \bar{A} \times \bar{B} \bullet \bar{C} \tag{7.17}$$

Equation (7.10) can be rewritten in the form

$$
\begin{aligned}
H_m &= \iint\limits_{\text{surface}} \bar{T}_m^{\text{curl}} \bullet \{\hat{n} \times \bar{H}^{\text{inc}}\} ds\, dt \\
&= \iint\limits_{\text{surface}} \left\{\bar{T}_m^{\text{curl}} \times \hat{n}\right\} \bullet \bar{H}^{\text{inc}} ds\, dt \\
&= \iint\limits_{\text{surface}} \bar{T}_m^{\text{div}} \bullet \bar{H}^{\text{inc}} ds\, dt
\end{aligned}
\tag{7.18}
$$

In (7.18), $\bar{T}_m^{\text{div}}$ denotes a divergence-conforming test function related to the specific curl-conforming test function under consideration by

$$
\bar{T}_m^{\text{div}} = -\hat{n} \times \bar{T}_m^{\text{curl}} \tag{7.19}
$$

In a similar manner, the second integral in (7.11) may be recast as

$$
\iint\limits_{\text{surface}} \bar{T}_m^{\text{curl}} \bullet \left\{\hat{n} \times \bar{H}_n^s\right\} ds\, dt = \iint\limits_{\text{surface}} \bar{T}_m^{\text{div}} \bullet \bar{H}_n^s\, ds\, dt \tag{7.20}
$$

The divergence-conforming functions are defined using the mapping of (5.77):

$$
\begin{bmatrix} T_x^{\text{div}} \\ T_y^{\text{div}} \\ T_z^{\text{div}} \end{bmatrix} = \frac{1}{D} \begin{bmatrix} \dfrac{\partial x}{\partial u} & \dfrac{\partial x}{\partial v} \\ \dfrac{\partial y}{\partial u} & \dfrac{\partial y}{\partial v} \\ \dfrac{\partial z}{\partial u} & \dfrac{\partial z}{\partial v} \end{bmatrix} \begin{bmatrix} R_u^{\text{div}} \\ R_v^{\text{div}} \end{bmatrix} = \frac{1}{D} \mathbf{J}^T \begin{bmatrix} R_u^{\text{div}} \\ R_v^{\text{div}} \end{bmatrix} \tag{7.21}
$$

where $\bar{R}^{\text{div}}$ denotes the testing function in the reference coordinates. (The normalization of the testing functions, as described in Section 5.6, is optional.) In order to construct the scalar vector products arising within (7.11), (7.18), and (7.20) by means of matrix manipulations, we observe that

$$
\begin{bmatrix} T_x^{\text{div}} \\ T_y^{\text{div}} \\ T_z^{\text{div}} \end{bmatrix}^{\mathrm{T}} = \frac{1}{D} \left[R_u^{\text{div}}\ R_v^{\text{div}} \right] \mathbf{J} \tag{7.22}
$$

and

$$\begin{bmatrix} T_x^{\text{curl}} \\ T_y^{\text{curl}} \\ T_z^{\text{curl}} \end{bmatrix}^{\text{T}} = \begin{bmatrix} R_u^{\text{curl}} & R_v^{\text{curl}} \end{bmatrix} \mathbf{J}^{-\text{T}} \tag{7.23}$$

Using (7.22), the integral in (7.18) may be expressed as

$$H_m = \iint_{\text{cell on surface}} \bar{T}_m^{\text{div}} \bullet \bar{H}^{\text{inc}} \, ds\, dt = \iint_{\text{reference cell}} \begin{bmatrix} R_u^{\text{div}} & R_v^{\text{div}} \end{bmatrix}_m \mathbf{J} \begin{bmatrix} H_x^{\text{inc}} \\ H_y^{\text{inc}} \\ H_z^{\text{inc}} \end{bmatrix} du\, dv \tag{7.24}$$

where the matrix $\mathbf{J}$ for the cell is defined in (5.53), and where we also used the differential surface area

$$ds\, dt = D\, du\, dv \tag{7.25}$$

The scale factors $1/D$ from (7.22) and D from (7.25) cancel in the integrand of (7.24). This integral is to be evaluated by numerical quadrature carried out in the reference coordinates u and v, with the incident field sampled at the desired quadrature points with the assistance of the mapping functions $x(u, v)$, $y(u, v)$, and $z(u, v)$.

The first integral in (7.11) can be expressed over a single cell, using (7.23) and (7.25), as

$$\iint_{\text{cell on surface}} \bar{T}_m^{\text{curl}} \bullet \bar{B}_n^{\text{curl}} \, ds\, dt = \iint_{\text{reference cell}} \begin{bmatrix} R_u^{\text{curl}} & R_v^{\text{curl}} \end{bmatrix}_m \mathbf{J}^{-\text{T}} \mathbf{J}^{-1} \begin{bmatrix} R_u^{\text{curl}} \\ R_v^{\text{curl}} \end{bmatrix}_n D\, du\, dv \tag{7.26}$$

The entries of the 3×2 matrix $\mathbf{J}^{-1}$, defined in (5.54), are not explicitly provided by the mapping functions $x(u, v)$, etc. These entries are obtained in practice by inverting the numerical 3×3 Jacobian matrix in Eq. (5.64) at each of the necessary quadrature points.

Using (7.20), the second integral in (7.11) can be written as

$$\iint_{\text{cell on surface}} \bar{T}_m^{\text{curl}} \bullet \left\{\hat{n} \times \bar{H}_n^s\right\} ds\, dt = \iint_{\text{cell on surface}} \bar{T}_m^{\text{div}} \bullet \bar{H}_n^s\, ds\, dt$$

$$= \iint_{\text{reference cell (observer)}} \begin{bmatrix} R_u^{\text{div}} & R_v^{\text{div}} \end{bmatrix}_m \mathbf{J}|_{\text{observer}} \begin{bmatrix} H_x^s \\ H_y^s \\ H_z^s \end{bmatrix} du\, dv \tag{7.27}$$

where, using (7.12),

$$\begin{bmatrix} H_x^s \\ H_y^s \\ H_z^s \end{bmatrix} = -\iint_{\text{reference cell (source)}} \begin{bmatrix} 0 & \frac{\partial G}{\partial z} & -\frac{\partial G}{\partial y} \\ -\frac{\partial G}{\partial z} & 0 & \frac{\partial G}{\partial x} \\ \frac{\partial G}{\partial y} & -\frac{\partial G}{\partial x} & 0 \end{bmatrix} \mathbf{J}^{-1}|_{\text{source}} \begin{bmatrix} R_u^{\text{curl}} \\ R_v^{\text{curl}} \end{bmatrix}_n D\, du\, dv \tag{7.28}$$

The integral in (7.27) is carried out over the observer cell, which is transformed to a reference cell using the Jacobian matrix $\mathbf{J}|_{\text{observer}}$ associated with that specific mapping. (As in (7.24), the scale factors $1/D$ and D cancel.) Equation (7.28) is embedded within (7.27), and involves an integral over the source cell, which is usually different from the observer cell. Since the cells are different, the mapping is different, and the transformation in (7.28) uses a matrix $\mathbf{J}^{-1}|_{\text{source}}$ that is not related to $\mathbf{J}|_{\text{observer}}$. These integrals must also be evaluated in the reference coordinates by numerical quadrature, with the derivatives of G in (7.28) determined from (7.13), and the entries of $\mathbf{J}^{-1}|_{\text{source}}$ obtained by the explicit numerical inversion of (5.64) at the quadrature points. The mapping functions $\{x(u, v), y(u, v), z(u, v)\}$ for both the source cell and the observer cell are required to compute the arguments of the derivatives of G. When the source and observer cells coincide, the quadrature procedure must be modified to handle the singularity in G as explained in Section 7.5.

7.4 NORMALIZATION OF CURL-CONFORMING BASIS FUNCTIONS

If interpolatory basis functions are used, and it is desired to interpret the coefficients $\{I_n\}$ as values of the surface current density, the curl-conforming basis functions used for $\bar{J}$ must be normalized so that the tangential component of each $\bar{B}_n$ has unity value at an appropriate location within each cell (in the curvilinear x–y–z space). This is accomplished by scaling each basis function by a constant equal to the magnitude of the appropriate base vector at that location, as explained in Section 5.4. For functions that straddle two cells (any of the "edge-based" functions of Chapter 4), the normalization constant may also contain a sign that ensures that the basis function points in a consistent tangential-vector direction from a global perspective. The normalization is incorporated into the matrix entries of the preceding section by multiplying the various components of the basis functions in the reference cell by the appropriate constants. These constants are cell-specific, and can be determined at the start of the analysis and stored in an array for convenient reference as required during the matrix construction phase of the MoM procedure, and any post-processing that involves the basis functions.

7.5 TREATMENT OF THE SINGULARITY OF THE GREEN'S FUNCTION

When the source and observer cells in (7.27) and (7.28) coincide, there will generally be points where the function R within G vanishes. In fact, the integral involves a singularity that must be treated with care.

For an observation point (s, t) located on a smooth portion of the source cell, the singularity can be treated in an analytical manner by replacing Eq. (7.12) with

$$\bar{H}_n^s(s,\ t) = \frac{\bar{B}_n(s,\ t)}{2} - \iint\limits_{\text{surface}-\varepsilon} \bar{B}_n(s',\ t') \times \nabla G\, ds'\, dt' \tag{7.29}$$

where the remaining integral is to exclude an infinitesimal region around the point (s, t) from the domain of integration [1]. The primary impact of this operation is the addition of the leading term on the right-hand side of (7.29), which effectively changes the first integral in (7.11) to one having only half its value:

$$Y_{mn} = \frac{1}{2} \iint\limits_{\text{surface}} \bar{T}_m \bullet \bar{B}_n\, ds\, dt - \iint\limits_{\text{surface}} \bar{T}_m \bullet \{\hat{n} \times \bar{H}_n^s\} ds\, dt \tag{7.30}$$

The calculation of $\bar{H}_n^s$ in the second term in (7.30) is modified in principle by the exclusion of an ε-neighborhood of the observation point from the domain.

However, in practice when evaluating this expression with numerical quadrature, as long as a node of the quadrature rule does not coincide with (s, t) the effect is the same whether or not the neighborhood is excluded.

Even after the exclusion of the ε-neighborhood of (s, t), the integrand in (7.29) still behaves as $1/R$ and will be difficult to evaluate accurately with standard quadrature rules. The $1/R$ behavior can be dealt with using the Duffy transformation approach described in Chapter 6. The reference cell can be divided into three or four triangular cells, each with a corner at the singularity, and the integrals over each cell can be evaluated after a change of variable that cancels the singularity as explained in Section 6.5. The result is that three or four integrals must be evaluated over square domains, and their results combined, to produce the result of the integral in (7.29) for each location of (s, t). Alternatives to the Duffy transformation are cited in Chapter 6.

7.6 RESULTS

A computer program was developed that implements the preceding expressions for the MoM matrix entries associated with $p = 0$ basis functions from Chapter 4 and quadratic cells mapped from triangular reference cells. For illustration, consider a perfectly conducting sphere of radius 0.5λ, where λ denotes the wavelength. Figure 7.1 depicts the current magnitude $|J_\theta|$ produced by a uniform plane wave incident on a sphere, obtained from the MFIE procedure with a model containing 300 cell edges (equivalently, 300 unknowns). The representation for J_θ along this cut is piecewise constant, as explicitly depicted in the figure. The sphere model is described in Section 6.7. The exact solution is shown for comparison.

Table 7.1 shows the percentage error in the current density, for several models of the sphere of radius 0.5λ. The MFIE performance is compared with that of the EFIE approach from Chapter 6. The error was determined by averaging over the difference between the current density coefficients and the exact solution for the appropriate current component (normal to an edge for the EFIE, tangential to an

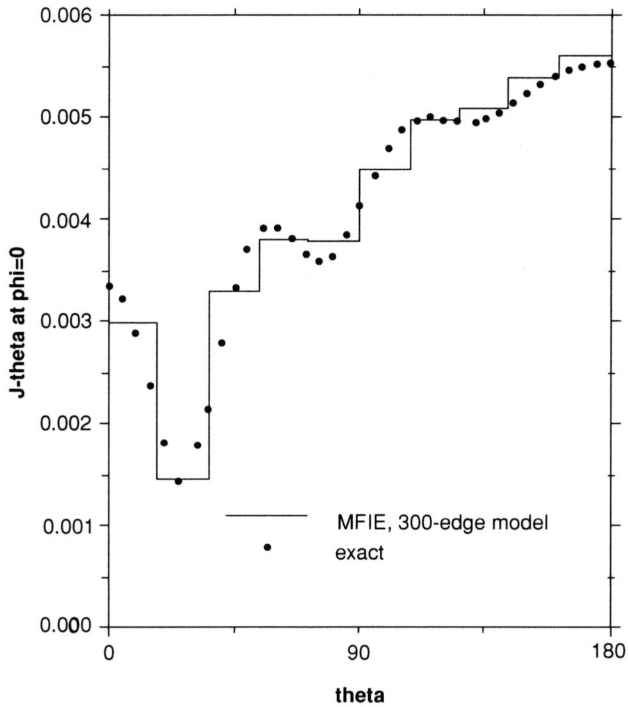

FIGURE 7.1: Magnitude of J_θ along the $\phi = 0$ cut, produced by a plane wave incident in the $\theta = 0$ direction with an x-component of electric field. The numerical solution obtained from the MFIE with curl-conforming basis functions is compared with the exact solution.

TABLE 7.1: Average Percent Error in the Current Density

NO. OF EDGES IN MODEL	LARGEST EDGE LENGTH IN MODEL (λ)	EFIE (DIV-CONF.), %	MFIE (CURL-CONF.), %
108	0.36	6.2	5.8
192	0.29	3.5	3.8
300	0.24	2.3	2.8
432	0.20	1.7	2.1

Note. Averaged over all edges at the edge center, for a sphere of radius 0.5λ modeled with curved quadratic cells mapped from triangles.

edge for the MFIE) at the edge centers. The error levels for the EFIE and MFIE decrease as the models are refined, and are similar for the two approaches.

REFERENCES

[1] A. J. Poggio and E. K. Miller, "Integral equation solutions of three-dimensional scattering problems," in *Computer Techniques for Electromagnetics*, R. Mittra, Ed. New York: Hemisphere, 1987.

Biography

Andrew F. Peterson received the B.S., M.S., and Ph.D. degrees in electrical engineering from the University of Illinois, Urbana-Champaign in 1982, 1983, and 1986, respectively. Since 1989, he has been a member of the faculty of the School of Electrical and Computer Engineering at the Georgia Institute of Technology, where he is now Professor and Associate Chair for Faculty Development. He teaches electromagnetic field theory and computational electromagnetics, and conducts research in the development of computational techniques for electromagnetic scattering, microwave devices, and electronic packaging applications. He is the principal author of the text *Computational Methods for Electromagnetics*, IEEE Press, 1998.

Dr. Peterson is a past recipient of the ONR Graduate Fellowship and the NSF Young Investigator Award. He has served as an Associate Editor of the *IEEE Transactions on Antennas and Propagation*, as an Associate Editor of the *IEEE Antennas and Wireless Propagation Letters*, as the General Chair of the 1998 IEEE AP-S International Symposium and URSI/USNC Radio Science Meeting, and as a member of IEEE AP-S AdCom. He also served for 6 years as a Director of ACES, and 2 years as Chair of the IEEE Atlanta Section. He will serve as the President of the IEEE AP-S during 2006. He is an IEEE Fellow and a recipient of the IEEE Third Millennium Medal. He is also a member of the Applied Computational Electromagnetics Society (ACES), the International Union of Radio Scientists (URSI) Commission B, the American Society for Engineering Education, and the American Association of University Professors.

MRTD (MULTI RESOLUTION TIME DOMAIN) METHOD IN ELECTROMAGNETICS

Nathan A. Bushyager

Manos M. Tentzeris
NSF-PRC Associate Director for RF Research School of ECE
Georgia Institute of Technology,
Atlanta, GA 30332-250 U.S.A.

MORGAN & CLAYPOOL PUBLISHERS

ABSTRACT

This book presents a method that allows the use of multiresolution principles in a time domain electromagnetic modeling technique that is applicable to general structures. The multiresolution time-domain (MRTD) technique, as it is often called, is presented for general basis functions. Additional techniques that are presented here allow the modeling of complex structures using a subcell representation that permits the modeling discrete electromagnetic effects at individual equivalent grid points. This is accomplished by transforming the application of the effects at individual points in the grid into the wavelet domain.

In this work, the MRTD technique is derived for a general wavelet basis using a relatively compact vector notation that both makes the technique easier to understand and illustrates the differences between MRTD basis functions. In addition, techniques such as the uniaxial perfectly matched layer (UPML) for arbitrary wavelet resolution and non-uniform gridding are presented. Using these techniques, any structure that can be simulated in Yee-FDTD can be modeled with in MRTD.

KEYWORDS

Adaptive Algorithms, MRTD, Wavelets, Adaptive Electromagnetic simulations, Time-/Space-adaptive gridding

Contents

CHAPTER 1

Introduction

The field of radio frequency (RF) and microwave design is growing at an astonishing rate due the confluence of several factors. Chief among these factors are the increased demand for RF and microwave consumer devices (cellular telephony and wireless data systems) and the increasing speed of digital devices, which has led to design limitations that previously applied only to traditional RF (radar and communications) circuits. There are also several quickly growing commercial RF applications such as automotive radar and RFID. As RF devices become more predominant in the consumer market, the design time for each generation decreases while performance demands increase. Design turnaround and performance gains traditional to the semiconductor device market are being expected of RF circuits. This in turn leads to increased demands on the RF designer and all associated tools, such as EM simulators.

Modern RF devices are built on a variety of technologies for a wide array of functionalities. In the attempt to reduce size and cost, multilayer substrates are commonplace. The constraints of each substrate are different, and the devices that can be fabricated with them often have no, or highly inaccurate, theoretical or empirical models. Design is usually performed using a top–down methodology, where the system is designed at the conceptual level and the details of the actual system are added as the design progresses. At the bottom of this design process, the actual physical layouts that represent the design blocks must be created. The characterization of these devices usually requires a full-wave electromagnetic simulator.

Full-wave simulators are often required to characterize effects that cannot be predicted, or properly accounted for, at higher levels of the design process. Examples of these effects are parasitic coupling, substrate modes, radiation, and package interference. These effects are often unique to the exact layout of a device and are too complicated to be treated theoretically. Any simulator that models the complete physics of electromagnetic interaction can be used to model these devices; however, time-domain techniques are popular and particularly well suited to these devices.

Time-domain techniques, as the name implies, determine the electromagnetic fields in a structure in response to a source condition. They are contrasted to frequency-domain methods, which determine the response to a harmonic source. There are several advantages to each class of methods, and the requirements of a particular problem determine which is best. The characteristic that is most often cited in recommending time-domain methods for the simulation of microwave devices is that a broadband response can be determined from a single simulation. In addition, time-domain simulators are usually derived using a differential form of Maxwell's equations, relating fields at points and allowing easy discretization of complex structures; they do not require the calculation of Green's functions or the inversion of a matrix. Furthermore, nonlinear effects can be simulated easily, as the field strength is measured as a function of time and these effects can be directly applied. The two main drawbacks of time-domain techniques are representing dispersive media and speed. The speed of these techniques, while comparable to other full-wave techniques, is not nearly fast enough for modern microwave design.

The most mature and widely used time-domain simulation technique is finite-difference time-domain (FDTD) (1). The Yee-FDTD scheme is one of the oldest, and one of the simplest, time-domain EM simulation techniques. It has remained popular due to this simplicity; not only is it relatively easy to apply, but it is also highly flexible. Because of its flexibility, the method has been applied to practically every type of electromagnetic analysis, from radar-cross-section investigation to optical characterization. A number of compatible techniques have been developed to model specific effects in the method. When developing other methods, the Yee-FDTD scheme is often used as a benchmark; most other techniques are designed to be faster.

While a number of techniques have been developed that are more efficient than FDTD for the same accuracy, these techniques add complexity and cannot be as generally applied. Several FDTD techniques using higher order basis functions (than the pulse functions used in FDTD) have been presented, but it is difficult to model general microwave structures because of the difficulty of representing boundary conditions (such as those required to model arbitrarily positioned metal structures). While the design community desires an improved method that offers several orders of magnitude improvement in simulation time, this is not possible. One promising method of decreasing design time is to identify those parts of the circuit that require full-wave simulation, and characterizing the remainder of the circuit with less computationally intensive techniques. However, as integration increases, the areas of the circuit that require full-wave simulation will increase. Thus, it is necessary to use the most efficient full-wave technique possible. One technique that has been suggested is multiresolution time-domain (MRTD) (2).

MRTD uses a wavelet-based discretization to represent the EM fields. The wavelets allow the resolution of the simulation to be changed as a function of both time and space. If the resolution is changed to respond to a propagating waveform, the number of operations that must be performed to characterize a structure can be minimized. The MRTD technique has been referred to as a generalization of the FDTD technique. The derivation of the method follows the same general scheme, but the relatively simple pulse basis functions used in the FDTD technique are replaced with the wavelet functions.

The purpose of this monograph is to present the MRTD technique to a reader unfamiliar with multiresolution (wavelet) analysis and little or no experience in numerical modeling. A more advanced reader will also benefit from the treatment presented here, especially the later chapters where newer techniques in MRTD analysis and examples are presented. In addition, the notation used to present

MRTD in this book attempts to unify the methods used to present all wavelet bases. By use of this method, a compact representation is presented that allows the characteristics of various wavelet schemes to be compared. This makes the method both easier to understand and provides a straightforward method to deal with the large number of nested coefficients that appear in a multiresolution scheme.

Since this is a relatively short book, it cannot cover all aspects of the complex multiresolution time domain technique. A multitude of papers have been written discussing the technique, using a wide variety of wavelet basis functions. The aspects of each wavelet scheme are well beyond the scope of this presentation. Because of this limitation, the lecture is organized such that the reader is presented with the general MRTD technique in the first chapter and overview of its properties and a discussion of its implementation in the second chapter. The remaining chapters focus on the implementation of the MRTD technique using Haar basis functions. The reason for this is twofold. First, the Haar wavelets are the simplest that can be used and thus can be covered in detail in the relatively short length of this book. Second, the Haar wavelets, like Yee-FDTD, are extremely general, and can thus be easily applied to complex structures while still keeping the advantages of the MRTD technique. The main advantage of the Haar wavelets is the ability to represent local features, such as perfect conductors, without the need to use image theory. A presentation using the Haar wavelets gets the reader familiar with the MRTD technique quickly, while still presenting the framework for a powerful, adaptive simulator.

CHAPTER 2

Background

This monograph presents techniques that can be used to model complex microwave structures in MRTD. In this section, the derivation of the MRTD technique in three dimensions for a general wavelet basis is presented. In addition, as the MRTD technique is presented as an adaptive alternative to FDTD, and several MRTD examples will be contrasted to FDTD, a brief overview of the FDTD method is presented. This chapter begins with a discussion of MRTD basics, including wavelet basis functions and the method of moments, followed by a discussion of two specific areas, PML and PEC modeling. After these topics are addressed, an overview of FDTD, focusing on its similarities and differences with MRTD, is presented.

2.1 MRTD BASICS

2.1.1 Wavelet Overview

The MRTD technique was first presented by Krumpholz and Katehi in 1996 [2]. At that time multiresolution analysis, the application of wavelet bases to numerical problems, was becoming popular in a number of fields as a way of increasing both the efficiency and accuracy of numerical methods. Their original MRTD paper provides a general discretization technique that can be used with any wavelet basis; however, the paper focused on the Battle–Lemarie wavelet scheme. Several other wavelet schemes have since been applied in the same manner, including Haar [3], Daubechies [4], and Cohen–Daubechies–Feauveau (CDF) [5].

While the above-mentioned wavelet schemes are all very different, they share a number of characteristics that make them suitable for numerical modeling. Most wavelet schemes used in for MRTD are orthonormal and are characterized by a scaling function and a mother wavelet. The mother wavelet is generated using the scaling function and is in turn used to generate all other wavelets that constitute the basis. The orthogonality that exists between the wavelets and scaling function, as well as all wavelets with other wavelets, make them natural choices for numerical discretization.

The key concept regarding wavelet expansions is that of *levels* of wavelet resolution. These levels correspond to sets of functions that can be added to the expansion to increase the accuracy of the wavelet discretization. In the following expressions, the scaling functions will be represented as $\varphi_i\,(x)$ [6], where

$$\varphi_i(x) = \varphi\left(\frac{x}{\Delta x} - i\right) \tag{1}$$

and $\varphi(x)$ represents the general scaling function. Likewise, the wavelets are offset throughout the grid, but are characterized with three indices. A typical wavelet coefficient, as used in MRTD, is represented as

$$\psi^r_{i,p}(x) = 2^{r/2}\psi_0\left(2^{r/2}\left(\frac{x}{\Delta x} - i\right) - p\right). \tag{2}$$

In this function, r represents the wavelet resolution and p can take any integer value between 0 and $2^r - 1$. Using these coefficients, each level of resolution, r, contains 2^r wavelets, offset by $\Delta x/2^r$.

For all wavelet schemes used in multiresolution analysis, the following hold [6]:

$$\int \varphi_i(x)\varphi_j(x) = \delta_{i,j} \tag{3}$$

$$\int \varphi_i(x)\psi^r_{j,p}(x) = 0 \quad \forall i, j, r, p \tag{4}$$

$$\int \psi^r_{i,p}(x)\psi^s_{j,q}(x) = \delta_{i,j}\delta_{r,s}\delta_{p,q}, \tag{5}$$

where $\delta_{i,j}$ is the Kronecker delta function

$$\delta_{i,j} = \begin{cases} 1 & \mathrm{i} = \mathrm{j} \\ 0 & \mathrm{i} \neq \mathrm{j} \end{cases}. \tag{6}$$

If the closed subspace represented by all wavelets of resolution r and higher is termed V_r, the following properties are required:

$$\ldots V_{-3} \subset V_{-2} \subset V_{-1} \subset V_0 \subset V_1 \subset V_2 \subset V_3 \ldots \tag{7}$$

$$\overline{\bigcup_{i \in Z} V_i} = L^2(R) \tag{8}$$

$$\overline{\bigcap_{i \in Z} V_i} = 0, \tag{9}$$

where Z is the set of all integers and $L^2(R)$ is the set of all square integrable functions. This is by no means a complete description of wavelets; it is only what is necessary to discuss the concepts in this monograph. Several books, such as [6], have been written on wavelet analysis, these can provide a much more broad appreciation of multiresolution analysis.

The consequence of these properties is that each addition of a level of wavelet resolution increases the accuracy of the representation, and arbitrary accuracy can be achieved using the appropriate wavelet resolution. This is in contrast to methods of improving accuracy in other basis functions, where increased accuracy is achieved by contracting the domain and increasing the number of basis functions. There are several basis functions that can be used to meet these requirements.

2.1.2 Haar Wavelets

The oldest and simplest wavelet basis is the Haar basis functions. When these wavelets are used in MRTD, a scheme equivalent to FDTD can result [7]. However, when used correctly, a time and space adaptive scheme that is more numerically efficient than FDTD results. As mentioned above, a number of other wavelet schemes have been used with the MRTD technique.

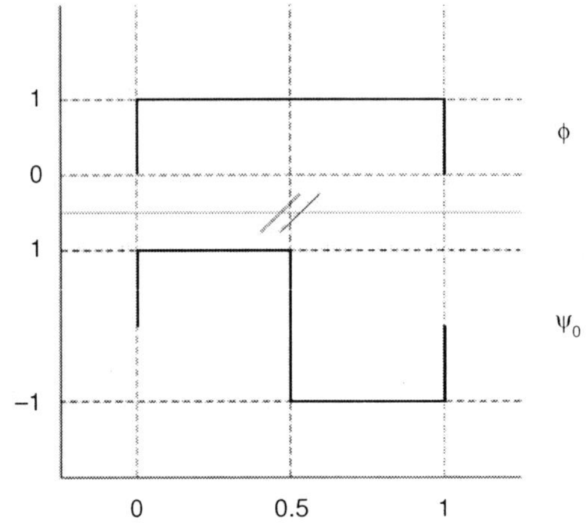

FIGURE 1: Haar scaling function (top) and mother wavelet (bottom).

An overview of the Haar wavelets is presented here. They are simple enough that they can be used to explain wavelet concepts relatively easily in one and two dimensions, while adequately demonstrating wavelet properties (three-dimensional (3-D) wavelets will be discussed, but are difficult to present in a two-dimensional (2-D) format). The examples presented in this work use Haar wavelets exclusively; they are preferred for a number of reasons that will be discussed. In addition, the composite cell technique presented in later chapters is practical only for Haar wavelets.

The Haar scaling function φ and mother wavelet ψ are presented in Fig. 1. The Haar scaling function is defined as $\chi_{[0,1)}$, the characteristic function from zero to one. The mother wavelet is based on this function. It is defined as

$$\psi_0(x) = \varphi(2x) - \varphi(2(x - 1/2)), \tag{10}$$

Higher level wavelets can be generated using (2). These wavelets for resolutions 1 and 2 are presented in Fig. 2.

As stated previously, and demonstrated in Fig. 2, there are 2^r wavelets for each level of resolution. Thus, there are 2 wavelets at $r = 1$ and 4 wavelets at $r = 2$. These functions represent scaled and translated mother wavelets. The scaling factor,

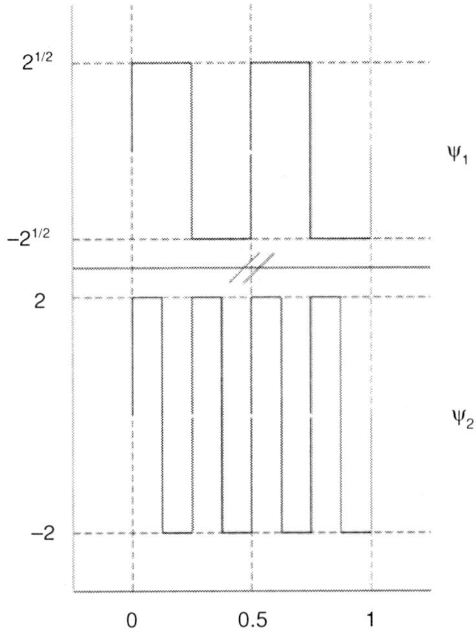

FIGURE 2: Haar wavelets for resolution 1 and 2.

$2^{r/2}$, presented in (2) is required to orthonormalize the functions, which guarantees (5). The domain of all of the wavelet functions for any level of resolution is identical to that of the mother wavelet and scaling function. A scheme that uses wavelets up to and including a resolution level r will use 2^{r+1} functions (including the scaling function).

There are several advantages to using Haar wavelets to model electromagnetic phenomena. The first is their finite domain nature. When Haar wavelets are used to expand an electric or magnetic field, the scaling functions and wavelets from one cell to the next do not overlap. This allows hard boundary conditions (setting discrete areas to a fixed value) to be easily applied. For example, a perfect-electrical-conductor (PEC) boundary condition can be applied by zeroing tangential electric fields. If the basis functions were to overlap, this condition could not be applied. The second advantage of using Haar wavelets is the relative ease of performing derivatives and integrals (although these integrals can be tabulated and thus calculated only once for any number of simulations). Because of their pulse nature, these operations become simple arithmetic. It will be seen that this eases the generation of the MRTD scheme. The major disadvantage of Haar wavelets is in the numerical dispersion of the resultant MRTD scheme. Compared to most other wavelet schemes, denser grids, and thus more coefficients, are required for Haar-MRTD schemes.

2.1.3 MRTD Scheme

The MRTD scheme is generated by first representing the electric and magnetic fields in terms of wavelets and then applying the method of moments to Maxwell's curl equations, (11) and (12), and the constitutive equations, (15) and (16). The differential form of Maxwell's equations in the time domain can be expressed as

$$\nabla \times \mathbf{E}(t) = -\frac{\partial \mathbf{B}(t)}{\partial t} - \mathbf{M}(t) \tag{11}$$

$$\nabla \times \mathbf{H}(t) = \frac{\partial \mathbf{D}(t)}{\partial t} + \mathbf{J}(t) \tag{12}$$

$$\nabla \cdot \mathbf{D}(t) = \rho_{\mathrm{e}}(t) \tag{13}$$

$$\nabla \cdot \mathbf{B}(t) = \rho_{\mathrm{m}}(t), \tag{14}$$

where $\mathbf{E}$ is the electric field (V/m), $\mathbf{H}$ is the magnetic field (A/m), $\mathbf{D}$ is the electric flux density (C/m^2), $\mathbf{B}$ is the magnetic flux density (Wb/m^2), ρ is charge density (C/m^3), $\mathbf{M}$ is the magnetic current density (V/m^2), and $\mathbf{J}$ is the electric current density (A/m^2). These equations relate the electric and magnetic fields at points in space. This analysis is limited to isotropic, nondispersive media, which have the constitutive equations

$$\mathbf{D}(\mathbf{r}) = \varepsilon(\mathbf{r})\,\mathbf{E}(\mathbf{r}) \tag{15}$$

$$\mathbf{B}(\mathbf{r}) = \mu(\mathbf{r})\,\mathbf{H}(\mathbf{r}), \tag{16}$$

where $\mathbf{r}$ is the position vector.

For a general material interface, the boundary conditions are

$$\hat{\mathbf{n}} \cdot (\mathbf{D}_2 - \mathbf{D}_1) = \rho_{\mathrm{s}} \tag{17}$$

$$\hat{\mathbf{n}} \cdot (\mathbf{B}_2 - \mathbf{B}_1) = 0 \tag{18}$$

$$\hat{\mathbf{n}} \times (\mathbf{E}_2 - \mathbf{E}_1) = 0 \tag{19}$$

$$\hat{\mathbf{n}} \times (\mathbf{H}_2 - \mathbf{H}_1) = \mathbf{J}_s, \tag{20}$$

where ρ_s, and $\mathbf{J}_s$ are the charge and electric current densities on the interface and $\hat{\mathbf{n}}$ is the normal unit vector to the interface.

Faraday's and Ampere's laws, (11) and (12) respectively, are vector equations and are expressed in Cartesian coordinates as

$$\frac{\partial B_x}{\partial t} = \frac{\partial E_y}{\partial z} - \frac{\partial E_z}{\partial y} - M_x \tag{21}$$

$$\frac{\partial B_y}{\partial t} = \frac{\partial E_z}{\partial x} - \frac{\partial E_x}{\partial z} - M_y \tag{22}$$

$$\frac{\partial B_z}{\partial t} = \frac{\partial E_x}{\partial y} - \frac{\partial E_y}{\partial x} - M_z. \tag{23}$$

$$\frac{\partial D_x}{\partial t} = \frac{\partial H_z}{\partial y} - \frac{\partial H_y}{\partial z} - J_x \tag{24}$$

$$\frac{\partial D_y}{\partial t} = \frac{\partial H_x}{\partial z} - \frac{\partial H_z}{\partial x} - J_y \tag{25}$$

$$\frac{\partial D_z}{\partial t} = \frac{\partial H_y}{\partial x} - \frac{\partial H_x}{\partial y} - J_z. \tag{26}$$

To generate the MRTD scheme, each E and H field in Faraday's and Ampere's laws must be expanded in terms of scaling and wavelet functions. A one-dimensional (1-D) field, for example E_x in a 1-D scheme, is expressed in scaling and wavelet functions up to resolution $r_{\max}$ as

$$E_x(x) = \sum_{n,i} h_n(t) \left[{}_nE_i^{x,\phi} \varphi_i(x) + \sum_{r=0}^{r_{\max}} \sum_{p=0}^{2^r-1} {}_nE_{i,r,p}^{x,\psi} \psi_{i,p}^{r}(x) \right], \tag{27}$$

where ${}_nE_i^{x,\phi}$ and ${}_nE_{i,r,p}^{x,\psi}$ are coefficients representing the magnitudes of the scaling and wavelet functions. This function is the discretization in both time and space. The time discretization, $h_n(t)$, is performed with simple pulse functions, equivalent to the Haar scaling functions presented in Fig. 1. The pulse functions are used in time to ensure causality. If the functions representing each time step overlapped, it would be possible for past events to be modified by future ones. The spatial index, i, indicates the position in cells along the x axis. The cells are defined by the domain of each scaling function, Δx. It is, technically, possible to use one scaling function the size of the entire domain. Practically, however, this is difficult to implement, and it removes some of the ability to locally increase resolution. In practice, several cells per the maximum excited wavelength are used. The criteria for choosing the cell size are discussed later, as a part of numerical dispersion.

In order to expand the fields in multiple dimensions, the products of all scaling/wavelet functions of each direction must be used. In two dimensions, this means four groups of coefficients, scaling-x/scaling-y, wavelet-x/scaling-y, scaling-x/wavelet-y, and wavelet-x/wavelet-y coefficients. As in (27), the wavelet terms require sums, and the wavelet/wavelet terms have nested x and y sums. For example, in a 2-D simulation, with $r_{\max} = 0$, the four functions used to represent the fields for the Haar wavelets are depicted in Fig. 3.

The maximum resolution is not required to be the same in each direction. For any wavelet basis, the number of coefficients required for a given maximum resolution is

$$\text{number of coefficients} = 2^{D + \sum\limits_{i=x,y,z} r_{\max,i}}, \tag{28}$$

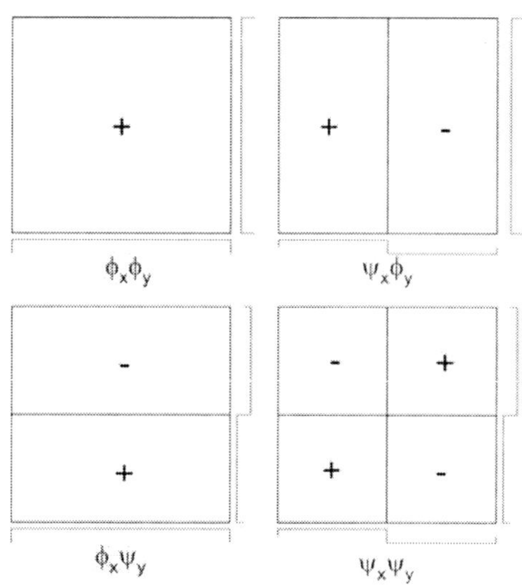

FIGURE 3: Two-dimensional Haar coefficients for $r_{\max} = 0$.

where D is the dimensionality of the simulator and the sum indicates that the maximum resolution may vary by direction. In a 3-D simulator, there are nine groups of coefficients needed to represent the field expansion, and an increase in resolution in every direction increases the number of coefficients needed by three powers of two.

The expansion of the E_x field in three dimensions, for any wavelet basis, is

$$E_x(x) = \sum_n h_{n(t)} \sum_{i,j,k} \left[\begin{array}{l} {}_nE^{x,\phi\phi\phi}_{i,j,k} \varphi_i(x)\varphi_j(y)\varphi_k(z) + \\ \sum_{r=0}^{r_{\max}} \sum_{p=0}^{2^r-1} {}_nE^{x,\psi\phi\phi}_{i,j,k,r,p} \psi^r_{i,p}(x)\varphi_j(y)\varphi_k(z) + \\ \sum_{r=0}^{r_{\max}} \sum_{p=0}^{2^r-1} {}_nE^{x,\phi\psi\phi}_{i,j,k,r,p} \varphi_i(x)\psi^r_{j,p}(y)\varphi_k(z) + \\ \sum_{r=0}^{r_{\max}} \sum_{p=0}^{2^r-1} {}_nE^{x,\phi\phi\psi}_{i,j,k,r,p} \varphi_i(x)\varphi_k(y)\psi^r_{k,p}(z) + \\ \sum_{r=0}^{r_{\max}} \sum_{p=0}^{2^r-1} \sum_{s=0}^{r_{\max}} \sum_{q=0}^{2^s-1} {}_nE^{x,\psi\psi\phi}_{i,j,k,r,p,s,q} \psi^r_{i,p}(x)\psi^s_{j,q}(y)\varphi_k(z) + \\ \sum_{r=0}^{r_{\max}} \sum_{p=0}^{2^r-1} \sum_{s=0}^{r_{\max}} \sum_{q=0}^{2^s-1} {}_nE^{x,\psi\phi\psi}_{i,j,k,r,p,s,q} \psi^r_{i,p}(x)\varphi_j(y)\psi^s_{k,q}(z) + \\ \sum_{r=0}^{r_{\max}} \sum_{p=0}^{2^r-1} \sum_{s=0}^{r_{\max}} \sum_{q=0}^{2^s-1} {}_nE^{x,\phi\psi\psi}_{i,j,k,r,p,s,q} \varphi_i(x)\psi^r_{j,p}(y)\psi^s_{k,q}(z) + \\ \sum_{r=0}^{r_{\max}} \sum_{p=0}^{2^r-1} \sum_{s=0}^{r_{\max}} \sum_{q=0}^{2^s-1} \sum_{v=0}^{r_{\max}} \sum_{w=0}^{2^v-1} {}_nE^{x,\psi\psi\psi}_{i,j,k,r,p,s,q,v,w} \psi^r_{i,p}(x)\psi^s_{j,q}(y)\psi^v_{k,w}(z) \end{array} \right]. \quad (29)$$

The eight groups of coefficients represent all of the combinations of scaling/wavelet functions in each direction. Using (28), for $r_{\max} = 0$ in all directions, there are eight coefficients per cell. For $r_{\max} = 2$, there are 512 coefficients per cell. The field coefficients are identified in this equation by the superscripts, denoting the field direction and scaling/wavelet component for each direction. For example, $E^{y,\phi\phi\psi}$ denotes a y-directed electric field coefficient, with a scaling function used in the x and y directions, and a wavelet in the z direction. The coefficients have a maximum of nine indices (for the $E^{dir,\psi\psi\psi}$ coefficient), which represent the position in space of the scaling function as well as the resolution and position of each wavelet.

The 3-D MRTD scheme is generated by first representing all fields (E, H, D, and B) in scaling/wavelet function expansions, as in (29), and setting the time and space steps. However, the fields are not collocated in space or time. The original MRTD scheme presented in [2] offsets electric fields one half a cell (the offset between scaling functions) along their coordinate axis (E_x by $\Delta x/2$ in x, for example), while magnetic fields are offset by half a cell in their two normal directions (H_x by $\Delta y/2$ in y and $\Delta z/2$ in z). However, a correction to this field arrangement [8,9] was later made so that that the magnitude of the offset depends on the maximum wavelet resolution used in the simulation. The corrected offset, s_d, is

$$s_d = \frac{\Delta d}{2^{r_{d,\max}+2}}, \tag{30}$$

where d denotes the direction (x, y, or z). The purpose of this offset is discussed in more detail as a consequence of numerical dispersion, but a brief graphical discussion is now presented which explains some of the purpose of the choice of offset.

Fig. 4 shows a 2-D Haar cell with $r_{\max} = 0$. The $r = 0$ wavelets are shown as a reference. The MRTD technique is characterized by cells that contain several basis functions. As a consequence, the field variation throughout the cell is not limited to the shape of a single function, but rather the sum of several. The position of the wavelet coefficients at $r_{\max}$, which can be calculated using (2), defines a number of *equivalent grid points* [9]. Specifically, two equivalent grid points are

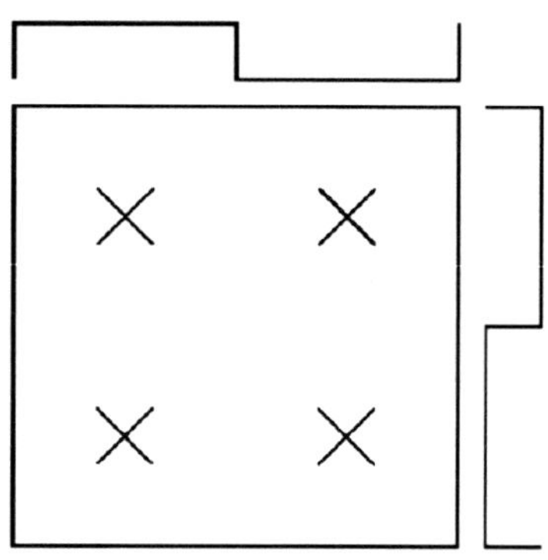

FIGURE 4: Two-dimensional Haar cell, $r_{\max} = 0$, equivalent grid points.

defined per $r_{\max}$ level wavelet positioned $\Delta x/(4 \times 2^{r_{\max}})$ and $3\Delta x/(4 \times 2^{r_{\max}})$ from the position specified in (2). It should be noted that the number of equivalent grid points is equal to the number of basis functions, (28). While the half cell offset used in [2] was chosen to give the cells an offset equal to Yee-FDTD [1], the offset presented in (30) arranges the equivalent grid points in the Yee manner. It should be noted that while Fig. 4 is shows the wavelets for the Haar scheme, the position of the equivalent grid points is valid for any wavelet basis. Furthermore, the cell size represents the domain of the scaling function in the Haar case, but it represents the spacing between the centers of the scaling functions in the general case. The equivalent grid points in Fig. 4 are marked with $\times$.

If a 2-D scheme with E_x, E_y, and H_z (TE$_z$ mode) is simulated with the $r_{\max} = 0$ cell of Fig. 4, and the fields are offset using (30) in the manner suggested above, the field arrangement shown in Fig. 5 results. This figure shows the equivalent grid points for all three fields and is valid for any wavelet basis. The arrangement of the grid points in this case is the same as the Yee-FDTD cell. If the same

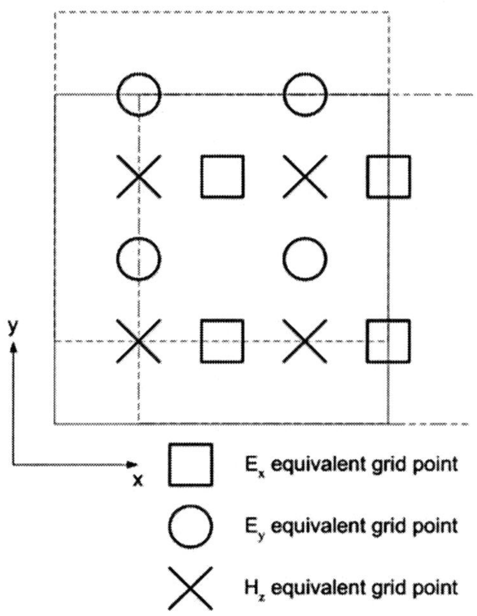

FIGURE 5: Equivalent grid points for a 2-D MRTD cell, $r_{\max} = 0$.

arrangement is used for a full 3-D scheme, the equivalent grid points are arranged in the same manner as Yee-FDTD.

Like FDTD, the update equations in MRTD are fully explicit. Thus, the time step must be chosen according to a stability requirement. This is discussed later. Also like FDTD, the electric and magnetic fields are offset in time. The time basis functions used for most MRTD schemes are pulses, as in FDTD. Similarly, the fields are offset by half a time step. Once the field offset is determined, the MRTD update scheme can be determined using the method of moments.

2.1.4 Method of Moments Overview

The method of moments is a general method used in a variety of numerical schemes. A brief overview of techniques that relate to MRTD is presented here. First, the method is applied by representing the unknown function as a sum of unknown coefficients multiplied by known basis functions. Next, a series of testing functions are chosen. These functions are multiplied into both sides of the equation and the integral is taken over the entire domain. This process is referred to as an inner product and the following notation will be used:

$$\langle f, g\rangle = \int f(\mathbf{r})\, g(\mathbf{r})\partial\mathbf{r}. \tag{31}$$

If the number of testing functions is equal to the number of unknown coefficients, the result is a series of linear equations that can be solved to determine the coefficients.

As a simple 1-D (in space) example, the equation

$$f(x) = \frac{\partial b(x)}{\partial x} \tag{32}$$

is presented, where $f(x)$ is unknown and $b(x)$ is a known function. The unknown expression can be expanded into a sum of basis functions

$$f(x, t) = \sum_{n=0}^{N} a_n c_n(x), \tag{33}$$

where a_n are unknown scalar coefficients and $c_n(x)$ are the known basis functions. A set of testing function $w_n(x)$ is chosen, and the inner product of (32) with each

$$\langle w_n, f\rangle = \langle w_n, b'\rangle \tag{34}$$

is taken. The result is a system of equations,

$$\begin{bmatrix} a_1\langle w_1, c_1\rangle & a_2\langle w_1, c_2\rangle & \cdots & a_N\langle w_1, c_N\rangle \\ a_1\langle w_2, c_1\rangle & a_2\langle w_2, c_2\rangle & & \\ \vdots & & \ddots & \\ a_1\langle w_N, c_1\rangle & & & a_N\langle w_N, c_N\rangle \end{bmatrix} = \begin{bmatrix} \langle w_1, b'\rangle \\ \langle w_2, b'\rangle \\ \vdots \\ \langle w_N, b'\rangle \end{bmatrix}, \tag{35}$$

which can be solved to yield the unknown coefficients, a_n. If the testing functions are chosen such that

$$\langle w_m, c_n\rangle = \delta_{m,n}, \tag{36}$$

then the matrix on the left-hand side of (35) is diagonal. In this case, the scheme is explicit; a matrix does not need to be inverted to solve the equations. Each coefficient can be determined directly,

$$a_n = \langle w_n, b'\rangle. \tag{37}$$

If an orthonormal set of basis functions is chosen

$$\langle w_m, w_n\rangle = \delta_{m,n}, \tag{38}$$

then the basis functions can be used as testing functions. This technique is called Galerkin's method.

2.2 MRTD UPDATE SCHEME

2.2.1 Time Localization

MRTD update equations are determined using Galerkin's method with wavelet discretizations of the electric and magnetic fields, such as (29). The half time step offset of the electric and magnetic field ensure that the fields are always updated from values at previous time steps, and thus the only unknowns are the updated values of the fields. Using (21)–(23), B fields are determined from previous E field values. Likewise, (24)–(26) are used to determine D fields from previous H fields. In many FDTD schemes the constitutive relationship, (15) and (16), is applied discretely at each cell, and thus H values are determined directly from E values, and vice-versa. For general MRTD basis functions, which can extend over the entire domain, this is not possible. Furthermore, even for non-overlapping, finite-domain wavelets, such as Haar wavelets, it is not practical because the dielectric constant cannot vary over the relatively large MRTD cell. Thus, the method of moments must also be applied to (15) and (16), to determine E fields from D fields and H fields from B fields. While this does add complexity to the method, it also allows the material to vary continuously throughout the cell.

The MRTD update equations are determined by localizing the coefficients in time by testing with the time basis functions (pulses) and then localizing in space by testing with the scaling/wavelet functions. The time derivative of the pulse function yields two Dirac Delta functions, located at the edges of the pulse,

$$\frac{\partial h_n(t)}{\partial t} = \delta(t - n\Delta t) - \delta(t - (n+1)\Delta t). \tag{39}$$

The time derivatives of the pulse functions that make up the time discretization are a Delta train, represented by the top line of Fig. 6. In (21)–(26), one field value is differentiated in time, while the other field values are differentiated in space. As stated previously, the E and H fields are offset in time by half a time step, which means that the spatial derivative terms are offset in time by half a time step from the time derivative terms.

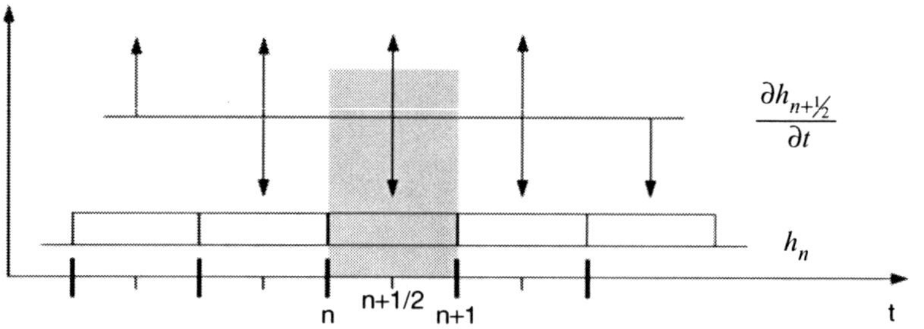

FIGURE 6: Time localization using pulse functions.

The time-testing function is chosen to overlap with the terms in the spatial derivative. If the B (and H) coefficients are located at the $n + 1/2$ time steps and the D (and E) functions are located at the n time steps, Fig. 6 represents the time discretization of (21)–(23). The delta train on the top of the figure represents the time derivative of the B fields, while the pulse functions on the bottom represent the E fields. The shaded area represents the domain of a testing function, which is collocated in time with the E fields. If the time and space basis functions that represent the E fields are separated

$$E(\mathbf{r}, t) = \sum_{n=0}^{N} h_n(t)\, {}_nE(\mathbf{r}), \tag{40}$$

where ${}_nE(\mathbf{r})$ is the wavelet/scaling space discretization for time step n, the quantities in the brackets of (29). Similarly,

$$B(\mathbf{r}, t) = \sum_{n=0}^{N} h_{n+1/2}(t)\, {}_{n+1/2}B(\mathbf{r}), \tag{41}$$

represents the B fields. Using these expressions, the inner products with h_n are

$$\langle h_n(t), E(\mathbf{r}, t)\rangle = \Delta t\, {}_nE(\mathbf{r}), \tag{42}$$

$$\left\langle h_n(t), \frac{\partial B(\mathbf{r}, t)}{\partial t}\right\rangle = {}_{n+1/2}B(\mathbf{r}) - {}_{n-1/2}B(\mathbf{r}). \tag{43}$$

As an example, (42) and (43) can be applied to (21). Ignoring the current terms and taking the inner product with h_n

$$\left\langle h_n, \frac{\partial B_x}{\partial t}\right\rangle = \left\langle h_n, \frac{\partial E_y}{\partial z} - \frac{\partial E_z}{\partial y}\right\rangle \tag{44}$$

yields

$${}_{n+1/2}B_x(\mathbf{r}) - {}_{n-1/2}B_x(\mathbf{r}) = \Delta t\left[\frac{\partial_n E_y(\mathbf{r})}{\partial z} - \frac{\partial_n E_z(\mathbf{r})}{\partial y}\right]. \tag{45}$$

Similar expressions can be found using (22) to (26). Solving for ${}_{n+1/2}B_x(\mathbf{r})$,

$${}_{n+1/2}B_x(\mathbf{r}) = {}_{n-1/2}B_x(\mathbf{r}) + \Delta t\left[\frac{\partial_n E_y(\mathbf{r})}{\partial z} - \frac{\partial_n E_z(\mathbf{r})}{\partial y}\right] \tag{46}$$

yields an update equation for ${}_{n+1/2}B_x(\mathbf{r})$ in terms of quantities from the previous time step. The next step in determining the MRTD update equations is to test with the spatial scaling/wavelet functions.

2.2.2 Space Localization

The 3-D basis functions are expressed as $\Gamma(x)\,\Gamma(y)\,\Gamma(z)$, where each Γ can be either φ_l or $\psi^r_{l,p}$, with l representing a directional index, i, j, or k. As stated in (28), for maximum resolution in the x, y, z direction $r_{\max,x}$, $r_{\max,y}$, $r_{\max,z}$, respectively, each cell (denoted by the triplet i, j, k) contains $2^{3+r_{\max,x}+r_{\max,y}+r_{\max,z}}$ basis functions. To determine an update formula for each coefficient, the testing functions are collocated in space with the time differentiated functions in Faraday's and Ampere's laws. In (46) this is equivalent to collocating the scaling/wavelet testing functions with the B_x fields.

A convenient notation was first introduced in [8], which represents the field discretization as a vector. Here, this notation is generalized for any wavelet basis, which allows the MRTD update equations to be written in a compact and easily understandable form. The wavelet/scaling coefficients for all wavelet resolutions can be represented as a vector, for example

$$
{}_{n+1/2}\mathbf{B}_{x,i,j,k} = \begin{bmatrix}
{}_{n+1/2}B^{x,\phi\phi\phi}_{i,j,k} \\
{}_{n+1/2}B^{x,\psi\phi\phi}_{i,j,k,0,0} \\
\vdots \\
{}_{n+1/2}B^{x,\phi\phi\psi}_{i,j,k,r_{\max,z},2^{r_{\max,z}}-1} \\
{}_{n+1/2}B^{x,\psi\psi\phi}_{i,j,k,0,0,0,0} \\
\vdots \\
{}_{n+1/2}B^{x,\phi\psi\psi}_{i,j,k,r_{\max,y},2^{r_{\max,y}}-1,r_{\max,z},2^{r_{\max,z}}-1} \\
{}_{n+1/2}B^{x,\psi\psi\psi}_{i,j,k,0,0,0,0,0,0} \\
\vdots \\
{}_{n+1/2}B^{x,\psi\psi\psi}_{i,j,k,r_{\max,x},2^{r_{\max,x}}-1,r_{\max,y},2^{r_{\max,y}}-1,r_{\max,z},2^{r_{\max,z}}-1}
\end{bmatrix}. \tag{47}
$$

If another vector Γ is defined

$$\Gamma_{i,j,k} = \begin{bmatrix} \varphi_i(x)\varphi_j(y)\varphi_k(z) \\ \psi^0_{i,0}(x)\varphi_j(y)\varphi_k(z) \\ \vdots \\ \varphi_i(x)\varphi_k(y)\psi^{r_{\max,z}}_{k,2^{r_{\max,z}}-1}(z) \\ \psi^0_{i,0}(x)\psi^0_{j,0}(y)\varphi_k(z) \\ \vdots \\ \varphi_i(x)\psi^{r_{\max y}}_{j,2^{r_{\max,y}}-1}(y)\psi^{r_{\max,z}}_{k,2^{r_{\max,z}}-1}(z) \\ \psi^0_{i,0}(x)\psi^0_{j,0}(y)\psi^0_{k,0}(z) \\ \vdots \\ \psi^{r_{\max x}}_{i,2^{r_{\max,x}}-1}(x)\psi^{r_{\max y}}_{j,2^{r_{\max,y}}-1}(y)\psi^{r_{\max,z}}_{k,2^{r_{\max,z}}-1}(z) \end{bmatrix}, \tag{48}$$

then

$$_{n+{}^1/_2}B_x(\mathbf{r}) = \sum_{i,j,k} \Gamma^T_{i,j,k}\,{}_{n+{}^1/_2}\mathbf{B}_{x,i,j,k}. \tag{49}$$

The update equation for each coefficient of **B** can then be determined by taking the inner product of (46) with each scaling/wavelet coefficient (every row of (48)).

Because the basis functions are orthonormal, a separate update equation is found for each scaling/wavelet coefficient. However, because of the offset of the B and E fields, the scaling/wavelet functions used for the B fields are not orthogonal with the functions used for the E fields. If the basis functions representing the B_x field are offset in the positive y and z directions from cell i, j, k (located at $i\Delta x$, $j\Delta y$, $k\Delta z$), while the E_y fields are offset in the y direction and E_z is offset in the z direction, then, using the vector notation of (47), the update for (46) becomes

$$\begin{aligned} _{n+{}^1/_2}\mathbf{B}_{x,i,j,k} = {}_{n-{}^1/_2}\mathbf{B}_{x,i,j,k} + \frac{\Delta t}{\Delta x \Delta y \Delta z}\Bigg[\sum_m \mathbf{U}^{B_x}_{E_y,m}\,{}_n\mathbf{E}_{y,i,j,k+m} \\ + \sum_m \mathbf{U}^{B_x}_{E_z,m}\,{}_n\mathbf{E}_{z,i,j+m,k}\Bigg]. \end{aligned} \tag{50}$$

In this expression, **U** represents an update matrix and consists of the inner products of the E basis functions with the B basis functions. The B and E fields are only offset in the direction of differentiation of the E field. Thus, m represents the position of all of the neighboring cells that the E and B fields overlap (depending on the domain of the chosen basis function).

The **U** matrices are $2^{3+r_{\max,x}+r_{\max,y}+r_{\max,z}} \times 2^{3+r_{\max,x}+r_{\max,y}+r_{\max,z}}$ in size and can be computed before simulation begins. They take the form

$$\mathbf{U}_{F_2,m}^{F_1} = \begin{bmatrix} \left\langle {}_{F_1}\Gamma_1, \frac{\partial}{\partial n}\, {}_{F_2}\Gamma_1\big|_m \right\rangle & \left\langle {}_{F_1}\Gamma_1, \frac{\partial}{\partial n}\, {}_{F_2}\Gamma_2\big|_m \right\rangle & \cdots & \left\langle {}_{F_1}\Gamma_1, \frac{\partial}{\partial n}\, {}_{F_2}\Gamma_L\big|_m \right\rangle \\ \left\langle {}_{F_1}\Gamma_2, \frac{\partial}{\partial n}\, {}_{F_2}\Gamma_L\big|_m \right\rangle & \left\langle {}_{F_1}\Gamma_2, \frac{\partial}{\partial n}\, {}_{F_2}\Gamma_2\big|_m \right\rangle & & \\ \vdots & & \ddots & \\ \left\langle {}_{F_1}\Gamma_L, \frac{\partial}{\partial n}\, {}_{F_2}\Gamma_L\big|_m \right\rangle & & & \left\langle {}_{F_1}\Gamma_L, \frac{\partial}{\partial n}\, {}_{F_2}\Gamma_L\big|_m \right\rangle \end{bmatrix}, \tag{51}$$

where F_1 is the field whose update is being found and F_2 is the field on which the update depends. In this case, ${}_{F_1}\Gamma_l$ is the lth row of Γ for the F_1 fields and ${}_{F_2}\Gamma_l\big|_m$ is the lth row of Γ for the F_2 fields, offset by m in the direction of differentiation. The $\partial/\partial n$ term represents the space derivative, in this case $n = y$ or z. If the basis functions used for each field were collocated, (51) would be diagonal. However, because of the offset scheme used, the functions are only collocated in two of the three Cartesian directions. For example, the B_x and E_y fields have the same location in the x and y directions but are offset in the z direction.

If the field offset in the x, y, and z directions is s_x, s_y, s_z, respectively, then the (2, 2) entry of (51) for $\mathbf{E}_{y,i,j,k}$ (same cell as the **B** vector, $m = 0$) is

$$\begin{aligned} \mathbf{U}_{E_y,m,2,2}^{B_{x1}} &= \left\langle \psi_{i,0}^{0}(x)\varphi_{j+s_y}(y)\varphi_{k+s_z}(z), \frac{\partial}{\partial z}\psi_{i,0}^{0}(x)\varphi_{j+s_y}(y)\varphi_k(z) \right\rangle \\ &= \iiint \psi_{i,0}^{0}(x)\varphi_{j+s_y}(y)\varphi_{k+s_z}(z)\frac{\partial}{\partial z}\left(\psi_{i,0}^{0}(x)\varphi_{j+s_y}(y)\varphi_k(z)\right) \partial x \partial y \partial z. \end{aligned} \tag{52}$$

This integral can be separated by direction

$$\begin{aligned}\iiint \psi_{i,0}^{0}(x)\varphi_{j+s_y}(y)\varphi_{k+s_z}(z)\frac{\partial}{\partial z}\left(\psi_{i,0}^{0}(x)\varphi_{j+s_y}(y)\varphi_k(z)\right)\partial x\partial y\partial z \\ = \int \psi_{i,0}^{0}(x)\psi_{i,0}^{0}(x)\partial x \int \varphi_{j+s_y}(y)\varphi_{j+s_y}(y)\partial y \int \varphi_{k+s_z}(z)\frac{\partial \varphi_k(z)}{\partial z}\partial z.\end{aligned} \tag{53}$$

Next, the orthogonality between the collocated basis functions can be exploited,

$$\begin{aligned}\int \psi_{i,0}^{0}(x)\psi_{i,0}^{0}(x)\partial x \int \varphi_{j+s_y}(y)\varphi_{j+s_y}(y)\partial y \int \varphi_{k+s_z}(z)\varphi_k(z)\partial z \\ = \Delta x \cdot \Delta y \cdot \int \varphi_{k+s_z}(z)\frac{\partial \varphi_k(z)}{\partial z}\partial z,\end{aligned} \tag{54}$$

and, finally, only a single 1-D integral must be evaluated. Depending on the functions involved, this can be done analytically or numerically. For either case, it can be tabulated, and does not have to be performed for each simulation. It should also be noted that due to the orthogonality of the basis functions, the majority of the entries in (51) are zero.

The procedure shown above can be applied to the remainder of (22)–(26), yielding update equations for all B and D fields. The remaining step in determining the MRTD update scheme is to find the updates for E and H fields from the D and B fields, respectively. As these fields are collocated in space and time, the procedure is less complicated, although potentially more computationally intensive.

2.2.3 Media Discretization

In a general structure, the material properties can vary as a function of position. For this work, only linear, isotropic, nondispersive media are considered. The relationship between the fields in this case is given in (15) and (16). Time localization in this case can be performed by taking the inner product with the time yielding

$$E_d(x, y, z) = \frac{1}{\varepsilon(x, y, z)} D_d(x, y, z), \tag{55}$$

$$H_d(x, y, z) = \frac{1}{\mu(x, y, z)} B_d(x, y, z) \quad d = x, y, \text{or } z. \tag{56}$$

The update for each component of E and H can be determined by taking the inner product with the scaling functions. However, because the offset of each field component is different, the update for each component is different. The update of a general E component takes the form

$$ {}_n\mathbf{E}_{d,i,j,k} = \sum_a \sum_b \sum_c \mathbf{U}^{E_d}_{D_d,i,j,k,a,b,c}\, {}_n\mathbf{D}_{d,i+a,j+b,k+c}, \tag{57} $$

where a, b, and d represent the relative positions of the surrounding D basis functions that overlap the $(i, j, k)E$ basis function. The number of entries in these sums is dependent on the basis functions. The coefficients of (57) take the form

$$ U^{E_d}_{D_d,a,b,c_{ij}} = \iiint \frac{1}{\varepsilon(x, y, z)}\, {}_{E_d}\Gamma_i\, {}_{D_d}\Gamma_j \Big|_{a,b,c} \partial x \partial y \partial z. \tag{58} $$

Unlike the **U** matricies that are used in the discretization of Faraday's and Ampere's laws, most of the entries in these update matricies are nonzero. Furthermore, the matrices vary by position. However, for regions of uniform media, they are diagonal for $l = m = n = 0$, and zero elsewhere.

These equations can be used together in a time-stepping scheme with the discretizations of Ampere's and Faraday's laws to create a general time marching MRTD scheme:

1. Determine B fields from E fields.
2. Determine H fields from B fields.
3. Determine D fields from H fields.
4. Determine E fields from D fields.
5. Repeat until simulation is complete.

2.2.4 Numerical Stability

The scheme that has been presented allows MRTD update schemes to be determined for any basis function. These schemes are fully explicit, the matrix expressions demonstrated only involve multiplication, and all numerical integration is performed before the start of simulation. For the majority of cases, a library of the update equations, **U**, can be constructed, so that the costly numerical integration does not have to be performed before each simulation. Most of the update matrices are very sparse, and a number of clever coding techniques and libraries can be used to reduce the processor and memory load of implementing these updates. Any basis function can be used by calculating the **U** matricies, which can be accomplished by solving a number of integrals, either analytically, or, more likely, numerically. To this point, however, no guidance has been given on choosing the time and space step values. To properly choose these values, the stability and dispersion of the scheme must be taken into consideration.

It is a known limitation of explicit techniques that the time step must limited as a function of the space step to maintain stability. If the time step is not below this limit, the resulting scheme quickly grows without bounds. One notable exception to this is the ADI-MRTD method [10], which is a semiimplicit scheme that remains stable (but not accurate) for any time step. This technique separates the time step into two substeps, and, as it does not alter the spatial grid, it is very similar to ADI-FDTD [11]. The discussion will focus on the traditional MRTD technique. A brief stability analysis of the MRTD technique is presented in [2], while additional studies are presented in [9, 12].

The stability condition for MRTD depends on the choice of the wavelet basis. The general dispersion analysis, however, is performed in the same manner as traditional FDTD techniques [13]. The time update portion (the discretization of the time derivative) and the space update portion (the discretization of the spatial derivative) are split into two separate problems. For numerical stability, the eigenvalues of the spatial problem must contain the eigenvalues of the temporal problem. For FDTD, this results in the condition [13]:

$$\Delta t \leq \frac{1}{\sqrt{\frac{1}{\mu\varepsilon}\left(\frac{1}{\Delta x^2}+\frac{1}{\Delta y^2}+\frac{1}{\Delta z^2}\right)}}. \tag{59}$$

For Battle–Lemarie S-MRTD (scaling function only) [12]

$$\Delta t \leq \frac{1}{\sum_{i=0}^{n_a-1} |a_i| \sqrt{\frac{1}{\mu\varepsilon}\left(\frac{1}{\Delta x^2} + \frac{1}{\Delta y^2} + \frac{1}{\Delta z^2}\right)}}, \tag{60}$$

where

$$s_{\mathrm{MRTD}} = \sum_{i=0}^{n_a-1} |a_i|, \tag{61}$$

the MRTD stability factor is the sum of the magnitudes of the inner products of the offset scaling functions from the spatial update equations (equivalent to the entries in the 1×1 **U** matricies using the notation presented here). In this equation, n_a is the *stencil size*, or the domain of the basis function in cells. The Battle–Lemarie basis functions are symmetric. Alternatively, the sum could be expressed as

$$s_{\mathrm{MRTD}} = \sum_{i=-n_b}^{n_a} |a_i|, \tag{62}$$

where n_a and n_b represent the stencil size of the function in both the positive and negative directions, respectively. In fact, if this expression is used, the stability condition is valid for any basis function. For the Battle–Lemarie scheme [12],

$$s_{\mathrm{MRTD}} = \frac{1}{\sum_{i=0}^{n_a-1} |a_i|} = 0.6371, \tag{63}$$

and the time step for the MRTD scheme is *smaller* than the time step for and FDTD grid with the same cell size. However, as will be shown next, the cells used in Battle–Lemarie MRTD are generally larger than those used in FDTD, and thus the technique is more efficient overall. It is demonstrated in [9] that, for any MRTD wavelet basis, the resolution of the scheme doubles for each addition of a level of wavelet resolution. This is equivalent to saying the time step of a scheme with one level of wavelet resolution in each direction will have a time step one-half of the value of S-MRTD. The resulting stability condition, valid for any MRTD basis, is

$$\Delta t \leq \frac{1}{\sum_{i=-n_b}^{n_a} |a_i| \sqrt{\frac{1}{\mu\varepsilon}\left(\frac{1}{\left(\frac{\Delta x}{2^{r_{\max,x}+1}}\right)^2} + \frac{1}{\left(\frac{\Delta y}{2^{r_{\max,y}+1}}\right)^2} + \frac{1}{\left(\frac{\Delta z}{2^{r_{\max,z}+1}}\right)^2}\right)}}. \tag{64}$$

The Haar MRTD technique can be analyzed using the same method. Several authors [3, 7–9] have noted the equivalence between Haar S-MRTD and FDTD, and indeed this relationship can be exploited to directly derive the general Haar MRTD stability requirement. Using this equivalence, and applying the resolution doubling argument to (59), the stability for Haar MRTD at any resolution level is [9]

$$\Delta t \leq \frac{1}{\sqrt{\frac{1}{\mu\varepsilon}\left(\frac{1}{\left(\frac{\Delta x}{2^{r_{\max,x}+1}}\right)^2} + \frac{1}{\left(\frac{\Delta y}{2^{r_{\max,y}+1}}\right)^2} + \frac{1}{\left(\frac{\Delta z}{2^{r_{\max,z}+1}}\right)^2}\right)}}. \tag{65}$$

This is a special case of (64), for the Haar scheme

$$s_{\mathrm{MRTD}} = 1. \tag{66}$$

2.2.5 Numerical Dispersion

Dispersion is defined as the variation of the wavelength, λ, with frequency, f. More commonly, it is discussed as the variation of the wavenumber, k, with angular frequency, ω, where

$$\omega = 2\pi f, \tag{67}$$

$$k = \frac{2\pi}{\lambda}. \tag{68}$$

By substituting the solution for a monochromatic plane wave into Maxwell's equations (assuming a linear, isotropic, nondispersive medium), the following dispersion relationship results:

$$k = \pm\frac{\omega}{c}, \tag{69}$$

where

$$c = \frac{1}{\sqrt{\mu\varepsilon}}. \tag{70}$$

In a 3-D representation, a wavevector, **k**, is defined, where, in Cartesian coordinates,

$$\mathbf{k} = k_x\hat{\imath} + k_y\hat{\jmath} + k_z\hat{k} \tag{71}$$

and

$$k = \sqrt{k_x^2 + k_y^2 + k_z^2}. \tag{72}$$

Using these parameters, a phase velocity, v_{p}, and group velocity, v_{g}, can be defined,

$$v_{\mathrm{p}} = \pm\frac{\omega}{k} = \pm c, \tag{73}$$

$$v_{\mathrm{g}} = \pm\frac{\partial\omega}{\partial k} = \pm c. \tag{74}$$

These parameters demonstrate that the wavelength and frequency have linear relationship, and that the phase and group velocities are independent of frequency.

These relationships are more complicated in a numerical scheme. A numerical scheme, by its nature, discretizes space into small, but finite, cells. In these cells, the waves cannot propagate in any direction, but rather in the directions defined by the grid. In addition, time is not continuous, but set as a multiple of a discrete time step. With these limitations it should not be surprising that the numerical wave propagation velocity can depend on direction, and, because of the discrete spatial and time steps, frequency. As it has been shown, Haar S-MRTD is equivalent to FDTD, and, as one of the simplest MRTD schemes, it will be used as the first example. For S-MRTD (and FDTD) [14],

$$\left[\frac{1}{c\Delta t}\sin\left(\frac{\omega\Delta t}{2}\right)\right]^2 = \left[\frac{1}{\Delta x}\sin\left(\frac{k_x\Delta x}{2}\right)\right]^2 + \left[\frac{1}{\Delta y}\sin\left(\frac{k_y\Delta y}{2}\right)\right]^2 + \left[\frac{1}{\Delta x}\sin\left(\frac{k_z\Delta z}{2}\right)\right]^2. \tag{75}$$

For numerical stability, the argument was advanced that increasing the resolution by one level effectively doubled the resolution of the scheme. This was represented in the time stability condition by dividing the space step by $2^{r_{\max}+1}$. For the dispersion analysis, this condition holds as well, thus [9],

$$\left[\frac{1}{c\,\Delta t}\sin\left(\frac{\omega\Delta t}{2}\right)\right]^2=\left[\frac{2^{r_{\max,x}+1}}{\Delta x}\sin\left(\frac{k_x\Delta x}{2^{r_{\max,x}+2}}\right)\right]^2+\left[\frac{2^{r_{\max,y}+1}}{\Delta y}\sin\left(\frac{k_y\Delta y}{2^{r_{\max,y}+2}}\right)\right]^2$$
$$+\left[\frac{2^{r_{\max,z}+1}}{\Delta x}\sin\left(\frac{k_z\Delta z}{2^{r_{\max,z}+2}}\right)\right]^2. \tag{76}$$

A similar dispersion analysis was performed in [12] for Battle–Lemarie S-MRTD, and, when made general for any wavelet basis and resolution level,

$$\left[\frac{1}{c\,\Delta t}\sin\left(\frac{\omega\Delta t}{2}\right)\right]^2=\left[\frac{2^{r_{\max,x}+1}}{\Delta x}\left(\sum_{i=n_b}^{n_a-1}a_i\sin\left(\frac{k_x\left(i+1/2\right)\Delta x}{2^{r_{\max,x}+2}}\right)\right)\right]^2$$
$$+\left[\frac{2^{r_{\max,y}+1}}{\Delta y}\left(\sum_{i=n_b}^{n_a-1}a_i\sin\left(\frac{k_y\left(i+1/2\right)\Delta y}{2^{r_{\max,y}+2}}\right)\right)\right]^2. \tag{77}$$
$$+\left[\frac{2^{r_{\max,z}+1}}{\Delta z}\left(\sum_{i=n_b}^{n_a-1}a_i\sin\left(\frac{k_x\left(z+1/2\right)\Delta z}{2^{r_{\max,z}+2}}\right)\right)\right]^2$$

This is the general dispersion relationship for MRTD. As (65) is a special case of (64), (76) is a special case of (77).

It is useful to express the wavenumber in terms of angular frequency. As (77) is a sum of sinusoids, an analytical solution cannot be found for the general case. It can, however, be solved numerically for specific wavelet bases. Several studies have been performed using these equations [4, 9, 12]. For reference, the solution for the wavenumber as a function of angular frequency for Haar MRTD in one of the grid major directions (along an axis), assuming a uniform grid size Δ, is [14]

$$k=\frac{2}{\Delta}\sin^{-1}\left(\frac{1}{S}\sin\left(\frac{\pi S}{N_\lambda}\right)\right), \tag{78}$$

and the phase velocity is

$$v_{\mathrm{p}} = \frac{\pi}{N_{\lambda} \sin^{-1}\left(\frac{1}{S}\sin\left(\frac{\pi S}{N_{\lambda}}\right)\right)} c, \tag{79}$$

using definitions

$$N_{\lambda} = \frac{\lambda_{\circ}}{\Delta} = \frac{c}{\omega \Delta}, \tag{80}$$

$$S = \frac{c\,\Delta t}{\Delta}. \tag{81}$$

Using this expression, the error between the numerical phase velocity and c can be determined, as well as what values of N_{λ} (the number of cells per wavelength) yield stable (nondamped) results. This equation is also valid for any wavelet resolution if Δ is the spacing between equivalent grid points. To keep phase error low, it is usual for these schemes to use more than 10 equivalent cells per smallest excited wavelength.

This condition is one of the major arguments for using the MRTD technique. Although it was demonstrated for Haar MRTD that the dispersion relationship, and thus the number of cells needed per wavelength, is the same for the MRTD scheme as in FDTD, other basis functions can use significantly fewer cells per wavelength. It was reported in [12] that a discretization of three to four cells per wavelength using Battle–Lemarie MRTD gave results comparable to FDTD with 10 cells per wavelength.

2.3 IMAGE THEORY FOR PEC MODELING

The techniques that have been presented thus far allow the modeling of arbitrary variation of the permeability and permittivity of the media under simulation. However, most structures of interest to microwave designers also contain metals. For many simulations in the time domain, these metals are treated as PECs, and this limitation will be used in this case as well. The treatment of a metal as a PEC is necessary, because otherwise the metal itself must be simulated (resulting in significantly increased number of grid points) or frequency dependent loss characteristics must be applied to the PEC.

In the prototypical time-domain technique, FDTD, PEC structures are simulated in a simple and straightforward manner. The PECs are located along grid boundaries, and the electric field values that are tangential to the PEC structures are zeroed each time step (after their updates are calculated). This explicitly enforces (17). The normal coefficients do not require any special processing, because they are offset from the PEC location. Because of the extended, and often overlapping, nature of MRTD basis functions, this simple processing is not possible in MRTD. Instead, PEC boundary conditions are traditionally applied to MRTD codes through image theory.

Several papers have been published that discuss the implementation of image theory to the modeling of PEC structures in MRTD [2, 9, 12, 15–17]. Using this method, PECs are modeled through the introduction of artificial image scaling and wavelet coefficients on the opposite side of the PEC. Tangential E field components and the normal H component are opposite their components across the image, while other components are identical. In this manner, when the real and image wavelet and scaling coefficients are summed at the PEC interface, the result is zero, the PEC boundary condition. This technique is difficult to apply in a general 3-D code, because a large amount of bookkeeping is required for each PEC in the grid. For any wavelet that contains a PEC within its stencil, a unique update scheme must be applied. In addition, multiple images, caused by the close proximity of multiple PECs, can exacerbate this problem further. While it has been stated in several publications that this technique can be applied to complex structures, no publications have been made detailing its use for a general 3-D structure with multiple PECs. At maximum, external boundaries have been presented.

A 1-D example of the application of image theory is presented in Fig. 7. This example uses biorthogonal scaling functions [5]. These functions are used because they have a relatively compact stencil size of four cells. The shaded area represents the width of the stencil. All magnetic field points within this distance of

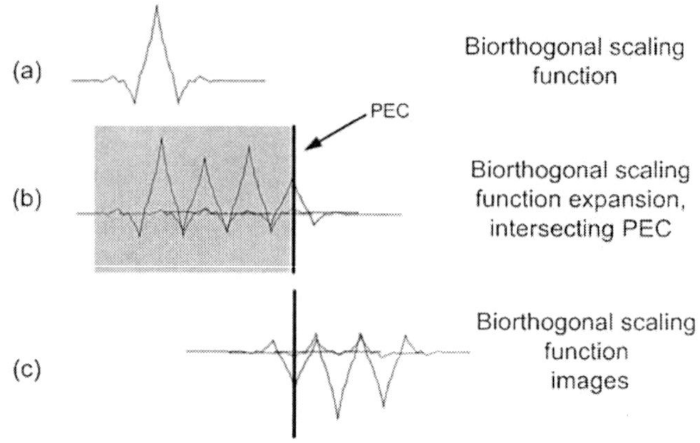

FIGURE 7: Image theory in one dimension, using biorthogonal scaling functions.

the PEC must use the images of the electric field on the opposite of the PEC when computing an update. If this PEC intersects the grid (the simulation is performed on both sides of the PEC), the actual field values on the opposite of the PEC are not used (the PEC effectively decouples these fields). In addition, when the fields are reconstructed, the images must be used in the reconstruction algorithm. This same method can be used to simulate perfect magnetic conductors (PMC) by imaging the magnetic fields across the boundary.

When performing simulations, using any wavelet basis other that the Haar basis functions, this treatment is necessary to terminate the space. In order to perform the leapfrog update scheme used in MRTD, the fields must be known on the outer boundary. Using image theory, the fields outside the computational space are computed as images of the internal fields. These field are not updated, and as they are the same (but different in magnitude) as the internal fields, they do not require additional memory. If the span of the basis functions extends more that the size of the computational domain across the boundary, additional images can be used. This has been termed the multiple image technique (MIT) [17].

2.4 FDTD

It has already been stated that FDTD is often used as a benchmark for other time domain schemes, such as MRTD. The chief advantage of the FDTD method is

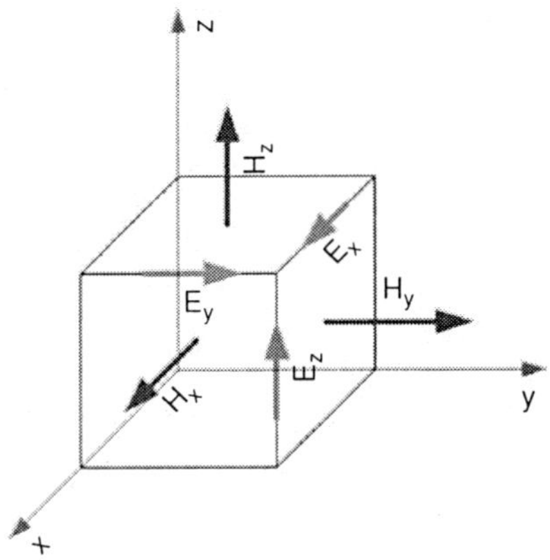

FIGURE 8: Yee cell in 3D.

its simplicity; it was originally derived using central differences with Maxwell's equations. However, the key aspect of the FDTD method does not come from its relatively simple derivation, but the arrangement of its grid points.

FDTD was originally developed in 1967 by Yee [1]. The feature of FDTD that makes it both simple and widely applicable is the Yee cell, which is presented in Fig. 8. A simple Cartesian grid is used, with the fields offset from the grid points in a specific manner. In Fig. 8 the electric field components are offset from the grid points half a cell in their coordinate direction (E_x in x, for example) and the magnetic fields are offset in the directions normal to their field components.

Using this simple field arrangement, and by offsetting the magnetic and electric fields by half a step in time, central differences can be applied to Faraday's and Ampere's law, creating a system of equations that can be used in a time marching scheme to update the fields. For example, the FDTD discretization of (21) is

$$\begin{aligned} &{}_{n+1/2}H_{x,i,j,k} \\ &= {}_{n-1/2}H_{x,i,j,k} + \frac{\Delta t}{\varepsilon}\left[\frac{{}_nE_{y,i,j,k} - {}_nE_{y,i,j,k-1}}{\Delta z} - \frac{{}_nE_{z,i,j,k} - {}_nE_{z,i,j-1,k}}{\Delta y}\right], \end{aligned} \quad (82)$$

if the convention that has been used in the MRTD case is applied, where all field components are offset in the positive direction, and the offsets are not explicitly shown in the field components ($E_{y,i,j,k}$ is located at $i\Delta x$, $(j+1/2)\Delta y$, $k\Delta z$, for example). In this equation, the constitutive relationship (15) is directly applied, as the updates are for actual field values at points instead of scaling/wavelet coefficients that span a much wider area.

It is useful to note that the FDTD update equations are exactly the same as Haar S-MRTD (scaling function only) update equations. The method-of-moments, applied with pulse basis functions, offset according to the scheme that has been presented for MRTD and FDTD, is equivalent to performing a central difference approximation. Furthermore, FDTD and Haar MRTD are equivalent for any $r_{\max}$ if all wavelets are used for all points in the grid. This can be demonstrated by transforming the MRTD scaling/wavelet scheme into a pointwise (where the actual field values are used) scheme, [18], or by noting that the Haar MRTD dispersion relationship for an arbitrary resolution level is the same as FDTD, if the FDTD cells are the same size as the MRTD equivalent grid points.

CHAPTER 3

Modeling of Practical Structures

The MRTD scheme presented in the previous chapter provides a framework for applying multiresolution principles to the simulation of modern microwave structures. Every paper published on MRTD discusses two major advantages of the MRTD technique:

1. The basis functions employed in the MRTD technique allow fewer cells to be used per maximum excited wavelength (compared to FDTD), allowing a more efficient overall simulation.
2. Wavelets can be added and subtracted dynamically during simulation, resulting in an adaptive scheme that automatically tailors its computational requirements to the complexity of propagating signals.

The number of cells required per wavelength is discussed in the previous chapter. The dispersion relationship, (77), can be solved numerically for any wavelet basis. It is shown in several papers [2, 4, 9, 20] that a variety of wavelet bases can be employed that allow less dense grids to be used. There are two distinct costs of applying these techniques. The first is determining the update coefficients, such as those used in (50) and (57), that result from Ampere's and Faraday's laws. However, these coefficients can be calculated before the start of simulation, and it is possible to build a library of the coefficients for several common cases. The second cost of applying the technique is the field updates themselves, which, due to the stencil of the basis functions, involve several surrounding field coefficients. Thus, while fewer cells are required per wavelength when compared to FDTD, more operations are required to calculate each field update. A comprehensive comparison of the benefits from using fewer cells in MRTD compared to the more simple updates used in FDTD has not been presented in literature.

Another difficulty of using general wavelet bases is that of simulating lossy media. The derivations of MRTD update equations presented in the previous chapter neglected loss. When Ohmic losses are added (considering both electric and magnetic loss), Faraday's and Ampere's laws become

$$\nabla \times \mathbf{E}(t) = -\frac{\partial \mathbf{B}(t)}{\partial t} - \sigma^* \mathbf{B}(t) \tag{83}$$

$$\nabla \times \mathbf{H}(t) = \frac{\partial \mathbf{D}(t)}{\partial t} + \sigma \mathbf{D}(t). \tag{84}$$

The procedure for determining the MRTD update procedures described in the previous chapter can be applied in this case as well. For example, the update for the B_x field is determined from

$$\frac{\partial B_x}{\partial t} = \frac{\partial E_y}{\partial z} - \frac{\partial E_z}{\partial y} - \sigma^* B_x. \tag{85}$$

To determine the time localized form of this equation, the inner product of both sides of (85) must be taken with $h_{n+1/2}(t)$. This process is slightly different than the case without loss, because both the time derivative of the B field and the B field itself are used in this equation. In Fig. 9, it is noted that the pulse used in the time localization overlaps the B field from two time steps. The inner product

$$\langle h_n(t), B_x \rangle = \frac{{}_{n+1/2}B_x(\mathbf{r}) + {}_{n-1/2}B_x(\mathbf{r})}{2} \Delta t \tag{86}$$

therefore includes terms from both the time step being updated and the previous time step. This is a statement of the semiimplicit approximation used in FDTD [14], but it is derived in this case as a direct result of the chosen basis functions.

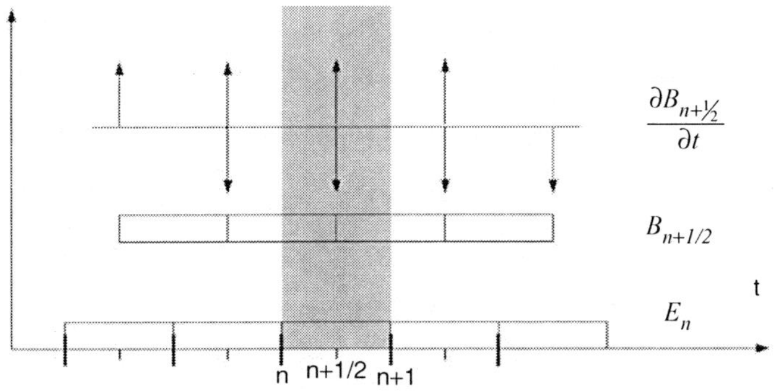

FIGURE 9: Time localization (pulse basis functions) for lossy case.

Using (86), the time localized form of (85) is

$$\begin{aligned} {}_{n+1/2}B_x(\mathbf{r}) &= {}_{n-1/2}B_x(\mathbf{r}) - \sigma^*\Delta t \frac{{}_{n+1/2}B_x(\mathbf{r}) + {}_{n-1/2}B_x(\mathbf{r})}{2} \\ &\quad + \Delta t \left[\frac{\partial_n E_y(\mathbf{r})}{\partial z} - \frac{\partial_n E_z(\mathbf{r})}{\partial y} \right]. \end{aligned} \tag{87}$$

Collecting terms

$$\begin{aligned} \left(1 + \frac{\sigma^*\Delta t}{2}\right) {}_{n+1/2}B_x(\mathbf{r}) &= \left(1 - \frac{\sigma^*\Delta t}{2}\right) {}_{n-1/2}B_x(\mathbf{r}) \\ &\quad + \Delta t \left[\frac{\partial_n E_y(\mathbf{r})}{\partial z} - \frac{\partial_n E_z(\mathbf{r})}{\partial y} \right] \end{aligned} \tag{88}$$

and finally solving for ${}_{n+1/2}B_x(\mathbf{r})$

$$\begin{aligned} {}_{n+1/2}B_x(\mathbf{r}) &= \left(\frac{2 - \sigma^*\Delta t}{2 + \sigma^*\Delta t}\right) {}_{n-1/2}B_x(\mathbf{r}) \\ &\quad + \left(\frac{2\Delta t}{2 + \sigma^*\Delta t}\right) \left[\frac{\partial_n E_y(\mathbf{r})}{\partial z} - \frac{\partial_n E_z(\mathbf{r})}{\partial y} \right]. \end{aligned} \tag{89}$$

When the space localization is performed to determine the updates for the individual wavelet coefficients, the following results:

$$\begin{aligned} {}_{n+1/2}\mathbf{B}_{x,i,j,k} &= \sum_a \sum_b \sum_c \mathbf{U}^{B_x}_{B_x,a,b,c\,n-1/2}\mathbf{B}_{x,i+a,j+b,k+c} \\ &\quad + \frac{\Delta t}{\Delta x \Delta y \Delta z} \left[\begin{array}{l} \sum_a \sum_b \sum_c \mathbf{U}^{B_x}_{E_y,a,b,c\,n-1/2}\mathbf{E}_{y,i+a,j+b,k+c} \\ + \sum_a \sum_b \sum_c \mathbf{U}^{B_x}_{E_z,a,b,c\,n-1/2}\mathbf{E}_{z,i+a,j+b,k+c} \end{array} \right], \end{aligned} \tag{90}$$

using the notation presented in the last chapter. In this case, however, it is noted that the basis functions can be offset in every direction (denoted by a, b, and c, in the x, y, and z directions, respectively), instead of the single direction that was denoted by m in Chapter 2. The entries of the $\mathbf{U}^{B_x}_{B_x,a,b,c}$ matrices take the form

$$U_{ij} = \left\langle {}_{B_x}\Gamma_i, \sigma^* \; {}_{B_x}\Gamma_j\big|_{a,b,c} \right\rangle = \iiint \left(\frac{2 - \sigma^* \Delta t}{2 + \sigma^* \Delta t} \right) {}_{B_x}\Gamma_i \; {}_{B_x}\Gamma_j\big|_{a,b,c} \partial x \partial y \partial z, \quad (91)$$

and the entries of the $\mathbf{U}^{B_x}_{E_y,a,b,c}$ matrices take the form

$$U_{ij} = \left\langle {}_{B_x}\Gamma_i, \sigma^* \; {}_{E}\Gamma_j\big|_{a,b,c} \right\rangle = \iiint \left(\frac{2\Delta t}{2 + \sigma^* \Delta t} \right) {}_{B}\Gamma_i \frac{\partial}{\partial z} \; {}_{E}\Gamma_j\big|_{a,b,c} \partial x \partial y \partial z, \quad (92)$$

where $\Gamma_j\big|_{a,b,c}$ denotes the scaling/wavelet function offset by a, b, and c (cells) in the x, y, z directions, respectively, from the coefficient update being calculated. In the general case of $\sigma^*(x, y, z)$ varying constantly in each direction, the $\mathbf{B_x}$ coefficients depend on surrounding $\mathbf{B_x}$ coefficients from the previous time step. This complicates the update of the coefficients, as when implemented on a computer, the array of updated values must be kept separate from previous values. In the lossless scheme, the fields being updated depend only on their previous value, and the updated fields can be stored in the same array as the previous fields. This doubles the amount of memory required to perform the scheme. In addition, it significantly increases the computational burden.

In the lossless case, the updates of any one coefficient depend on the previous value of the coefficient and the values of two other fields (the normal E fields in the B update case, and the normal H fields in the D case). For these fields, a sum of coefficients (multiplied by the entries in the $\mathbf{U}$ matrix) from neighboring grid points is required, and the number of grid points depends on the stencil of the basis function. For example, a stencil of 12 (6 in the positive, 6 in the negative direction) cells is normally used for Battle–Lemarie wavelets [16]. However, because the fields are offset only in one direction, the sum must only be performed in one direction (as in (50)). In the case including loss, however, the addition of σ into the inner products destroys the orthogonality in the inner product integrals, and the sums must be performed in all three directions. For a stencil size n, the result is a summation over n^3 elements. As an example, in the lossless case, for Battle–Lemarie wavelets using a stencil size of 12, the total number of neighboring cells required for the update is

Previous value + normal field 1 + normal field 2

$$1 + n + n = 2n + 1, \tag{93}$$

$$1 + 12 + 12 = 25, \tag{94}$$

while for the lossy case

$$n^3 + n^3 + n^3 = 3n^3, \tag{95}$$

$$1728 + 1728 + 1728 = 5184. \tag{96}$$

This significantly complicates the method. However, for stencil sizes of 1 (such as Haar basis functions) the requirements are the same in the lossy and lossless case. In addition, there are several classes of wavelets and wavelet like functions that allow this condition to be reduced. One example is the Daubechies' wavelets with vanishing moments [19, 20].

The final difficulty of simulating complex structures using general wavelet bases is that of representing PEC structures within an MRTD grid. In the previous chapter image theory was discussed; the primary example presented, as well as the examples shown in the referenced literature, was for the case of a PEC wall terminating the structure. A general method for simulating arbitrarily placed internal PEC structures has not been presented in literature. This makes simulation of most microwave structures difficult. One notable exception is [20] which utilizes an FDTD/Haar-MRTD interface to simulate highly detailed structure in FDTD, and MRTD to simulate open areas. However, this technique cannot be readily expanded to other wavelet bases. In this chapter a method is presented that allows arbitrarily placed PEC structures to be simulated; however, it is shown that it is practical only for Haar-MRTD schemes.

As this monograph is an introduction to the MRTD method, and Haar wavelets are used to keep the examples relatively simple, in this chapter a technique is presented that allows subcell structures to be modeled within Haar MRTD cells. This technique is shown to be a bridge between pointwise field updates, such as those used in FDTD and the wavelet/scaling updates used in MRTD. This technique allows the adaptive resolution characteristics of MRTD to be exploited for the simulation of any structure.

3.1 GENERAL, EXPLICIT, SUBCELL FIELD MODIFICATION

One challenge in MRTD modeling is representing arbitrary structures in such a way that the time-space adaptive grid can be used effectively. To accomplish this, it is necessary to determine a method of representing PEC structures that are smaller than an MRTD cell. Using the Haar basis functions, it is possible to apply the PEC boundary condition in a similar method as FDTD grids. It is shown that the method effectively provides a bridge between pointwise electromagnetic effects and the distributed wavelet field representation used in MRTD.

In FDTD, which is equivalent to Haar S-MRTD, PEC structures are explicitly represented by zeroing electric field values that are tangential to PEC field locations. The update equations for FDTD can be determined in the same manner as the MRTD equations in the previous chapter. In fact, the B_x update equation is a special case of (50). Similarly, the D_x update equation is [14]

$$
\begin{aligned}
{}_{n+1}D_{x,i,j,k} = {}_{n}D_{x,i,j,k} + \Delta t \Bigg[& \frac{{}_{n+1/2}H_{z,i,j,k} - {}_{n+1/2}H_{z,i,j-1,k}}{\Delta y} \\
& - \frac{{}_{n+1/2}H_{y,i,j,k} - {}_{n+1/2}H_{y,i,j,k-1}}{\Delta z} \Bigg],
\end{aligned}
\tag{97}
$$

in the lossless case (for the purpose of applying PEC boundary conditions, loss is irrelevant). In FDTD, the constitutive relationships are applied at each cell, so

$$
\begin{aligned}
{}_{n+1}E_{x,i,j,k} = {}_{n}E_{x,i,j,k} + \frac{\Delta t}{\varepsilon} \Bigg[& \frac{{}_{n+1/2}H_{z,i,j,k} - {}_{n+1/2}H_{z,i,j-1,k}}{\Delta y} \\
& - \frac{{}_{n+1/2}H_{y,i,j,k} - {}_{n+1/2}H_{y,i,j,k-1}}{\Delta z} \Bigg].
\end{aligned}
\tag{98}
$$

Note that in this notation, the spatial offsets from the grid points are not written explicitly, similar to the MRTD case in the previous chapter (but follow the same scheme as MRTD). In this case, the electric fields are located half a cell from point $i\Delta x$, $j\Delta y$, $k\Delta z$ in the direction of their field component, and the magnetic fields are located half a cell from $i\Delta x$, $j\Delta y$, $k\Delta z$ in the two directions normal to the field component that they represent. A 2-D cross-section of an FDTD grid intersected by a PEC is presented in Fig. 10.

In FDTD, the PEC is placed along the locations of the electric field points. The practical result is that the size of the structure being simulated is constrained by the grid. To represent field variation caused by the PECs, several cells are usually placed across each PEC. The PEC condition is applied by first updating the electric fields, and then setting the field values that overlap with the PEC locations to zero. This is possible because the electric field update equations, such as (98), use only previous field values.

The time update scheme used in MRTD takes the same form as FDTD (the same basis functions are used for the time discretization). Instead of using

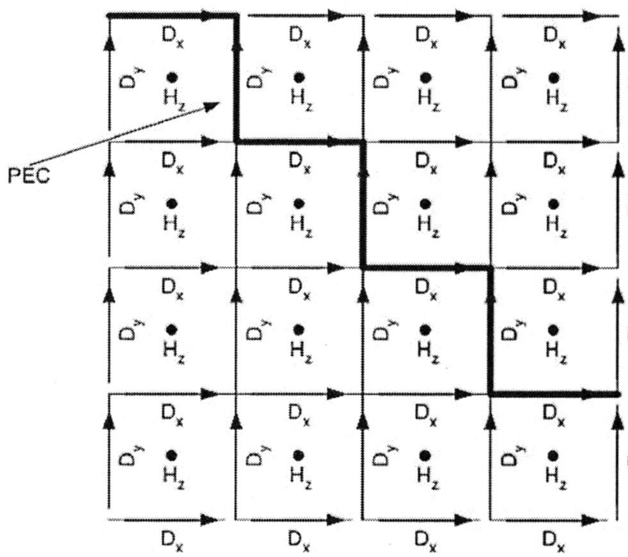

FIGURE 10: FDTD grid intersected by PEC.

image theory, which creates artificial scaling/wavelet coefficients to apply PEC conditions, it would be convenient to explicitly enforce the boundary condition of zero tangential electric field on the MRTD grid.

For a general wavelet basis, using the notation presented in the previous chapter, the electric field is reconstructed as

$$ {}_{n}E_{\text{dir}}\left(\mathbf{r}\right)=\sum_{i}\sum_{j}\sum_{k}\Gamma^{T}_{i,j,k}\,{}_{n}\mathbf{E}_{\text{dir},i,j,k}. \tag{99} $$

This function is used to give the fields at every point in the grid. It is useful to note that the MRTD scheme never updates the field values themselves, as in FDTD, but only updates the scaling/wavelet coefficients that then must be reconstructed to determine field values. This is important to note when probing the fields during simulation, as the fields must be reconstructed at points of interest. When probing fields during simulation, (99) is unwieldy; reconstructing the fields for the entire grid for each time step is computationally prohibitive. Instead, the fields for an individual cell can be reconstructed using

$$ {}_{n}E_{\text{dir},i,j,k}\left(\mathbf{r}\right)=\sum_{a=-n_a}^{n_b}\sum_{b=-n_a}^{n_b}\sum_{c=-n_a}^{n_b}\Gamma^{T}_{i+a,j+b,k+c}\,{}_{n}\mathbf{E}_{\text{dir},i+a,j+b,k+c}, \tag{100} $$

where n_a and n_b represent the size of the stencil (the overlap of the wavelet/scaling function in cells).

Once the fields are reconstructed, a localized PEC can be applied by multiplying the reconstructed electric field by a function $P(\mathbf{r})$, where

$$ P(\mathbf{r})=\begin{cases}0 & \text{PEC location}\\ 1 & \text{Elsewhere}\end{cases}. \tag{101} $$

Once the PEC boundary condition has been applied, the fields must be transformed back into the wavelet domain. This can be accomplished by applying a wavelet transform

$$ \mathbf{E}_{\text{dir},i,j,k}=\left\langle \Gamma_{i,j,k},\ P(\mathbf{r})\,{}_{n}E_{\text{dir},i,j,k}(\mathbf{r})\right\rangle. \tag{102} $$

Of course, the new components must be found at all locations whose stencils include a PEC location. By combining (99) and (102), the wavelet coefficient with the PEC condition applied can be found directly from other scaling/wavelet coefficients using

$$\mathbf{E}_{\mathrm{dir},i,j,k} = \left\langle \Gamma_{i,j,k},\ P(\mathbf{r}) \sum_i \sum_j \sum_k \Gamma^T_{i,j,k\,n} \mathbf{E}_{\mathrm{dir},i,j,k} \right\rangle . \tag{103}$$

In the case of a general wavelet resolution, there are several difficulties in applying this technique. The first is that, for arbitrarily placed PECs, the reconstruction, and subsequent wavelet decomposition, of the entire grid is required for each time step. This procedure likely requires more computation time than the field updates and is therefore not practical. In addition, the implicit assumption in the above procedure is that the fields, when deconstructed back into the wavelet domain after the application of the PEC boundary condition, reconstruct to the same values at all non-PEC locations, and zero at PEC locations. While all of the wavelet bases discussed in this book are complete, this condition is true only for an infinite level of wavelet resolution. For a limited wavelet resolution, which is kept the same before and after the application of the PEC, the application of the PEC as outlined here will result in modified fields outside the PEC region, and nonzero fields inside the PEC region. This is not consistent with Maxwell's equations.

For this technique to be applied successfully, the wavelet basis must satisfy two conditions:

1. The scaling/wavelet functions for one cell must not overlap with a neighboring cell.
2. The application of a PEC boundary condition (zeroing the field) over a range within an MRTD cell must not affect neighboring field values.

The first condition allows the PEC boundary condition to be applied locally. This means that only the scaling/wavelet coefficients in a single cell must be modified to apply the PEC boundary condition. The second ensures that the fields will not undergo nonphysical modifications when the PEC condition is applied. Both of these conditions are satisfied by the Haar wavelet basis.

3.2 PROPERTIES OF WAVELET DISCRETIZATIONS

3.2.1 HAAR Wavelets

To demonstrate that the Haar basis can be used to explicitly apply subcell PEC boundary conditions, it is first useful to study how the Haar basis functions discretize field values. The Haar scaling and wavelet functions are defined in the previous chapter. When the Haar wavelets are reconstructed, they are constantly valued over discrete regions. For example, Fig. 3 presents the 2-D Haar scaling function and wavelets for $r_{\mathrm{max}} = 0$ in both directions. When these functions are summed, they yield four independent regions of constant field value. These regions are centered at the equivalent grid points. As was stated previously, the number of equivalent grid points is the same as the number of basis functions. In the 3-D case, the number of equivalent grid points (and basis functions) is

$$\text{number of equivalent grid points} = 2^{3+r_{\mathrm{max},x}+r_{\mathrm{max},y}+r_{\mathrm{max},z}}, \tag{104}$$

which is equivalent to (28) for the 3-D case.

As another example, the $r_{\mathrm{max}} = 1$, in all directions, wavelets for the 2-D case are presented in Fig. 11. In this case there are $2^{2+r_{\mathrm{max},x}+r_{\mathrm{max},y}} = 2^{2+1+1} = 16$ coefficients per cell. In this figure, each of the wavelets has two possible, identical in magnitude, values, with opposite sign. These are represented in the figure by different shading. For the wavelet/wavelet case, these represent four distinct regions. For the highest level wavelet/wavelet terms (the $\psi_p^{r_{\mathrm{max}}}(x)\,\psi_p^{r_{\mathrm{max}}}(y)$ terms), the equivalent grid points are at the center of the areas of constant magnitude. Each of the highest level wavelet/wavelet terms represents four equivalent grid points in the 2-D scheme.

This discussion is easily extended to 3D. In the $r_{\mathrm{max}} = 1$ case an additional four scaling/wavelet coefficients are required to describe the fields in the z direction. The total number of basis functions in this case is $2^{3+r_{\mathrm{max},x}+r_{\mathrm{max},y}+r_{\mathrm{max},z}} = 2^{3+1+1+1} = 64$. In this case, the regions of constant value are rectangular solids, and for the wavelet/wavelet terms there are eight areas of constant value. Again, the center of these regions for the highest resolution wavelet/wavelet terms represents the

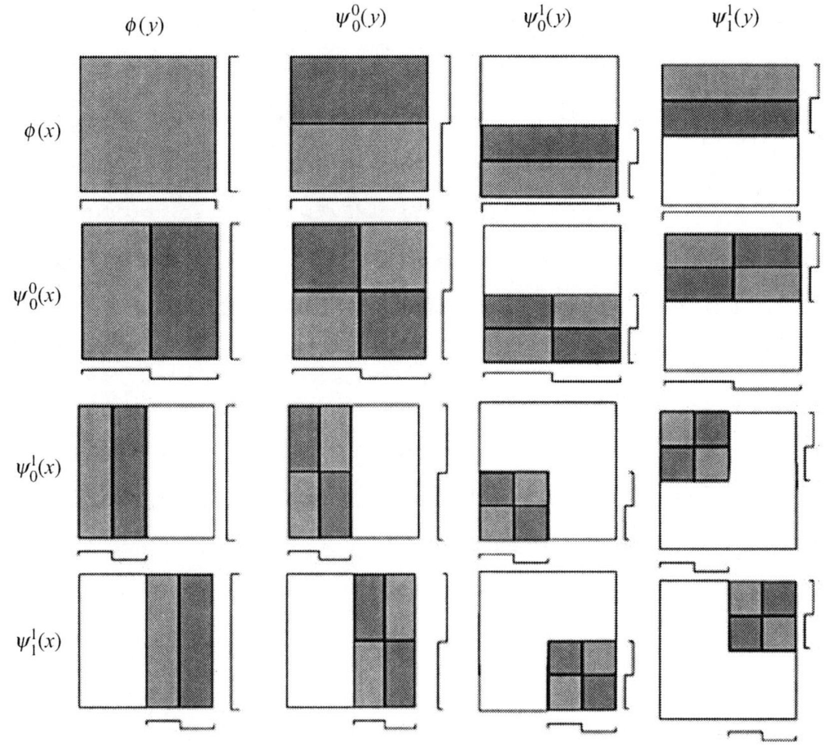

FIGURE 11: Haar scaling and wavelet functions in 2-D, $r_{max} = 1$.

equivalent grid points. When the fields are reconstructed they add to independent values, located at the equivalent grid points. The domain of each value is

$$L = \frac{\Delta D}{2^{1+r_{\max,D}}}, \tag{105}$$

in each dimension, D. In the $r_{\max} = 1$ case, the value is constant at each equivalent grid point over a range of $\Delta x/4 \times \Delta y/4 \times \Delta z/4$.

One important note is that, for any cell, the fields can be represented in an equivalent manner by scaling functions alone. In this case the scaling functions must be the size of the equivalent grid points. The advantage of using the wavelets is that the resolution can be varied on a cell by cell basis. In cells containing complex structures, high resolution cells can be used. In surrounding cells, lower resolution can be used; Additionally, the wavelet resolution can be varied during simulation. For complex field variation, high resolution wavelets can be used; for less variation, lower resolution can be used. To apply a similar method with scaling functions only (FDTD), an interpolation scheme must be used to interface the high and low resolution areas. In the MRTD scheme, variable resolution is automatically applied by zeroing high resolution wavelet coefficients when they are not needed.

The practicality of using Haar wavelet addition/subtraction as a method of representing fields varying at different rates can be represented analytically and graphically. For example, a 1-D Haar wavelet system with $r_{\max} = 0$ is presented. In this scheme there is only one scaling and one wavelet function. These functions can be used to represent any two values. Fig. 12(a) shows the Haar scaling and wavelet function and Fig. 12(b) shows an example of a dual valued piecewise constant function that can be represented using these values. The function $f(x)$ in Fig. 12(b) has value c in the range (0, 0.5) and the value d in the range (0.5, 1). The scaling

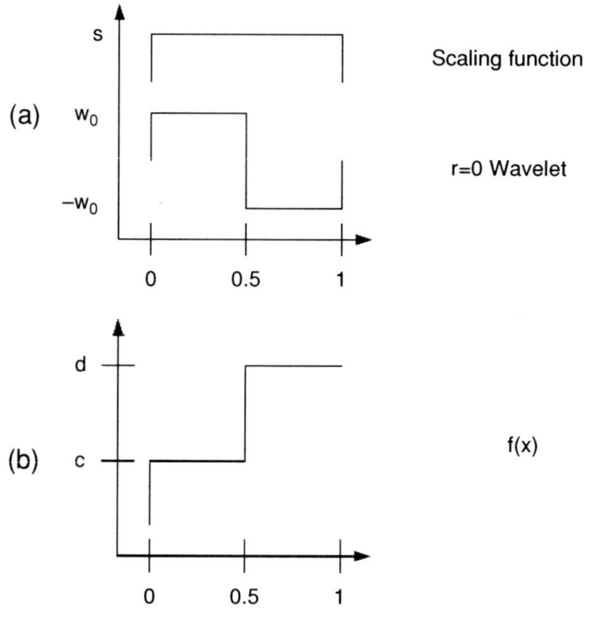

FIGURE 12: (a) Haar scaling and $r_{\max} = 0$ wavelet and (b) sample function that can be represented using (a).

function has value s in the range (0, 1) and 0 elsewhere. The wavelet function is valued w_0 in the range (0, 0.5) and $-w_0$ in the range (0.5, 1).

The magnitudes of the scaling/wavelet functions can be found using the system

$$\begin{aligned} s + w_0 &= c, \\ s - w_0 &= d. \end{aligned} \tag{106}$$

This system can be easily solved to yield

$$s = \frac{c+d}{2} \qquad w_0 = \frac{c-d}{2}. \tag{107}$$

While this is an extremely basic example it demonstrates an important property of Haar discretizations. The scaling function is the average of the discrete values of $f(x)$. For this two-valued function example, if the values are identical, $w_0 = 0$, and the wavelet coefficient is not needed. In practical simulations, w_0 can be neglected if it is significantly smaller than the scaling function.

This same scheme holds for higher level wavelets. For contrast, a similar example is presented for $r_{\max} = 1$. In Fig. 13 the addition of the level one wavelets (the tails are removed for $r_{\max} = 1$ to demonstrate that the wavelets represent two independent functions) allows the piecewise constant valued function to have a maximum of four independent values. The wavelets have two values, equal in magnitude and opposite in sign, and for simplicity the magnitude is indicated in the center of each function. In this case a system of four equations can be constructed to determine the scaling/wavelet function magnitudes. The solution of this system,

$$\begin{aligned} s &= \frac{c+d+e+f}{4} & w_0 &= \frac{c+d-e-f}{4} \\ w_{0,0} &= \frac{c-d}{2} & w_{0,1} &= \frac{e-f}{2} \end{aligned} \tag{108}$$

shows that the scaling function still represents the average of the function. The sum of the scaling term and the 0th level wavelet represents the average on either half of the domain. If the variation of the field on either half of the domain is small, the high-level wavelets can be ignored.

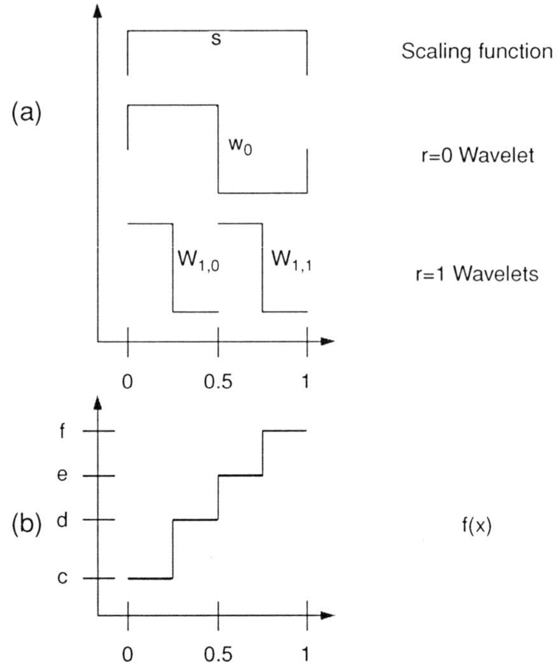

FIGURE 13: Haar example with (a) scaling through $r_{\max} = 1$ wavelets and (b) an example function that can be represented with these functions.

One other interesting property of Haar wavelet expansions related to the representation of PEC structures can be demonstrated using this example. The values of the scaling/wavelet functions presented in (108) can be determined regardless of the values of $f(x)$. If $f(x)$is zeroed over one of the ranges, the remainder of the values can still be represented using the scaling/wavelet basis. The values of all of the basis functions in the cell change, but the values obtained when the functions are summed remain the same.

In the two examples that have been presented of Haar wavelet decompositions, the values of the Haar wavelets were found using a system of linear equations. In the general case of a continuous function, the method of moments can be applied. In the special cases that have been presented of piecewise constant functions with constant domains equal to the equivalent grid point dimensions, the method of moments discretization reduces to a system of linear equations. If the values of the function over these constant areas are represented as a vector **F** and the scaling/wavelet function magnitudes are represented as a vector $\mathbf{F_w}$, then they are related by a reconstruction matrix **R** where

$$\mathbf{F} = \mathbf{R}\mathbf{F_w}. \tag{109}$$

This representation is also valid for the 2-D and 3-D cases. Because there are the same number of equivalent grid points as scaling/wavelet functions, **R** is square. In the two and three dimensional cases, any ordering of the wavelet coefficients and equivalent grid points can be used, it only affects the positioning of the coefficients in the **R** matrix. **R** can be generated quickly by examining the contribution of each wavelet/scaling coefficient to each equivalent grid point.

Once the reconstruction matrix is determined, the wavelet transform can be easily performed for arbitrary values of **F**. A wavelet transform matrix, **W** is defined as

$$\mathbf{W} = \mathbf{R}^{-1} \tag{110}$$

and

$$\mathbf{F_w} = \mathbf{W}\mathbf{F}. \tag{111}$$

These matrices provide a quick and easy transition between scaling/wavelet coefficients and allow the PEC boundary condition to be explicitly applied at any equivalent grid point in the Haar MRTD scheme without affecting neighboring field values. The only restriction is that the metals must be the size of an equivalent grid point. By increasing the resolution to the appropriate level, arbitrary structures can be represented.

3.3 HAAR SUBCELL PEC APPLICATION

The discussion in the previous section demonstrates that Haar wavelets can be used to apply the explicit PEC method that has been presented. The Haar wavelets are nonoverlapping, and thus a modification of the fields in one cell does not affect neighboring cells. In addition, the reconstruction matrix can be used to quickly determine the field values within one cell, which can be subsequently modified and transformed back into the wavelet domain. The MRTD update algorithm that was presented in the previous chapter can be modified (the added steps are indicated in boldface):

1. Determine B fields from E fields.
2. Determine H fields from B fields.
3. Determine D fields from H fields.
4. **Reconstruct D fields in PEC grid locations.**
5. **Zero fields tangential to PECs.**
6. **Transform D fields back to wavelet domain.**
7. Determine E fields from D fields.
8. Repeat until simulation is complete.

The PEC condition is applied directly to the D fields after they are updated. Mathematically, these steps are relatively simple. First, the fields are reconstructed in the cell where the PEC is to be applied,

$$\mathbf{D}_{\mathrm{dir},i,j,k} = \mathbf{R}\mathbf{D}_{\mathbf{w},\mathrm{dir},i,j,k}. \tag{112}$$

In this case the subscripts indicate the fields at cell i, j, k. The vector $\mathbf{D}_{\mathrm{dir},i,j,k}$ contains the reconstructed field values in the cell. The entries in this vector that correspond to positions of PECs are zeroed. This can also be represented mathematically if a matrix $\mathbf{I}_{\mathbf{P}}$ is defined, where $\mathbf{I}_{\mathbf{P}}$ is the identity matrix with the rows that correspond to the PEC locations zeroed. The application of the PEC to the field becomes

$$\mathbf{D}_{\mathbf{PEC},\mathrm{dir},i,j,k} = \mathbf{I}_{\mathbf{P}}\mathbf{D}_{\mathrm{dir},i,j,k}. \tag{113}$$

To continue the MRTD field updates, the $\mathbf{D}$ fields must be transformed back to the wavelet domain

$$\mathbf{D}_{\mathbf{w},\mathrm{dir},i,j,k} = \mathbf{R}^{-1}\mathbf{D}_{\mathbf{PEC},\mathrm{dir},i,j,k}. \tag{114}$$

The procedure that has been presented requires that the **D** vector be multiplied by three matrices. These matrices are $2^{3+r_{\max,x}+r_{\max,y}+r_{\max,z}} \times 2^{3+r_{\max,x}+r_{\max,y}+r_{\max,z}}$ in size, for the $r_{\max} = 1$ case the matrix is 64×64, containing 4096 entries. While many of the entries in these matrices are zero, and thus the multiplication can be significantly simplified, this procedure still adds nonnegligible computational overhead. There are two methods that can be used to apply this process in a more efficient manner.

First, the three steps (112)–(114) can be combined. Each step is a matrix multiplication, so an alternate method is

$$\mathbf{D}_{\mathbf{w},\mathrm{dir},i,j,k} = \mathbf{P}_{\mathrm{dir},i,j,k}\mathbf{D}_{\mathbf{w},\mathrm{dir},i,j,k} = \mathbf{R}^{-1}\mathbf{I}_{\mathbf{P}}\mathbf{R}\mathbf{D}_{\mathbf{w},\mathrm{dir},i,j,k}. \tag{115}$$

Before simulation begins, the $\mathbf{P}_{\mathrm{dir},i,j,k}$ matrices can be calculated, and then the PEC condition can be quickly applied with a single matrix multiplication. This method is a simple way to modify an existing Haar MRTD code to add local PEC modeling. An even more efficient method is to directly modify the MRTD update equations.

In Haar MRTD the D field update equations can be calculated using the same procedure as (50), yielding

$${}_{n+1}\mathbf{D}_{x,i,j,k} = {}_{n}\mathbf{D}_{x,i,j,k} + \frac{\Delta t}{\Delta x \Delta y \Delta z}\begin{bmatrix} \mathbf{U}^{D_x}_{H_y,1}{}_{n+1/2}\mathbf{H}_{y,i,j,k+1} + \mathbf{U}^{D_x}_{H_y,0}{}_{n+1/2}\mathbf{H}_{y,i,j,k} \\ + \mathbf{U}^{D_x}_{H_z,1}{}_{n+1/2}\mathbf{H}_{z,i,j+1,k} + \mathbf{U}^{D_x}_{H_z,0}{}_{n+1/2}\mathbf{H}_{z,i,j,k} \end{bmatrix}, \tag{116}$$

$${}_{n+1}\mathbf{D}_{y,i,j,k} = {}_{n}\mathbf{D}_{y,i,j,k} + \frac{\Delta t}{\Delta x \Delta y \Delta z}\begin{bmatrix} \mathbf{U}^{D_y}_{H_x,1}{}_{n+1/2}\mathbf{H}_{x,i,j,k+1} + \mathbf{U}^{D_y}_{H_x,0}{}_{n+1/2}\mathbf{H}_{x,i,j,k} \\ + \mathbf{U}^{D_y}_{H_z,1}{}_{n+1/2}\mathbf{H}_{z,i+1,j,k} + \mathbf{U}^{D_y}_{H_z,0}{}_{n+1/2}\mathbf{H}_{z,i,j,k} \end{bmatrix}, \tag{117}$$

$${}_{n+1}\mathbf{D}_{z,i,j,k} = {}_{n}\mathbf{D}_{z,i,j,k} + \frac{\Delta t}{\Delta x \Delta y \Delta z}\begin{bmatrix} \mathbf{U}^{D_z}_{H_x,1}{}_{n+1/2}\mathbf{H}_{x,i,j+1,k} + \mathbf{U}^{D_z}_{H_x,0}{}_{n+1/2}\mathbf{H}_{x,i,j,k} \\ + \mathbf{U}^{D_z}_{H_y,1}{}_{n+1/2}\mathbf{H}_{y,i+1,j,k} + \mathbf{U}^{D_z}_{H_y,0}{}_{n+1/2}\mathbf{H}_{y,i,j,k} \end{bmatrix}. \tag{118}$$

In the Haar MRTD case, the sums are not required, because the overlap of the D and B basis functions extends only to the nearest neighbors. All of these equations have the same format; the only differences between the equations are the field values involved and the entries of the **U** matrices. It should be noted that the **U** matrices are the same size as the reconstruction/wavelet decomposition matrices. Neglecting unneeded subscripts, any of (116)–(118) can be represented as

$$\mathbf{D}_{\mathrm{W}\mathrm{dir1}} = \mathbf{D}_{\mathrm{W}\mathrm{dir1}} + \frac{\Delta t}{\Delta x \Delta y \Delta z}\begin{bmatrix} \mathbf{U}_1\mathbf{H}_{\mathrm{W}\mathrm{dir2,1}} + \mathbf{U}_2\mathbf{H}_{\mathrm{W}\mathrm{dir2,2}} \\ + \mathbf{U}_3\mathbf{H}_{\mathrm{W}\mathrm{dir3,1}} + \mathbf{U}_4\mathbf{H}_{\mathrm{W}\mathrm{dir3,2}} \end{bmatrix}, \tag{119}$$

where the **w** subscript indicates that the vector is scaling/wavelet magnitudes. The updated **D** vector can be reconstructed to give field values at specific points by multiplication with **R**,

$$\mathbf{D}_{\mathrm{dir1}} = \mathbf{R}\mathbf{D}_{\mathrm{W}\mathrm{dir1}} = \mathbf{R}\mathbf{D}_{\mathrm{W}\mathrm{dir1}} + \frac{\Delta t}{\Delta x \Delta y \Delta z}\begin{bmatrix} \mathbf{R}\mathbf{U}_1\mathbf{H}_{\mathrm{W}\mathrm{dir2,1}} + \mathbf{R}\mathbf{U}_2\mathbf{H}_{\mathrm{W}\mathrm{dir2,2}} \\ + \mathbf{R}\mathbf{U}_3\mathbf{H}_{\mathrm{W}\mathrm{dir3,1}} + \mathbf{R}\mathbf{U}_4\mathbf{H}_{\mathrm{W}\mathrm{dir3,2}} \end{bmatrix}. \tag{120}$$

The $\mathbf{D}_{\mathrm{dir1}} = \mathbf{R}\mathbf{D}_{\mathrm{W}\mathrm{dir1}}$ term represents the field values at the equivalent grid points, not the scaling/wavelet coefficients. Thus, (120) is an equation that gives updated **D** field values from **H** scaling/wavelet coefficients. If the PEC condition is applied,

$$\begin{aligned}\mathbf{D}_{\mathrm{dir1,PEC}} &= \mathbf{I}_{\mathrm{P}}\mathbf{R}\mathbf{D}_{\mathrm{W}\mathrm{dir1}} = \mathbf{I}_{\mathrm{P}}\mathbf{R}\mathbf{D}_{\mathrm{W}\mathrm{dir1}} \\ &\quad + \frac{\Delta t}{\Delta x \Delta y \Delta z}\begin{bmatrix} \mathbf{I}_{\mathrm{P}}\mathbf{R}\mathbf{U}_1\mathbf{H}_{\mathrm{W}\mathrm{dir2,1}} + \mathbf{I}_{\mathrm{P}}\mathbf{R}\mathbf{U}_2\mathbf{H}_{\mathrm{W}\mathrm{dir2,2}} \\ + \mathbf{I}_{\mathrm{P}}\mathbf{R}\mathbf{U}_3\mathbf{H}_{\mathrm{W}\mathrm{dir3,1}} + \mathbf{I}_{\mathrm{P}}\mathbf{R}\mathbf{U}_4\mathbf{H}_{\mathrm{W}\mathrm{dir3,2}} \end{bmatrix}.\end{aligned} \tag{121}$$

Transforming back into the wavelet domain,

$$\begin{aligned}\mathbf{D}_{\mathrm{W}\mathrm{dir1,PEC}} &= \mathbf{R}^{-1}\mathbf{I}_{\mathrm{P}}\mathbf{R}\mathbf{D}_{\mathrm{W}\mathrm{dir1}} = \mathbf{R}^{-1}\mathbf{I}_{\mathrm{P}}\mathbf{R}\mathbf{D}_{\mathrm{W}\mathrm{dir1}} \\ &\quad + \frac{\Delta t}{\Delta x \Delta y \Delta z}\begin{bmatrix} \mathbf{R}^{-1}\mathbf{I}_{\mathrm{P}}\mathbf{R}\mathbf{U}_1\mathbf{H}_{\mathrm{W}\mathrm{dir2,1}} + \mathbf{R}^{-1}\mathbf{I}_{\mathrm{P}}\mathbf{R}\mathbf{U}_2\mathbf{H}_{\mathrm{W}\mathrm{dir2,2}} \\ + \mathbf{R}^{-1}\mathbf{I}_{\mathrm{P}}\mathbf{R}\mathbf{U}_3\mathbf{H}_{\mathrm{W}\mathrm{dir3,1}} + \mathbf{R}^{-1}\mathbf{I}_{\mathrm{P}}\mathbf{R}\mathbf{U}_4\mathbf{H}_{\mathrm{W}\mathrm{dir3,2}} \end{bmatrix}.\end{aligned} \tag{122}$$

If a new matrix $\mathbf{U}_{\mathrm{PEC}}$ is defined

$$\mathbf{U}_{\mathrm{PEC}} = \mathbf{R}^{-1}\mathbf{I}_{\mathrm{P}}\mathbf{R}\mathbf{U}, \tag{123}$$

then a new update equation can be defined

$$\mathbf{D}_{\mathrm{Wdir1,PEC}} = \mathbf{D}_{\mathrm{Wdir,PEC1}} + \frac{\Delta t}{\Delta x \Delta y \Delta z}\begin{bmatrix} \mathbf{U}_{\mathrm{PEC1}}\mathbf{H}_{\mathrm{Wdir2,1}} + \mathbf{U}_{\mathrm{PEC2}}\mathbf{H}_{\mathrm{Wdir2,2}} \\ + \mathbf{U}_{\mathrm{PEC3}}\mathbf{H}_{\mathrm{Wdir3,1}} + \mathbf{U}_{\mathrm{PEC4}}\mathbf{H}_{\mathrm{Wdir3,2}} \end{bmatrix}. \tag{124}$$

The $\mathbf{R}^{-1}\mathbf{I}_{\mathrm{P}}\mathbf{R}\mathbf{D}_{\mathrm{Wdir1}}$ condition does not have to be explicitly enforced because $\mathbf{D}_{\mathrm{Wdir1}}$ represents the scaling/wavelet coefficients from the previous time step, where the condition has already been applied.

This technique provides a method that can be used to automatically apply the PEC boundary condition within the Haar-MRTD update equations. In this method, the $\mathbf{U}_{\mathrm{PEC}}$ matrices can be calculated before the simulation begins, and therefore do not add significant overhead to the field updates. In fact, this method can be expanded to allow other subcell effects to be automatically applied in the MRTD update scheme.

3.4 GENERAL SUBCELL EFFECTS IN HAAR-MRTD: COMPOSITE CELLS

In the previous section, the reconstruction/wavelet transform matrices were used to apply the PEC boundary condition directly in the MRTD update. If the reconstruction/deconstruction is applied to all of the fields in the update equation,

$$\begin{aligned} \mathbf{D}_{\mathrm{dir1}} = \mathbf{R}\mathbf{D}_{\mathrm{Wdir1}} = \mathbf{R}\mathbf{D}_{\mathrm{Wdir1}} \\ + \frac{\Delta t}{\Delta x \Delta y \Delta z}\begin{bmatrix} \mathbf{R}\mathbf{U}_1\mathbf{R}^{-1}\mathbf{R}\mathbf{H}_{\mathrm{Wdir2,1}} + \mathbf{R}\mathbf{U}_2\mathbf{R}^{-1}\mathbf{R}\mathbf{H}_{\mathrm{Wdir2,2}} \\ +\mathbf{R}\mathbf{U}_3\mathbf{R}^{-1}\mathbf{R}\mathbf{H}_{\mathrm{Wdir3,1}} + \mathbf{R}\mathbf{U}_4\mathbf{R}^{-1}\mathbf{R}\mathbf{H}_{\mathrm{Wdir3,2}} \end{bmatrix}, \end{aligned} \tag{125}$$

and the substitutions

$$\mathbf{H} = \mathbf{R}\mathbf{H}_{\mathrm{W}}, \tag{126}$$

$$\mathbf{U}_{\mathrm{L}} = \mathbf{R}\mathbf{U}\mathbf{R}^{-1}, \tag{127}$$

where the subscript **L** is used to refer to the update equation for local, pointwise, fields, is made, then

$$\mathbf{D}_{\mathrm{dir1}} = \mathbf{D}_{\mathrm{dir1}} + \frac{\Delta t}{\Delta x \Delta y \Delta z}\begin{bmatrix} \mathbf{U}_{\mathrm{L1}}\mathbf{H}_{\mathrm{dir2,1}} + \mathbf{U}_{\mathrm{L2}}\mathbf{H}_{\mathrm{dir2,2}} \\ + \mathbf{U}_{\mathrm{L3}}\mathbf{H}_{\mathrm{dir3,1}} + \mathbf{U}_{\mathrm{L4}}\mathbf{H}_{\mathrm{dir3,2}} \end{bmatrix}. \tag{128}$$

This equation is an update for the field values at the equivalent grid points from field values at equivalent grid points. It has been repeatedly stated that MRTD and FDTD are equivalent schemes, and (128) is the conversion from MRTD to FDTD. It should be noted, however, that the MRTD scheme can still be used to vary the wavelet resolution, and thus the number of equivalent grid points, on a cell-by-cell basis. This is an inherent property of MRTD that is not available in FDTD. The pointwise field update representation (128) provides the ability to manipulate the fields at individual equivalent grid points and then transform the fields back to the scaling/wavelet domain.

One application of this technique, and indeed the motivation for this technique, is the application of the PEC boundary condition at individual equivalent grid points that has already been presented. Another simple application of this technique is the addition of a current source at equivalent grid points. This method is equivalent to wavelet transforming a spatial source condition.

In the pointwise update scheme, a source, **J**, can be added at each equivalent grid point:

$$\mathbf{D}_{\mathrm{dir1}} = \mathbf{D}_{\mathrm{dir1}} + \frac{\Delta t}{\Delta x \Delta y \Delta z}\begin{bmatrix} \mathbf{U}_{\mathrm{L1}}\mathbf{H}_{dir2,1} + \mathbf{U}_{\mathrm{L2}}\mathbf{H}_{\mathrm{dir2,2}} \\ + \mathbf{U}_{\mathrm{L3}}\mathbf{H}_{\mathrm{dir3,1}} + \mathbf{U}_{\mathrm{L4}}\mathbf{H}_{\mathrm{dir3,2}} \end{bmatrix} + \mathbf{J}. \tag{129}$$

Multiplying (129) by $\mathbf{R}^{-1}$ converts the equation back to the wavelet domain, yielding

$$\begin{aligned} \mathbf{D}_{\mathrm{Wdir1}} &= \mathbf{R}^{-1}\mathbf{D}_{\mathrm{dir1}} + \frac{\Delta t}{\Delta x \Delta y \Delta z}\mathbf{R}^{-1}\begin{bmatrix} \mathbf{U}_{\mathrm{L1}}\mathbf{H}_{\mathrm{dir2,1}} + \mathbf{U}_{\mathrm{L2}}\mathbf{H}_{\mathrm{dir2,2}} \\ + \mathbf{U}_{\mathrm{L3}}\mathbf{H}_{\mathrm{dir3,1}} + \mathbf{U}_{\mathrm{L4}}\mathbf{H}_{\mathrm{dir3,2}} \end{bmatrix} + \mathbf{R}^{-1}\mathbf{J} \\ &= \mathbf{D}_{\mathrm{Wdir1}} + \frac{\Delta t}{\Delta x \Delta y \Delta z}\begin{bmatrix} \mathbf{U}_{1}\mathbf{H}_{\mathrm{Wdir2,1}} + \mathbf{U}_{2}\mathbf{H}_{\mathrm{Wdir2,2}} \\ + \mathbf{U}_{3}\mathbf{H}_{\mathrm{Wdir3,1}} + \mathbf{U}_{4}\mathbf{H}_{\mathrm{Wdir3,2}} \end{bmatrix} + \mathbf{R}^{-1}\mathbf{J}. \end{aligned} \tag{130}$$

This equation can be used to apply an arbitrarily placed source into the MRTD cell.

One of the advantages of the Yee-FDTD scheme is that its popularity and longevity has led to the development of a large number of techniques that can extend its use. Some examples of techniques that have been developed for Yee-FDTD are the modeling of thin wires [21], narrow slots [22], curved structures (with a locally conformal grid) [23], thin material sheets [24], dispersive surfaces (such as thin metals) [25], SPICE circuits [26], local field correction [27], and lumped elements [28]. The result of all of these techniques is modified FDTD update equations for the areas where the effect is applied. The surrounding fields are updated normally. Using the technique that has been presented here, it is possible to directly bridge the modified FDTD update equations to the MRTD technique.

An example of how this technique can be applied to bridge the pointwise FDTD modifications to the MRTD technique is presented here with lumped elements. The procedure for representing lumped elements in FDTD is presented in [28], and a brief overview is given here.

A lumped element, here specified to be a resistor, capacitor, or inductor, can be represented in Ampere's law as a current source $\mathbf{J}_{\mathrm{L}}$,

$$\nabla \times \mathbf{H}(t) = \frac{\partial \mathbf{D}(t)}{\partial t} + \mathbf{J}_{\mathrm{L}}(t). \tag{131}$$

In this case currents due to loss and impressed currents are neglected for simplicity. For each of the lumped elements listed above, a relationship can be determined for the current as a function of voltage. The $\mathbf{D_z}$ component update will be shown here, although this procedure can be easily modified for any direction. Including the lumped element current, the z component of (131) is

$$\frac{\partial D_z}{\partial t} = \frac{\partial H_y}{\partial x} - \frac{\partial H_x}{\partial y} - \frac{I_z}{\Delta x \Delta y}, \tag{132}$$

noting that

$$J_{L,z} = \frac{I_z}{\Delta x \Delta y}. \tag{133}$$

I_z can be determined for each element. For resistors,

$$I_{z,R} = \frac{V}{R} = \frac{1}{R}\int\limits_{\Delta z} E\partial z; \tag{134}$$

for capacitors,

$$I_{z,C} = C\frac{\partial V}{\partial t} = C\frac{\partial \int\limits_{\Delta z} E\partial z}{\partial t}; \tag{135}$$

and for inductors,

$$I_{z,C} = \frac{1}{I}\int V\partial t = \frac{1}{I}\int\int\limits_{\Delta z} E\partial z\partial t. \tag{136}$$

For any of these cases, the MRTD update equations can be determined by inserting the current relationship into Ampere's law and applying the method of moments. However, this method is somewhat difficult. An example is presented for the resistor.

The voltage across a single equivalent grid point can be simply calculated as

$$V_{i,j,k} = E_{i,j,k}\Delta z. \tag{137}$$

However, for Ampere's law, the update is performed at the $n + 1/2$ time step, while the E field is known only for the n and $n + 1$ time steps. Similar to the method that is used for Ohmic losses, the semiimplicit approximation can be applied. In this case,

$$V_{i,j,k} = \frac{{}_{n+1}E_{i,j,k} +_n E_{i,j,k}}{2}\Delta z. \tag{138}$$

When the method of moments is applied to (132), with the voltage–current relationship (138), the resulting update equation is

$$\begin{aligned} {}_{n+1}\mathbf{D}_{z,i,j,k} &= {}_n\mathbf{D}_{z,i,j,k} + \mathbf{L}\left({}_{n+1}\mathbf{E}_{z,i,j,k} + {}_n\mathbf{E}_{z,i,j,k}\right) \\ &+ \frac{\Delta t}{\Delta x\Delta y\Delta z}\begin{bmatrix} \mathbf{U}^{D_z}_{H_x,1\,n+1/2}\mathbf{H}_{x,i,j+1,k} + \mathbf{U}^{D_z}_{H_x,0\,n+1/2}\mathbf{H}_{x,i,j,k} \\ + \mathbf{U}^{D_z}_{H_y,1\,n+1/2}\mathbf{H}_{y,i+1,j,k} + \mathbf{U}^{D_z}_{H_y,0\,n+1/2}\mathbf{H}_{y,i,j,k} \end{bmatrix}. \end{aligned} \tag{139}$$

The entries of **L** have the form,

$$L_{ij} = \int_{\mathbf{r}} L(\mathbf{r})\Gamma_i\Gamma_j\partial\mathbf{r}, \tag{140}$$

with

$$L(\mathbf{r}) = \begin{cases} \dfrac{\Delta z}{2R\Delta x\Delta y} \text{ resistor location} \\ 0 \text{ elsewhere} \end{cases}, \tag{141}$$

if Δx, Δy, and Δz are the size of the equivalent grid points. Two difficulties are immediately apparent with the equation. First, it includes both **D** and **E** coefficients, where before **E** coefficients were updated directly from the **D** fields. Secondly, **D** fields are updated using **E** fields at the same time step.

The first of these difficulties can be easily accounted for using the constitutive relationship. In (57), the electric field is determined from the D field. In the Haar case, this relationship becomes

$${}_n\mathbf{E}_{d,i,j,k} = \mathbf{U}^{E_d}_{D_d,i,j,k,a,b,c}\,{}_n\mathbf{D}_{d,i,j,k} = \varepsilon_{d,i,j,k}\,{}_n\mathbf{D}_{d,i,j,k}. \tag{142}$$

Because no neighboring field values are required to determine the **E** coefficients from the **D** coefficients, the transformation can be applied directly to the update equation. In this case,

$$\begin{aligned}
&{}_{n+1}\mathbf{E}_{z,i,j,k} \\
&\quad = \varepsilon_{d,i,j,k}\,{}_{n+1}\mathbf{D}_{z,i,j,k} \\
&\quad = \varepsilon_{d,i,j,k}\,{}_n\mathbf{D}_{z,i,j,k} \\
&\qquad + \frac{\Delta t}{\Delta x\Delta y\Delta z}\varepsilon_{d,i,j,k}\begin{bmatrix} \mathbf{U}_{H_x,1}\,{}_{n+1/2}\mathbf{H}_{x,i,j+1,k} + \mathbf{U}_{H_x,0}\,{}_{n+1/2}\mathbf{H}_{x,i,j,k} \\ + \mathbf{U}_{H_y,1}\,{}_{n+1/2}\mathbf{H}_{y,i+1,j,k} + \mathbf{U}_{H_y,0}\,{}_{n+1/2}\mathbf{H}_{y,i,j,k} \end{bmatrix} \qquad (143) \\
&\quad = \mathbf{E}_{z,i,j,k} + \frac{\Delta t}{\Delta x\Delta y\Delta z}\varepsilon_{d,i,j,k}\begin{bmatrix} \mathbf{U}_{H_x,1}\,{}_{n+1/2}\mathbf{H}_{x,i,j+1,k} + \mathbf{U}_{H_x,0}\,{}_{n+1/2}\mathbf{H}_{x,i,j,k} \\ + \mathbf{U}_{H_y,1}\,{}_{n+1/2}\mathbf{H}_{y,i+1,j,k} + \mathbf{U}_{H_y,0}\,{}_{n+1/2}\mathbf{H}_{y,i,j,k} \end{bmatrix}.
\end{aligned}$$

Using this update, (139) becomes

$$
{}_{n+1}\mathbf{E}_{z,i,j,k} = {}_{n}\mathbf{E}_{z,i,j,k} + \mathbf{L}\left({}_{n+1}\mathbf{E}_{z,i,j,k} + {}_{n}\mathbf{E}_{z,i,j,k}\right)
+ \frac{\Delta t}{\Delta x \Delta y \Delta z}\varepsilon \begin{bmatrix} \mathbf{U}_{H_x,1\,n+1/2}\mathbf{H}_{x,i,j+1,k} + \mathbf{U}_{H_x,0\,n+1/2}\mathbf{H}_{x,i,j,k} \\ +\mathbf{U}_{H_y,1\,n+1/2}\mathbf{H}_{y,i+1,j,k} + \mathbf{U}_{H_y,0\,n+1/2}\mathbf{H}_{y,i,j,k} \end{bmatrix}. \quad (144)
$$

Collecting terms

$$
(\mathbf{I} - \mathbf{L})_{n+1}\mathbf{E}_{z,i,j,k} = (\mathbf{I} + \mathbf{L})_{n}\mathbf{E}_{z,i,j,k} + \frac{\Delta t}{\Delta x \Delta y \Delta z}\varepsilon
\times \begin{bmatrix} \mathbf{U}_{H_x,1\,n+1/2}\mathbf{H}_{x,i,j+1,k} + \mathbf{U}_{H_x,0\,n+1/2}\mathbf{H}_{x,i,j,k} \\ +\mathbf{U}_{H_y,1\,n+1/2}\mathbf{H}_{y,i+1,j,k} + \mathbf{U}_{H_y,0\,n+1/2}\mathbf{H}_{y,i,j,k} \end{bmatrix}, \quad (145)
$$

and solving for ${}_{n+1}\mathbf{E}_{x,i,j,k}$

$$
{}_{n+1}\mathbf{E}_{z,i,j,k} = (\mathbf{I} - \mathbf{L})^{-1}(\mathbf{I} + \mathbf{L})_{n}\mathbf{E}_{z,i,j,k} + \frac{\Delta t}{\Delta x \Delta y \Delta z}(\mathbf{I} - \mathbf{L})^{-1}\varepsilon
\times \begin{bmatrix} \mathbf{U}_{H_x,1\,n+1/2}\mathbf{H}_{x,i,j+1,k} + \mathbf{U}_{H_x,0\,n+1/2}\mathbf{H}_{x,i,j,k} \\ +\mathbf{U}_{H_y,1\,n+1/2}\mathbf{H}_{y,i+1,j,k} + \mathbf{U}_{H_y,0\,n+1/2}\mathbf{H}_{y,i,j,k} \end{bmatrix}. \quad (146)
$$

A similar procedure can be used to determine the updates for capacitors and inductors. This procedure is not difficult and most of the operations involved are matrix multiplications that can be performed before the start of simulation. The only integration required is to determine the **L** matrix. However, there is another method that can be used in this case that removes the need to calculate the **L** matrix.

The E_z update equation in FDTD for a resistor is [28]

$$
{}_{n+1}E_{z,i,j,k} = \left(\frac{1 - \frac{\Delta t \Delta z}{2R\varepsilon \Delta x \Delta y}}{1 + \frac{\Delta t \Delta z}{2R\varepsilon \Delta x \Delta y}}\right){}_{n}E_{z,i,j,k} \quad (147)
$$

$$
+ \left(\frac{\frac{\Delta t}{\varepsilon}}{1 + \frac{\Delta t \Delta z}{2R\varepsilon \Delta x \Delta y}}\right)\begin{bmatrix} \frac{{}_{n+1/2}H_{y,i+1,j,k} - {}_{n+1/2}H_{y,i,j,k}}{\Delta x} \\ -\frac{{}_{n+1/2}H_{x,i,j+1,k} - {}_{n+1/2}H_{x,i,j,k}}{\Delta y} \end{bmatrix}. \quad (148)
$$

Using (128), two additional lumped element matrices can be created that apply the lumped element formulation in (147). The two lumped element matrices $\mathbf{L_1}$ and

$\mathbf{L}_2$ are diagonal with

$$L_{1,i,i} = \begin{cases} \varepsilon \left(\dfrac{1 - \frac{\Delta t \Delta z}{2R\varepsilon \Delta x \Delta y}}{1 + \frac{\Delta t \Delta z}{2R\varepsilon \Delta x \Delta y}} \right) & \text{resistor equivalent grid points} \\ 1 & \text{elsewhere} \end{cases}, \tag{149}$$

$$L_{2,i,i} = \begin{cases} \left(\dfrac{1}{1 + \frac{\Delta t \Delta z}{2R\varepsilon \Delta x \Delta y}} \right) & \text{resistor equivalent grid points} \\ 1 & \text{elsewhere} \end{cases}. \tag{150}$$

$\mathbf{L}_1$ and $\mathbf{L}_2$ can be inserted into (128)

$$\mathbf{D}_{\text{dir1}} = \mathbf{L}_1 \mathbf{D}_{\text{dir1}} + \frac{\Delta t}{\Delta x \Delta y \Delta z} \mathbf{L}_2 \begin{bmatrix} \mathbf{U}_{\mathbf{L1}} \mathbf{H}_{\text{dir2},1} + \mathbf{U}_{\mathbf{L2}} \mathbf{H}_{\text{dir2},2} \\ + \mathbf{U}_{\mathbf{L3}} \mathbf{H}_{\text{dir3},1} + \mathbf{U}_{\mathbf{L4}} \mathbf{H}_{\text{dir3},2} \end{bmatrix}, \tag{151}$$

where again, for simplicity, the subscripts have been dropped. By rotating the values of Δx, Δy, and Δz, $\mathbf{L}$ can be used for a resistor in any direction. Converting back to the wavelet domain,

$$\begin{aligned} \mathbf{D}_{\mathbf{W}\text{dir1}} = {} & \mathbf{R}^{-1} \mathbf{L}_1 \mathbf{R}^{-1} \mathbf{D}_{\mathbf{W}\text{dir1}} + \frac{\Delta t}{\Delta x \Delta y \Delta z} \mathbf{R}^{-1} \mathbf{L}_2 \\ & \times \begin{bmatrix} \mathbf{U}_{\mathbf{L1}} \mathbf{R} \mathbf{H}_{\mathbf{W}\text{dir2},1} + \mathbf{U}_{\mathbf{L2}} \mathbf{R} \mathbf{H}_{\mathbf{W}\text{dir2},2} \\ + \mathbf{U}_{\mathbf{L3}} \mathbf{R} \mathbf{H}_{\mathbf{W}\text{dir3},1} + \mathbf{U}_{\mathbf{L4}} \mathbf{R} \mathbf{H}_{\mathbf{W}\text{dir3},2} \end{bmatrix}. \end{aligned} \tag{152}$$

These matrices can be combined so that each field is only multiplied by one matrix for each update

$$\mathbf{D}_{\mathbf{W}\text{dir1}} = \mathbf{U}_{\mathbf{R}} \mathbf{D}_{\mathbf{W}\text{dir1}} + \frac{\Delta t}{\Delta x \Delta y \Delta z} \begin{bmatrix} \mathbf{U}_{\mathbf{R1}} \mathbf{H}_{\mathbf{W}\text{dir2},1} + \mathbf{U}_{\mathbf{R2}} \mathbf{H}_{\mathbf{W}\text{dir2},2} \\ + \mathbf{U}_{\mathbf{R3}} \mathbf{H}_{\mathbf{W}\text{dir3},1} + \mathbf{U}_{\mathbf{R4}} \mathbf{H}_{\mathbf{W}\text{dir3},2} \end{bmatrix}. \tag{153}$$

Using this method, resistors can be inserted into the MRTD grid at any equivalent grid point, and modified update equations can be found before simulation through simple matrix multiplications. One other important note is that this formulation introduced a new update matrix for the current **D** field. This does add one extra matrix multiplication to the field update, but, as is shown in the next chapter, the general MRTD update including UPML includes this matrix.

Similar updates can be derived for capacitors and inductors. Capacitors are modeled in FDTD using

$$ {}_{n+1}E_{z,i,j,k} = {}_{n}E_{z,i,j,k} + \left(\frac{\frac{\Delta t}{\varepsilon}}{1 + \frac{C\Delta z}{\varepsilon \Delta x \Delta y}} \right) \left[\begin{array}{c} \dfrac{{}_{n+1/2}H_{y,i+1,j,k} - {}_{n+1/2}H_{y,i,j,k}}{\Delta x} \\ -\dfrac{{}_{n+1/2}H_{x,i,j+1,k} - {}_{n+1/2}H_{x,i,j,k}}{\Delta y} \end{array} \right], \tag{154} $$

so, in this case, only one **L** matrix needs to be determined,

$$ L_{2,i,i} = \begin{cases} \left(\dfrac{1}{1 + \frac{C\Delta z}{\varepsilon \Delta x \Delta y}} \right) & \text{capacitor equivalent grid points} \\ 1 & \text{elsewhere} \end{cases}. \tag{155} $$

In FDTD, inductors are represented using,

$$ \begin{aligned} {}_{n+1}E_{z,i,j,k} = {}_{n}E_{z,i,j,k} &+ \frac{\Delta t}{\varepsilon} \left[\begin{array}{c} \dfrac{{}_{n+1/2}H_{y,i+1,j,k} - {}_{n+1/2}H_{y,i,j,k}}{\Delta x} \\ -\dfrac{{}_{n+1/2}H_{x,i,j+1,k} - {}_{n+1/2}H_{x,i,j,k}}{\Delta y} \end{array} \right] \\ &- \frac{\Delta z (\Delta t)^2}{\varepsilon L \Delta x \Delta y} \sum_{m=1}^{n} {}_{m}E_{z,i,j,k}, \end{aligned} \tag{156} $$

where it is noted that the sum can be represented as a single value that is augmented at each time step. In this case, both $\mathbf{L}_1$ and $\mathbf{L}_2$ are the identity matrix, but another matrix must be created. In this case, the sum term is similar to the previously presented current source. If (128) is modified

$$ \mathbf{D}_{\text{dir1}} = \mathbf{D}_{\text{dir1}} + \frac{\Delta t}{\Delta x \Delta y \Delta z} \left[\begin{array}{c} \mathbf{U}_{\text{L1}}\mathbf{H}_{\text{dir2,1}} + \mathbf{U}_{\text{L2}}\mathbf{H}_{\text{dir2,2}} \\ + \mathbf{U}_{\text{L3}}\mathbf{H}_{\text{dir3,1}} + \mathbf{U}_{\text{L4}}\mathbf{H}_{\text{dir3,2}} \end{array} \right] + \mathbf{L} \sum_{m=0}^{n} \mathbf{D}, \tag{157} $$

with

$$ L_{i,i} = \begin{cases} \dfrac{\Delta z (\Delta t)^2}{\varepsilon L \Delta x \Delta y} & \text{inductor equivalent grid points} \\ 0 & \text{elsewhere} \end{cases}, \tag{158} $$

then, using the same procedure as above, (157) can be converted into the wavelet domain. In the case of the inductor, an additional vector, $\sum_{m=0}^{n} \mathbf{D}_\mathrm{W}$ must be used to calculate the updates. It is noted that it is only necessary to store this vector in cells that contain inductors.

CHAPTER 4

Other Techniques Necessary for Simulation: UPML, Variable Gridding, Source Excitation and Time/Space Adaptive Gridding

In the previous chapter, a method is presented that allows a variety of elements that are smaller than a single MRTD cell to be simulated using the Haar MRTD method. Unlike the method presented in [22], this technique treats the entire space as an MRTD grid, using scaling and wavelet functions exclusively to represent the field. The method was created to model subcell PEC structures, as most microwave structures are simulated using only PEC and dielectric media. In addition, it is shown that the method can be generalized to allow the modeling of several other subcell effects; a specific example is provided for lumped elements. The MRTD method that includes subcell elements and varying dielectrics within individual

MRTD cells is termed the composite-cell method. This is because, instead of homogenous cells that are used in most MRTD simulations, these cells can include complex structures. The advantage of this method is that these subcell structures can be treated within the MRTD framework, allowing the adaptive MRTD grid to be used to effectively model these structures.

In order to perform simulations using the MRTD method, there are a few more topics that must be addressed. The first is grid truncation. While a variety of absorbing boundary conditions can be implemented, the treatment in this book focuses exclusively on the UPML. It is one of the best absorbers (if computationally intensive), and can be easily implemented using the MRTD framework previously presented. In addition, grid excitation is necessary to inject signals into the computational space. The treatment in this chapter follows on the impressed current source formulation presented in the previous chapter. Finally, a method of varying grid size is presented, which allows the modeling of complex structures more accurately and efficiently. Like in the previous chapter, this section discusses many techniques that were originally developed for the FDTD method.

4.1 ARBITRARY WAVELET RESOLUTION UPML

In order to perform simulations of microwave structures, a method of truncating the computational space with a reflectionless (or low reflection) condition is necessary. These absorbing boundary conditions allow the computational space to closely surround the simulated structure. Without such a condition, signals that reach the outer boundary (represented as a PEC or PMC) will reflect into the space. This renders the field values observed in the simulated structure useless; there is no way to differentiate between the reflected and incident signal. The most widely used method of truncating the computational space is the perfectly matched layer, which has the properties of extremely high loss while matching arbitrary incident signals.

The UPML [19] was first developed as an alternative to the Berenger PML [18] that did not require nonphysical field splitting. The UPML is expressed as a material with a carefully designed permittivity and permeability tensor [14]

$$\bar{\bar{\varepsilon}}(\omega) = \bar{\bar{\mu}}(\omega) = \begin{bmatrix} \frac{s_y s_z}{s_x} & 0 & 0 \\ 0 & \frac{s_x s_z}{s_y} & 0 \\ 0 & 0 & \frac{s_x s_y}{s_z} \end{bmatrix}, \tag{158}$$

where

$$s_n = k_n + \frac{\sigma_n}{j\omega\varepsilon} \qquad n = x, y, z. \tag{159}$$

The parameter σ_n represents the loss inside the UPML and k_n is a matching parameter that can be used to fine tune the UPML performance.

The permittivity and permeability tensors inside the UPML are frequency dependent, and are therefore difficult to represent in a time domain scheme. If the constitutive relationship

$$D_x(\omega) = \varepsilon \frac{s_z}{s_x} E_x(\omega) \qquad D_y(\omega) = \varepsilon \frac{s_x}{s_y} E_y(\omega) \qquad D_z(\omega) = \varepsilon \frac{s_y}{s_z} E_z(\omega) \tag{160}$$

is defined, then a coupled set of time domain equations can be defined to update the electric and magnetic fields. The derivation for the electric fields is shown here, the magnetic field updates are found in the same manner.

The time domain differential equation for the D field is

$$\begin{bmatrix} \frac{\partial H_z}{\partial y} - \frac{\partial H_y}{\partial z} \\ \frac{\partial H_x}{\partial z} - \frac{\partial H_z}{\partial x} \\ \frac{\partial H_y}{\partial x} - \frac{\partial H_x}{\partial y} \end{bmatrix} = \begin{bmatrix} k_y & 0 & 0 \\ 0 & k_z & 0 \\ 0 & 0 & k_z \end{bmatrix} \frac{\partial}{\partial t} \begin{bmatrix} D_x \\ D_y \\ D_z \end{bmatrix} + \frac{1}{\varepsilon} \begin{bmatrix} \sigma_y & 0 & 0 \\ 0 & \sigma_z & 0 \\ 0 & 0 & \sigma_z \end{bmatrix} \begin{bmatrix} D_x \\ D_y \\ D_z \end{bmatrix}. \tag{161}$$

This system is very similar to Ampere's law when Ohmic loss is included (the only difference is the constant that precedes the time derivative). It can be discretized using the procedure outlined in Chapter 2. For example, the D_x component of

(161) is

$$\frac{\partial H_z}{\partial y} - \frac{\partial H_y}{\partial z} = k_y \frac{\partial D_x}{\partial t} + \frac{\sigma_y}{\varepsilon} D_x, \tag{162}$$

and the time localized form is

$$_{n+1}D_x = {}_nD_x - \frac{\Delta t \sigma_y}{k_y \varepsilon} \left(\frac{_{n+1}D_x + {}_nD_x}{2} \right) + \frac{\Delta t}{k_y} \left(\frac{\partial_{n+1/2} H_z}{\partial y} - \frac{\partial_{n+1/2} H_y}{\partial z} \right). \tag{163}$$

Combining D terms,

$$_{n+1}D_x = {}_nD_x \left(\frac{2k_y \varepsilon - \Delta t \sigma_y}{2k_y \varepsilon + \Delta t \sigma_y} \right) + \left(\frac{2k_y \varepsilon \Delta t}{2k_y \varepsilon + \Delta t \sigma_y} \right) \left(\frac{\partial_{n+1/2} H_z}{\partial y} - \frac{\partial_{n+1/2} H_y}{\partial z} \right). \tag{164}$$

Next, space localization can be performed, and the update matrices can be computed. The $\mathbf{D_x}$ update is

$$\begin{aligned} _{n+1}\mathbf{D}_{x,i,j,k} &= \mathbf{U}_{D_x,}^{D_x,}\, {}_n\mathbf{D}_{x,i,j,k} + \frac{1}{\Delta x \Delta y \Delta z} \\ &\times \left[\sum_l \sum_m \mathbf{U}_{H_y,m}^{D_x}\, {}_{n+1/2}\mathbf{H}_{y,i,j+l,k+m} + \sum_m \mathbf{U}_{H_z,m}^{D_x}\, {}_{n+1/2}\mathbf{H}_{z,i,j+m,k} \right], \end{aligned} \tag{165}$$

where

$$U_{D_x,i,j}^{D_x} = \left\langle {}_{D_x}\Gamma_i, \left(\frac{2k_y \varepsilon - \Delta t \sigma_y}{2k_y \varepsilon + \Delta t \sigma_y} \right)_{D_x} \Gamma_j \right\rangle, \tag{166}$$

$$U_{H_y,m,i,j}^{D_x} = \left\langle {}_{D_x}\Gamma_i, -\left(\frac{2k_y \varepsilon \Delta t}{2k_y \varepsilon + \Delta t \sigma_y} \right) \frac{\partial}{\partial z}\, {}_{H_y}\Gamma_j \big|_m \right\rangle, \tag{167}$$

$$U_{H_z,m,i,j}^{D_x} = \left\langle {}_{D_x}\Gamma_i, \left(\frac{2k_y \varepsilon \Delta t}{2k_y \varepsilon + \Delta t \sigma_y} \right) \frac{\partial}{\partial y}\, {}_{H_z}\Gamma_j \big|_m \right\rangle. \tag{168}$$

Similar equations can be found for the $\mathbf{D_y}$ and $\mathbf{D_z}$ components.

The relationship between the E and D fields inside the PML is also more complicated than in the isotropic nondispersive case that has already been presented.

When the D_x relationship of (160) is converted to the time domain [19],

$$\frac{\partial}{\partial t}(k_x D_x) + \frac{\sigma_x}{\varepsilon} D_x = \varepsilon \left[\frac{\partial}{\partial t}(k_z E_x) + \frac{\sigma_z}{\varepsilon} E_x \right] \tag{169}$$

must be solved to determine the updated E fields from D fields.

The time localized form of (169) is

$$\begin{aligned} k_x \left({}_{n+1}D_x - {}_{n}D_x \right) + \frac{\Delta t \sigma_x}{\varepsilon} \frac{{}_{n+1}D_x + {}_{n}D_x}{2} \\ = \varepsilon \left[k_z \left({}_{n+1}E_x - {}_{n}E_x \right) + \frac{\Delta t \sigma_z}{\varepsilon} \frac{{}_{n+1}E_x + {}_{n}E_x}{2} \right]. \end{aligned} \tag{170}$$

Collecting terms and solving for ${}_{n+1}E_x$,

$${}_{n+1}E_x = \left(\frac{2\varepsilon k_z - \sigma_z \Delta t}{2\varepsilon k_z + \sigma_z \Delta t} \right) {}_{n}E_x + \left[\frac{1}{(2\varepsilon k_z + \sigma_z \Delta t)\,\varepsilon} \right] \left[\begin{array}{l} (2\varepsilon k_x + \sigma_x \Delta t)\, {}_{n+1}D_x \\ -\,(2\varepsilon k_x - \sigma_x \Delta t)\, {}_{n}D_x \end{array} \right]. \tag{171}$$

Solving for the $\mathbf{E_x}$ coefficients,

$${}_{n+1}\mathbf{E}_{x,i,j,k} = \sum_m \mathbf{U}^{E_x,}_{E_x,i,j,m}\, {}_{n}\mathbf{E}_{x,i,j,k+m} + \sum_a \sum_b \sum_c \mathbf{U}^{E_x}_{D_x,i,j,k,a,b,c}\, {}_{n}\mathbf{D}_{x,i+a,j+b,k+c}, \tag{172}$$

where

$$U^{E_x}_{E_x i,j,m} = \left\langle {}_{E_x}\Gamma_i, \left(\frac{2\varepsilon k_z - \sigma_z \Delta t}{2\varepsilon k_z + \sigma_z \Delta t} \right) {}_{E_x}\Gamma_j \Big|_m \right\rangle, \tag{173}$$

$$U^{E_x}_{D_{x+}i,j} = \left\langle {}_{E_x}\Gamma_i, \frac{1}{\varepsilon} \left(\frac{2\varepsilon k_x + \sigma_x \Delta t}{2\varepsilon k_z + \sigma_z \Delta t} \right) {}_{E_x}\Gamma_j \Big|_{a,b,c} \right\rangle, \tag{174}$$

$$U^{E_x}_{D_x i,j} = \left\langle {}_{E_x}\Gamma_i, -\frac{1}{\varepsilon} \left(\frac{2\varepsilon k_x - \sigma_x \Delta t}{2\varepsilon k_z + \sigma_z \Delta t} \right) {}_{E_x}\Gamma_j \Big|_{a,b,c} \right\rangle, \tag{175}$$

and, again, similar equations can be found for the other directions. A process similar to the one presented here can be used to find the updates for the magnetic fields.

It has been left until this point to discuss the values of σ and k that should be used with this scheme. The technique for implementing the UPML that has been presented here is derived in an equivalent manner to the FDTD UPML [19] (until the last step, where the updates for the scaling/wavelet coefficients are found). The values of σ and k can be chosen in exactly the same manner as an FDTD implementation. In this case, $\sigma_a(n)$, where a is x, y,or z, is nonzero only within a predetermined distance d of the a normal outer boundaries. Usually, a width of 10 cells is sufficient for the UPML. Inside this boundary, $\sigma_a(n)$ is varied from zero to a maximum value $\sigma_{\max}$. A polynomial grading works well,

$$\sigma_a(n) = \left(n/_d\right)^m \sigma_{q,\max}. \tag{176}$$

Similiarly, k can be varied from 1 in the non-UPML region to $k_{\max}$ at the outer edge of the grid.

There is one important difference between the FDTD and MRTD implementations of the UPML. In FDTD, the σ values are discretized and applied only at FDTD grid points. In MRTD, however, the function $\sigma_a(n)$ is used when calculating the update equations, and thus the variation in $\sigma_a(n)$ across a cell is accounted for. This method could also be applied to FDTD, by determining FDTD coefficients as in S-MRTD.

It is important to note the similarity between the field updates for the PML and the field updates for fields outside the PML. If $\sigma = 0$ and $k = 1$, the equations are the same as for the PML free case. For implementation of the PML in a computer, it is convenient to use the PML formulation everywhere, and to use only nonzero σ values inside the PML region. As an added convenience, by using this formulation, the PML does not need to cover an entire cell. For example, in the $r_{\max} = 1$ case, there are four equivalent grid points per direction. If at least 10 equivalent cells of PML coverage are required, then this scheme permits exactly 10 to be used, instead of 12 (for three cells entirely covered with PML).

4.2 NONUNIFORM GRID IN MRTD

In the FDTD method, one common technique for conformal meshing is nonuniform gridding [29]. For a fixed grid size, it is easy to pick a structure (for example, parallel transmission lines) where a uniform grid (constant Δx, Δy, Δz) cannot align with the structure at all points. If, instead, the grid spacing can vary with position, many more structures can be accurately represented in the FDTD grid. Of course, curved structures must still be stair-stepped, but even with these structures, a more accurate grid using fewer grid points is possible. The FDTD algorithm depends on neighboring cells having identical dimensions at intersections, and if the grid is varied so that the grid size is a function of all three coordinate directions, the grid becomes incompatible with FDTD. Therefore, FDTD nonuniform grids must vary the spacing as a function of individual directions only. This means

$$\Delta x = \Delta x(i) \qquad \Delta y = \Delta y(j) \qquad \Delta z = \Delta z(k). \tag{177}$$

One example of a grid that uses this scheme is presented in Fig. 14.

The 2-D grid that is presented shows the positions of the E_x, E_y, and H_z fields. The value of Δx or Δy must be computed based on the field update. For the space derivatives of fields, the value of Δx or Δy is the separation between the field components. Thus, for electric field space derivatives, the Δx or Δy value is simply the grid spacing. However, for magnetic field spatial derivatives, the value is instead the average of the Δ values for the cells containing each magnetic field. For example, if the H_z coefficient centered at Δx_3, Δy_3 in Fig. 14 is updated, the E_y field coefficients required are the ones centered at Δy_3 and located on either end of Δx_3. For this update, $\Delta y = \Delta y_3$. But when the E_y field component centered at Δy_3 and between Δx_3 and Δx_4 is updated, the H_z fields required are those centered at Δx_3, Δy_3 and Δx_4, Δy_3. In this case, $\Delta x = (\Delta x_3 + \Delta x_4)/2$.

It would be convenient to have a similar capability for MRTD. While the MRTD adaptive grid does allow the resolution to vary on a cell-by-cell basis, the equivalent grid size is always

$$\Delta n_{equiv} = \frac{\Delta n}{2^{1+r_n}}. \tag{178}$$

Using this grid size, it is difficult to align the grid with realistic structures. Furthermore, increasing the resolution until an equivalent grid point is close to a feature

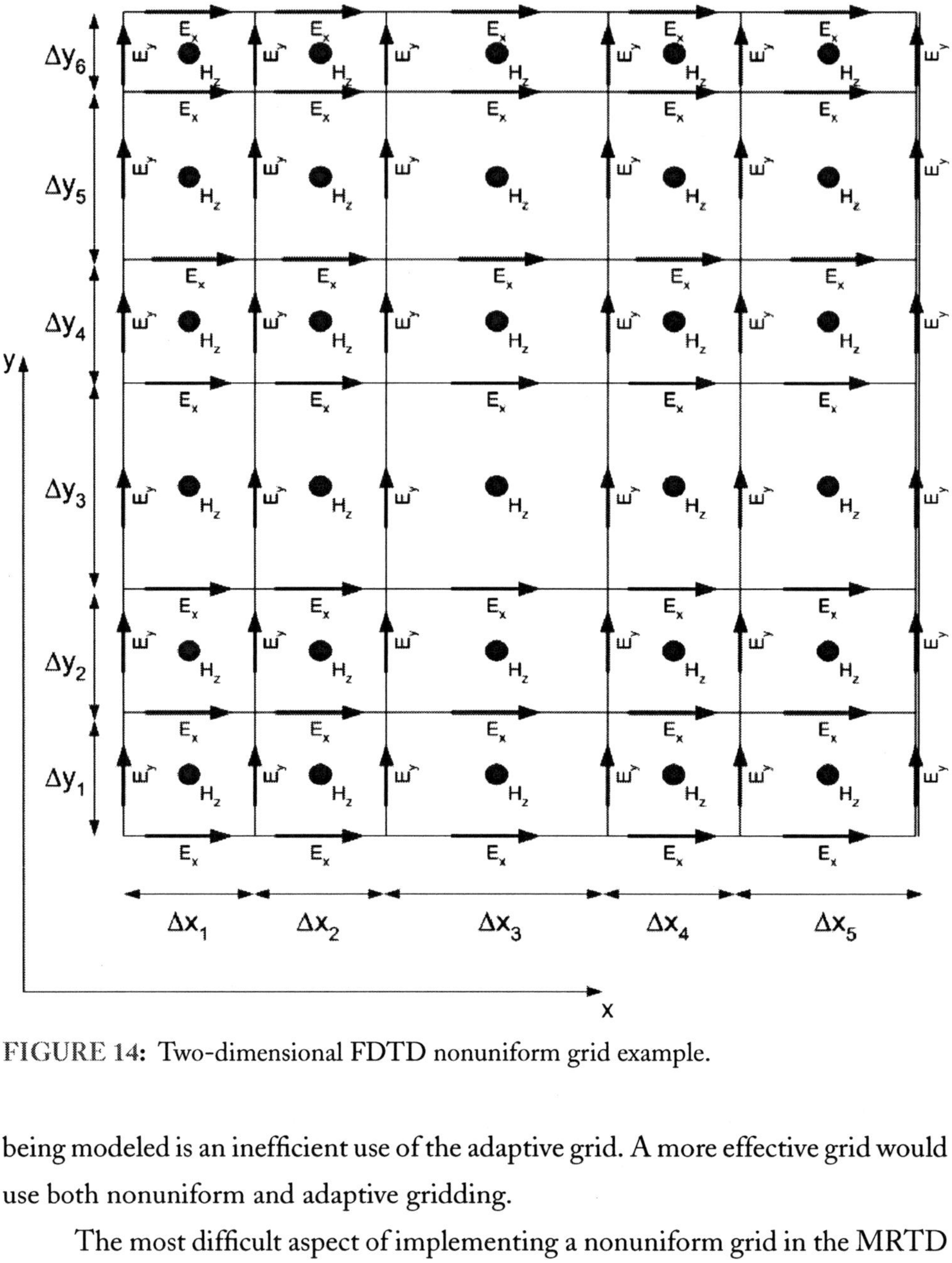

FIGURE 14: Two-dimensional FDTD nonuniform grid example.

being modeled is an inefficient use of the adaptive grid. A more effective grid would use both nonuniform and adaptive gridding.

The most difficult aspect of implementing a nonuniform grid in the MRTD method is determining the correct Δx, Δy, and Δz values to use for each field. In MRTD, the cell size does not appear explicitly in the field update (for the formulation presented here), but is an implicit part of determining the update coefficients. In the derivation presented in Chapter 2, a formula for the offset between the electric and magnetic fields is given as

$$s_d = \frac{\Delta_d}{2^{r_{d,\max}+2}}. \tag{179}$$

In this formula, $r_{d,\max}$ is the maximum resolution in the given direction. It was discussed that in the general 3-D scheme, the fields being updated (the time derivative terms in Faraday's and Ampere's laws) are offset from the other fields in the update (the spatial derivative terms) only in the direction of their spatial derivative. If this scheme is applied consistently for all electric and magnetic field coefficients, there is a specific arrangement of grid points that must be used. This arrangement is outlined here.

The convention chosen for this work is that all fields are offset relative to a grid that is defined by (177). The field components can be indexed to each of these grid points, but the actual domain of each function, defined by the Haar-MRTD scaling function (this discussion can be easily extended to general wavelet basis by defining the domain as the spacing between the centers of the scaling functions, centered on each scaling function), begins at the offset from the grid point. The grid is consistent with the conditions above if the electric fields are offset by s in their coordinate directions, and the magnetic fields are offset by s_d in the two directions normal to their coordinate direction. An example of a grid that meets these criteria is demonstrated in Fig. 15. It should be noted that while the offset can be made relative to another point (for example, the magnetic fields could be offset in their coordinate directions and the electric field in the normals) the resulting arrangement of the fields relative to each other must be maintained.

The offset of (177) denotes the start of the scaling function for the electric or magnetic field, but does not specify the size of the scaling function. If the size of each scaling function is chosen to simply as $\Delta x(i) \times \Delta y(j) \times \Delta z(k)$, then the basis functions for a single field can overlap. In Fig. 15, a 2-D view of an MRTD grid for $r_{\max} = 0$ (in both directions) is used to demonstrate variable gridding. For $r_{\max} = 0$, $s_x = s_y = 1/4$. In the figure, the positions of the E_z and H_z equivalent grid

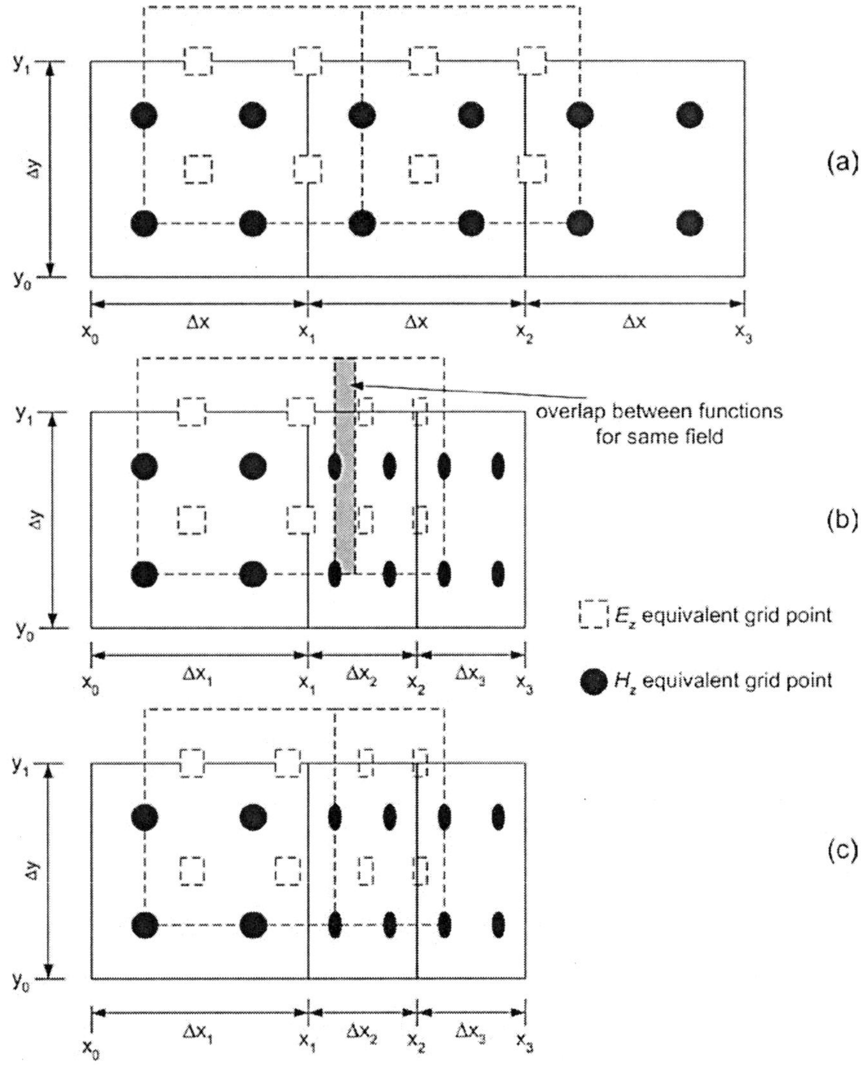

FIGURE 15: Offset between electric and magnetic fields in MRTD (a) fixed grid, (b) nonuniform grid (implemented incorrectly), and (c) nonuniform grid (implemented correctly).

points and the domain of the scaling functions are shown. The E_z field is offset in the z direction (normal to the page), while the H_z field is offset in the x and y directions.

Fig. 15(a) demonstrates a uniform MRTD grid. In this case, both scaling functions have dimensions $\Delta x \times \Delta y$. The scaling functions for each field start at the same point relative to the grid, and form a nonoverlapping basis covering the entire space. Fig. 15(b) demonstrates a nonuniform MRTD grid. In this case, the Δx_1 is twice Δx_2. The E_z field is not offset in the x direction, and thus for each cell $\Delta x = \Delta x_i$. The H_z field, however, is offset in the x direction. When $\Delta x = \Delta x_i$ is used for this field, the H_z field for the first cell continues to the center of the neighboring cell. The H_z field for the second cell, however, begins one quarter of the cell width from x_1. This causes the field coefficients that represent the H_z field to overlap. While it may be possible to use this field arrangement for MRTD, the overlap between neighboring field components would render the scheme implicit, and the wavelet basis nonorthogonal.

In order to keep the wavelet scheme nonoverlapping, and still allow for a variable grid, the width of the offset scaling functions must be set so that it begins s_d from the current grid point, and ends s from the next grid point (note that s_d is proportional to the cell spacing). This is demonstrated in Fig. 15(c). A scheme for determining the grid size for the different basis functions can then be quantified:

1. For nonoffset directions (directions normal to field component for electric fields, or direction of field component for magnetic fields),

$$\Delta n = \Delta n(l), \tag{180}$$

   where l is an index denoting the position along the axis.

2. For offset directions (direction of field component for electric fields, or normal to field component for magnetic fields),

$$\Delta n = \Delta n(l) + \frac{\Delta n(l+1) - \Delta n(l)}{4}. \tag{181}$$

4.3 MRTD GRID EXCITATION

In Chapter 3, a method was presented that allows the addition of an impressed field to the MRTD field update equations. A modified $\mathbf{D_x}$ update was presented, (129), that allows a source function to be applied at specific equivalent grid points and then be transformed back into the wavelet domain. While this equation can be used to impress a field at any (or every) point in the MRTD grid, it is likely that only a relatively small number of grid points will be modified to excite the grid. In this case, the fields can be excited using

$$\mathbf{Dw}_{\mathrm{dir1}} = \mathbf{Dw}_{\mathrm{dir1}} + \mathbf{R}^{-1}\mathbf{J}, \tag{182}$$

after the fields are updated, if **J** is a vector of field values that is only nonzero at desired equivalent grid points. In this case, (129) is simply split into two steps.

One of the most common types of excitations for microwave circuit simulation is the microstrip line. One method of exciting a microstrip line is to impress a constant electric field in a plane normal to the microstrip metal, directly under the microstrip. For any wavelet resolution, it is necessary to use (182). This is because, when wavelets are applied, there is more than one equivalent grid point in each direction. This relatively simple excitation is demonstrated in Fig. 16.

In this example, $r_{\max} = 0$ in all directions; therefore, there are two equivalent grid points per cell in each direction (eight total equivalent grid points). In this example, only the fields directly under the microstrip are excited. The field strength is represented by arrows, demonstrating that the impressed field at all points is identical in magnitude and direction. The arrows are located at the E_y equivalent grid points, and the cells are represented by the alternating shading. This is meant

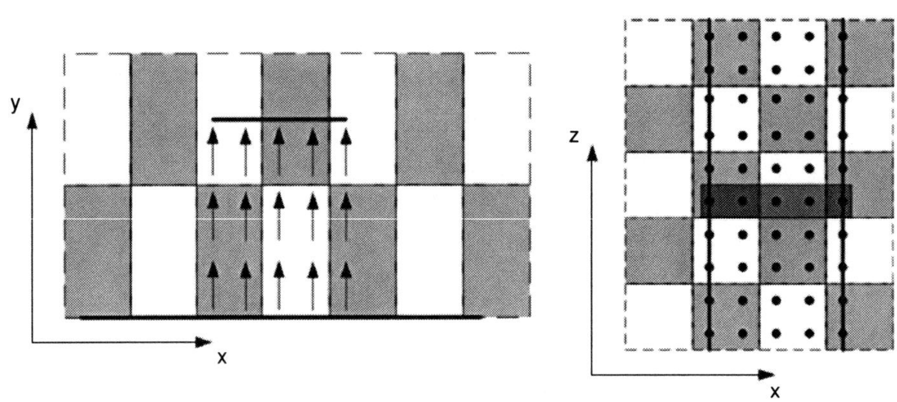

FIGURE 16: Excitation in MRTD, $r_{\max} = 0$ cells demonstrated by alternate shading.

as simple approximation of the microstrip mode; as the field propagates through the grid it quickly matches the physical microstrip field pattern. In this example, five E_y field points fall under the microstrip line. As there are two of the field points in the y direction per cell, one cell is split in the x direction, where only half is excited. Similarly, all cells that contain the PEC are split in the y direction.

As in all other directions, there are two E_y field points in the z direction per cell. These field points are represented by the dots in Fig. 16, and the boundaries of the top metallization are the dark lines. For the excitation, however, only a single set of fields is excited. If both are excited, the fields resemble two identical, slightly offset pulses, needlessly complicating analysis. This is similar to subcell PEC analysis, where PECs can be simulated with an even number of cells in their tangential directions, but without subcell modeling, the PECs are several equivalent cells in thickness.

Similar limitations apply to coplanar waveguide (CPW) excitations; this case is presented in Fig. 17. For CPW excitations, the field is guided between the center

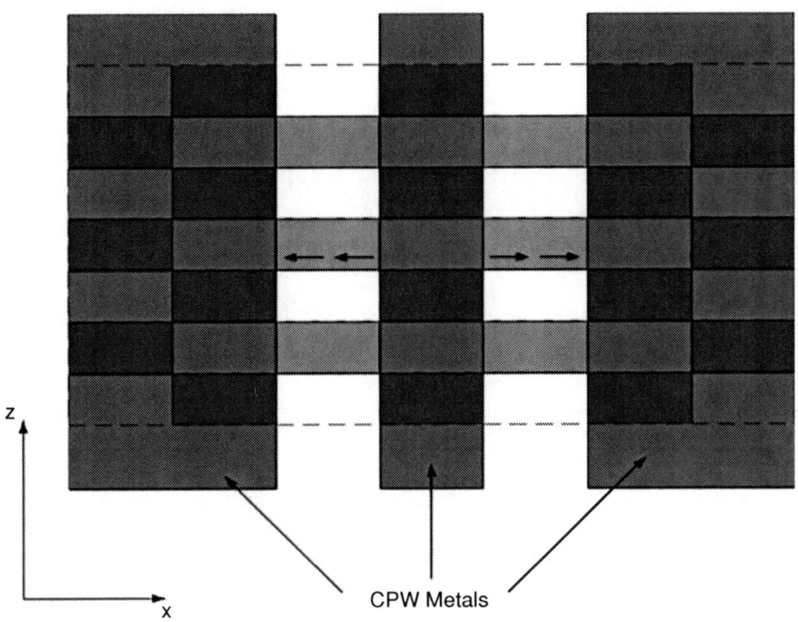

FIGURE 17: CPW excitation, demonstrating subcell excitation.

conductor and the surrounding ground planes. In this case, the field is excited as a voltage between the center conductor and the outer ground planes. In the direction of propagation (the *z* direction in the diagram) the field is only one cell thick. The impressed voltage along a line one equivalent cell in thickness can also used for wire antennas and probe feeds (in coaxial fed microstrip antennas or waveguides, for example).

4.4 TIME/SPACE ADAPTIVE GRIDDING

At the beginning of Chapter 3, the two main advantages of the MRTD method are presented. The first of these advantages is that many of the wavelet basis functions that can be used in MRTD allow the use of fewer basis functions per wavelength than FDTD, resulting in fewer coefficient updates per time step. However, it can be very difficult to apply PEC boundary conditions and the UPML boundary with these cases. In this book, the focus is placed on Haar wavelet basis functions, and while it is true that the Haar cells are larger than FDTD cells, the total number of scaling/wavelet coefficients needed per wavelength is the same as in the FDTD case. The advantage of the Haar scheme is the time-and-space adaptive grid.

The methods that have been presented allow for the modeling of general structures using Haar MRTD and are derived using an equivalence between FDTD and Haar MRTD. If the Haar MRTD method is applied, and the maximum wavelet resolution is used in all cells, the scheme is equivalent to FDTD. However, the chief advantage of the MRTD method is that the wavelet resolution can be varied from cell to cell. Most simulations are not constrained by the dispersion requirement, but instead by the structures that are simulated. For accurate simulation, several grid points must be used across each feature. Using Haar MRTD, the resolution can be locally increased to model complex structures, and reduced elsewhere for computational efficiency. This technique can be applied statically before the start of simulation, and automatically during simulation. When applied during simulation, this feature is usually referred to as the MRTD time-and-space adaptive grid [30].

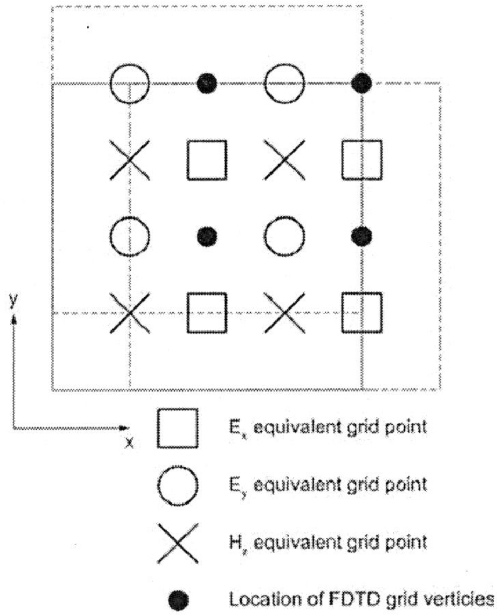

FIGURE 18: Haar MRTD cell (2D) showing FDTD grid points, $r_{max} = 0$.

The concept of the equivalent grid point was introduced in Chapter 2 to demonstrate the offset of the electric and magnetic fields in MRTD. In Haar MRTD the equivalent grid points are functionally identical to FDTD grid points (regions of constant field value). When wavelets are added and subtracted from a Haar representation, the effect is to add and remove equivalent grid points. This allows dense discretizations to be used in the area of complex structures, and coarse discretizations to be used elsewhere. Fig. 18 shows the arrangement of the fields (equivalent grid points) in a single $r_{max} = 0$ Haar MRTD cell, and also shows the positions of the FDTD grid points if the Yee-cell in Chapter 2 is used. These FDTD equivalent grid points can be used to demonstrate the FDTD equivalent grid for Haar MRTD adaptive resolution.

As an example of the capabilities of the Haar MRTD adaptive grid, consider the grid presented in Fig. 19. In this example, three wavelet resolutions are used from $r_{max} = -1$ (scaling function only) to $r_{max} = 1$. In the $r_{max} = -1$ case, there is one equivalent FDTD cell per MRTD cell; in the $r_{max} = 1$ case, there are eight equivalent FDTD cells per MRTD cell (for a 2-D example). Similar to Fig. 18, the

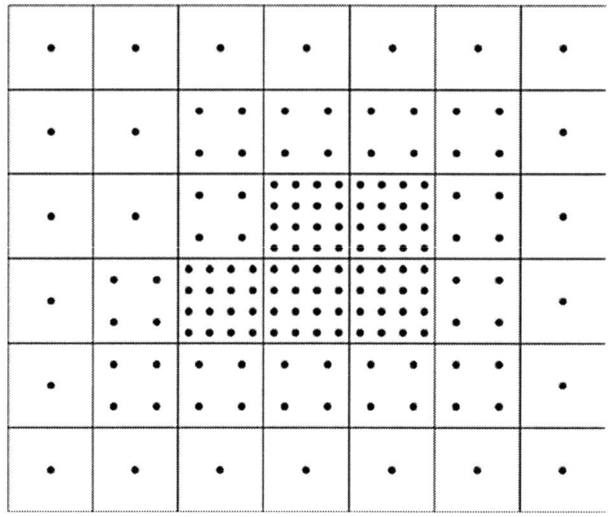

FIGURE 19: Adaptive grid example.

dots represent the equivalent FDTD cells, and the surrounding squares represent the MRTD grid boundaries.

In Fig. 19, the adaptive grid is used to place a dense discretization in the center, and a coarse discretization in the surrounding area. Using this grid, a highly detailed structure can be modeled in the center, while the surrounding, homogeneous area is simulated with low resolution. This technique is more powerful than the nonuniform mesh that is presented earlier in this chapter because the spacing of the equivalent grid points is a function of all three grid dimensions, and thus can be increased locally to represent complex features. One important note is that, while this example demonstrates the grid when the resolution is the same in all dimensions, this is not a requirement.

While the grid that is presented in Fig. 19 can be fixed for the entire simulation, the MRTD method also provides the ability to change the resolution during simulation. In Chapter 3, it is demonstrated that at any point in the grid, the highest resolution wavelet represents the deviation from the average. If the field is not changing rapidly, the high resolution wavelets can be neglected during the simulation. In [30], it is suggested that an absolute and relative threshold be used to determine whether a wavelet is needed. This concept forms the basis for the adaptive grid.

For the examples presented in the next chapter, the thresholds were chosen as in [30]. The relative threshold is a fraction of the value of the scaling function. If the value of the wavelet is less than this fraction of the scaling function, it can be neglected. The absolute threshold is required for low-field values, and does not change during simulation. If the fields are near zero, the difference between the scaling and wavelet coefficients will not be large. However, the wavelets are not required because the field does not vary rapidly. In this case, the wavelets will be lower than the absolute threshold, and can be neglected.

This scheme is not effective if it is applied every time step. For the test to be applied, the wavelet value must be calculated, which defeats the purpose of the adaptive grid. In this work a scheme was used where the wavelets were tested at a user defined period. Wavelets that are below either threshold are neglected until the next testing period. Using this scheme, only the highest level wavelets are tested. If the wavelets are removed, at the next testing period both the currently highest level wavelets are tested (one lower than the previous level) and the next highest level is tested. In this manner, it can be determined if wavelets need to be reintroduced into the grid.

In the code that was used in this simulation, both time-adaptive grids and static (but variable in space) grids are used. Using this scheme, high-resolution grids can be used to specify complex structures, while the resolution around the structure can be predefined to a low value. In this manner, the adaptive grid can be applied to vary the resolution in response to field propagation, while the resolution is not higher than necessary at any point in the structure.

CHAPTER 5

MRTD Simulation Examples

The techniques that are presented in this book can be used to simulate a wide variety of structures in MRTD. Using the subcell modeling of arbitrary structures, any structure that can be simulated in FDTD can be simulated in MRTD. If a structure can be represented within a cell, then the cells used in an MRTD simulation can be significantly larger. Using the method presented here, any structure that can be simulated in FDTD can be simulated with MRTD. In addition, the MRTD time-and-space adaptive grid can be applied to allow fewer grid points in the MRTD case.

In order to verify the method, several simulations were run using the MRTD code that was developed. The code utilizes all of the features that have been discussed, including the UPML, nonuniform gridding, and lumped element modeling. For the practical cases, the MRTD results were verified by an FDTD code.

To measure the effectiveness of the MRTD code, both simulation time and number of equivalent grid points required to simulate the structure were recorded. The number of equivalent grid points, which is equal to the total number of grid points in the FDTD case, is useful because it directly relates to the number of calculations required. This is the best measure of how one MRTD grid compares to another. In the cases where the time-adaptive grid was used, the number of equivalent grid points provided a method to measure the effectiveness of the adaptive scheme.

5.1 ANALYZING MRTD OUTPUT

The microstrip structures that were evaluated in the following examples were measured by computing the voltage on the microstrip lines. The voltage is calculated using

$$V_{ab} = -\int_{ab} \mathbf{E}(\mathbf{r}) \cdot \partial \mathbf{l}, \tag{183}$$

where a is the ground plane and b is the microstrip conductor. This is the same method that is used for FDTD simulations [14]. Similarly, current is calculated using

$$I = \oint_{C} \mathbf{H}(\mathbf{r}) \cdot \partial \mathbf{l}, \tag{184}$$

where C is the contour that the current passes through. These are relatively simple calculations, and are solved in the discrete case by summing the electric or magnetic field values along the path and multiplying by the space step. Of course, in MRTD, the field must be reconstructed to perform the sum, and the space step used must be the equivalent cell size. In this case it is useful to note that while the reconstruction is a matrix operation, the field at any one point in the cell can be reconstructed by multiplying the coefficient vector by the row of the reconstruction matrix that represents the desired point.

Most structures are not characterized in the frequency domain; therefore, the time domain voltages must be transformed into the frequency domain using a discrete Fourier transform (any DFT method will suffice). Once the frequency domain voltages are determined, they can be used to calculate device parameters such as S parameters and characteristic impedance. S parameters are calculated using [14]

$$S_{mn} = \frac{V_m(\omega)}{V_n(\omega)} \sqrt{\frac{Z_{o,n}(\omega)}{Z_{o,m}(\omega)}}, \tag{185}$$

where the V_m is the voltage out of port m and V_n is the voltage into port n. In most simulations, one port is excited and the output is measured at the other ports. In this case the total voltage measured at the output ports can be used directly in (185). The voltage recorded at the input, however, is the total of the input waveform and reflections from the structure being characterized. In this case, a separate simulation of the input line alone can be used to determine the input. This "thru" case can be used as V_n in (185). For the special case of S_{nn}, the following can be used:

$$S_{nn} = \frac{V_{n,total}(\omega) - V_{n,thru}(\omega)}{V_{n,thru}(\omega)}. \tag{186}$$

For the microstrip lines used here, the characteristic impedance can be calculated simply by dividing the voltage between the ground and the conductor at one point with the current through the conductor at the same point. However, the electric and magnetic fields are offset. Both the half-time step and half equivalent grid point offset must be accounted for. Using [31]

$$Z_o(\omega) = \frac{V_z(\omega)\, e^{-j\omega\Delta t/2}}{\sqrt{I_{z-1/2}(\omega)\, I_{z+1/2}(\omega)}}, \tag{187}$$

the characteristic impedance can be determined. In this case, the line runs in the z direction and the $z - 1/2$ and $z + 1/2$ subscripts indicate that the current values are offset from the voltage values by one half of a space step.

5.2 PML ABSORPTION USING MICROSTRIP LINE

The first test that was run of the MRTD code was that of a simple microstrip line. Beyond a simple check to see if the code is working, this test allows the performance of the PML to be evaluated. As has already been presented, the PML is a numerical representation of a nonphysical (at least at the time of this printing) high-loss material that is perfectly matched for incoming waves at any angle. When implemented numerically, however, the material exhibits small reflections. The reflections are generated in two ways. First, the medium is highly lossy, but not infinitely lossy; some of the incoming wave will propagate through the material, reflect off of the outer PEC boundary, and reenter the simulation space. In addition, when implemented in a discrete space, the PML is not perfectly matched and small reflections occur. The parameters that can be used to finely tune the PML are σ, k, and the degree of the polynomial that is used to grade them. Using these parameters, it is possible to characterize the PML for a number of applications and values of the parameters [14].

The purpose of the test used here was not to completely characterize the PML, but rather to verify that the PML was functioning as expected. The test structure is a microstrip line terminated at one end with PML. The microstrip line is excited using the method presented in the previous chapter. The time shape of the pulse is Gaussian, with time duration chosen such that the 3 dB point is at 30 GHz. Two structures are simulated. In the first case, the line is chosen to be sufficiently long so that reflections from the far wall do not return until the initial pulse (and its reflection from the near wall) has passed. In the second case, the line is terminated with PML very close to the excitation.

The simulation was performed using $r_{\max} = 1$ in all directions (64 equivalent grid points per cell). The microstrip PEC is 452-μm wide and the substrate is 200-μm thick with $\varepsilon_r = 2.2$. The predicted Z_0 is 50 Ω. The grid size is 452 μm across the microstrip line, so that four equivalent grid points are used to model the line. The cell height is 200 μm; the entire height of the substrate is modeled with one cell (four equivalent grid points). Using this grid, subcell PEC modeling is necessary to restrict the microstrip conductor to only one cell in height.

For this case, S_{11} was calculated to quantify the reflection from the PML. The reflection, as seen in Fig. 20, is lower that −60 dB at 10 GHz, and slowly increases with frequency. This is expected behavior; the reflection from the PML is caused primarily by the effect of the discrete grid. As the frequency increases, the relative size of the cells also increases. The PML could most likely be tuned by manipulating σ, k, their variation across the PML, and the grid size, but it is not required for the structures that are characterized in the further examples. The response of the featured antenna is significantly higher than this noise floor, making the reflection irrelevant.

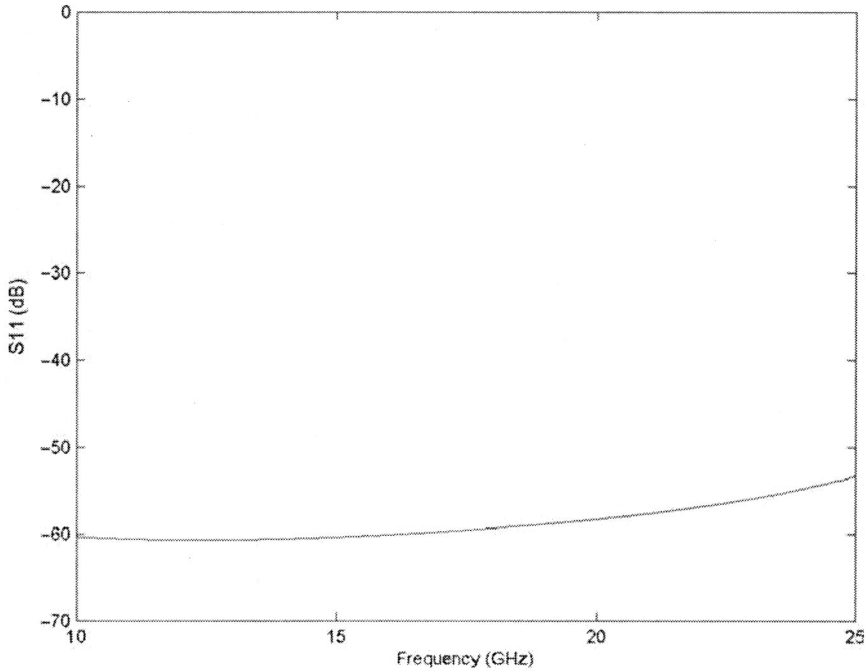

FIGURE 20: Reflection from PML.

5.3 LUMPED ELEMENT VERIFICATION: RESISTOR-TERMINATED MICROSTRIP LINE

In order to test the lumped element algorithm that is presented in Chapter 3, a very simple case was simulated. A microstrip line, the same as in the antenna example, was terminated with resistors. This technique is commonly used to simulate measurement environments. As was shown in the previous example, the PML has low reflections across a wide band. In measurement, the ports are often terminated with a load of fixed impedance. If the input line is not matched to the load, or the characteristic impedance of the line changes with frequency, then reflections will occur from the termination. By terminating the simulation with resistors, the measurement environment is simulated.

In this example, the microstrip line of the previous example is terminated with a PML at one end and five parallel resistors at the other. The PEC that represents the microstrip conductor is four equivalent grid points wide (5 E_z field locations).

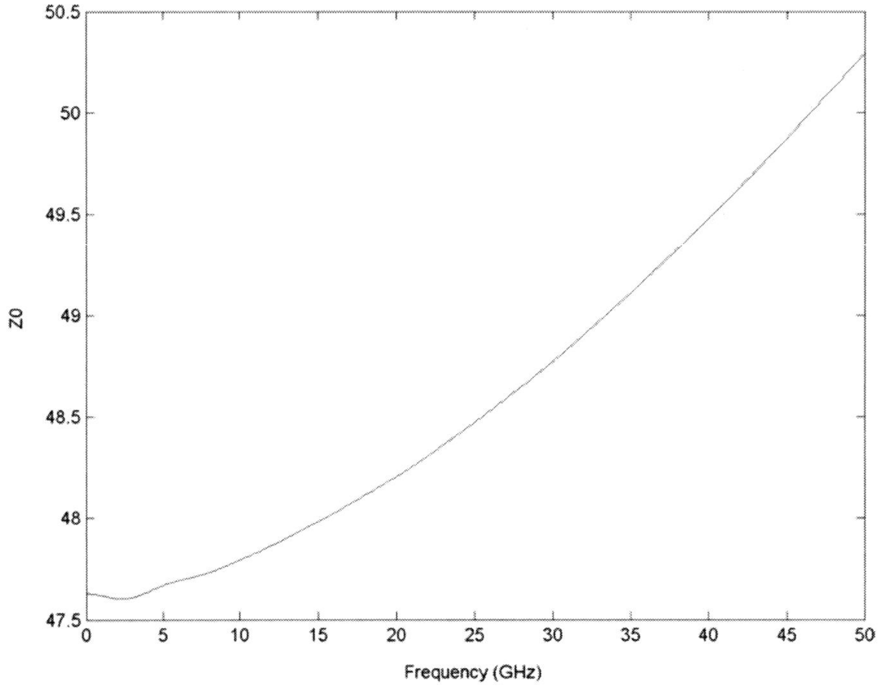

FIGURE 21: Characteristic impedance of microstrip line.

The impedance of the line was first calculated using (188), and is presented in Fig. 21. The impedance can be seen to vary slightly over the 50 GHz band. The resistors were chosen to represent a 49 Ω load (each resistor is 245 Ω), the impedance at the center of the band.

The time domain response is given in Fig. 22. The time domain is presented instead of the S parameters because, in this case, the reflection from the resistor can be seen. The initial pulse is a Gaussian derivative, with the 3 dB frequency at 50 GHz. The reflection occurs at approximately 175 ps, and is the small ripple in the response. In the case where both ends of the line are terminated with PML, the ripple does not appear. This response demonstrates that the resistor is matched for part of the band, but not all. This is verified by S_{11}, which is presented in Fig. 23, demonstrating that the best match occurs at approximately 37 GHz, the same frequency that $Z_0 = 49\Omega$.

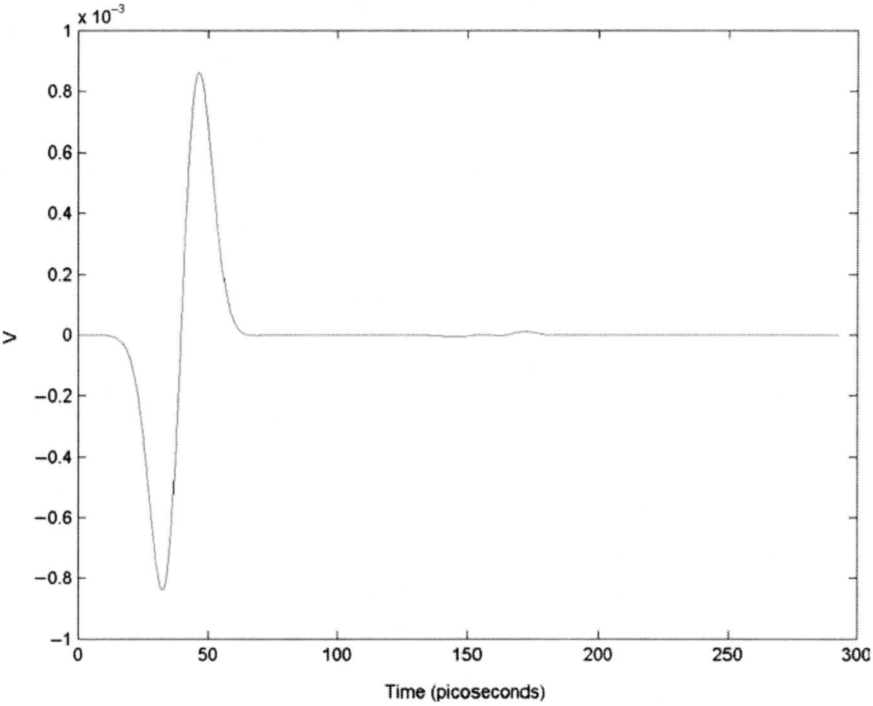

FIGURE 22: Time domain reflection from resistor.

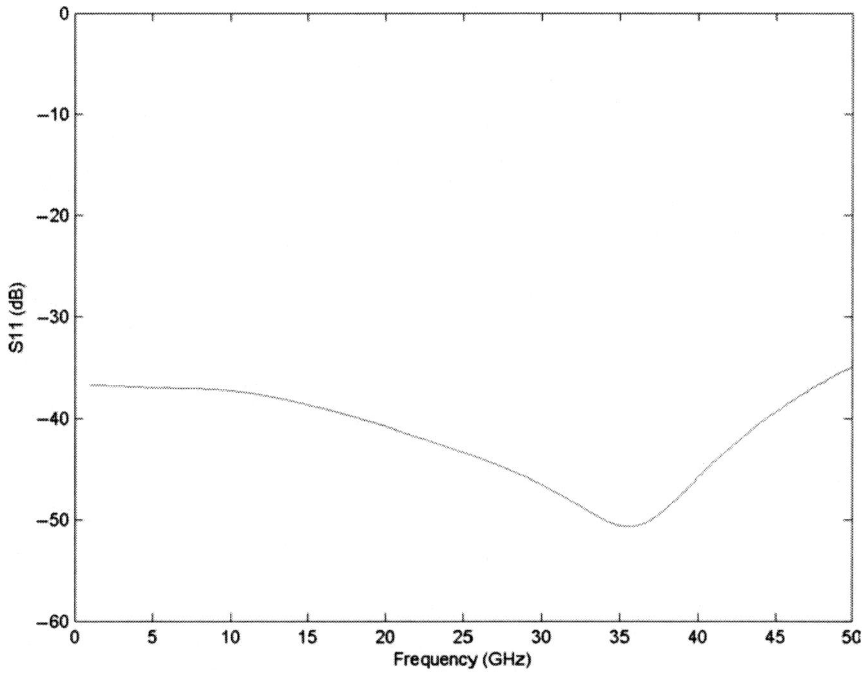

FIGURE 23: Frequency domain reflection from resistor.

5.4 MICROSTRIP PATCH ANTENNA

The PML example that was presented effectively validates the code and the techniques that have been presented. Both the microstrip line and the resistors used to terminate it in the previous example are smaller than a cell in size, validating the subcell PEC procedure. However, in order to show the advantage of Haar MRTD simulations over FDTD simulations (and verify that the code gives results similar to FDTD), another example is presented. In this example, a 30 GHz microstrip patch antenna is simulated. Several simulations are performed, and the effectiveness of the adaptive grid over a variety of threshold values is evaluated. In addition, a static multiresolution grid is used, as well as a mix of static and adaptive grids.

The antenna that was simulated is presented in Fig. 24. The antenna is based on a design for a 30GHz patch antenna built on a thin polyimide film [32].

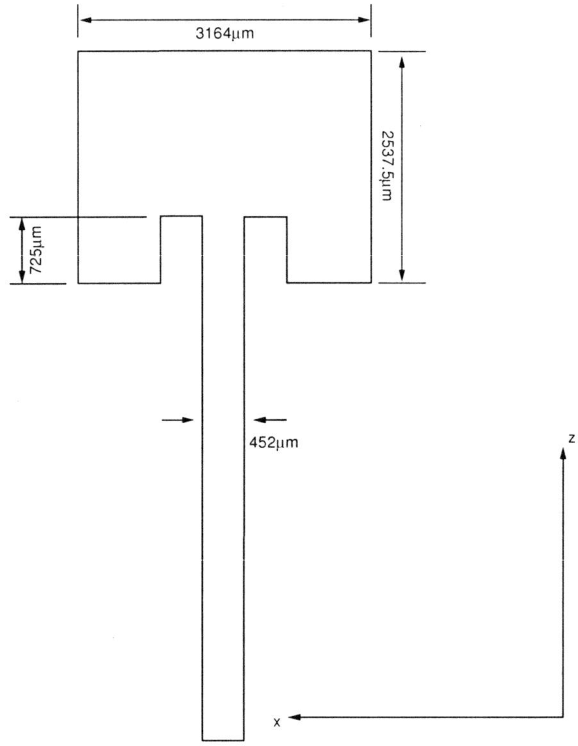

FIGURE 24: Microstrip fed patch antenna.

The microstrip line is identical to the line used in the previous two examples. In the simulation, $r_{\max} = 1$ is used in all directions. A uniform grid is used with $\Delta x = 452\ \mu\text{m}$, $\Delta y = 200\ \mu\text{m}$, and $\Delta z = 725\ \mu\text{m}$. The uniform grid is used to ensure that the FDTD grid and the MRTD grid are identical. When a nonuniform grid is used in FDTD, a change in cell size affects only adjacent cells. However, in the MRTD case, the change in cell size affects every equivalent cell within the MRTD cell. Using the offset method presented here, an FDTD grid cannot be generated from the MRTD grid by simply subdividing the grid size. While it is possible to match an MRTD and FDTD grid (by applying a variable grid to every cell in the FDTD grid), the purpose of this example is to measure the effect of the MRTD adaptive grid.

To avoid interaction of the antenna with the boundary, a space of three times the antenna width is placed between the PML and the antenna on all sides (three times the substrate height on the top). Including the antenna, the surrounding space, and the absorber, the simulation space is 40 cells $\times$ 7 cells $\times$ 23 cells. In equivalent grid points (and FDTD cells) the grid is $160 \times 28 \times 92$. To model the antenna, a 50 GHz Gaussian derivative pulse is used to excite the microstrip line. As a reference to the adaptive grid case, an FDTD simulation and MRTD simulation with constant grid were run. The time domain voltage on the feed line is presented in Fig. 25.

In this figure, excellent agreement is observed between the MRTD and FDTD simulations. However, there are small differences between these time domain plots and these are revealed in the plot of S_{11}, Fig. 26. It can be seen from the S_{11} plot that there is a small difference between the simulations (approx 1.5%) in the resonance frequency of the antenna. When the time response is examined more closely, Fig. 27, the differences between the FDTD and MRTD simulations are easier to observe. The primary reason for this difference is the modeling of the varying dielectric in the different techniques. In the FDTD simulation, the ε_r value used in the update equation is applied locally. For field values on the air/substrate interface, the average of the two values is used. In the MRTD case, the discretization

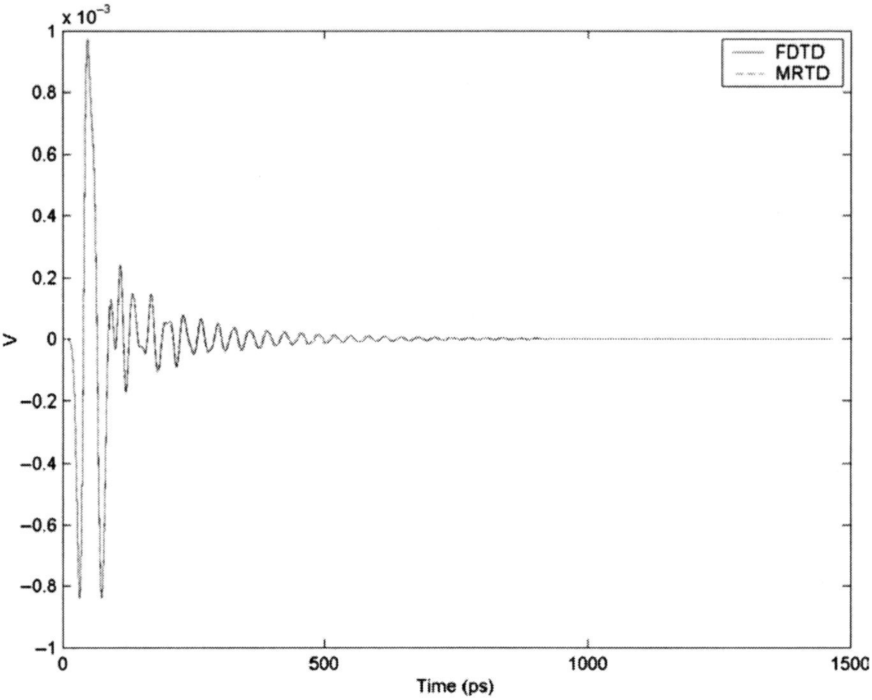

FIGURE 25: Time domain response of antenna, FDTD and MRTD, $r_{\max} = 1$.

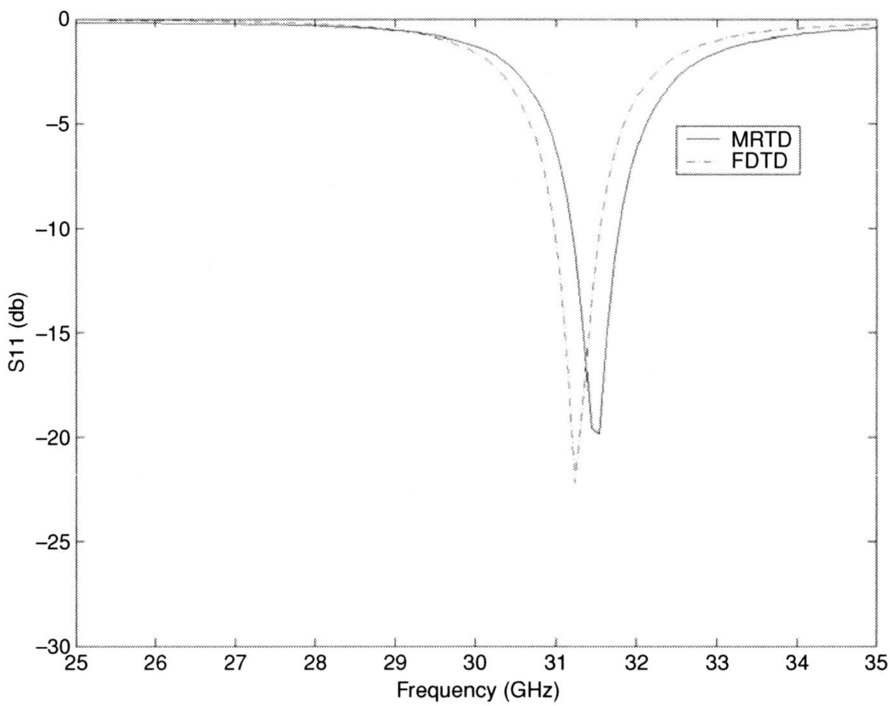

FIGURE 26: S_{11} of patch antenna, MRTD/FDTD comparison.

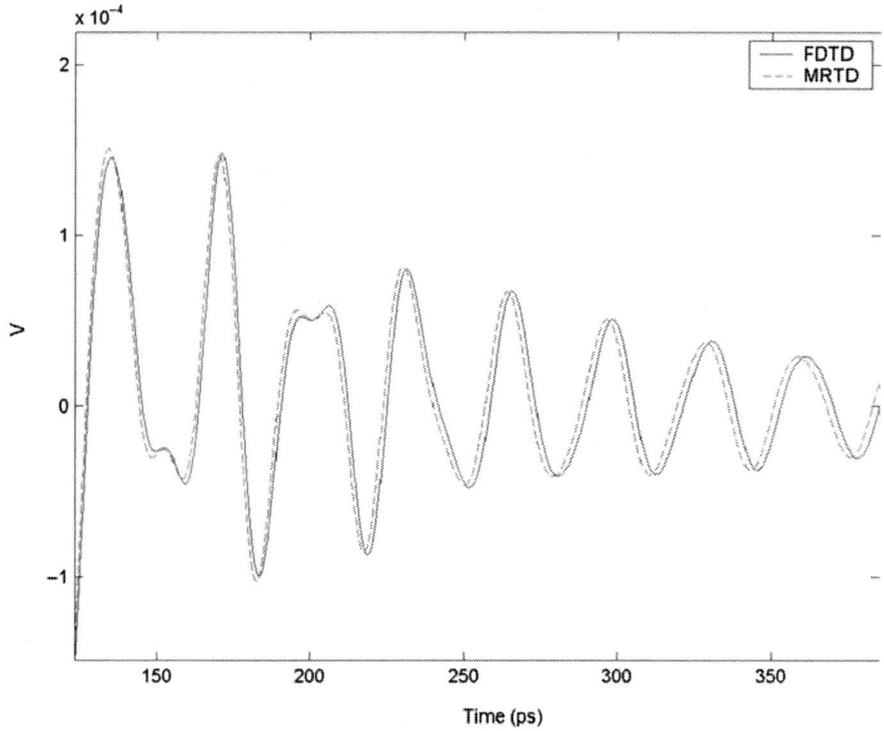

FIGURE 27: Close-up of differences between MRTD/FDTD, time domain.

of the constitutive relationship involves field values across the entire cell, leading to the small difference between the two techniques.

To verify that the differences in the methods that are used to model material interfaces is responsible for the frequency shift in Fig. 26, another experiment was run where ε_r was kept constant across the entire space. For this experiment, the ε_{eff} (2.47) of the microstrip line was chosen. For this case, S_{11} is presented in Fig. 28. In this example, the MRTD and FDTD results overlap; there is no frequency shift.

The remainder of the cases that were simulated using this technique are experiments with the grid resolution. The first tests that were run used a static grid, but lowered the resolution in the area surrounding the antenna. This is a very important test because it demonstrates that the technique can be used to locally increase the grid resolution to simulate a complex structure (the antenna), while the

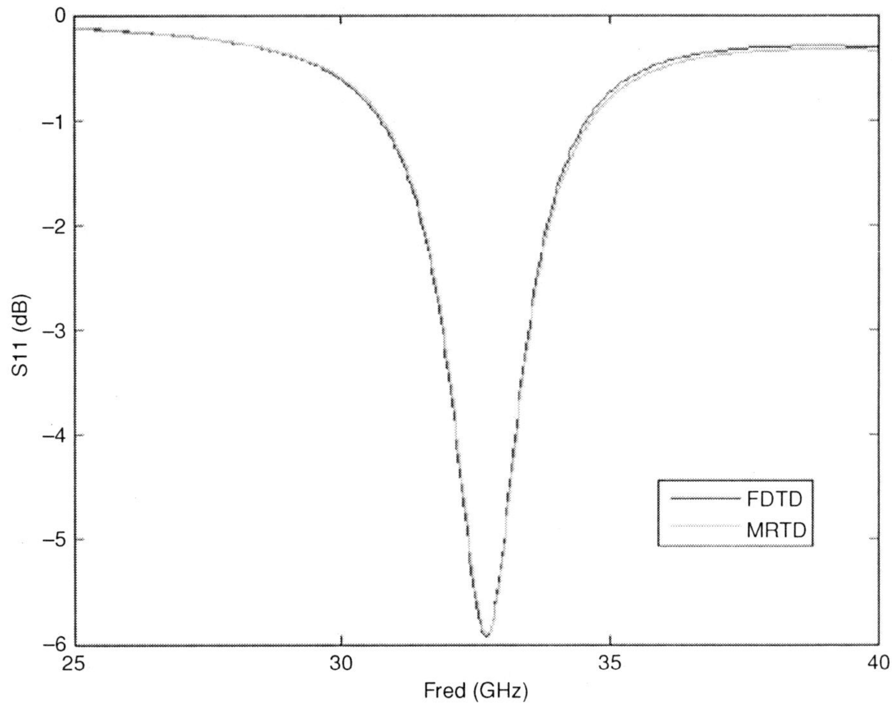

FIGURE 28: FDTD/MRTD comparison, constant ε_r.

surrounding area can use a coarser grid. In this case, the smallest grid dimension used in the x direction (tangential to the antenna surface) is $\lambda/53.1$. The grid is sampled significantly higher than required for accuracy in free space; however, this grid is necessary to accurately model the feed. When the resolution is reduced in the area surrounding the antenna, the spacing is $\lambda/26.5$ for $r = 0$ and $\lambda/13.3$ for $r = -1$ (scaling function only).

The results for the $r = -1$ case are presented in Fig. 29. The grid in this case has 12.4% fewer grid points than in the grid that uses the same resolution everywhere. If the PML cells are neglected for this calculation, the number of cells is reduced by 34%. This modest improvement is due to the relatively small structure that is simulated. In this structure, the only area that can be reduced in resolution is the area surrounding the antenna, and it uses relatively small number of cells. In this case, the grid has 6440 cells (with 64 equivalent grid points per cell for a total

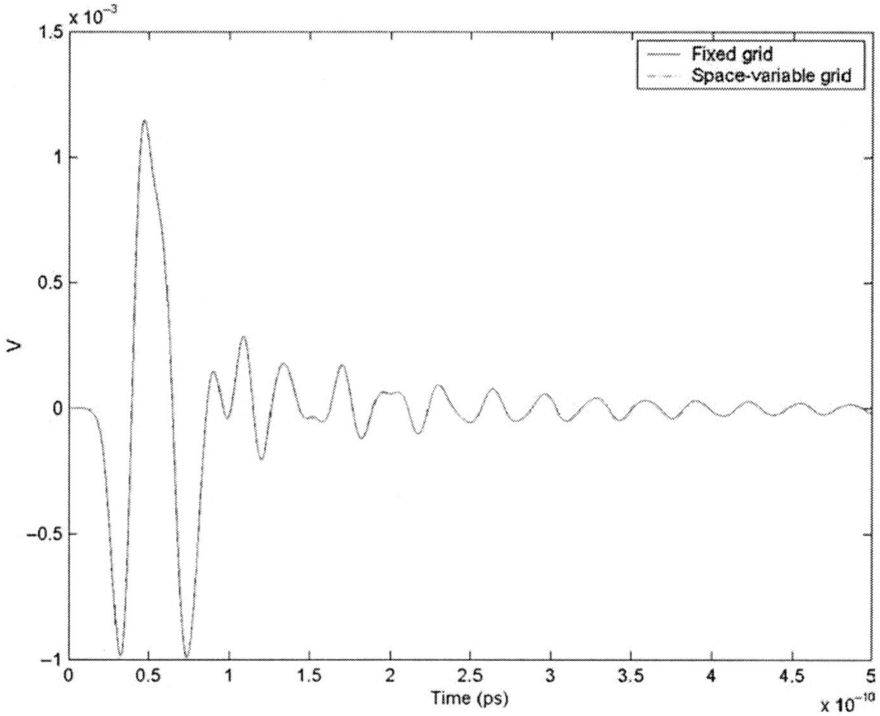

FIGURE 29: Comparison of $r_{\max} = 1$ MRTD with surround grid fixed at $r = -1$.

of 412,160 equivalent grid points), and 1077 of these cells surround the antenna and are reduced in resolution (cells in the PML are not reduced). In addition, the cells on the bottom of the grid intersect the ground plane, and the resolution is not lowered in the y direction. The results for the $r = 0$ case are presented in Fig. 30, the number of non-PML grid points is reduced by 28.1%.

The remaining cases that are presented demonstrate the adaptive grid. In these cases the resolution was set to one at the beginning of the simulation, and relative and adaptive thresholds, t_r and t_a respectively, were used to in the manner indicated in the previous chapter to automatically determine if wavelets were needed. The t_a values that are presented are relative to the maximum value of the pulse for the entire simulation (this can be determined *a-priori* from the excitation). A number of cases were run to test the technique; representative cases are presented

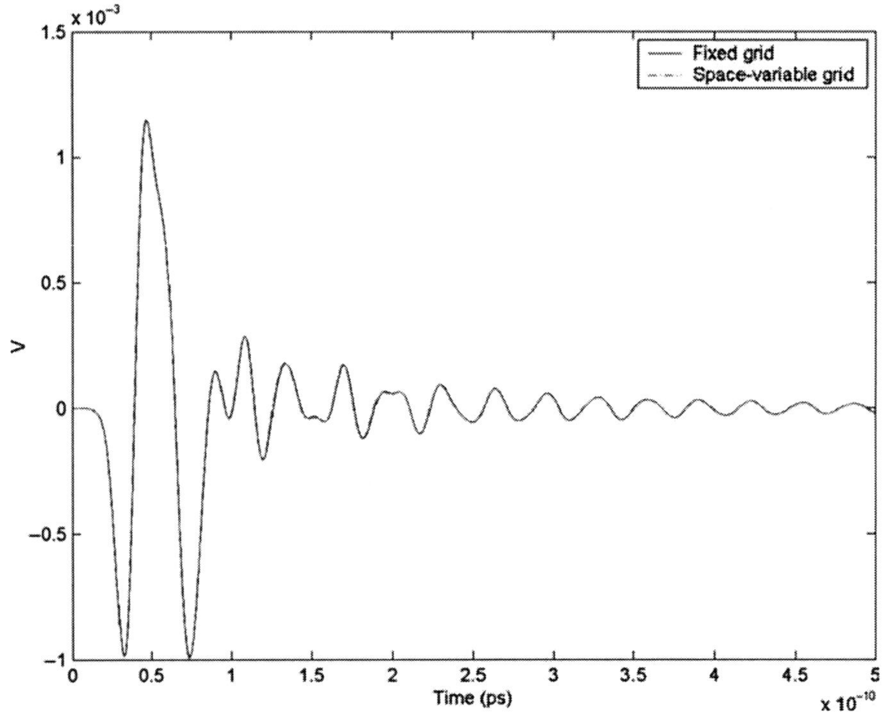

FIGURE 30: Comparison of $r_{\max} = 1$ MRTD with surround grid fixed at $r = 0$.

here. In all of these cases, the thresholding is performed every 110 time steps. This value is chosen because it represents the Nyquist limit for the resonant frequency of the antenna. When the thresholding is performed significantly more often than this rate, the effectiveness is largely unchanged, and for values higher than this limit, the scheme is slightly less efficient.

In the first two cases that are presented, the threshold values are too high. In Fig. 31, the simulation uses $t_r = 0.1$ $t_a = 0.01$. In the time domain output, it can be seen that the oscillations of the wave are quickly damped. This is because the adaptive grid algorithm removes wavelets that have significant values. A similar result is presented in Fig. 32. In this case, both thresholds are reduced by an order of magnitude from the previous case. It can be seen that the damping takes a longer time to occur. It is interesting to note that in both of these cases the results at the time before the results are abruptly damped closely match the fixed grid case.

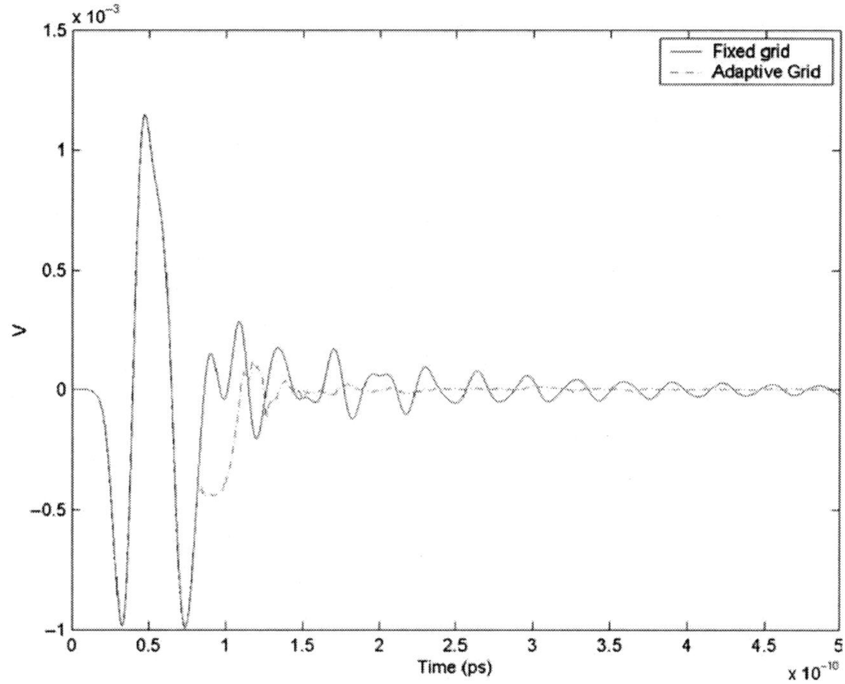

FIGURE 31: Comparison of fixed grid MRTD with adaptive grid $t_r = 0.1$ and $t_a = 0.01$.

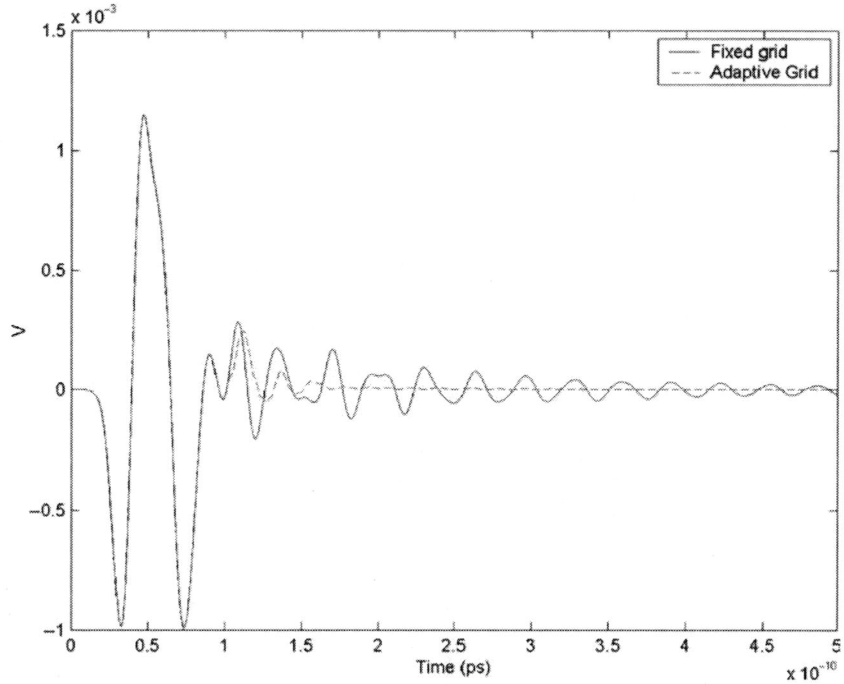

FIGURE 32: Comparison of fixed grid MRTD with adaptive grid $t_r = 0.01$ and $t_a = 0.001$.

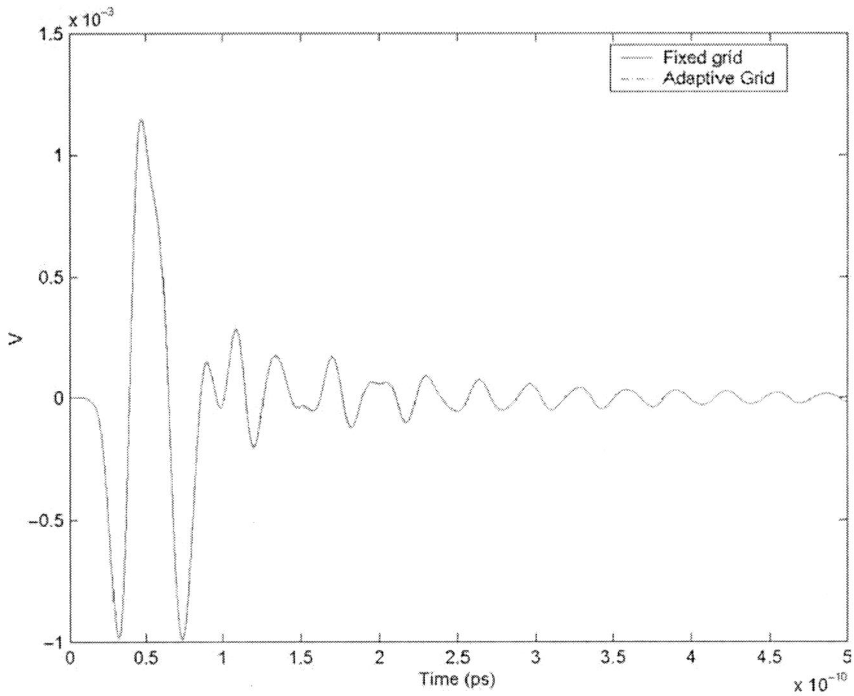

FIGURE 33: Comparison of fixed grid MRTD with adaptive grid $t_r = 1 \times 10^{-4}$ and $t_a = 1 \times 10^{-5}$.

In the final case that is presented the results are very close to the fixed grid case (0.73% error). The error in this case was calculated by integrating the magnitude of the difference between the uniform and variable grid cases and dividing by the integral of the magnitude of the uniform case. In Fig. 33, the simulation uses $t_r = 1 \times 10^{-4}$ and $t_a = 1 \times 10^{-5}$. In this case the adaptive algorithm does not remove wavelets that are needed. However, results exactly the same as the fixed grid case could easily be achieved by setting the thresholds to an extremely low value. For the method to be effective, it must also reduce the number of calculations required to update the fields.

In Fig. 34, the number of the equivalent grid points, normalized to the maximum, at each time step is presented for the $t_r = 1 \times 10^{-4}$, $t_a = 1 \times 10^{-5}$ case. This plot can be used to calculate the total number of grid points required

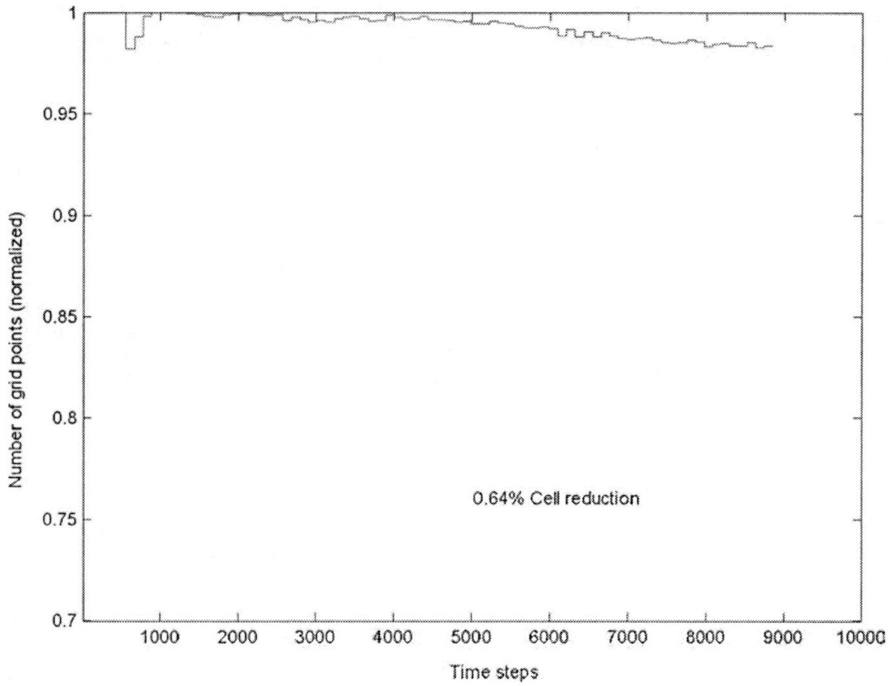

FIGURE 34: Number of gridpoints used in simulation vs. time, $t_r = 1 \times 10^{-4}$ and $t_a = 1 \times 10^{-5}$.

for the simulation. For the case that is presented, the reduction in grid points is only 0.6%. This is extremely low, and is most likely because the structure being simulated occupies most of the grid. The adaptive algorithm will most likely yield higher efficiency in grids with finely detailed structures that are dispersed in a larger grid. In the next section, one of these structures is explored.

5.5 DUAL MICROSTRIP PATCH ANTENNAS

The previous example served to demonstrate the subcell technique as applied to a relatively simple structure and showed the benefit of both fixed subcell grids and adaptive grids. The adaptive gridding method that has been presented allows the grid to be varied as a function of position and provides modest gains for many structures. This technique is even more effective for highly detailed structures with

FIGURE 35: Dual microstrip patch antenna (shaded area is PML).

large homogeneous areas. For this example, the antenna from the previous section is employed. Another similar antenna, this one tuned for 43 GHz, is placed on the same substrate, separated by approximately three wavelengths. An example of a situation where this configuration may be used is in a device that has separate send and receive bands. In this case, it is useful to know the isolation between the antennas and the change in the radiation due to the proximity of the antennas. This structure is presented in Fig. 35.

For this example, the feed for the 31GHz antenna is labeled port one, and the feed for the 43GHz antenna is labeled port two. For this structure, S_{11} and S_{22} are presented in Figs. 36 and 37, respectively. The return loss for the 31 GHz antenna is almost exactly the same as in the case where the antenna was evaluated alone. This can be further explained by examining S_{21}, the coupling between the antennas. In this case, the maximum coupling occurs, as expected, at 43 GHz (the resonant frequency of the second antenna), but it is very low; the isolation between the antennas is 48 dB.

This simulation was evaluated with the same metrics as the previous example. A fixed subcell grid was employed, and the difference between the two simulations was examined. For the fixed grid, high resolution was used to model the antenna, while low resolution was used around the antenna. The antenna was simulated with $r = 1$, and the surrounding area used $r = -1$ (scaling function only). It is important

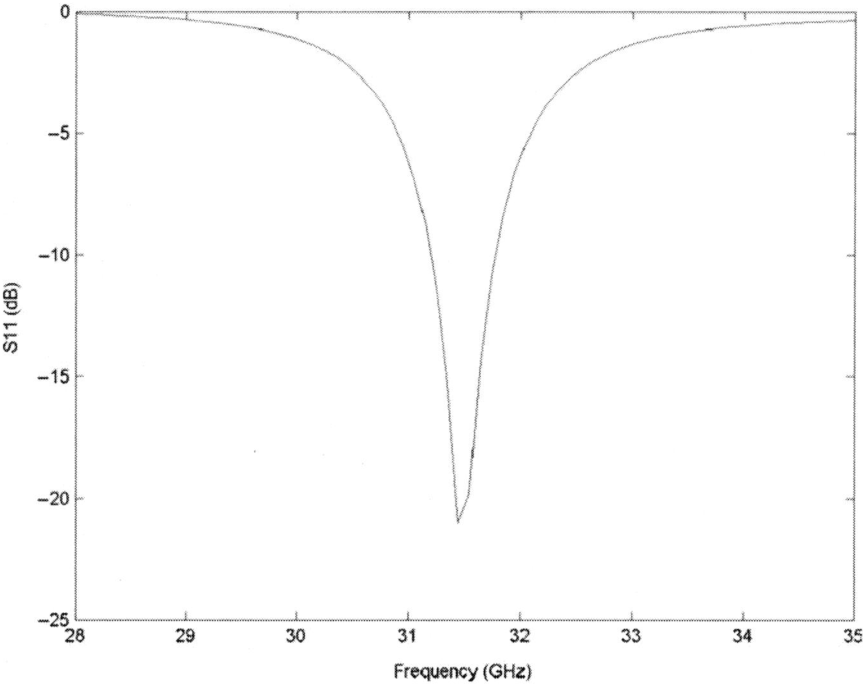

FIGURE 36: S_{11} for dual antennas.

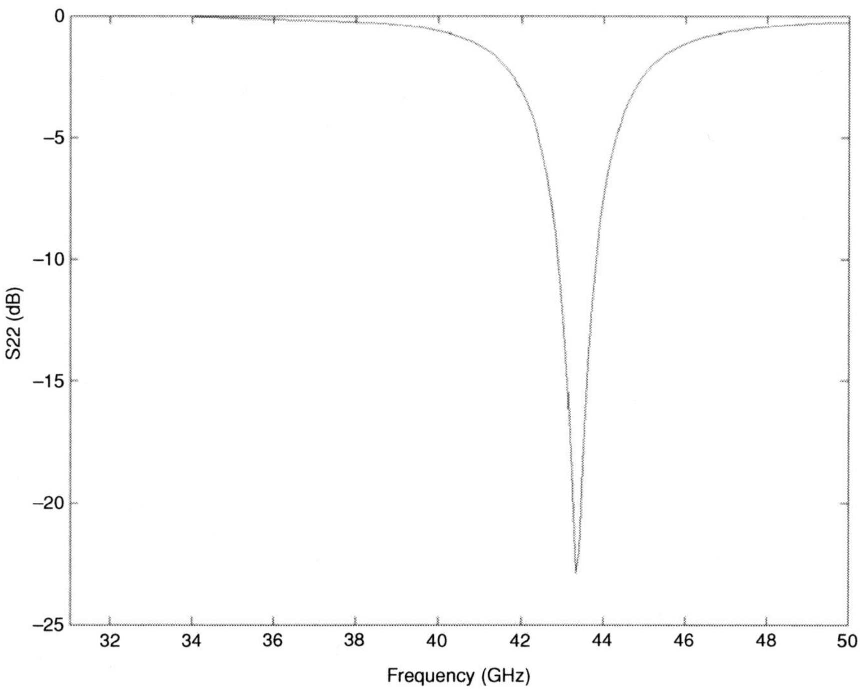

FIGURE 37: S_{22} for dual antennas.

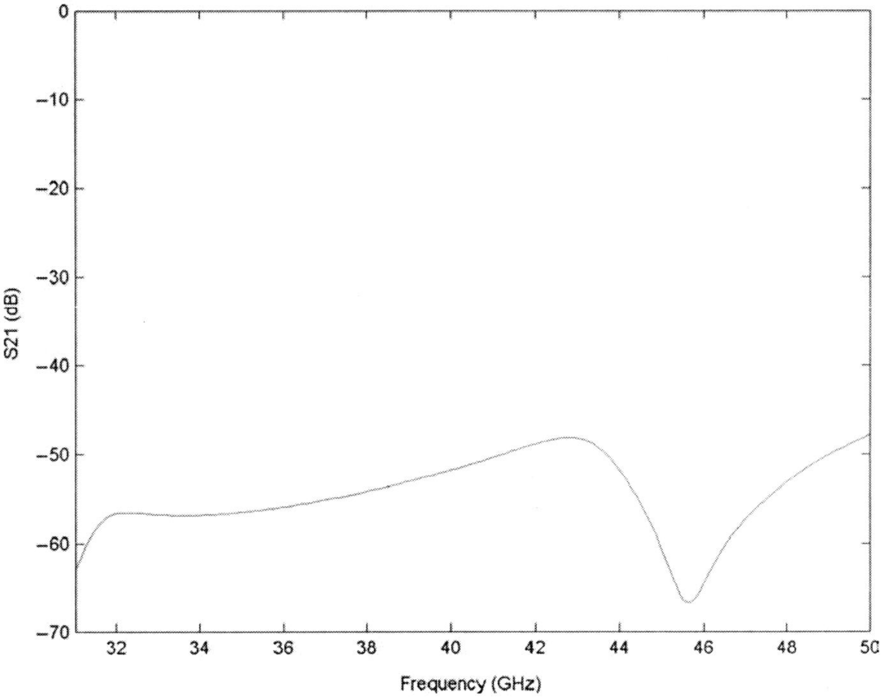

FIGURE 38: S_{21} for dual antennas.

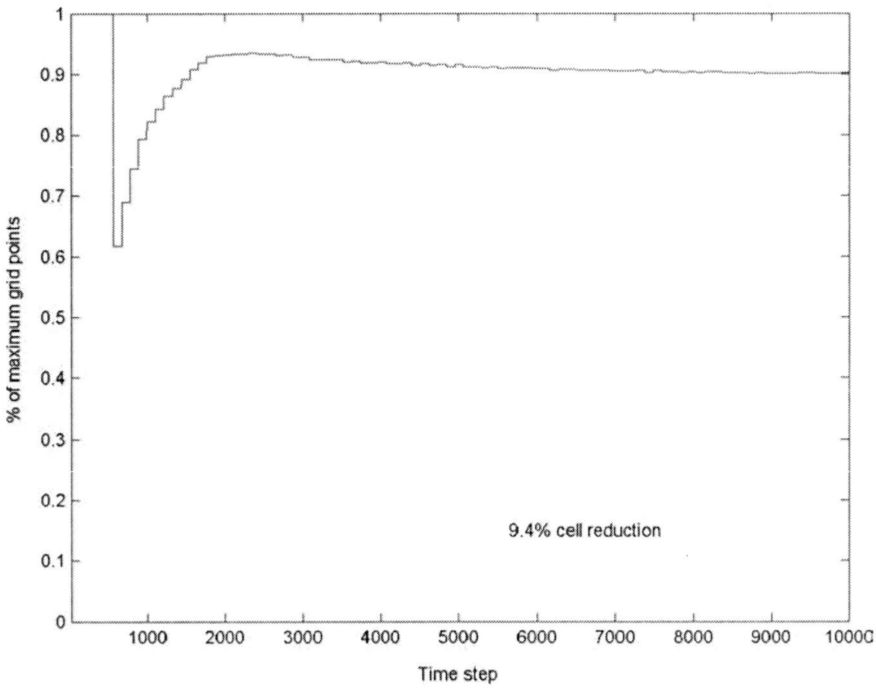

FIGURE 39: Adaptive grid for initial $r = 1$ everywhere.

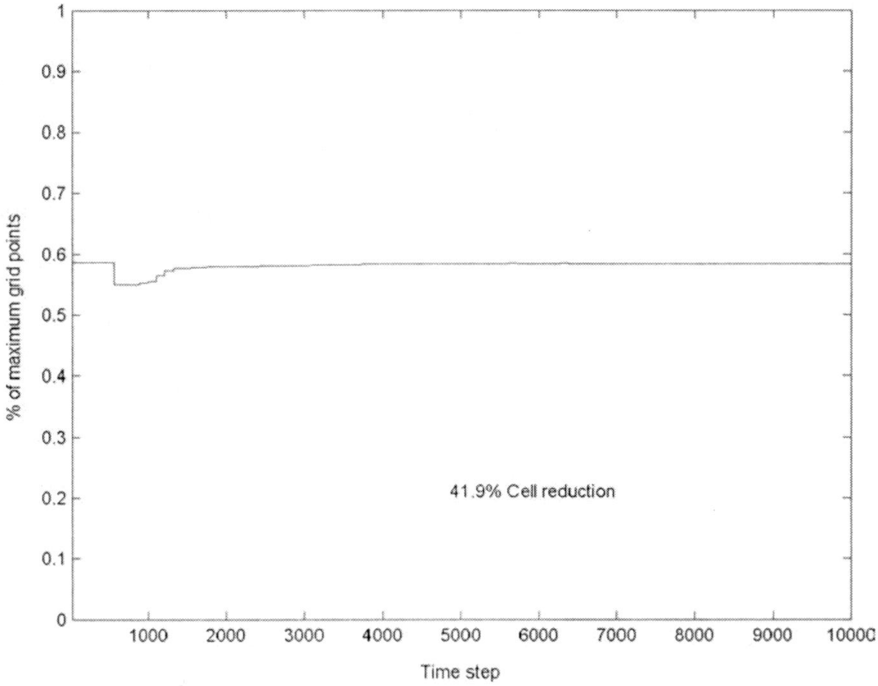

FIGURE 40: Adaptive grid with resolution surrounding antenna fixed at $r = -1$.

to note that even in the $r = -1$ case, there are still more than 12 equivalent grid points per wavelength. In the previous example, it was noted that the fixed subcell grid used less overall grid points than the adaptive grid case (where the resolution at the beginning was $r = 1$ everywhere). In this example, a simulation was run where the adaptive grid was used in the area of the antenna, while the surrounding resolution was still fixed at $r = -1$. The goal of this simulation was to determine what improvement, if any, could be obtained through the application of both the adaptive and fixed subcell grids.

For the fixed subcell grid case, where the resolution between and surrounding the antennas was reduced to -1, the efficiency when compared to the uniform grid case is 41.4%. The two other cases of interest are the adaptive grid for the entire grid and the adaptive grid for the antenna area only (with the surrounding grid at -1 as in the fixed subcell case). For the case where only adaptive gridding is used, there is a 9.4% reduction in cells. This is significantly lower than the reduction using the fixed subcell grid, but it does demonstrate that the adaptive grid can be used to achieve nontrivial grid reduction while maintaining accuracy. In the case where the adaptive grid is combined with the fixed grid, the total cell reduction is 41.9%. This is only 0.6% higher than the fixed grid case, showing that the adaptive grid does not cause much improvement when the grid is already fairly sparse.

References

[1] K. S. Yee, "Numerical solution of initial boundary value problems involving Maxwell's equations in isotropic media," *IEEE Trans. Antennas Propagat.*, vol. AP-14, pp. 302–307, Mar. 1966.

[2] M. Krumpholz and L. P. B. Katehi, "New time domain schemes based on multiresolution analysis," *IEEE Trans. Microwave Theory Tech.*, vol. 44, pp. 555–561, April 1996.

[3] M. Fujii and W. J. R. Hoefer, "A three-dimensional Haar-wavelet-based multiresolution analysis similar to the FDTD method—derivation and application," *IEEE Trans. Microwave Theory Tech.*, vol. 46, pp. 2463–2475, Dec. 1998.

[4] Y. Tretiakov and G. W. Pan, "On Daubechies wavelet based time domain scheme," *Proc. Antennas Prop. Sym.*, vol. 4, pp. 810–813, July 2001.

[5] T. Dogaru and L. Carin, "Multiresolution time-domain using CDF biorthogonal wavelets," *IEEE Trans. Microwave Theory Tech.*, vol. 49, pp. 902–912, May 2001.

[6] I. Daubechies, *10 Lectures on Wavelets*, Society for Industrial and Applied Mathematics, Philadelphia, 1992.

[7] G. Carat, R. Gillard, J. Citerne, and J. Wiart, "An efficient analysis of planar microwave circuits using a DWT-based Haar MRTD scheme," *IEEE Trans. Microwave Theory Tech.*, vol. 48, pp. 2261–2270, Dec. 2000.

[8] T. Dogaru and L. Carin, "Application of Haar-wavelet-based multiresolution time-domain schemes to electromagnetic scattering problems," *IEEE Trans. Antennas Propagat.*, vol. 50, no. 6, pp. 774–784, June 2002.

[9] C. Sarris and L. P. B. Katehi, "Fundamental gridding-related dispersion effects in multiresolution time-domain schemes," *IEEE Trans. Microwave Theory Tech.*, vol. 49, no. 12, pp. 2248–2257, Dec. 2001.

[10] Z. Chen and J. Zhang, "An unconditionally stable 3-D ADI-MRTD method free of the CFL stability condition," *IEEE Microwave Wireless Comp. Lett.*, vol. 11, pp. 349–351, Aug. 2001.

[11] F. Zheng, Z. Chen and J. Zhang, "Toward the development of a three dimensional unconditionally stable finite-difference time-domain method," *IEEE Trans. Microwave Theory Tech.*, vol. 48, pp. 1550–1558, Sept. 2000.

[12] E. Tentzeris, R. L. Robertson, J. F. Harvey, and L. P. B. Katehi, "Stability and dispersion analysis of Battle-Lemarie-Based MRTD schemes," *IEEE Trans. Microwave Theory Tech.*, vol. 47, pp. 1004–1013, July 1999.

[13] A. Taflove and M. E. Brodwin, "Numerical solution for steady-state electromagnetic scattering problems using the time-dependent Maxwell's equations," *IEEE Trans. Microwave Theory Tech.*, vol. 23, pp. 623–630, Aug. 1975.

[14] A. Taflove and S. Hagness, *Computational Electromagnetics, the Finite-Difference Time-Domain Technique*, 2nd ed. Artech House, Boston, 2003.

[15] R. Robertson, E. Tentzeris, M. Krumpholz, and L. P. B. Katehi, "MRTD analysis of dielectric cavity structures," *Proc. IEEE MTT-S*, vol. 3, pp. 1861–1864, June 1996.

[16] E. Tentzeris, M. Krumpholz, and L. P. B. Katehi, "Application of MRTD to printed transmission lines," *Proc. IEEE MTT-S*, vol. 2, pp. 573–576, June 1996.

[17] Q. Cao, Y. Chen, and R. Mittra, "Multiple image technique (MIT) and anistropic perfectly matched layer (APML) in implementation of MRTD scheme for boundary truncations of microwave structures," *IEEE Trans. Microwave Theory Tech.*, vol. 50, pp. 1578–1589, 2002.

[18] N. Bushyager, J. Papapolymeroy, and M. M. Tentzeris, "A composite-cell multiresolution time-domain technique for the design of antenna systems including electromagnetic band gap and via-array structures," *IEEE Trans. Antennas Prop.*, vol. 53, pp. 2700–2710, Aug. 2005.

[19] Y. W. Cheong et al., "Wavelet-Galerkin scheme of time-dependent inhomogeneous electromagnetic problems," *IEEE Microwave Guided Wave Lett.*, vol. 9, pp. 297–299, Aug. 1999.

[20] M. Fujii and W. J. R. Hoefer, "Dispersion of time domain wavelet galerkin method based on daubechies' compactly supported scaling functions with three and four vanishing moments," *IEEE Microwave Guided Wave Lett.*, vol. 10, pp. 125–127, April 2000.

[21] K. R. Umashankar, A. Taflove, and B. Becker, "Calculation and experimental validation of induced currents on coupled wires in an arbitrary shaped cavity," *IEEE Trans. Antennas Propagat.*, vol. 35, pp. 1248–1257, Nov. 1987.

[22] A. Taflove et al., "Detailed FDTD analysis of electromagnetic fields penetrating narrow slots and lapped joints in thick conducting screens," *IEEE Trans. Antennas Propagat.*, vol. 36, pp. 247–257, Feb. 1988.

[23] S. Dey and R. Mittra, "A modified locally-conformal finite-difference time-domain algorithm for modeling three-dimensional perfectly conducting objects," *Microwave Optical Tech. Lett.*, vol. 17, pp. 349–352, April 1998.

[24] J. G. Maloney and G. S. Smith, "The efficient modeling of thein material sheets in the FDTD method," *IEEE Trans. Antennas Propagat.*, vol. 40, pp. 323–330, March 1992.

[25] J. G. Maloney, G. S. Smith, and W. R. Scott, Jr. "Accurate computation of the radiation from simple antennas using the finite-difference time-domain method," *IEEE Trans. Antennas Propagat.*, vol. 38, pp. 1059–1068, July 1990.

[26] J. G. Maloney and G. S. Smith, "The use of surface impedance concepts in the finite-difference time-domain method," *IEEE Trans. Antennas Propagat.*, vol. 40, pp. 38–48, Jan. 1992.

[27] N. Bushyager, E. Dalton, and E. M. Tentzeris, "Modeling of complex RF/wireless structures using computationally optimized time-domain techniques," *Int. J. Num. Modeling*, vol. 17, no. 3, pp. 223–236, April 2004.

[28] M. J. Piket-May, A. Taflove, and J. Baron, "FDTD modeling of digital signal propagation in 3-D circuits with passive and active loads," *IEEE Trans. Microwave Theory Tech.*, vol. 42, pp. 1514–1523, Aug. 1994.

[29] P. Monk and E. Suli, "Error estimates for Yee's method on non-uniform grids," *IEEE Trans. Magnetics*, vol. 30, pp. 3200–3203, Sept. 1994.

[30] E. M. Tentzeris, A. Cangellaris, L. P. B. Katehi, and J. Harvey, "Multiresolution time-domain (MRTD) adaptive schemes using arbitrary resolutions of wavelets," *IEEE Trans. Microwave Theory Tech.*, vol. 50, pp. 501–516, Feb. 2002.

[31] J. Fang and D. Xue, "Numerical errors in the computation of impedances by FDTD method and ways to eliminate them," *IEEE Microwae Guided Wave Lett.*, vol. 5, pp. 6–8, Jan. 1995.

[32] E. Y. Tsai, A. M. Bacon, M. Tentzeris, and J. Papapolymerou, "Design and development of novel micromachined patch antennas for wireless applications" in *Proc. Asian-Pacific Microwave Symposium*, Nov. 2002, pp. 821–824.

Biographies

Dr. Nathan A. Bushyager received a B.S. in Engineering Science from The Pennsylvania State University in 1999 and an M.S. and Ph.D. from The Georgia Institute of Technology in 2003 and 2004, respectively. In addition to this book he has authored a book chapter, six journal papers, and has presented more than 30 conference papers. In 2001, he received the Best Student Paper award at the Applied Computational Electromagnetics Society symposium. Dr. Bushyager is currently a Senior RF Engineer at Northrop Grumman in Linthicum Maryland.

Professor Manos M. Tentzeris received the Diploma Degree in Electrical and Computer Engineering from the National Technical University of Athens ("Magna Cum Laude") in Greece and the M.S. and Ph.D. degrees in Electrical Engineering and Computer Science from the University of Michigan, Ann Arbor, MI and he is currently an Associate Professor with School of ECE, Georgia Tech, Atlanta, GA. He has published more than 200 papers in refereed Journals and Conference Proceedings and 8 book chapters and he is in the process of writing 3 books. Dr. Tentzeris has helped develop academic programs in Highly Integrated/Multilayer Packaging for RF and Wireless Applications, Microwave MEM's, SOP-integrated antennas and Adaptive Numerical Electromagnetics (FDTD, MultiResolution Algorithms) and heads the ATHENA research group (15 researchers). He is the Georgia Tech NSF-Packaging Research Center Associate Director for RF Research and the RF Alliance Leader. He is also the leader of the RFID Research Group of the Georgia Electronic Design Center (GEDC) of the State of Georgia. He was the recipient of the 2006 IEEE MTT Outstanding Young Engineer Award, the 2004 IEEE Transactions on Advanced Packaging Commendable Paper Award, the 2003 NASA Godfrey "Art" Anzic Collaborative Distinguished Publication Award for his activities in the area of finite-ground low-loss low-crosstalk coplanar waveguides, the 2003 IBC International Educator of the Year Award, the 2003 IEEE CPMT Outstanding Young Engineer Award for his work on 3D multilayer integrated RF modules, the 2002 International Conference on Microwave and Millimeter-Wave Technology Best Paper Award (Beijing, CHINA) for his work on Compact/SOP-integrated RF components for low-cost high-performance wireless front-ends, the 2002 Georgia Tech-ECE Outstanding Junior Faculty Award, the 2001 ACES Conference Best Paper Award and the 2000 NSF CAREER Award for his work on the development of MRTD technique that allows for the system-level simulation of RF integrated modules and the 1997 Best Paper Award of the International Hybrid Microelectronics and Packaging Society for the development of design rules for low-crosstalk finite-ground embedded transmission lines. He was also the 1999 Technical Program Co-Chair of the 54th ARFTG Conference, Atlanta, GA and the Chair of the 2005 IEEE CEM-TD Workshop and he is the Vice-Chair of the RF Technical Committee (TC16) of the IEEE CPMT Society.

He has organized various sessions and workshops on RF/Wireless Packaging and Integration in IEEE ECTC, IMS and APS Symposia in all of which he is a member of the Technical Program Committee in the area of "Components and RF". He is the Associate Editor of IEEE Transactions on Advanced Packaging. Dr.Tentzeris was a Visiting Professor with the Technical University of Munich, Germany for the summer of 2002, where he introduced a course in the area of High-Frequency Packaging. He has given more than 40 invited talks in the same area to various universities and companies in Europe, Asia and America. He is a Senior Member of IEEE, a member of URSI-Commission D, an Associate Member of EuMA and a member of the Technical Chamber of Greece.

— Prof. Manos M. Tentzeris
NSF-PRC Associate Director for RF Research School of ECE
Georgia Institute of Technology
Atlanta, GA 30332-250 U.S.A.

Higher Order FDTD Schemes for Waveguide and Antenna Structures

Nikolaos V. Kantartzis and Theodoros D. Tsiboukis
Applied and Computational Electromagnetic Laboratory
Department of Electrical and Computer Engineering
Aristotle University of Thessaloniki
Thessaloniki, Greece

SYNTHESIS LECTURES ON COMPUTATIONAL ELECTROMAGNETICS #3

MORGAN & CLAYPOOL PUBLISHERS

ABSTRACT

This publication provides a comprehensive and systematically organized coverage of higher order *finite-difference time-domain* or FDTD schemes, demonstrating their potential role as a powerful modeling tool in computational electromagnetics. Special emphasis is drawn on the analysis of contemporary waveguide and antenna structures. Acknowledged as a significant breakthrough in the evolution of the original Yee's algorithm, the higher order FDTD operators remain the subject of an ongoing scientific research. Among their indisputable merits, one can distinguish the enhanced levels of accuracy even for coarse grid resolutions, the fast convergence rates, and the adjustable stability. In fact, as the fabrication standards of modern systems get stricter, it is apparent that such properties become very appealing for the accomplishment of elaborate and credible designs.

KEYWORDS

finite-difference time-domain methods, FDTD, computational electromagnetics, Yee's algorithm, waveguide and antenna structure and analysis, electromagnetic modeling,

Contents

Preface

The present book provides a comprehensive and systematically organized coverage of higher order FDTD schemes, demonstrating their potential role as a powerful modeling tool in computational electromagnetics. Special emphasis is drawn on the analysis of contemporary waveguide and antenna structures. Acknowledged as a significant breakthrough in the evolution of the original Yee's algorithm, the higher order FDTD operators remain the subject of an ongoing scientific research. Among their indisputable merits, one can distinguish the enhanced levels of accuracy even for coarse grid resolutions, the fast convergence rates, and the adjustable stability. In fact, as the fabrication standards of modern systems get stricter, it is apparent that such properties become very appealing for the accomplishment of elaborate and credible designs.

In Chapter 1, the book begins with a brief, yet informative, introduction on time-domain algorithms, the most characteristic shortcomings of Yee's technique, and the classification of higher order FDTD schemes in conventional and nonstandard counterparts. The investigation of the former – both explicit and implicit – along with a thorough analysis of interface treatment and dispersion-error optimization is conducted in Chapter 2. Chapter 3 covers the theoretical formulation of higher order nonstandard operators and introduces a generalized curvilinear covariant/contravariant methodology for broadband simulations. The topic of absorbing boundary conditions for open-region termination is carefully examined in Chapter 4, which also presents several techniques for the manipulation of the inevitably widened spatial stencils near dissimilar material interfaces or lattice ends. Chapter 5 discusses diverse structural extensions and temporal integration approaches focusing on lossy, frequency-dependent and dispersion-reduction approaches. Moreover, the hybridization of higher order FDTD forms with other efficient algorithms and the possibility of increasing the order of accuracy of several alternative methods are explored in Chapter 6. Finally, Chapters 7 and 8 are devoted to the numerical modeling of up-to-date waveguides and antennas, respectively. A lot of indicative examples are presented and various complicated configurations are successfully solved. All applications are realistic, while most of the results are compared with measurements or reference data. It is stressed that every chapter closes with an extensive list of references that offer additional evidence and details on the competence of higher order schemes to the more interested reader.

Conclusively, the flavor of the book is the profitable link between the underlying physical background and the computational practice of the higher order FDTD method. Therefore, it is the hope of the authors that it will be contributive as a basis to the researcher pursuing effective substitutes to traditional ideas or serve as a state-of-the-art review for this rapidly growing numerical technique.

Thessaloniki | N. V. Kantartzis
April 2006 | T. D. Tsiboukis

CHAPTER 1
Introduction

1.1 TIME-DOMAIN MODELING IN COMPUTATIONAL ELECTROMAGNETICS

The consistent numerical analysis of complex electromagnetic problems in the time domain is strongly connected to the combined representation of fundamental vector quantities via the appropriate mathematical schemes, usually in the form of partial differential equations. Their precise theoretical formulation along with the problem's intrinsic properties guarantees the well posedness of a simulation and achieves fast algorithmic convergence, if several constraints are satisfied. As opposed to the theory of other quantitative models, the analytical solution of a partial differential equation is rather laborious to be extracted and when accomplished, it rarely helps much in answering questions important for an engineer. Also, from the applications viewpoint, a researcher is primarily interested in data that satisfy certain subsidiary requirements, such as boundary or initial conditions. It is well-known that any continuous function can, under the proper assumptions, be accurately approximated by a set of weighted polynomials. Such an issue would be valid if the approximation of boundary values was always exact. However, this does not hold for the majority of numerical approaches. Furthermore, the correct interpretation of a problem is normally a difficult task, since the initial data are incomplete and discontinuous, indicating the weakness of the particular procedure. Consequently, to obtain acceptable solutions, time-domain methods should be carefully designed.

The necessity to develop such techniques is reflected in the serious scientific research dealing with the pertinent discretization strategies for the construction of the computational lattice. Over the last two decades, the rapid evolution of modern devices has turned this interest into an intensive demand. Therefore, several popular methodologies have been introduced, the most prominent of which is the *finite-difference time-domain (FDTD)* algorithm [1–6]. Due to its ability to model problems in an elegant and simple way, it has become a very powerful solver. Time-domain modeling, however, can be alternatively conducted by other constantly growing schemes such as the finite integration technique (FIT) [7, 8], the transmission line modeling (TLM) [9], the finite-element time-domain (FETD) [10–12], the pseudospectral time-domain (PSTD) [13], and the multiresolution time-domain (MRTD) [14] method. Despite their differences, all of them share an important issue: the establishment of an optimal tessellation that yields accurate and trustworthy results.

1.2 THE FDTD METHOD FOR WAVEGUIDE AND ANTENNA ANALYSIS

1.2.1 Essential Principles and Evolution Aspects

Originally proposed by K. S. Yee in 1966 [1], the FDTD technique has emerged as a foremost numerical tool for the solution of the time-dependent Maxwell's equations in either *differential* or *integral* form. As computing facilities became faster and more widely accessible, its premises found tremendous applications and proved their superiority in radiation and scattering problems. Motivated by these advances, the scientific community introduced numerous enhancements to the initial idea that greatly broadened its applicability.

Amid the most active areas, one can distinguish the study of waveguide and antenna structures [15–24]. Implementing field theory issues and employing no potentials, the FDTD algorithm exploits existing sets of electromagnetic rules that facilitate the calculation of significant quantities, such as S-parameters, radiation patterns, return loss, resonances, gain, or directivity. The advantages of Yee's concepts are *manifold*. First of all, it provides a unified discipline for diverse real-world arrangements and a multitude of phenomena. Spatial and temporal variations are well resolved by the sampling process to limit errors and assure the self-consistency of the meshes. Secondly, by biasing the discretization on the structural properties of the theoretical analysis, the FDTD method provides efficient models that preserve many important attributes of the original configuration. Additionally, the fact that it updates *both* electric and magnetic components (rather than one as in the wave equation) leads to the rigorous treatment of several interactions. Not to mention that the manipulation of tangential or radial singularities near geometric oddities is easier to realize when both fields are available. Owing to its intuitive mathematical notions, the main kernel remains open for ongoing development, something that may be experienced by the large amount of scientific publications. Finally, FDTD schemes handle the unwanted spurious solutions encountered in waveguide problems in a more mature way. Although affected by their appearance, the disruption of the results is not so noticeable. The explicit fulfillment of Maxwell's equations in accordance with the advantageous leapfrog process are some of its features that lead toward this direction. By discretizing the continuous space with the staggered cells, the FDTD approach enables, in most of the cases, the rigorous satisfaction of divergence equations.

1.2.2 Inherent Shortcomings and Efficiency Enhancement

As competent as the FDTD method is, its limitations are not so difficult to identify. Except for its *second-order* accuracy, restraining the electrical size and duration of the problems one can solve, the *staircase* approximation generates some serious drawbacks. These are the dispersion, anisotropy, and dissipation errors.

Numerical *dispersion* is an undesired nonphysical effect inherently present in the FDTD technique. Generally, this artificially created deficiency affects the accurate update of the simulated waves in the computational space by enforcing a *strict* dependence of phase velocity on frequency, propagation direction, and grid discretization. The consequences of dispersion are several and sometimes – along with the problem under investigation – may become prohibitive for the accomplishment of the analysis. Two of them are the cumulative delay or phase errors and the unnatural refraction. According to the former, the inevitable discrepancies from the original speed of light arouse a series of phenomena such as amplification of single-pulse waveforms, generation of spurious modes, and mesh anisotropy. So, if a device is based on phase canceling, even an apparently small error in wave velocity is likely to accumulate to unacceptable levels. In addition, this sort of error is principally responsible for the dislocation of resonances in the frequency domain, an issue that is critical for the performance of Yee's scheme at waveguide applications. Conversely, *anisotropy* errors are present when the cell shape varies over the lattice, namely in curvilinear, nonorthogonal, or unstructured cell complexes. Under these circumstances, a wave experiences a diverse numerical speed in different sections of the grid. Such a shortcoming corresponds to an inhomogeneous medium, which causes refraction to occur. The above remarks indicate that these errors comprise a crucial obstacle in FDTD modeling able to confine the accuracy of the algorithm and if neglected or ineffectively suppressed, it can easily spoil the entire solution. The list of errors is completed by the numerical *dissipation*, which causes unnatural wave dampening. Since its origins are associated with those of dispersion, their properties are comprehensively explored.

The decisive role of the aforementioned mechanisms in the development of the FDTD technique has been the subject of a considerable research [24–39]. As it cannot be totally eliminated (after all it is an *intrinsic* attribute of the discrete state), efforts have concentrated on its mitigation or, given the suitable conditions, on its compensation. Basically, every new idea aims at the improvement of the dispersion relation, governing the propagation of plane waves. In fact, up to this moment, *four* distinct directions have been established for the reduction of numerical dispersion. These are the use of finer grids, the embedding of artificial anisotropy directly in Yee's algorithm, the construction of optimized spatial/temporal operators, and the use of higher order schemes. The notion of the first procedure is quite simple. If the resolution increases, dispersion errors will tend to diminution. It is often convenient to preestimate the effect of this error and choose the proper spatial increment. Unfortunately, a denser lattice is not always viable since it implies excessive memory requirements. Bearing in mind that the problems where dispersion is more prominent are heavily discretized due to their structural details, it is easy to realize the insufficiency of this attempt. Inarguably, the *higher order FDTD formulations* constitute one of the most promising approaches for the drastic suppression of the prior errors. Evidence of this deduction is the ongoing development of efficient contributions that attempt to make these concepts a state-of-the-art technique in computational electromagnetics.

1.3 THE HIGHER ORDER FDTD FORMULATION

1.3.1 Merits and Performance Issues

Higher order time-domain schemes are characterized by their ability to rigorously resolve wave propagation over fairly long distances, involving only a *few* points per wavelength. The enhanced features of these forms are confirmed not only by their application as differential operators inside infinitesimal spaces, but also from their unperturbed validity in macroscopic regions. They are, hence, natural candidates for the construction of spatial approximants and time integrators in the context of the FDTD method. Such a statement, however, requires that the very first stage in the discretization of a problem should be the punctual fulfillment of specific relations. Given the initial profile of higher order abstractions, Maxwell's equations can now be accurately expressed as *algebraic links* between field vectors from which the respective differential counterparts are derived. On the other hand, this framework is compatible with the separation of discrete and continuous analogs of physical terms. A very instructive way to realize the previous remarks is the comprehension of the fundamental difference that distinguishes quantities related to points from those associated with noninfinitesimal regions. Assuming that all update equations are meant to be a function of the latter terms – directly connected to the cells – it is reasonable to prefer higher order schemes for our simulations.

Furthermore, since the above concepts receive a general manipulation rather than a purely spatial one, it would be very effective if one adopted a truly *space-time* approach. For every quantity appearing in the modeling expressions, there must be predicted that the lattice has the appropriate type of cell. Differently speaking, the simple collection of nodes of the traditional finite-difference treatment is proven to be insufficient for this task. Not to mention that, owing to the duality of electric and magnetic fields, two individual ensembles should be considered, namely the primary and the secondary mesh. Such structures might be geometrically coincident – as in the Cartesian FDTD formulation – and only logically different. Observe that as soon as the space-time quantities of the particular problem are properly discerned and the correct strategy has been selected, the type of time integration is *uniquely* determined. This is opposed to other existing approaches that conduct two separate actions for the solution of a problem; i.e., they first discretize the domain in space and then form the set of differential equations that are approximated in the time variable. The topological interpretation of the higher order FDTD method has an additional advantage apart from the already enumerated ones: it reveals the *pitfalls* of the classical staircase outlook. Therefore, through the suitable discretization, arbitrary curvatures and media discontinuities can be adequately represented. As a matter of fact, these properties will be readily clarified in the relevant chapters, where a systematic analysis, accompanied by selected numerical results, is elaborately conducted.

1.3.2 Basic Classification

Depending on the nature of their operators, higher order FDTD algorithms are classified in two categories throughout this book: the *conventional* and the *nonstandard* ones. In the former class, spatial and temporal differentiation is performed by means of approximants *directly* derived from the truncation of Taylor series [40]. Hence, the accuracy order and convergence rate increase, if more terms are considered in the above expansion. The benefit of this abstraction lies on the fact that the resulting forms offer credible simulations with coarser resolutions. Conventional higher order techniques are, mainly, limited to Cartesian coordinates. Furthermore, given that lengthy propagation distances are associated with multiple interactions from metal or dielectric walls, the extraction of stable three-dimensional boundary formulas has a crucial effect on their usage. Besides, their application to the metrics of a nonorthogonal transformation is often mandatory and yields notable speed/storage penalties, as body-fitted discretizations require special care.

On the other hand, the construction of higher order nonstandard schemes arises from the idea of *exact* counterparts, which guarantee that consistency, stability, and convergence issues do not arise. By properly modifying spatial and temporal derivative approximations, via the suitable correction functions, this formulation focuses on the derivation of a *fully isotropic Laplacian*, with an adjustable accuracy order in a wideband sense [41–44]. A major difference with the conventional counterparts is their *extension* to generalized curvilinear coordinates. Moreover for electrically large applications, this translates into greatly suppressed dispersion errors and remarkable reductions in computation resources, allowing simulations of realistic complexity. Hence, in the case of arbitrarily curved media interfaces, not coinciding with the grid axes, or boundaries separating media with dissimilar constitutive parameters, the higher order nonstandard techniques are very efficient, as they circumvent artificial oscillations and spurious vector parasites.

Obviously, both higher order FDTD methodologies sacrifice simplicity at the expense of accuracy. Indeed, they are mathematically more involved than the second-order analogs, due to their extended spatial stencils and the additional terms during derivative approximation. Nevertheless, these concerns are abundantly compensated by their enhanced modeling profile that provides notable capabilities to the researchers. Furthermore, their modular structure encourages hybridizations with other approaches, a fact that may lead to robust tools, powerful and flexible enough to cope with future electromagnetic problems.

REFERENCES

[1] K. S. Yee, "Numerical solution of initial boundary value problems involving Maxwell's equations in isotropic media," *IEEE Trans. Antennas Propag.*, vol. AP-14, no. 3, pp. 302–307, May 1966.

[2] A. Taflove and M. E. Brodwin, "Numerical solution of steady-state electromagnetic scattering problems using the time-dependent Maxwell's equations," *IEEE Trans. Microw. Theory Tech.*, vol. 23, no. 8, pp. 623–630, Aug. 1975.

[3] K. S. Kunz and R. J. Luebbers, *The Finite Difference Time Domain Method for Electromagnetics.* Boca Raton, FL: CRC Press, 1993.

[4] A. Taflove and S. C. Hagness, *Computational Electrodynamics: The Finite-Difference Time-Domain Method*, 3rd ed. Norwood, MA: Artech House, 2005.

[5] A. C. Cangellaris, "Time-domain finite methods for electromagnetic wave propagation and scattering," *IEEE Trans. Magn.*, vol. 27, no. 5, pp. 3780–3785, Sep. 1991.

[6] A. Taflove, Ed., *Advances in Computational Electrodynamics: The Finite-Difference Time-Domain Method.* Norwood, MA: Artech House, 1998.

[7] T. Weiland, "A discretization method for the solution of Maxwell's equations for six-component fields," *Electron. Commun. (AEÜ)*, vol. 31, pp. 116–120, 1977.

[8] T. Weiland, "Time domain electromagnetic field computation with finite difference methods," *Int. J. Numer. Model.*, vol. 9, no. 4, pp. 295–319, July 1996.

[9] C. Christopoulos, *The Transmission-Line Modeling Method: TLM.* Piscataway, NJ: IEEE Press, 1995.

[10] J.-F. Lee, R. Lee, and A. C. Cangellaris, "Time-domain finite-element methods," *IEEE Trans. Antennas Propag.*, vol. 45, no. 3, pp. 430–442, Mar. 1997.

[11] M. Feliziani and F. Maradei, "Mixed finite-difference/Whitney elements time domain (FD/WE-TD) method," *IEEE Trans. Magn.*, vol. 34, no. 5, pp. 3222–3227, Sep. 1998.

[12] D. Jiao, A. A. Ergin, B. Shanker, E. Michielssen, and J.-M. Jin, "A fast higher-order time-domain finite element boundary integral method for 3-D electromagnetic scattering analysis," *IEEE Trans. Antennas Propag.*, vol. 50, pp. 1192–1202, Sep. 2002.

[13] Q. Liu, "The PSTD algorithm: A time-domain method requiring only two cells per wavelength," *Microw. Optical Technol. Lett.*, vol. 15, no. 3, pp. 158–165, 1997.

[14] E. M. Tentzeris, A. C. Cangellaris, L. P. B. Katehi, and J. F. Harvey, "Multiresolution time-domain (MRTD) adaptive schemes using arbitrary resolutions of wavelets," *IEEE Trans. Microw. Theory Tech.*, vol. 50, no. 2, pp. 501–516, Feb. 2002.

[15] C. A. Balanis, *Antenna Theory: Analysis and Design.* New York: IEEE Press and Wiley Interscience, 2005.

[16] R. D. Graglia, R. J. Luebbers, and D. R. Wilton, Eds, *Special Issue on Advanced Numerical*

Techniques in Electromagnetics, IEEE Trans. Antennas Propag., vol. 45, no. 3, pp. 313–572. Mar. 1997.

[17] T. Itoh and B. Houshmand, Eds., *Time-Domain Methods for Microwave Structures: Analysis and Design*. Piscataway, NJ: IEEE Press, 1998.

[18] A. F. Peterson, S. L. Ray, and R. Mittra, *Computational Methods for Electromagnetics*. Piscataway, NJ: IEEE Press/OUP Series, 1998.

[19] J.-F. Lee and R. Lee, Eds., *Special Issue on Computational Electromagnetics, Comput. Methods Appl. Mech. Engrg.*, vol. 169, 1999.

[20] D. H. Werner and R. Mittra, *Frontiers in Electromagnetics*, Piscataway, NJ: IEEE Press, 2000.

[21] D. M. Sullivan, *Electromagnetic Simulation Using the FDTD Method*. Piscataway, NJ: IEEE Press, 2000.

[22] F. L. Teixeira, Ed., *Geometric Methods for Computational Electromagnetics (Progress in Electromagnetics Research Series – PIER 32)*. Cambridge, MA: EMW Publishing, 2001.

[23] W. Sui, *Time-Domain Computer Analysis of Nonlinear Hybrid Systems*. Boca Raton, FL: CRC Press, 2002.

[24] J. S. Hesthaven, "High-order accurate methods in time-domain computational electromagnetics: A review," in *Advances in Imaging and Electron Physics*, P. Hawkes, Ed. New York: Academic Press, 2003, vol. 127, pp. 59–123.

[25] S. L. Ray, "Numerical dispersion and stability characteristics of time-domain methods on nonorthogonal meshes," *IEEE Trans. Antennas Propag.*, vol. 41, no. 2, pp. 233–235, Feb. 1993.

[26] K. L. Shlager, J. G. Maloney, S. L. Ray, and A. F. Peterson, "Relative accuracy of several finite difference time domain methods in two and three-dimensions," *IEEE Trans. Antennas Propag.*, vol. 41, no. 12, pp. 1732–1737, Dec. 1993.

[27] C. K. W. Tam and J. C. Webb, "Dispersion-relation-preserving finite difference schemes for computational acoustics," *J. Comput. Phys.*, vol. 107, no. 2, pp. 262–281, Aug. 1993.

[28] S. Castillo and S. Omick, "Suppression of dispersion in FDTD solutions of Maxwell's equations," *J. Electromagn. Waves Applicat.*, vol. 8, nos. 9–10, pp. 1193–1221, 1994.

[29] Y. Liu, "Fourier analysis of numerical algorithms for Maxwell equations," *J. Comput. Phys.*, vol. 124, pp. 396–406, Mar. 1996.

[30] M. Okoniewski, M. Mrozowski, and M. A. Stuchly, "Simple treatment of multi-term dispersion in FDTD," *IEEE Microw. Guided Wave Lett.*, vol. 7, no. 5, pp. 121–123, May 1997.

[31] J. W. Nehrbass, J. O. Jevtić, and R. Lee, "Reducing the phase error for finite-difference methods without increasing the order," *IEEE Trans. Antennas Propag.*, vol. 46, no. 8, pp. 1194–1201, Aug. 1998.

[32] J. B. Schneider and C. L. Wagner, "FDTD dispersion revisited: Faster-than-light propagation," *IEEE Microw. Guided Wave Lett.*, vol. 9, no. 2, pp. 54–56, Feb. 1999.

[33] J. S. Juntunen and T. D. Tsiboukis, "Reduction of numerical dispersion in FDTD method through artificial anisotropy," *IEEE Trans. Microw. Theory Tech.*, vol. 48, no. 4, pp. 582–588, Apr. 2000.

[34] J. B. Schneider and R. J. Kruhlak, "Dispersion of homogeneous and inhomogeneous waves in the Yee finite difference time-domain grid," *IEEE Trans. Microw. Theory Tech.*, vol. 49, no. 2, pp. 280–287, Feb. 2001.

[35] K. Suzuki, T. Kashiwa, and Y. Hosoya, "Reducing the numerical dispersion in the FDTD analysis by modifying anisotropically the speed of light," *Electron. Commun. Japan (Part II: Electronics)*, vol. 85, no. 1, pp. 50–58, Jan. 2002.

[36] E. A. Forgy and W. C. Chew, "A time-domain method with isotropic dispersion and increased stability on an overlapped lattice," *IEEE Trans. Antennas Propag.*, vol. 50, no. 7, pp. 983–996, July 2002.

[37] K. L. Shlager and J. B. Schneider, "Comparison of the dispersion properties of several low-dispersion finite-difference time-domain algorithms," *IEEE Trans. Antennas Propag.*, vol. 51, no. 3, pp. 642–653, Mar. 2003.

[38] S. Wang and F. L. Teixeira, "Dispersion-relation-preserving FDTD algorithms for large-scale three-dimensional problems," *IEEE Trans. Antennas Propag.*, vol. 51, no. 8, pp. 1818–1828, Aug. 2003.

[39] A. P. Zhao, "Rigorous analysis of the influence of the aspect ratio of Yee's unit cell on the numerical dispersion property of the 2-D and 3-D FDTD methods," *IEEE Trans. Antennas Propag.*, vol. 52, no. 7, pp. 1630–1637, July 2004.

[40] J. Fang, *Time Domain Finite Difference Computation for Maxwell's Equations*, Ph.D. thesis, California Univ., Berkeley, 1989.

[41] R. E. Mickens, *Nonstandard Finite Difference Models of Differential Equations*. Singapore: World Scientific, 1994.

[42] B. Gustafsson, H.-O. Kreiss, and J. Oliger, *Time Dependent Problems and Difference Methods*. New York: Pure and Applied Mathematics, Wiley, 1995.

[43] J. B. Cole, "Application of nonstandard finite differences to solve the wave equation and Maxwell's equations," in *Applications of Nonstandard Finite Difference Schemes*, R. E. Mickens, Ed. Singapore: World Scientific, 2000, ch. 3, pp. 109–154.

[44] N. V. Kantartzis and T. D. Tsiboukis, "A generalised methodology based on higher-order conventional and nonstandard FDTD concepts for the systematic development of enhanced dispersionless wide-angle absorbing perfectly matched layers," *Int. J. Numer. Model.*, vol. 13, no. 5, pp. 417–440, Sep.–Oct. 2000.

CHAPTER 2

Conventional Higher Order FDTD Differentiation

2.1 INTRODUCTION

The consistent analysis of complex electromagnetic field problems via the FDTD method is strongly connected to the accurate spatial and temporal approximation of the corresponding partial derivatives in the discrete computational domain. Despite its great popularity, the second-order algorithm, as occurs with all numerical approaches, suffers from dispersion and anisotropy errors, especially when modeling generally curved or electrically large applications [1, 2]. These artificial lattice defects influence the accurate update of the simulated waves by enforcing a nonphysical relation of phase velocity with frequency, propagation direction, and grid discretization. Observing the strong dependence of dispersion anomalies on spatial sampling, one possible solution to their mitigation, up to an acceptable level, could be a finer mesh resolution. However, in the prior problems that extend to several hundreds of wavelengths and demand prolonged time intervals, full-vector analysis entails inhibiting overheads even for the most powerful computing machines.

An alternative to cell division comes from the higher order FDTD schemes, whose *conventional* version is recognized as an efficient means of achieving satisfactorily lower reflection errors for a given node density. Since their early stages of development [3–12], they received a notable attention and soon became one of the most promising research topics in time-domain computational electromagnetics. Consequently, a wide variety of robust higher order algorithms – both *explicit* and *implicit* – have been, hitherto, proposed. Among them, one can discern rigorous boundary [13–23] and material interface [24–28] formulations, curvilinear [29–31] or conformal [32–34] realizations, stable temporal integration procedures [35–40], hybrid implementations [41–43], and fast approaches in lossy and dispersive media [44–46]. Special attention has also been drawn to low-dispersion methods based on modified stencil structure concepts [47–52], angle-optimized filtered processes [53, 54], dispersion-preserving relations [55, 56], and controllable error estimators [57–59]. Due to their potential expediency, the performance of the most popular higher order schemes has been the subject of a very comprehensive and elaborate study [60] as well as several instructive reviews [23, 61–63].

It is the objective of this chapter to investigate the evolution of these schemes and systematically describe, apart from their key concepts and implementation aspects, various types of arrangements. Starting with some fundamental issues of electromagnetic fields, the basic form of the conventional higher order FDTD algorithm and its generalizations are considered next. Then, analysis continues with the treatment of dielectric or metal boundaries and media interfaces which, in the case of incorrect modeling, lead to serious instabilities and loss of convergence that contaminate the simulation outcomes. Moreover, the critical issue of phase- and dispersion-error reduction is elaborately examined through diverse optimally designed methods, while the chapter closes with the extension of higher order FDTD concepts in orthogonal curvilinear lattices.

2.2 FUNDAMENTALS

Let us suppose a medium in which all electromagnetic field vectors $\mathbf{F}$ are continuous, bounded, single-valued functions of position and time $\mathbf{F} = \mathbf{F}(x, y, z, t)$ and incorporate continuous derivatives as well. Then, the differential and integral form of Maxwell's equations can be expressed as

Faraday's law:

$$\nabla \times \mathbf{E} = -\mathbf{M}_s - \mathbf{M}_c - \frac{\partial \mathbf{B}}{\partial t} \Leftrightarrow \oint_C \mathbf{E} \cdot d\mathbf{l} = -\iint_S \mathbf{M}_s \cdot d\mathbf{S} - \iint_S \mathbf{M}_c \cdot d\mathbf{S} - \frac{\partial}{\partial t} \iint_S \mathbf{B} \cdot d\mathbf{S} \tag{2.1}$$

Ampère's law:

$$\nabla \times \mathbf{H} = \mathbf{J}_s + \mathbf{J}_c + \frac{\partial \mathbf{D}}{\partial t} \Leftrightarrow \oint_C \mathbf{H} \cdot d\mathbf{l} = \iint_S \mathbf{J}_s \cdot d\mathbf{S} + \iint_S \mathbf{J}_c \cdot d\mathbf{S} + \frac{\partial}{\partial t} \iint_S \mathbf{D} \cdot d\mathbf{S} \tag{2.2}$$

Gauss' law (electric field):

$$\nabla \cdot \mathbf{D} = \rho_e \Leftrightarrow \oiint_S \mathbf{D} \cdot d\mathbf{S} = \iiint_V \rho_e \, dV \tag{2.3}$$

Gauss' law (magnetic field):

$$\nabla \cdot \mathbf{B} = \rho_m \Leftrightarrow \oiint_S \mathbf{B} \cdot d\mathbf{S} = \iiint_V \rho_m \, dV \tag{2.4}$$

where

$\mathbf{E}$ is the electric field intensity,
$\mathbf{D}$ the electric flux density,
$\mathbf{J}_s$ the imposed (source) electric current density,
$\mathbf{J}_c$ the conduction (losses) electric current density,
ρ_e the electric charge density (source), and
$\mathbf{B}$ the magnetic flux density,
$\mathbf{H}$ the magnetic field intensity,
$\mathbf{M}_s$ the imposed (source) magnetic current density,
$\mathbf{M}_c$ the conduction (losses) magnetic current density,
ρ_m the magnetic charge density (source).

Furthermore, S is an arbitrary surface (with the infinitesimal surface element vector $d\mathbf{S}$ pointing in the direction of the outward normal unit vector $\hat{\mathbf{n}}$, $d\mathbf{S} = \hat{\mathbf{n}}\, dS$) which, when open, has the closed contour C (with the infinitesimal line element vector $d\mathbf{l}$ pointing in the tangential direction of C) as its boundary, while V is the volume enclosed by the closed surface S [2]. It must be mentioned that the temporal derivatives of electric and magnetic flux densities at the right-hand side of the differential form of (2.1) and (2.2) denote the displacement magnetic and electric current densities, respectively. To complete our analysis, the associated boundary conditions must also be taken into account. The differential form is the most prevalent representation for the treatment of boundary-value electromagnetic field problems. However, the physical meanings of Maxwell's equations are easier to understand if expressed in their integral form. Electromagnetic field vectors are additionally related by a set of equations, which justify the presence of charged particles contained in the material. Such particles interact with the fields, generate currents, and normally modify the propagation of electromagnetic waves in the medium under study as compared to that in free space. These expressions are the constitutive relations

$$\mathbf{D} = \varepsilon\,\mathbf{E} \quad \text{and} \quad \mathbf{B} = \mu\mathbf{H}. \tag{2.5}$$

Finally, the conduction electric and magnetic current densities are related to the corresponding intensities as

$$\mathbf{J}_c = \sigma\mathbf{E} \quad \text{and} \quad \mathbf{M}_c = \sigma^*\mathbf{H}. \tag{2.6}$$

In (2.5) and (2.6), ε, μ, σ, and σ^* are the constitutive parameters that characterize the electric and magnetic properties of the material. Hence, ε is the electric permittivity, μ the magnetic permeability, σ the electric conductivity, and σ^* the equivalent magnetic resistivity. The variation of the constitutive parameters as a function of diverse field characteristics (such as intensity, position, direction, and frequency) leads to their classification according to structure and behavior.

A very convenient form for the time-dependent Maxwell's equations can be derived in terms of *flux tensor notation*. Hence, if in (2.1)–(2.4) $\mathbf{J}_s = \mathbf{M}_s = \mathbf{M}_c = \mathbf{0}$ and $\rho_e = \rho_m = 0$, one gets

$$\frac{\partial \mathbf{U}}{\partial t} + \nabla \cdot \mathbf{W} = \frac{\partial \mathbf{U}}{\partial t} + \frac{\partial \mathbf{Q}_x}{\partial x} + \frac{\partial \mathbf{Q}_y}{\partial y} + \frac{\partial \mathbf{Q}_z}{\partial z} = -\mathbf{S}, \tag{2.7}$$

where the solution vector $\mathbf{U} = [B_x, B_y, B_z, D_x, D_y, D_z]^T$ and $\mathbf{W}$ is the point flux tensor $\mathbf{Q}_x \hat{\mathbf{x}} + \mathbf{Q}_y \hat{\mathbf{y}} + \mathbf{Q}_z \hat{\mathbf{z}}$ in Cartesian coordinates. The rest of the component vectors are given by

$$\mathbf{Q}_x = \begin{bmatrix} 0 \\ -D_z/\varepsilon \\ D_y/\varepsilon \\ 0 \\ B_z/\mu \\ -B_y/\mu \end{bmatrix}, \quad \mathbf{Q}_y = \begin{bmatrix} D_z/\varepsilon \\ 0 \\ -D_x/\varepsilon \\ -B_z/\mu \\ 0 \\ B_x/\mu \end{bmatrix}, \quad \mathbf{Q}_z = \begin{bmatrix} -D_y/\varepsilon \\ D_x/\varepsilon \\ 0 \\ B_y/\mu \\ -B_x/\mu \\ 0 \end{bmatrix}, \quad \mathbf{S} = \begin{bmatrix} 0 \\ 0 \\ 0 \\ \sigma E_x \\ \sigma E_y \\ \sigma E_z \end{bmatrix}. \tag{2.8}$$

On the other hand, the structure of (2.7) in a curvilinear coordinate system is extracted by the necessary transformation, which in its most general formulation launches a bidirectional relation between the two sets of independent temporal and spatial variables. Nonetheless, in the majority of practical applications, the existence of a transformation that couples only the latter variables is assumed to be adequate, namely

$$u = u(x, y, z), \qquad v = v(x, y, z), \qquad w = w(x, y, z). \tag{2.9}$$

This new form is acquired through the division of the successively differentiated equations by the Jacobian of the transformation, $\mathcal{V}$, and the analogous use of the metrical identities. Therefore,

$$\frac{\partial \mathbf{U}}{\partial t} + \nabla \cdot \mathbf{W} = \frac{\partial \mathbf{U}}{\partial t} + \frac{\partial \mathbf{Q}_u}{\partial u} + \frac{\partial \mathbf{Q}_v}{\partial v} + \frac{\partial \mathbf{Q}_w}{\partial w} = -\mathbf{S}, \tag{2.10}$$

in which the transformed variables remain unchanged with the exception of their normalization by $\mathcal{V}$, as

$$\mathbf{U} = \frac{1}{\mathcal{V}} \left[B_x, \; B_y, \; B_z, \; D_x, \; D_y, \; D_z \right]^T. \tag{2.11}$$

The flux tensor components $\mathbf{Q}_u$,$\mathbf{Q}_v$, and $\mathbf{Q}_w$ are the products of the coordinate transformation metrics by the corresponding Cartesian quantities $\mathbf{Q}_u = u_x \mathbf{Q}_x + u_y \mathbf{Q}_y + u_z \mathbf{Q}_z$, $\mathbf{Q}_v = v_x \mathbf{Q}_x + v_y \mathbf{Q}_y + v_z \mathbf{Q}_z$, and $\mathbf{Q}_w = w_x \mathbf{Q}_x + w_y \mathbf{Q}_y + w_z \mathbf{Q}_z$. In particular and after some algebra,

$$\mathcal{Q}_{\zeta} = \begin{bmatrix} 0 & 0 & 0 & 0 & -\mathcal{F}_z^{\zeta} & \mathcal{F}_y^{\zeta} \\ 0 & 0 & 0 & \mathcal{F}_z^{\zeta} & 0 & -\mathcal{F}_x^{\zeta} \\ 0 & 0 & 0 & -\mathcal{F}_y^{\zeta} & \mathcal{F}_x^{\zeta} & 0 \\ 0 & \mathcal{G}_z^{\zeta} & -\mathcal{G}_y^{\zeta} & 0 & 0 & 0 \\ -\mathcal{G}_z^{\zeta} & 0 & \mathcal{G}_x^{\zeta} & 0 & 0 & 0 \\ \mathcal{G}_y^{\zeta} & -\mathcal{G}_x^{\zeta} & 0 & 0 & 0 & 0 \end{bmatrix} \begin{bmatrix} B_x \\ B_y \\ B_z \\ D_x \\ D_y \\ D_z \end{bmatrix}, \quad \text{with} \begin{array}{l} \mathcal{F}_p^{\zeta} = \frac{\zeta_p}{\varepsilon \mathcal{V}} \\ \mathcal{G}_p^{\zeta} = \frac{\zeta_p}{\mu \mathcal{V}} \end{array} \quad \text{and} \begin{array}{l} \zeta = u, v, w \\ p = x, y, z \end{array}. \tag{2.12}$$

It is noteworthy to observe that the mapping between the physical and the transformed space is unique if Jacobian $\mathcal{V}$ is nonzero everywhere in the computational domain.

2.3 DEVELOPMENT OF THE BASIC CONVENTIONAL ALGORITHM

This section describes the main conventional higher order FDTD technique, presented in [3, 4, 8]. Retaining the usual notation, the members of the family are hereafter designated as (N, M), with the numbers in parentheses signifying the formal accuracy of temporal and spatial differentiation, respectively. For example, the simplest and most broadly implemented members are the (2, 4) and (4, 4) schemes.

Let us express Maxwell's equations (2.1) and (2.2) in matrix form as

$$\frac{\partial}{\partial t}\begin{bmatrix} \mathbf{E} \\ \mathbf{H} \end{bmatrix} = \begin{bmatrix} 0 & \varepsilon^{-1} \\ -\mu^{-1} & 0 \end{bmatrix}\begin{bmatrix} \nabla\times & 0 \\ 0 & \nabla\times \end{bmatrix}\begin{bmatrix} \mathbf{E} \\ \mathbf{H} \end{bmatrix} - \begin{bmatrix} \sigma\varepsilon^{-1} & 0 \\ 0 & \sigma^{*}\mu^{-1} \end{bmatrix}\begin{bmatrix} \mathbf{E} \\ \mathbf{H} \end{bmatrix}$$
$$\Rightarrow \frac{\partial \mathbf{U}}{\partial t} = \mathbf{CD}\left[\mathbf{U}\right] - \mathbf{LU}, \tag{2.13}$$

where it has been presumed that $\mathbf{J}_c$ and $\mathbf{M}_c$ are given by (2.6), $\mathbf{J}_s = \mathbf{M}_s = \mathbf{0}$, $\rho_e = \rho_m = 0$, and $\mathbf{D}[.]$ is the matrix curl operator acting on vector $\mathbf{U} = [\mathbf{E}\ \mathbf{H}]$. Moreover, f at a three-dimensional (3-D) FDTD Cartesian lattice is represented by $f(x, y, z, t) = f(i\Delta x, j\Delta y, k\Delta z, n\Delta t) = f|_{i,j,k}^{n}$ with Δx, Δy, Δz, and Δt the spatial and temporal increments, respectively. Expanding vector $\mathbf{U}$ in Taylor series with respect to time, one obtains

$$\left.\frac{\partial \mathbf{U}}{\partial t}\right|_{i,j,k}^{n} = \frac{\mathbf{U}|_{i,j,k}^{n+1/2} - \mathbf{U}|_{i,j,k}^{n-1/2}}{\Delta t} - \frac{(\Delta t)^2}{24}\left.\frac{\partial^3 \mathbf{U}}{\partial t^3}\right|_{i,j,k}^{n} + O\left[(\Delta t)^4\right], \tag{2.14}$$

which controls the accuracy order of temporal differentiation. Indeed, if the terms with the third-order temporal derivatives on the right-hand side of (2.14) are ignored, the $(2, M)$ family of schemes with Mth-order spatial accuracy is extracted. If these terms are considered, the resulting approximation is fourth-order accurate in time; that is, the $(4, M)$ family is acquired.

In the latter case, substitution of (2.14) into (2.13) yields

$$\varepsilon\, \mathbf{E}|_{i,j,k}^{n+1} = \varepsilon\, \mathbf{E}|_{i,j,k}^{n} + \Delta t\ \nabla \times \mathbf{H}|_{i,j,k}^{n+1/2} - \Delta t\sigma\, \mathbf{E}|_{i,j,k}^{n+1/2} + \frac{\varepsilon(\Delta t)^3}{24}\, \frac{\partial^3 \mathbf{E}}{\partial t^3}\bigg|_{i,j,k}^{n+1/2}, \tag{2.15}$$

$$\mu \mathbf{H}|_{i,j,k}^{n+1/2} = \mu \mathbf{H}|_{i,j,k}^{n-1/2} - \Delta t\ \nabla \times \mathbf{E}|_{i,j,k}^{n} - \Delta t\sigma^*\, \mathbf{H}|_{i,j,k}^{n} + \frac{\mu(\Delta t)^3}{24}\, \frac{\partial^3 \mathbf{H}}{\partial t^3}\bigg|_{i,j,k}^{n}. \tag{2.16}$$

As can be observed, the direct calculation of the $\partial^3/\partial t^3$ derivatives is rather laborious because they require extra time levels. To circumvent this difficulty, they are converted into spatial analogs, through repeated differentiation and consecutive use of Maxwell's equations, as

$$\frac{\partial^3}{\partial t^3}\begin{bmatrix} \varepsilon \mathbf{E} \\ \mu \mathbf{H} \end{bmatrix} = \frac{1}{\mu\varepsilon}\nabla \times \nabla^2 \begin{bmatrix} \mathbf{H} \\ -\mathbf{E} \end{bmatrix} - \nabla^2 \begin{bmatrix} \frac{1}{\mu}\left(\frac{\sigma^*}{\mu} + 2\frac{\sigma}{\varepsilon}\right)\mathbf{E} \\ \frac{1}{\varepsilon}\left(\frac{\sigma}{\varepsilon} + 2\frac{\sigma^*}{\mu}\right)\mathbf{H} \end{bmatrix}$$
$$- \left(\frac{\sigma}{\varepsilon} + \frac{\sigma^*}{\mu}\right)\nabla \times \begin{bmatrix} -\frac{\sigma^*}{\mu}\mathbf{H} \\ \frac{\sigma}{\varepsilon}\mathbf{E} \end{bmatrix} + \frac{\partial}{\partial t}\begin{bmatrix} \frac{\sigma^2}{\varepsilon}\mathbf{E} \\ \frac{(\sigma^*)^2}{\mu}\mathbf{H} \end{bmatrix}. \tag{2.17}$$

It is important to state that the derivation of (2.17) requires the medium to be *locally* homogeneous; i.e., it is assumed that all constitutive parameters are *spatially independent*. The small-valued $(\Delta t)^3/24$ term in (2.15) and (2.16) permits further simplifications. Since all the derivatives encountered on the right-hand side of (2.17) are multiplied by it, they can be amply discretized by the common second-order finite differences. Therefore, one avoids ambiguous mathematical complexities and at the same time manages to construct a significantly improved FDTD approach, presenting lower phase and propagation errors.

To attain the fourth-order spatial discretization of (2.15) and (2.16), the following central finite-difference schemes are employed. The former is the Yee's *staggered-grid* arrangement, while the latter is the *collocated-mesh* configuration, where component f and its derivatives are positioned at the same node. So,

$$\textit{Staggered grid:}\ \frac{\partial f}{\partial \zeta}\bigg|_{\mathrm{ps}}^{t} \cong \frac{\partial f}{\partial \zeta}\bigg|_{l}^{n} = \frac{9}{8}\left(\frac{f|_{l+1/2}^{n} - f|_{l-1/2}^{n}}{\Delta\zeta}\right) - \frac{1}{24}\left(\frac{f|_{l+3/2}^{n} - f|_{l-3/2}^{n}}{\Delta\zeta}\right) \tag{2.18}$$

$$\textit{Collocated grid:}\ \frac{\partial f}{\partial \zeta}\bigg|_{\mathrm{ps}}^{t} \cong \frac{\partial f}{\partial \zeta}\bigg|_{l}^{n} = \frac{2}{3}\left(\frac{f|_{l+1}^{n} - f|_{l-1}^{n}}{\Delta\zeta}\right) - \frac{1}{12}\left(\frac{f|_{l+2}^{n} - f|_{l-2}^{n}}{\Delta\zeta}\right) \tag{2.19}$$

which calculates the ζ-derivative of f at point ps, with a stencil of $l\Delta\zeta$ toward the direction of differentiation (the only displayed), for $\zeta \in (x, y, z)$ and $l \in (i, j, k)$. This computation involves field values at symmetrically located nodes around point ps on a straight line parallel to ζ-axis that crosses the point of interest. Increasing the order of spatial approximation for the (2, M) or (4, M) schemes, the general expression of (2.18) for a spatial operator that spans in $M-1$

cells and takes into account M nodes, is given by

$$\left.\frac{\partial f}{\partial \zeta}\right|_{\mathrm{ps}}^{t} \simeq \left.\frac{\partial f}{\partial \zeta}\right|_{l}^{n} = \frac{1}{\Delta\zeta}\sum_{s=1}^{M/2} C_s \left[f\big|_{l+(2s-1)/2}^{n} - f\big|_{l-(2s-1)/2}^{n} \right]. \tag{2.20}$$

Coefficients C_s are determined via the Taylor expansion of the right-hand side of (2.20) and the requirement for minimum truncation error. Application of this process results in the $M/2 \times M/2$ system

$$\begin{bmatrix} 1 & 3 & \cdots & M-1 \\ 1 & 3^3 & \cdots & (M-1)^3 \\ \vdots & \vdots & \ddots & \vdots \\ 1 & 3^{M-1} & \cdots & (M-1)^{M-1} \end{bmatrix} \begin{bmatrix} C_1 \\ C_2 \\ \vdots \\ C_{M/2} \end{bmatrix} = \begin{bmatrix} 1 \\ 0 \\ \vdots \\ 0 \end{bmatrix} \tag{2.21}$$

a closed-form solution of which can be written as

$$C_s = \frac{(-1)^{s+1}\,(M-1)!!^2}{2^{M-2}\left(\frac{1}{2}M+s-1\right)!\left(\frac{1}{2}M-s\right)!\,(2s-1)^2} \tag{2.22}$$

with $N!! = N(N-2)(N-4)\ldots$. The order of truncation error

$$\frac{\Delta^M}{2^M(M+1)!}\sum_{s=1}^{M/2} C_s\,(2s-1)^{M+1}\, f^{(M+1)}\big|_{l}^{n} \tag{2.23}$$

specifies the order of spatial approximation, whose maximum value depends on operator spanning. Table 2.1 provides the most frequently used coefficients C_s, also found in the comparative study of [60], along with the corresponding truncation errors up to tenth order. A useful indicator regarding the performance of (2.20) is the dispersion relation that demonstrates the variation of wavenumber versus frequency. Hence, for the (2, M) schemes, by presuming plane-wave

TABLE 2.1: Explicit Central-Difference Spatial Approximations of Various Orders

ORDER M	C_1	C_2	C_3	C_4	C_5	TRUNCATION ERROR
2	1	0	0	0	0	$-\frac{(\Delta\zeta)^2}{24} f^{(3)}\big\|_l^n$
4	$\frac{9}{8}$	$-\frac{1}{24}$	0	0	0	$\frac{3(\Delta\zeta)^4}{640} f^{(5)}\big\|_l^n$
6	$\frac{75}{64}$	$-\frac{25}{384}$	$\frac{3}{640}$	0	0	$-\frac{5(\Delta\zeta)^6}{7168} f^{(7)}\big\|_l^n$
8	$\frac{1225}{1024}$	$-\frac{245}{3072}$	$\frac{49}{5120}$	$-\frac{5}{7168}$	0	$\frac{35(\Delta\zeta)^8}{294912} f^{(9)}\big\|_l^n$
10	$\frac{19845}{16384}$	$-\frac{735}{8192}$	$\frac{567}{40960}$	$-\frac{405}{229376}$	$\frac{35}{294912}$	$-\frac{63(\Delta\zeta)^{10}}{2883584} f^{(11)}\big\|_l^n$

propagation in a 3-D Cartesian space, one derives

$$\left[\frac{\sin(\omega\Delta t/2)}{\upsilon\Delta t}\right]^2 = \sum_{\zeta=x,y,z}\frac{1}{(\Delta\zeta)^2}\left[\sum_{s=1}^{M/2} C_s \sin\left(\frac{2s-1}{2}k_\zeta^{\text{num}}\Delta\zeta\right)\right]^2, \qquad (2.24)$$

where $\upsilon = 1/\sqrt{\mu\varepsilon}$ and $\mathbf{k}^{\text{num}} = k_x^{\text{num}}\hat{\mathbf{x}} + k_y^{\text{num}}\hat{\mathbf{y}} + k_z^{\text{num}}\hat{\mathbf{z}}$ is the numerical wavenumber.

Moreover, through the von Neumann method along with the use of discrete Fourier modes [1], the stability condition for any member of the (2, *M*) family is extracted. For instance, the (2, 4) case has

$$\upsilon\Delta t \leq \frac{6}{7}\left[\sqrt{\frac{1}{(\Delta x)^2}+\frac{1}{(\Delta y)^2}+\frac{1}{(\Delta z)^2}}\right]^{-1}, \qquad (2.25)$$

indicating that the solution attained by the prior discretization will converge to the exact one as increments $\Delta\zeta$ become smaller. However, since the orders of spatial and temporal approximation are different, the (2, *M*) schemes are not anticipated to demonstrate their best accuracy, when Δt is equal or, at least, very close to the maximum one. Moreover, even in the case of a mesh refinement process, their convergence rate will not exceed that of the common Yee's algorithm unless the time-step becomes adequately small, so that the error of temporal integration is not dominant.

To indicate the advantages of the conventional (2, *M*) FDTD method, Figure 2.1(a) presents the relative error of the normalized phase velocity versus spatial frequency $k\Delta h$, where Δh corresponds to a uniform grid. Evidently, the second-order operator has a limited spectral

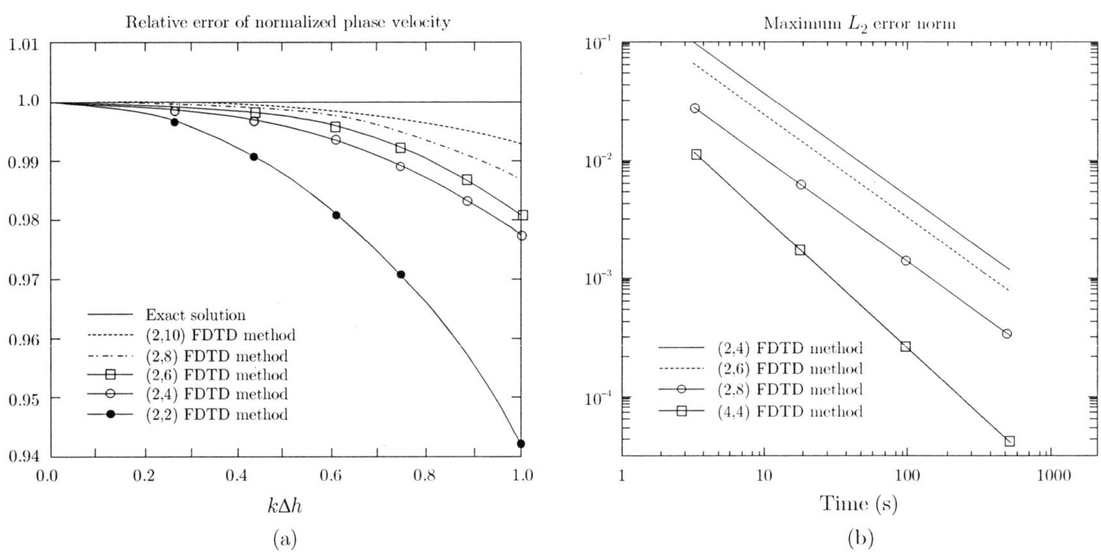

FIGURE 2.1: (a) Relative error of the normalized phase velocity and (b) maximum L_2 error norm for various conventional higher order FDTD schemes

bandwidth when compared to its higher order analogs. Furthermore, the behavior of the L_2 (global error) norm with respect to time is fairly sufficient as shown in Figure 2.1(b), which proves the relatively high convergence rate of the (2, *M*) constituents.

From the analysis of this section, it is straightforward to identify several reasons for the successful application of the above higher order FDTD schemes, but their most appealing attribute is perhaps their *increased accuracy*. Also, the use of staggered lattices enhances the correct handling of several geometries and concurrently preserves the divergence of initial conditions in homogeneous areas. However, the weaknesses of these schemes are equally easy to recognize. In particular, their inherent discretization errors *still* leave some unresolved issues, regarding radiation boundary conditions and material discontinuities, which obstruct their applicability to several practical problems. As prolonged propagation distances are normally associated with multiple interactions with media interfaces, the specific schemes may have a considerable influence on the efficiency of the simulation. Actually, these defects have triggered the development of alternative techniques with improved capabilities, low dispersion errors or expansions to other coordinate systems.

2.4 HIGHER ORDER FDTD MODELING OF BOUNDARIES AND MATERIAL INTERFACES

The principal intention of a consistent numerical boundary approach for the manipulation of demanding media interfaces is the satisfaction of the proper physical jump conditions (far field, impedance, or metal). Their role is deemed crucial, since they can guarantee the stability of the time-domain simulation and assure the desired order of convergence over the entire domain. Evidence of this is the evolution of various state-of-the-art algorithms, whose foremost properties and characteristics are described in the following paragraphs.

2.4.1 Maximum Accuracy and Optimized Compact Schemes

This robust higher order finite-difference method, originally presented in [10, 13, 25], develops a *seven-point* spatial operator along with an explicit *six-stage* time-advancing technique of the Runge–Kutta form. For the former operator, two central-difference suboperators are required: a) an antisymmetric

$$\delta_\zeta^{\mathrm{A}}\left[f|_l^n\right] = \frac{1}{\Delta\zeta}\left[p_1\left(f|_{l+1}^n - f|_{l-1}^n\right) + p_2\left(f|_{l+2}^n - f|_{l-2}^n\right) + p_3\left(f|_{l+3}^n - f|_{l-3}^n\right)\right], \quad (2.26)$$

and b) a symmetric one, used to add a small amount of numerical dissipation,

$$\delta_\zeta^{\mathrm{S}}\left[f|_l^n\right] = \frac{1}{\Delta\zeta}\left[q_0 f|_l^n + q_1\left(f|_{l+1}^n + f|_{l-1}^n\right) + q_2\left(f|_{l+2}^n + f|_{l-2}^n\right) + q_3\left(f|_{l+3}^n + f|_{l-3}^n\right)\right], \quad (2.27)$$

for $\zeta = x, y, z$. As can be observed, (2.26) has a maximum formal order that fluctuates as $(\Delta\zeta)^6$ when $p_1 = \frac{3}{4}$, $p_2 = -\frac{3}{20}$, and $p_3 = \frac{1}{60}$. Furthermore, the maximum formal order of (2.27) varies as $(\Delta\zeta)^5$ for $q_0 = \frac{1}{10}$, $q_1 = -\frac{3}{4}q_0$, $q_2 = \frac{3}{10}q_0$, and $q_3 = -\frac{1}{20}q_0$. If the preceding concepts are applied to a hyperbolic system of equations, like the Maxwell's one, a certain splitting algorithm is employed. For example in the case of $\frac{\partial \mathbf{U}}{\partial t} + \mathbf{R}\frac{\partial \mathbf{U}}{\partial x} = 0$, written according to (2.10), the term $\mathbf{R}\frac{\partial \mathbf{U}}{\partial t}$ is approximated by

$$\mathbf{R}\left.\frac{\partial \mathbf{U}}{\partial x}\right|_l = \mathbf{R}\delta_x^{\mathrm{A}}\left[\mathbf{U}|_l^n\right] + |\mathbf{R}|\,\delta_x^{\mathrm{S}}\left[\mathbf{U}|_l^n\right], \quad \text{where } |\mathbf{R}| = \mathbf{K}\,|\mathbf{A}|\,\mathbf{K}^{-1} \tag{2.28}$$

with $\mathbf{K}$ the matrix of right eigenvectors and $\mathbf{A}$ the eigenvalue matrix of $\mathbf{R}$. In a similar manner and without any additional modifications, the technique treats derivatives along y- and z-axis.

On the other hand, the temporal update of differential equation $d\mathbf{U}/dt = f(u,t)$ is conducted in terms of

$$\begin{aligned}
u_1\big|_l^{n+\tau_1} &= u\big|_l^n + \tau_1 \Delta t f\big|_l^n, & u_2\big|_l^{n+\tau_2} &= u\big|_l^n + \tau_2 \Delta t f_1\big|_l^{n+\tau_1},\\
u_3\big|_l^{n+\tau_3} &= u\big|_l^n + \tau_3 \Delta t f_2\big|_l^{n+\tau_2}, & u_4\big|_l^{n+\tau_4} &= u\big|_l^n + \tau_4 \Delta t f_3\big|_l^{n+\tau_3},\\
u_5\big|_l^{n+\tau_5} &= u\big|_l^n + \tau_5 \Delta t f_4\big|_l^{n+\tau_4}, & u\big|_l^{n+1} &= u\big|_l^n + \Delta t f_5\big|_l^{n+\tau_5},
\end{aligned} \tag{2.29}$$

for $f_m|_l^{n+\tau} = f\left(u_m|_l^{n+\tau}, (n+\tau)\Delta t\right)$. To obtain the sixth-order accuracy for our time integration strategy, parameters τ_i $(i = 1, 2, \ldots, 5)$ are selected as $\tau_1 = \frac{1}{6}$, $\tau_2 = \frac{1}{5}$, $\tau_3 = \frac{1}{4}$, $\tau_4 = \frac{1}{3}$, and $\tau_5 = \frac{1}{2}$. For the stability of (2.29), the Lax–Richtmyer analysis is utilized in order to yield the correct condition for the system's amplification matrix $\mathbf{G}$. So, for a fully discrete finite-difference scheme applied to hyperbolic problems, the necessary and sufficient condition is $\left\|(\mathbf{G}(\Delta\zeta, \Delta t))^n\right\| \leq \vartheta$ for all $n \geq 0$ up to a preset upper time limit.

Proceeding to the manipulation of interfaces that have dissimilar dielectric properties, each region is modeled as a *separate domain* and the outcomes are coupled via the pertinent continuity conditions. This idea enables the distinct generation of diverse grids with the desired resolution. More specifically, at the interface between two dielectric media or a perfect electric conducting (PEC) boundary, a local flux-vector formulation of Maxwell's equations is considered. Therefore, as discussed in [25] for incoming waves, derivatives are approximated by

$$\begin{aligned}
\delta_\zeta\left[f|_{\text{boundary}}^n\right] = \frac{1}{60\Delta\zeta}&\left[-3f|_0^n - 119f|_1^n + 255f|_2^n - 240f|_3^n\right.\\
&\left.+155f|_4^n - 57f|_5^n + 9f|_6^n\right],
\end{aligned} \tag{2.30a}$$

$$\begin{aligned}
\delta_\zeta\left[f|_{\text{boundary}}^n\right] = \frac{1}{60\Delta\zeta}&\left[9f|_0^n - 66f|_1^n + 70f|_2^n - 60f|_3^n\right.\\
&\left.+75f|_4^n - 34f|_5^n + 6f|_6^n\right],
\end{aligned} \tag{2.30b}$$

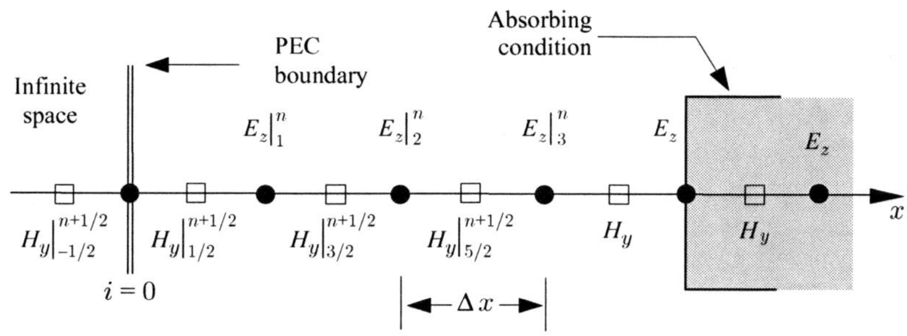

FIGURE 2.2: Treatment of E_z components at a lattice terminated by a PEC boundary and an absorbing boundary condition

while for outgoing waves, the difference operators for the last three points in the mesh are formed through a regular fifth-order space extrapolation in conjunction with the incoming differencing practice.

Another effective means to confine the stencil size in the case of PEC boundaries are the *central compact operators* [5, 7], implemented for the higher order FDTD method in [14], and expressed as

$$a_2 \left.\frac{\partial f}{\partial x}\right|_{i+1}^{n} + a_1 \left.\frac{\partial f}{\partial x}\right|_{i}^{n} + a_2 \left.\frac{\partial f}{\partial x}\right|_{i-1}^{n} = b\frac{f|_{i+1/2}^{n} - f|_{i-1/2}^{n}}{\Delta x}. \tag{2.31}$$

Coefficients a_1, a_2 and b are obtained by the Fourier analysis and the relatively rapid solution of the resulting tridiagonal system of equations, due to the implicit nature of (2.31). A typical set is $a_1 = 22$, $a_2 = 1$, and $b = 24$. To comprehend their function, let us observe Figure 2.2 that assumes the computation of $\partial H_y/\partial x$ and $\partial E_z/\partial x$ at $i = 0$. For the first case, constraint $E_y = E_z = H_x = 0$ at $i = 0$ indicates that $\partial H_y/\partial x$ (likewise for all $\mathbf{H}$ derivatives) must also be zero. In the second case, to calculate $\partial E_z/\partial x$ at $i = \frac{1}{2}$, one needs its values at $i = -\frac{1}{2}, \frac{1}{2}, \frac{3}{2}$. Nonetheless, point $i = -\frac{1}{2}$ is outside the domain and to find a reliable value for the tridiagonal matrix, the explicit, sixth-order central-difference scheme is selected

$$\begin{aligned}\left.\frac{\partial f}{\partial x}\right|_{i+1/2}^{n} = \frac{1}{1920\Delta x}\Big[&-1627f|_{i}^{n} + 633f|_{i+1}^{n} + 2360f|_{i+2}^{n} - 2350f|_{i+3}^{n} \\ &+1365f|_{i+4}^{n} - 443f|_{i+5}^{n} + 62f|_{i+6}^{n}\Big].\end{aligned} \tag{2.32}$$

The properties of techniques (2.26)–(2.29), and (2.31) are, indeed, quite instructive. Their amplitude errors are less than the phase ones, for any wavenumber. Moreover, time integration generates drastically smaller errors than the spatial discretization, and so allowing the phase discrepancies to be independent of the Courant limit. The above schemes are able to perform even when propagation direction is not aligned with the mesh axes.

2.4.2 Implicit Methods for Complex Cartesian Domains

The basic operator for the construction of these competent *implicit* finite-difference schemes has been proposed in [15]. Their staggered form, for $\zeta = x, y, z$, is denoted as

$$\frac{1}{24}\left(\delta_\zeta\left[f|_{i+1}^n\right]+\delta_\zeta\left[f|_{i-1}^n\right]\right)+\frac{11}{12}\delta_\zeta\left[f|_i^n\right]=\frac{1}{\Delta\zeta}\left(f|_{i+1/2}^n-f|_{i-1/2}^n\right). \tag{2.33}$$

Hence, spatial derivatives are approximated to fourth-order through the matrix arrangement of

$$\mathcal{A}=\frac{1}{24}\begin{bmatrix}26 & -5 & 4 & -1 & 0 & \dots & 0\\ 1 & 22 & 1 & 0 & 0 & \dots & 0\\ 0 & 1 & 22 & 1 & 0 & \dots & 0\\ \vdots & \vdots & \ddots & \ddots & \ddots & \vdots & \vdots\\ 0 & 0 & \dots & 1 & 22 & 1 & 0\\ 0 & 0 & \dots & 0 & 1 & 22 & 1\\ 0 & 0 & \dots & -1 & 4 & -5 & 26\end{bmatrix}$$
$$\mathcal{A}\frac{\partial}{\partial\zeta}\begin{bmatrix}f|_{1/2}^n\\ f|_{3/2}^n\\ \vdots\\ f|_{L-3/2}^n\\ f|_{L-1/2}^n\end{bmatrix}=\frac{1}{\Delta\zeta}\left(\begin{bmatrix}f|_1^n\\ f|_2^n\\ \vdots\\ f|_{L-1}^n\\ f|_L^n\end{bmatrix}-\begin{bmatrix}f|_0^n\\ f|_1^n\\ \vdots\\ f|_{L-2}^n\\ f|_{L-1}^n\end{bmatrix}\right) \tag{2.34}$$

in which L is the size of the domain. It is mentioned that the only difference between the FDTD algorithm and (2.34) is the multiplication by $\mathcal{A}^{-1}$. Actually, this matrix is never calculated but *decomposed* as $\mathcal{A} = \mathcal{LU}$ along every direction, with $\mathcal{L}$ and $\mathcal{U}$ being bidiagonal matrices. The preceding method is, also, second-order accurate in time and enables the selection of coarser grids without any loss of convergence.

However at the boundaries, (2.34) should be modified in order to account for nonuniform spacings and evade erroneous instabilities [19, 24]. Let us consider the one-dimensional (1-D) grid of Figure 2.3(a) with components H_y and E_z. Extension to 2- and 3-D problems follows

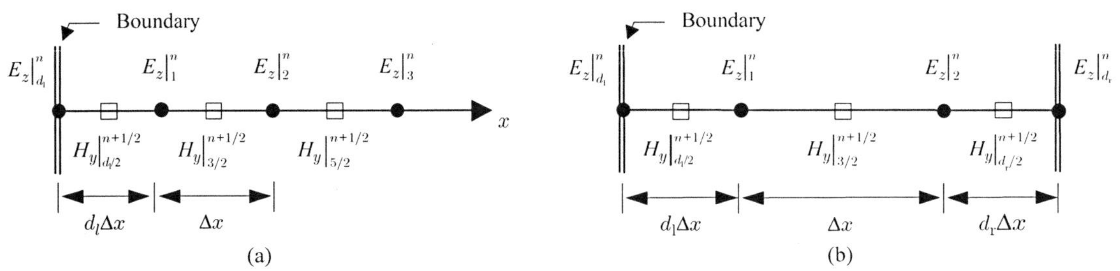

FIGURE 2.3: Treatment of a general lattice for (a) five nodes or more and (b) four nodes

the same process. Since the boundary points may not coincide with lattice nodes, the focal idea is to use H_y at $i = d_l/2$ instead of H_y at $i = 1/2$, where $d_l \Delta x$ is the distance between E_z at $i = 1$ and E_z at $i = d_l$. So, (2.34) becomes

$$\mathcal{A}_b^{\mathrm{E}} = \begin{bmatrix} a_{\mathrm{l}}^{\mathrm{E}} & 0 & \dots & \dots & \dots & \dots & 0 \\ a_{\mathrm{l1}}^{\mathrm{E}} & b_{\mathrm{l1}}^{\mathrm{E}} & c_{\mathrm{l1}}^{\mathrm{E}} & 0 & \dots & \dots & 0 \\ 0 & 1 & 22 & 1 & 0 & \dots & 0 \\ \vdots & \vdots & \ddots & \ddots & \ddots & \vdots & \vdots \\ 0 & \dots & 0 & 1 & 22 & 1 & 0 \\ 0 & \dots & \dots & 0 & c_{\mathrm{r1}}^{\mathrm{E}} & b_{\mathrm{r1}}^{\mathrm{E}} & a_{\mathrm{r1}}^{\mathrm{E}} \\ 0 & \dots & \dots & \dots & \dots & 0 & a_{\mathrm{r}}^{\mathrm{E}} \end{bmatrix} \quad \mathcal{A}_b^{\mathrm{H}} = \begin{bmatrix} a_{\mathrm{l}}^{\mathrm{H}} & b_{\mathrm{l}}^{\mathrm{H}} & 0 & \dots & \dots & \dots & 0 \\ 1 & 22 & 1 & 0 & \dots & \dots & 0 \\ 0 & 1 & 22 & 1 & 0 & \dots & 0 \\ \vdots & \vdots & \ddots & \ddots & \ddots & \vdots & \vdots \\ 0 & \dots & 0 & 1 & 22 & 1 & 0 \\ 0 & \dots & \dots & 0 & 1 & 22 & 1 \\ 0 & \dots & \dots & \dots & 0 & b_{\mathrm{r}}^{\mathrm{H}} & a_{\mathrm{r}}^{\mathrm{H}} \end{bmatrix}, \tag{2.35}$$

with supscripts "l" and "r" representing the left and right boundary of the domain, respectively. In addition and by means of the appropriate Taylor expansions, one gets

$$a_s^{\mathrm{E}} = 24 d_s, \quad a_{s1}^{\mathrm{E}} = \frac{8}{(1+d_s)(3+d_s)}, \quad b_{s1}^{\mathrm{E}} = \frac{4(5+6d_s)}{1+d_s}, \quad c_{s1}^{\mathrm{E}} = \frac{4}{3+d_s}, \tag{2.36}$$

$$a_s^{\mathrm{H}} = 3(3 + 4d_s + d_s^2), \qquad b_s^{\mathrm{H}} = 3(1 - d_s), \quad \text{for } s = \mathrm{r}, \mathrm{l}, \tag{2.37}$$

while the matrix configurations of (2.35)–(2.37) are

$$\begin{aligned} \mathcal{A}_b^{\mathrm{E}} \frac{\partial}{\partial x} \begin{bmatrix} E_z|_{d_l/2}^n \\ E_z|_{3/2}^n \\ \vdots \\ E_z|_{L-3d_{\mathrm{r}}/2}^n \\ E_z|_{L-d_{\mathrm{r}}/2}^n \end{bmatrix} &= \frac{24}{\Delta x} \begin{bmatrix} E_z|_1^n - E_z|_{d_l}^n \\ E_z|_2^n - E_z|_1^n \\ \vdots \\ E_z|_{L-1}^n - E_z|_{L-2}^n \\ E_z|_L^n - E_z|_{L-d_{\mathrm{r}}}^n \end{bmatrix} \\ \mathcal{A}_b^{\mathrm{H}} \frac{\partial}{\partial x} \begin{bmatrix} H_y|_1^{n+1/2} \\ H_y|_2^{n+1/2} \\ \vdots \\ H_y|_{L-2}^{n+1/2} \\ H_y|_{L-1}^{n+1/2} \end{bmatrix} &= \frac{24}{\Delta x} \begin{bmatrix} H_y|_{3/2}^{n+1/2} - H_y|_{d_l/2}^{n+1/2} \\ H_y|_{5/2}^{n+1/2} - H_y|_{3/2}^{n+1/2} \\ \vdots \\ H_y|_{L-3/2}^{n+1/2} - H_y|_{L-5/2}^{n+1/2} \\ H_y|_{L-d_{\mathrm{r}}/2}^{n+1/2} - H_y|_{L-3/2}^{n+1/2} \end{bmatrix}. \end{aligned} \tag{2.38}$$

Note that the previous formulation holds for more than five grid nodes in any direction. Assuming that there are four grid nodes, as in Figure 2.3(b), (2.35)–(2.38) are given by

$$\mathcal{A}_b^{\mathrm{E}} = \begin{bmatrix} 24d_1 & 0 & 0 \\ a_1 & a_2 & a_3 \\ 0 & 0 & 24d_{\mathrm{r}} \end{bmatrix} \quad \text{and} \quad \mathcal{A}_b^{\mathrm{H}} = \begin{bmatrix} 3(3+4d_1+d_1^2) & 3(1-d_1^2) \\ 3(1-d_{\mathrm{r}}^2) & 3(3+4d_{\mathrm{r}}+d_{\mathrm{r}}^2) \end{bmatrix}, \tag{2.39}$$

where

$$a_1 = \frac{24}{3(1+d_1)(2+d_1+d_{\mathrm{r}})}, \qquad a_2 = \frac{24(2+3d_1+3d_{\mathrm{r}}+3d_1d_{\mathrm{r}})}{3(1+d_1)(1+d_{\mathrm{r}})},$$

$$a_3 = \frac{24}{3(2+d_1+3d_{\mathrm{r}}+d_1d_{\mathrm{r}}+d_{\mathrm{r}}^2)}, \qquad \mathcal{A}_b^{\mathrm{E}} \frac{\partial}{\partial x} \begin{bmatrix} E_z|_{d_1/2}^n \\ E_z|_{3/2}^n \\ E_z|_{d_{\mathrm{r}}/2}^n \end{bmatrix} = \frac{24}{\Delta x} \begin{bmatrix} E_z|_1^n - E_z|_{d_1}^n \\ E_z|_2^n - E_z|_1^n \\ E_z|_{d_{\mathrm{r}}}^n - E_z|_2^n \end{bmatrix},$$

$$\text{and} \quad \mathcal{A}_b^{\mathrm{H}} \frac{\partial}{\partial x} \begin{bmatrix} H_y|_1^{n+1/2} \\ H_y|_2^{n+1/2} \end{bmatrix} = \frac{24}{\Delta x} \begin{bmatrix} H_y|_{3/2}^{n+1/2} - H_y|_{d_1/2}^{n+1/2} \\ H_y|_{d_{\mathrm{r}}/2}^{n+1/2} - H_y|_{3/2}^{n+1/2} \end{bmatrix}. \tag{2.40}$$

For the time-marching procedure, the leapfrog integration may be substituted by the fourth-order Runge–Kutta one, which staggers the variables in space but not in time. Thus, for $\partial U/\partial t = f(U)$, it is derived

$$\begin{aligned} U_1 &= U|^n + \tfrac{\Delta t}{4} f(U|^n), & U_2 &= U|^n + \tfrac{\Delta t}{3} f(U_1), \\ U_3 &= U|^n + \tfrac{\Delta t}{2} f(U_2), & U|^{n+1} &= U|^n + \Delta t f(U_3). \end{aligned} \tag{2.41}$$

The corresponding stability criteria for the (2, 4) and (4, 4) schemes are

$$(2,4): \quad \upsilon \Delta t \le \frac{\Delta x}{1.2\sqrt{2}} \quad \text{and} \quad (4,4): \ \upsilon \Delta t \le \frac{5\Delta x}{6}. \tag{2.42}$$

In general, boundary conditions are imposed at the end of each stage of (2.41) or the leapfrog time-step. Finally, in the case of absorbing boundary conditions, all derivatives are computed by the implicit algorithm across the entire domain including its interior and the absorber. Then, each system is updated by (2.41).

The performance of the above schemes can be validated by the results of Figures 2.4(a) and 2.4(b), where the variation of the maximum L_2 error norm, i.e., the global error, as a function of the $\Delta t/\Delta h$ number and computational time are, respectively, illustrated. It becomes apparent that the implicit higher order schemes outperform Yee's technique, achieving fast convergence in terms of a reasonable CPU overhead.

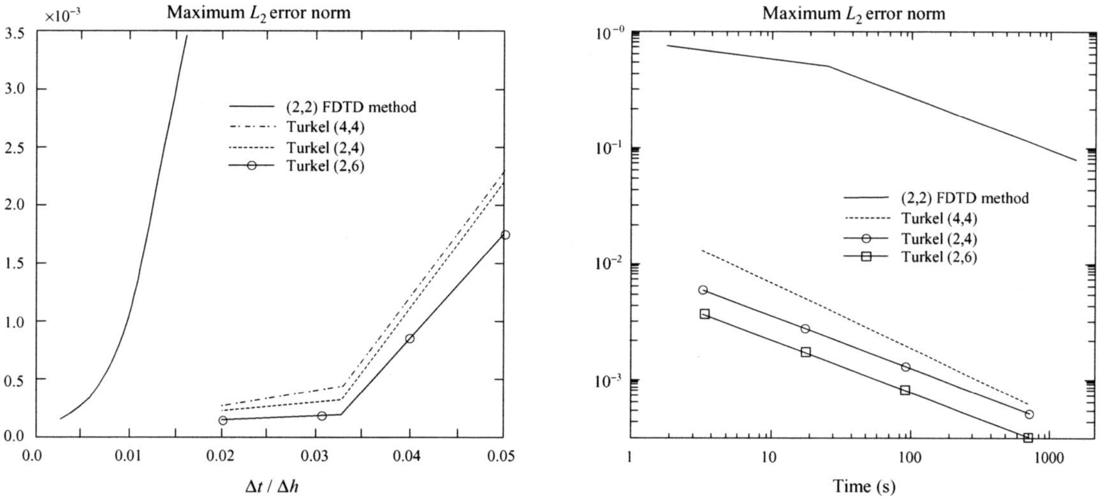

(a) (b)

FIGURE 2.4: Behavior of the maximum L_2 error norm versus (a) $\Delta t/\Delta h$ and (b) computational time

2.4.3 One-Sided Difference Operators for Homogeneous Dielectric Boundaries

Implementing a domain-decomposition concept, accurate *one-sided difference* operators [26] are combined with extrapolation/interpolation techniques to treat dielectric interfaces as boundary points between separate subdomains. Spatial derivatives in each region are computed via the typical (2, 4) FDTD scheme.

Let us focus on a two dimensional (2-D) $L \times L$ grid bounded by a homogeneous dielectric or PEC wall, and apply (2.18) at all nodes except at the first and last, where the modified boundary conditions are to be imposed. These interior lattice points (one electric and one magnetic) are directly adjacent to the first and last electric field nodes of the domain. For the global approximation of spatial derivatives at such locations, the ensuing set of third and fourth-order one-sided approximations are utilized:

$$\left.\frac{\partial f}{\partial y}\right|_{i,l}^{n} = \frac{1}{24\Delta y}\left(\mp 23 f\big|_{i,l\mp 1/2}^{n} \pm 21 f\big|_{i,l\pm 3/2}^{n} \pm 3 f\big|_{i,l\pm 5/2}^{n} \mp f\big|_{i,l\pm 7/2}^{n}\right), \quad \text{for } l = 1,\, L-1, \tag{2.43}$$

$$\left.\frac{\partial f}{\partial y}\right|_{i,l}^{n} = \frac{1}{24\Delta y}\left(\mp 22 f\big|_{i,l\mp 1/2}^{n} \pm 17 f\big|_{i,l\pm 1/2}^{n} \pm 9 f\big|_{i,l\pm 3/2}^{n} \mp 5 f\big|_{i,l\pm 5/2}^{n} \pm f\big|_{i,l\pm 7/2}^{n}\right), \quad \text{for } l = 1/2,\, L-1/2, \tag{2.44}$$

where the signs "±" and "∓" are selected according to the first or second value of l. A similar practice holds for the derivative toward the x-direction. Next, as in (2.34) and (2.35), one can define

$$
\mathcal{B}_b^{\mathrm{H}} = \begin{bmatrix} -23 & 21 & 3 & -1 & \dots & \dots & 0 \\ 1 & -27 & 27 & -1 & \dots & \dots & 0 \\ 0 & 1 & -27 & 27 & -1 & \dots & 0 \\ \vdots & \vdots & \ddots & \ddots & \ddots & \vdots & \vdots \\ 0 & \dots & \dots & 1 & -27 & 27 & -1 \\ 0 & \dots & \dots & 1 & -3 & -21 & 23 \end{bmatrix}
$$
$$
\mathcal{B}_b^{\mathrm{E}} = \begin{bmatrix} -22 & 17 & 9 & -5 & 1 & \dots & 0 \\ 1 & -27 & 27 & -1 & \dots & \dots & 0 \\ 0 & 1 & -27 & 27 & -1 & \dots & 0 \\ \vdots & \vdots & \ddots & \ddots & \ddots & \vdots & \vdots \\ 0 & \dots & \dots & 1 & -27 & 27 & -1 \\ 0 & \dots & \dots & 5 & -9 & -17 & 22 \end{bmatrix} \tag{2.45}
$$

to obtain the matrix form of the $\partial f/\partial y$ approximation at the midpoint between mesh nodes and at the mesh nodes, respectively, as

$$
\frac{\partial}{\partial y}\begin{bmatrix} f|_{i,1/2}^{n} \\ f|_{i,3/2}^{n} \\ \vdots \\ f|_{i,L-3/2}^{n} \\ f|_{i,L-1/2}^{n} \end{bmatrix} = \frac{1}{24\Delta y}\mathcal{B}_b^{\mathrm{E}}\begin{bmatrix} f|_{i,0}^{n} \\ f|_{i,1}^{n} \\ \vdots \\ f|_{i,L-1}^{n} \\ f|_{i,L}^{n} \end{bmatrix} \quad \frac{\partial}{\partial y}\begin{bmatrix} f|_{i,1}^{n} \\ f|_{i,2}^{n} \\ \vdots \\ f|_{i,L-2}^{n} \\ f|_{i,L-1}^{n} \end{bmatrix} = \frac{1}{24\Delta y}\mathcal{B}_b^{\mathrm{H}}\begin{bmatrix} f|_{i,1/2}^{n} \\ f|_{i,3/2}^{n} \\ \vdots \\ f|_{i,L-3/2}^{n} \\ f|_{i,L-1/2}^{n} \end{bmatrix}. \tag{2.46}
$$

In this manner, the modified explicit (2, 4) FDTD scheme for the TM case is given by

$$
E_z|_{i,j}^{n+1} = E_z|_{i,j}^{n} + \frac{\Delta t}{24\varepsilon\Delta x}\left[\mathcal{B}_b^{\mathrm{H}}\right] H_y|_{i+1/2,j}^{n+1/2} - \frac{\Delta t}{24\varepsilon\Delta y} H_x|_{i,j+1/2}^{n+1/2}\left[\overline{\mathcal{B}}_b^{\mathrm{H}}\right] \tag{2.47}
$$

$$
H_x|_{i,j+1/2}^{n+1/2} = H_x|_{i,j+1/2}^{n-1/2} - \frac{\Delta t}{24\mu\Delta y} E_z|_{i,j}^{n}\left[\overline{\mathcal{B}}_b^{\mathrm{E}}\right] \tag{2.48}
$$

$$
H_y|_{i+1/2,j}^{n+1/2} = H_x|_{i+1/2,j}^{n-1/2} + \frac{\Delta t}{24\mu\Delta x}\left[\mathcal{B}_b^{\mathrm{E}}\right] E_z|_{i,j}^{n} \tag{2.49}
$$

with $\overline{\mathcal{B}}_b^{\mathrm{H}}$ and $\overline{\mathcal{B}}_b^{\mathrm{E}}$ denoting the matrices that correspond to stencil treatment along j and the brackets representing the appropriate elements for the (i, j) position of f inside the grid.

Conversely, if a PEC wall is located on a tangential magnetic field node, then due to the homogeneous Neumann condition, $\partial H^{\tan}/\partial n = 0$, the above procedure should be pertinently

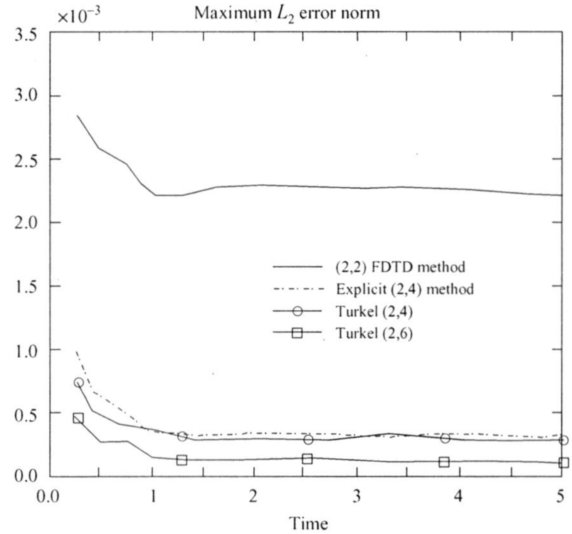

FIGURE 2.5: Maximum L_2 error norm as a function of time

altered. Selecting the 1-D case, i.e., H_y and E_z, for clarity, our attention is drawn to boundary condition $\partial H_y/\partial x = 0$ at $i = 1/2$. Since the electric field node at $i = 1$ can now be viewed as a boundary node, an update expression in terms of the condition at $i = 1/2$ has to be derived. Therefore, one may evaluate $\partial H_y/\partial x$ at $i = 1$ by a fourth-order arrangement of the fluxes at adjacent electric field nodes 2, 3, 4, and the boundary 1 node as

$$\left.\frac{\partial H_y}{\partial x}\right|_1^{n+1/2} = \left.\frac{\partial H_y}{\partial x}\right|_2^{n+1/2} - \frac{3}{5}\left.\frac{\partial H_y}{\partial x}\right|_3^{n+1/2} + \frac{1}{7}\left.\frac{\partial H_y}{\partial x}\right|_4^{n+1/2} \tag{2.50}$$

with the first derivative on the right hand side calculated using (2.43) and those at $i = 3, 4$ by (2.18). This practice leads to the time-advancing relation of

$$E_z|_1^{n+1} = E_z|_1^n + \frac{\Delta t}{\varepsilon}\left.\frac{\partial H_y}{\partial x}\right|_1^{n+1/2}. \tag{2.51}$$

It is mentioned that extension of (2.43)–(2.51) to 3-D applications follows along analogous lines.

As a simple example to test the efficiency of the one-sided difference operators, consider a rectangular cavity with a small metallic stub on its lower base. Figure 2.5 gives the temporal evolution of the global error (L_2 norm) and Table 2.2 summarizes the dispersion

TABLE 2.2: Dispersion Error of One-sided Operators

METHOD	GRID RESOLUTION	DISPERSION ERROR	CONVERGENCE RATE
FDTD	1/15	0.00980	
FDTD	1/45	0.00217	2.0046
FDTD	1/75	5.1743×10^{-4}	1.9961
Turkel (2, 4)	1/15	0.00124	
Turkel (2, 4)	1/45	3.8642×10^{-4}	1.8936
Turkel (2, 4)	1/75	7.1234×10^{-5}	1.9613
Explicit (2, 4)	1/15	0.00125	
Explicit (2, 4)	1/45	3.8645×10^{-4}	1.8852
Explicit (2, 4)	1/75	7.1249×10^{-5}	1.9607

error and convergence rate of diverse implementations including the operators of Section 2.4.2. Results demonstrate that both higher order schemes are equivalently accurate, retaining also a satisfactory convergence rate for all values of grid resolution.

The stability criterion of the technique is extracted by means of the ordinary von Neumann method as

$$\frac{\Delta t}{\sqrt{\min[\mu\varepsilon]}} \leq 2\frac{\Delta h}{\rho^{\infty}} \tag{2.52}$$

in which ρ^{∞} is the spectral radius of the matrices utilized for the approximation of spatial derivatives and Δh the respective increment. Note that because of the dielectric walls the maximum Δh must be determined by employing the $\min[\mu\varepsilon]$ over the domain under study, and the maximum resolved frequency with a prefixed number of points per wavelength. Evidently, (2.52) has the form of the Courant condition (2.25), thus maintaining the fourth-order character of the method.

2.4.4 Treatment of Inhomogeneous Dielectric Boundaries

In the case of an inhomogeneous dielectric, serving as a boundary or an intermediate layer, one-sided difference operators have to be reformulated in order to circumvent possible instabilities. The key concept for these amendments, which lies on the efficient analysis of [20, 23], presumes an explicit (2, 4) FDTD approach in the homogeneous areas of the computational space and

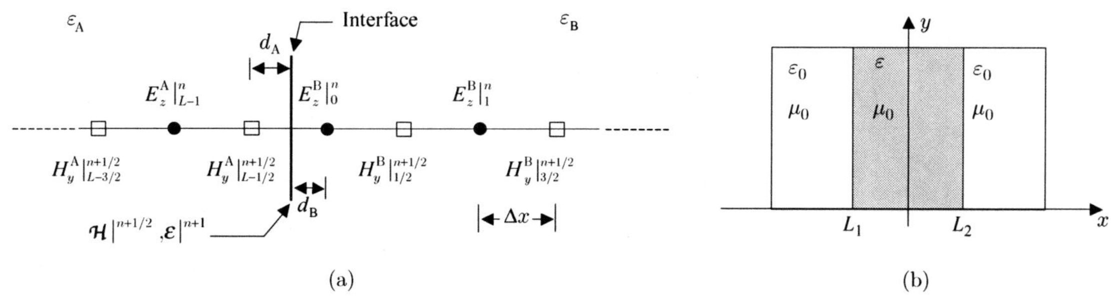

FIGURE 2.6: (a) A 1-D interface separating two different media and the auxiliary distances d_{A}, d_{B}. (b) A vertical dielectric slab in a 2-D computational domain

a *locally* modified scheme near its boundaries or interfaces. Actually, it is the second part of this combined method that permits the mitigation of staircase effects and the fulfillment of the physically correct jump conditions at the problematic areas. Moreover, owing to its local profile, the preceding abstraction retains the merits of the initial algorithm, since the majority of the extra overhead is performed in a preprocessing phase.

Consider the 1-D graphical depiction of Figure 2.6(a), where the two different dielectrics are designated as A and B, respectively. All field components located at nodes inside the domain are updated by (2.18), while E_z^{A} at $i = L-1$ and H_y^{B} at $i = 1/2$ are evaluated via the one-sided third-order scheme of (2.43). The most significant issue, nevertheless, is the computation of H_y^{A} at $i = L-1/2$ and E_z^{B} at $i = 0$ next to the interface. This is accomplished by the extrapolated terms $\mathcal{H}$ and $\mathcal{E}$ on the left- and right-hand side of the discontinuity and their consecutive combination with the physical continuity conditions as

$$\begin{aligned}\mathcal{H}|^{n+1/2} &= \frac{(7-2d_{\mathrm{B}})(5-2d_{\mathrm{B}})}{16}\left[\frac{3-2d_{\mathrm{B}}}{3}\,H_y^{\mathrm{A}}\Big|_{L-1/2}^{n+1/2} - (1-2d_{\mathrm{B}})\,H_y^{\mathrm{A}}\Big|_{L-3/2}^{n+1/2}\right]\\ &+\frac{(3-2d_{\mathrm{B}})(1-2d_{\mathrm{B}})}{16}\left[(7-2d_{\mathrm{B}})\,H_y^{\mathrm{A}}\Big|_{L-5/2}^{n+1/2} - \frac{5-2d_{\mathrm{B}}}{3}\,H_y^{\mathrm{A}}\Big|_{L-7/2}^{n+1/2}\right], \end{aligned} \tag{2.53}$$

$$\begin{aligned}\mathcal{E}|^{n} &= \frac{(7-2d_{\mathrm{A}})(5-2d_{\mathrm{A}})}{16}\left[\frac{3-2d_{\mathrm{A}}}{3}\,E_z^{\mathrm{B}}\Big|_{0}^{n} - (1-2d_{\mathrm{A}})\,E_z^{\mathrm{B}}\Big|_{1}^{n}\right]\\ &+\frac{(3-2d_{\mathrm{A}})(1-2d_{\mathrm{A}})}{16}\left[(7-2d_{\mathrm{A}})\,E_z^{\mathrm{B}}\Big|_{2}^{n} - \frac{5-2d_{\mathrm{A}}}{3}\,E_z^{\mathrm{B}}\Big|_{3}^{n}\right], \end{aligned} \tag{2.54}$$

where $d_{\mathrm{A}} + d_{\mathrm{B}} = 1/2$. Hence, the resulting expressions for H_y^{A} at $i = L-1/2$ and E_z^{B} at

$i = 0$ are

$$\left.\frac{\partial H_y^{\mathrm{A}}}{\partial t}\right|_{L-1/2}^{n} = C_1(d_{\mathrm{B}})\,\mathcal{E}|^n - C_2(d_{\mathrm{B}})\,E_z^{\mathrm{A}}\Big|_{L-1}^{n} + C_3(d_{\mathrm{B}})\,E_z^{\mathrm{A}}\Big|_{L-2}^{n} - C_4(d_{\mathrm{B}})\,E_z^{\mathrm{A}}\Big|_{L-3}^{n}, \quad (2.55a)$$

$$\left.\varepsilon^{\mathrm{B}}\frac{\partial E_z^{\mathrm{B}}}{\partial t}\right|_{0}^{n+1/2} = -\,C_1(d_{\mathrm{A}})\,\mathcal{H}|^{n+1/2} + C_2(d_{\mathrm{A}})\,H_y^{\mathrm{B}}\Big|_{1/2}^{n+1/2} - C_3(d_{\mathrm{A}})\,H_y^{\mathrm{B}}\Big|_{3/2}^{n+1/2} + C_4(d_{\mathrm{A}})\,H_y^{\mathrm{B}}\Big|_{5/2}^{n+1/2}, \quad (2.55b)$$

with

$$C_1(s) = \frac{46}{\Delta x(1+2s)(3+2s)(5+2s)}, \qquad C_2(s) = \frac{15-16s}{4\Delta x(1+2s)},$$

$$C_3(s) = \frac{5-12s}{2\Delta x(3+2s)}, \qquad C_4(s) = \frac{3-8s}{4\Delta x(5+2s)}, \qquad \text{for } s = \mathrm{A}, \mathrm{B}.$$

It is stressed that the stencils of (2.55) do not generate any artificial oscillations even if the interface is in very close proximity to a lattice node, guaranteeing so a uniformly bounded time-step approach.

Let us now take into account a 2-D vertical dielectric slab that divides the computational domain into *three* regions: two filled with air, namely $\varepsilon = \varepsilon_0$ for $i < L_1$ and $i > L_2$, and one with the dielectric ε for $L_1 \leq i \leq L_2$, as shown in Figure 2.6(b). The chief feature of this material is that its permittivity is a piecewise constant function of x solely, with the points of material discontinuity restricted to coincide with electric field nodes. Across this dielectric interface, tangential components E_z and H_y are continuous, whereas all second-order and higher order derivatives of both fields are discontinuous. Thus, for H_y at L_1 and L_2, one may potentially use the fifth-order extrapolation of [26] based on data from the air regions,

$$H_y\Big|_{l,j}^{n+1/2} = \frac{1}{128}\left(325 H_y\Big|_{l\mp1/2,j}^{n+1/2} + 35\,H_y\Big|_{l\mp9/2,j}^{n+1/2}\right) - \frac{1}{64}\left(210\,H_y\Big|_{l\mp3/2,j}^{n+1/2} - 189\,H_y\Big|_{l\mp5/2,j}^{n+1/2} + 90\,H_y\Big|_{l\mp7/2,j}^{n+1/2}\right) \quad (2.56)$$

with the "$-$" and "$+$" sign for $l = L_1$ and $l = L_2$, respectively. To circumvent differentiation across the interfaces, the value of $\partial H_y/\partial x$ at these nodes is extracted by

$$\left.\frac{\partial H_y}{\partial x}\right|_{l,j}^{n+1/2} = \mp\frac{1126}{315\Delta x}\,H_y\Big|_{l,j}^{n+1/2} \pm \frac{1}{16\Delta x}\left(\frac{315}{2}\,H_y\Big|_{l\pm1/2,j}^{n+1/2} \mp 35\,H_y\Big|_{l\pm3/2,j}^{n+1/2} \pm\frac{189}{10}\,H_y\Big|_{l\pm5/2,j}^{n+1/2} \mp \frac{45}{7}\,H_y\Big|_{l\pm7/2,j}^{n+1/2} \pm \frac{35}{36}\,H_y\Big|_{l\pm9/2,j}^{n+1/2}\right). \quad (2.57)$$

The next step involves the calculation of E_z at $l = L_1, L_2$ as

$$E_z\big|_{l,j}^{n+1} = E_z\big|_{l,j}^{n} - \frac{9\Delta t}{8\varepsilon\Delta y}\left(H_x\big|_{l,j+1/2}^{n+1/2} - H_x\big|_{l,j-1/2}^{n+1/2}\right)$$
$$+\frac{\Delta t}{24\varepsilon\Delta y}\left(H_x\big|_{l,j+3/2}^{n+1/2} - H_x\big|_{l,j-3/2}^{n+1/2}\right) + \frac{\Delta t}{\varepsilon}\frac{\partial H_y}{\partial x}\bigg|_{l,j}^{n+1/2}. \qquad (2.58)$$

Should the dielectric interface be positioned at $L_1 + 1/2$, i.e., at a magnetic field node, the role of E_z and H_y is reversed with the former provided by

$$E_z\big|_{L_1+1/2,j}^{n} = \frac{1}{128}\left(315E_z\big|_{L_1,j}^{n} + 35\ E_z\big|_{L_1-4,j}^{n}\right) - \frac{1}{64}\left(210\ E_z\big|_{L_1-1,j}^{n}\right.$$
$$\left.-189\ E_z\big|_{L_1-2,j}^{n} + 90\ E_z\big|_{L_1-3,j}^{n}\right) \qquad (2.59)$$

and, through the suitable data from the corresponding region,

$$\frac{\partial E_z}{\partial x}\bigg|_{L_1+1/2,j}^{n} = -\frac{1126}{315\Delta x}\ E_z\big|_{L_1+1/2,j}^{n} + \frac{1}{16\Delta x}\left(\frac{315}{2}\ E_z\big|_{L_1+1,j}^{n} - 35\ E_z\big|_{L_1+2,j}^{n}\right.$$
$$\left.+\frac{189}{10}E_z\big|_{L_1+3,j}^{n} - \frac{45}{7}E_z\big|_{L_1+4,j}^{n} + \frac{35}{36}E_z\big|_{L_1+5,j}^{n}\right). \qquad (2.60)$$

Therefore, the update of H_y at $L_1 + 1/2$ leads to

$$H_y\big|_{L_1+1/2,j}^{n+1/2} = H_y\big|_{L_1+1/2,j}^{n-1/2} + \frac{\Delta t}{\mu}\frac{\partial E_z}{\partial x}\bigg|_{L_1+1/2,j}^{n}. \qquad (2.61)$$

As a conclusive remark, it is stated that the fairly sufficient performance of (2.55)–(2.61) deteriorates with the increase of the contrast between the dielectric constants. This problem is, essentially, attributed to the faster loss of smoothness at the dielectric interface and the consequent augmentation of the dispersion error.

2.4.5 Derivative Matching with Fictitious Points

The use of *fictitious points* as a means of locally modifying the differential stencils near laborious media interfaces in finite-difference simulations has been initially developed in [17, 18] and extended in [21, 28]. The specific method, which matches the problematic boundaries with physical derivative conditions, enhances the flexibility of higher order FDTD schemes and facilitates the discretization of difficult geometries.

Starting from the 1-D flux tensor form of Maxwell's equations $\partial\mathbf{Q}/\partial t = \mathbf{R}\,\partial\mathbf{Q}/\partial x$ with $\mathbf{Q} = [E_z\ H_y]^T$, it is assumed that the interface between media A and B is located at $i = L$. As a consequence, the constitutive parameter matrix $\mathbf{R}$ becomes $\mathbf{R} = \mathbf{R}_\mathrm{A}$ for $i < L$ and $\mathbf{R} = \mathbf{R}_\mathrm{B}$

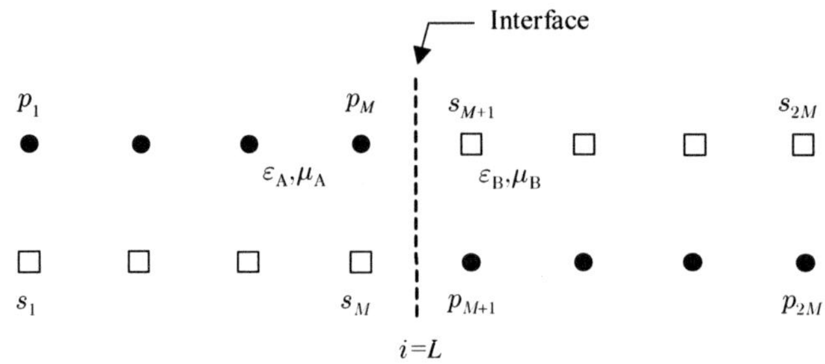

FIGURE 2.7: Positions of the real and fictitious lattice points in the vicinity of a material interface

for $i > L$. Since both fields are continuous, their temporal derivatives also are continuous, a fact which yields the one-sided quantities

$$\mathbf{R}_\mathrm{A}^m \left.\frac{\partial^m \mathbf{Q}}{\partial x^m}\right|_{L^-}^n = \mathbf{R}_\mathrm{B}^m \left.\frac{\partial^m \mathbf{Q}}{\partial x^m}\right|_{L^+}^n, \tag{2.62}$$

with $m = 0, 1, 2, \ldots$ the differentiation order and the signs denoting location on the left or right of the interface. To continue the implicit derivative matching analysis, E_z and H_y components are placed at real, p_i, and fictitious, s_i, mesh nodes for $i = 1, 2, \ldots, 2M$, according to Figure 2.7. The values of s_i are obtained in terms of p_i ones at every time-step via the representation of

$$s_i = \sum_{j=1}^{2M} w_{i,j} p_j \quad \text{for } i = 1, 2, \ldots, 2M, \text{ producing the system of } \mathbf{S} = \mathbf{WP}, \tag{2.63}$$

where $w_{i,j}$ are certain coefficients, $\mathbf{S} = [s_1 \, s_2 \ldots s_{2M}]^T$, and $\mathbf{P} = [p_1 p_2 \ldots p_{2M}]^T$. The resulting unknown $2M \times 2M$ matrix $\mathbf{W}$ is determined by considering its rows $\mathbf{W}_i = [w_{i,1} w_{i,2} \ldots w_{i,2M}]$ as new variables and solving for a vector $\bar{\mathbf{W}} = [\mathbf{W}_1 \mathbf{W}_2 \ldots \mathbf{W}_{2M}]^T$ from an algebraic system $(2M)^2 \times (2M)^2$. In a similar context, a set of $2M$ vectors $\mathbf{I}_i$ as the rows of a $2M \times 2M$ identity matrix $\mathbf{I}$ are defined, namely $\mathbf{I}_1 = [100\ldots 0]^T$, $\mathbf{I}_2 = [010\ldots 0]^T$ up to $\mathbf{I}_{2M} = [000\ldots 1]^T$. So, it is derived that $s_i = \mathbf{W}_i\mathbf{P}$ and $p_i = \mathbf{I}_i\mathbf{P}$.

For the discretization of the $2M$ jump conditions, the algorithm uses central finite-difference schemes at each mesh node to construct $2M \times 2M$ algebraic equations. For instance, let us consider jump condition

$$\frac{1}{\mu_\mathrm{A}} \left.\frac{\partial^2 E_z}{\partial x^2}\right|_{L^-}^n = \frac{1}{\mu_\mathrm{B}} \left.\frac{\partial^2 E_z}{\partial x^2}\right|_{L^+}^n \Rightarrow \frac{1}{\mu_\mathrm{A}} \left(\sum_{j=1}^{M} q_{2,j} s_j + \sum_{j=M+1}^{2M} q_{2,j} p_j \right)$$
$$= \frac{1}{\mu_\mathrm{B}} \left(\sum_{j=1}^{M} q_{2,j} p_j + \sum_{j=M+1}^{2M} q_{2,j} s_j \right) \tag{2.64}$$

with $q_{2,j}$, $j = 1, 2, \ldots, 2M$, the suitable differencing weights acquired through the fast method of [18]. Separating the known values in (2.64) and substituting s_i,p_i, the discrete version of the subsequent $2M$ algebraic equations of the prior jump condition are extracted

$$\frac{1}{\mu_{\mathrm{A}}}\sum_{j=M+1}^{2M} q_{2,j}\mathbf{R}_j^T - \frac{1}{\mu_{\mathrm{B}}}\sum_{j=1}^{M} q_{2,j}\mathbf{R}_j^T = \underbrace{\frac{1}{\mu_{\mathrm{B}}}\sum_{j=M+1}^{2M} q_{2,j}\mathbf{I}_j^T - \frac{1}{\mu_{\mathrm{A}}}\sum_{j=1}^{M} q_{2,j}\mathbf{I}_j^T}_{\text{known values}}, \tag{2.65}$$

which are independent of field values. Observe that for the remaining computational domain, operator (2.18) is utilized for spatial discretization and the fourth-stage Runge–Kutta integrator (2.41) for the time update.

Extension to the 2-D TM case opts for the vector form of $\partial\mathbf{Q}/\partial t = \mathbf{R}\,\partial\mathbf{Q}/\partial x + \mathbf{Z}\,\partial\mathbf{Q}/\partial y$ with $Q = [H_x H_y E_z]^T$. If the material interface is again placed at $i = L$, constitutive parameter matrices $\mathbf{R}$ and $\mathbf{Z}$ receive the appropriate values at media A and B, as in the 1-D problem. Herein, derivative matching is performed for $\partial E_z/\partial x$ and $\partial H_y/\partial x$ with the ensuing jump conditions across the interface

$$\textit{zero-order:}\ \mathbf{Q}\rfloor_{L^-}^{n} = \mathbf{Q}\rfloor_{L^+}^{n}, \tag{2.66}$$

$$\textit{higher order:}\left(\mathbf{R}_1\frac{\partial}{\partial x} + \mathbf{Z}_1\frac{\partial}{\partial y}\right)^m \mathbf{Q}\rfloor_{L^-}^{n} = \left(\mathbf{R}_2\frac{\partial}{\partial x} + \mathbf{Z}_2\frac{\partial}{\partial y}\right)^m \mathbf{Q}\rfloor_{L^+}^{n}. \tag{2.67}$$

As can be deduced, for $m \geq 2$, expression (2.67) leads to cross derivatives by x and y, whose evaluation is rather cumbersome. To alleviate this difficulty, only one fictitious point can be considered at each side of the interface and hence only the zero- and first-order jump conditions are implemented. While this notion gives reliable solutions, an alternative quasi-fourth-order strategy has been presented in [28] for the consideration of higher order conditions and cross-derivative computation. A fairly interesting feature of the derivative matching method is that it encompasses various schemes with different orders that permit its hybridization with other high-accuracy time-domain approaches.

2.5 DISPERSION-OPTIMIZED HIGHER ORDER FDTD TECHNIQUES

The potential to exploit the advanced features of higher order FDTD schemes in the construction of algorithms, with reduced dispersion error at specific parts of the frequency spectrum, is indeed a challenging task. This section presents several robust methodologies that attain this critical issue from different perspectives.

2.5.1 Modified (2, 4) Scheme with Improved Phase Accuracy

The modified (2, 4) FDTD algorithm [47] is constructed by means of the integral form of Maxwell's equations (2.1) and (2.2) for $\mathbf{J}_c = \mathbf{J}_s = \mathbf{M}_c = \mathbf{M}_s = \mathbf{0}$ and $\rho_e = \rho_m = 0$. Focusing on a uniform 2-D mesh ($\Delta h = \Delta x = \Delta y$) and considering the standard (2, 4) update equation for the TM mode, rewritten as

$$\begin{aligned}\varepsilon \frac{\partial E_z}{\partial t}\bigg|_{i,j}^{n+1/2} &= \frac{9}{8(\Delta h)^2}\Delta h\left(H_y\big|_{i+1/2,j}^{n+1/2} - H_y\big|_{i-1/2,j}^{n+1/2} - H_x\big|_{i,j+1/2}^{n+1/2} + H_x\big|_{i,j-1/2}^{n+1/2}\right)\\ &\quad -\frac{1}{8(3\Delta h)^2}3\Delta h\left(H_y\big|_{i+3/2,j}^{n+1/2} - H_y\big|_{i-3/2,j}^{n+1/2} - H_x\big|_{i,j+3/2}^{n+1/2} + H_x\big|_{i,j-3/2}^{n+1/2}\right)\\ &\cong \frac{9}{8(\Delta h)^2}\oint_{L_1}\mathbf{H}\cdot d\mathbf{l} - \frac{1}{8(3\Delta h)^2}\oint_{L_2}\mathbf{H}\cdot d\mathbf{l},\end{aligned} \tag{2.68}$$

analysis attempts to extract a modified expression in conjunction with Figure 2.8(a). Owing to the fact that

$$\oint_{L_1}\mathbf{H}\cdot d\mathbf{l} = \frac{\partial}{\partial t}\iint_{S_1}\mathbf{D}\cdot d\mathbf{S} \cong \varepsilon(\Delta h)^2\frac{\partial E_z}{\partial t}, \tag{2.69a}$$

$$\oint_{L_2}\mathbf{H}\cdot d\mathbf{l} = \frac{\partial}{\partial t}\iint_{S_2}\mathbf{D}\cdot d\mathbf{S} \cong 9\varepsilon(\Delta h)^2\frac{\partial E_z}{\partial t}, \tag{2.69b}$$

the initial time-advancing scheme becomes

$$\left(\varepsilon\frac{\partial E_z}{\partial t}\right)^{\text{FDTD}} = -\frac{1}{8}\left(\varepsilon\frac{\partial E_z}{\partial t}\right)_{L_2} + \frac{9}{8}\left(\varepsilon\frac{\partial E_z}{\partial t}\right)_{L_1}. \tag{2.70}$$

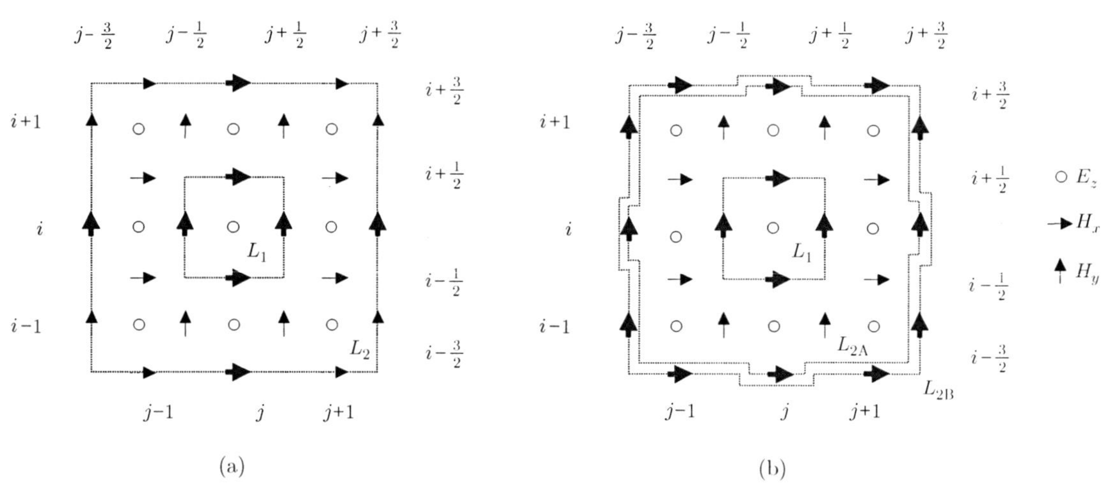

FIGURE 2.8: (a) The inner and outer loops for the modified 2-D (2, 4) FDTD scheme. (b) Bisplitting of the outer loop and assignment of the eight diagonal field nodes

Equation (2.70) demonstrates that (2.68) constitutes a weighted sum of the results obtained by the implementation of Ampère's law on the two loops L_1 and L_2. Moreover, (2.70) suggests an easy way to incorporate *all 12* nodes along the outer loop in the FDTD approach. It must be mentioned, herein, that such a process will alter the values of the two right-hand side coefficients whose sum must always be unity. For this purpose and to enable a more adaptable minimization of the dispersion error, L_2 loop is divided into two different loops $L_{2\mathrm{A}}$ and $L_{2\mathrm{B}}$, as depicted in Figure 2.8(b). Thus, (2.70) receives the more consistent form of

$$\left(\varepsilon\frac{\partial E_z}{\partial t}\right)^{\mathrm{FDTD}} = N_1\left(\varepsilon\frac{\partial E_z}{\partial t}\right)_{L_{2\mathrm{A}}} + N_2\left(\varepsilon\frac{\partial E_z}{\partial t}\right)_{L_{2\mathrm{B}}} + (1-N_1-N_2)\left(\varepsilon\frac{\partial E_z}{\partial t}\right)_{L_1}, \tag{2.71}$$

which leads to the optimized (2, 4) TM FDTD equations

$$\begin{aligned}\varepsilon\frac{\partial E_z}{\partial t}\bigg|_{i,j}^{n+1/2} &= \frac{N_1}{3\Delta h}\left(H_y\big|_{i+3/2,j}^{n+1/2} - H_y\big|_{i-3/2,j}^{n+1/2} - H_x\big|_{i,j+3/2}^{n+1/2} + H_x\big|_{i,j-3/2}^{n+1/2}\right)\\ &+ \frac{N_2}{6\Delta h}\left(H_y\big|_{i+3/2,j-1}^{n+1/2} - H_y\big|_{i-3/2,j-1}^{n+1/2} - H_y\big|_{i-3/2,j+1}^{n+1/2} + H_y\big|_{i+3/2,j+1}^{n+1/2}\right.\\ &\left. + H_x\big|_{i-1,j-3/2}^{n+1/2} - H_x\big|_{i-1,j+3/2}^{n+1/2} - H_x\big|_{i+1,j+3/2}^{n+1/2} + H_x\big|_{i+1,j-3/2}^{n+1/2}\right)\\ &- \frac{1-N_1-N_2}{\Delta h}\left(H_y\big|_{i+1/2,j}^{n+1/2} - H_y\big|_{i-1/2,j}^{n+1/2} - H_x\big|_{i,j+1/2}^{n+1/2} + H_x\big|_{i,j-1/2}^{n+1/2}\right),\end{aligned} \tag{2.72}$$

$$\mu\frac{\partial H_x}{\partial t}\bigg|_{i,j}^{n} = \frac{N_1}{3\Delta h}\left(E_z\big|_{i,j+3/2}^{n} - E_z\big|_{i,j-3/2}^{n}\right) + \frac{1-N_1}{\Delta h}\left(E_z\big|_{i,j+1/2}^{n} - E_z\big|_{i,j-1/2}^{n}\right), \tag{2.73}$$

$$\mu\frac{\partial H_y}{\partial t}\bigg|_{i,j}^{n} = \frac{N_1}{3\Delta h}\left(E_z\big|_{i+3/2,j}^{n} - E_z\big|_{i-3/2,j}^{n}\right) + \frac{1-N_1}{\Delta h}\left(E_z\big|_{i+1/2,j}^{n} - E_z\big|_{i-1/2,j}^{n}\right), \tag{2.74}$$

with $N_1 < 0$ and $N_2 < 1.5 - 2N_1$. In order to compute the optimal value of tuning parameters N_1 and N_2, the dispersion relation of (2.71), provided with those of the other higher order schemes in Section 2.4, is solved for the numerical wavenumber k^{num} as a function of N_1 and N_2 at every propagation angle φ. The outcomes of this procedure are then substituted in the definition of the 2-D dispersion error

$$e^{\mathrm{disp}} = \frac{1}{2\pi}\int_0^{2\pi}\left[\frac{k-k^{\mathrm{num}}(\varphi)}{k}\right]^2 d\varphi, \tag{2.75a}$$

with k being the physical wavenumber. Finally, (2.75a) is minimized in terms of N_1 and N_2 to get the required values. Note that when $N_1 = -1/8$ and $N_2 = 0$, the standard (2, 4) FDTD scheme is derived.

The stability of the update equations (2.72)–(2.74) is ensured by the modified Courant criterion of

$$\upsilon \Delta t \le \frac{3\Delta h}{\tau\sqrt{2(3-4N_1)(3-4N_1-2N_2)}} \quad \text{for } \tau \ge 1. \tag{2.75b}$$

The fourth-order FDTD method, so described, accomplishes low-phase errors even at relatively coarse resolutions, while its structural flexibility permits hybridization with the second-order Yee's technique. Also, the algorithm exhibits a relatively acceptable behavior for wideband applications.

On the other hand, extension to 3-D problems can be achieved by changing the contributions of L_1 and L_2 loops in the time-dependent Maxwell's equations, so that phase-velocity error is zeroed at some prefixed propagation angles [48, 51]. According to this approach, the time-domain update of electric and magnetic components is the combination of two separate calculations with a (2, 4) framework. Therefore, each field quantity is obtained from the values of its surrounding dual components on an inner and outer loop (N_1 and N_2) in the principal plane as well as from the corresponding values on another inner and outer loop in each of the two adjacent parallel plains (Figure 2.9). The latter modification introduces two additional tuning parameters N_3 and N_4. Obviously, the field in the principal plane represents an average for the field extending to the distance of the adjacent planes, and so correcting the numerical phase velocity to a larger extent. As anticipated, the complexity of the (2, 4) FDTD expressions is increased but it is still affordable and relatively straightforward to implement. For instance, the

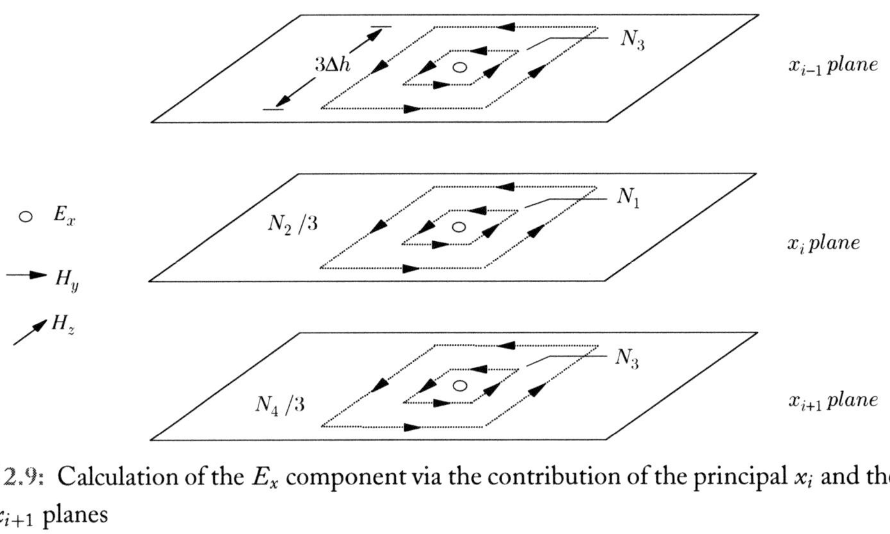

FIGURE 2.9: Calculation of the E_x component via the contribution of the principal x_i and the adjacent x_{i-1} and x_{i+1} planes

E_x component in a 3-D domain becomes

$$\varepsilon \frac{\partial E_x}{\partial t}\bigg|_{i,j,k}^{n+1/2} = \frac{1}{\Delta h}\left[N_1\,\mathcal{A}|_{1/2} + \frac{1}{3}N_2\,\mathcal{B}|_{1/2} + N_3\,\mathcal{A}|_{3/2} + \frac{1}{3}N_4\,\mathcal{B}|_{3/2}\right], \tag{2.76}$$

with

$$\mathcal{A}|_s = H_z\big|_{i,j+s,k}^{n+1/2} - H_z\big|_{i,j-s,k}^{n+1/2} - H_y\big|_{i,j,k+s}^{n+1/2} + H_y\big|_{i,j,k-s}^{n+1/2},$$

$$\begin{aligned}\mathcal{B}|_s = {} & H_z\big|_{i+1,j+s,k}^{n+1/2} - H_z\big|_{i+1,j-s,k}^{n+1/2} + H_z\big|_{i-1,j+s,k}^{n+1/2} - H_z\big|_{i-1,j-s,k}^{n+1/2} - H_y\big|_{i+1,j,k+s}^{n+1/2} \\ & + H_y\big|_{i+1,j,k-s}^{n+1/2} - H_y\big|_{i-1,j,k+s}^{n+1/2} + H_y\big|_{i-1,j,k-s}^{n+1/2},\end{aligned}$$

and $N_1 + N_2 + 2(N_3 + N_4) = 1$. The above facts make the algorithm suitable for large-scale problems, without any stability or convergence defects, on condition that is applied away from material discontinuities.

2.5.2 Angle-Optimized FDTD Algorithms

The principal concept of this efficient dispersion-reduction method, developed in [53, 54], originates from the necessity to annihilate lattice reflection errors around a *preassigned angular span*. Such a need is particularly present in electrically large waveguides or high-directivity antenna problems.

Considering the 2-D TE case first, the semidiscrete equations for E_x, E_y, and H_z may be written as

$$\varepsilon \frac{\partial E_x}{\partial t}\bigg|_{i+1/2,j}^{n+1/2} = \mathcal{C}(\Delta y)\left(H_z\big|_{i+1/2,j+1/2}^{n+1/2} - H_z\big|_{i+1/2,j-1/2}^{n+1/2}\right), \tag{2.77}$$

$$\varepsilon \frac{\partial E_y}{\partial t}\bigg|_{i,j+1/2}^{n+1/2} = \mathcal{C}(\Delta x)\left(H_z\big|_{i-1/2,j+1/2}^{n+1/2} - H_z\big|_{i+1/2,j+1/2}^{n+1/2}\right), \tag{2.78}$$

$$\begin{aligned}\mu \frac{\partial H_z}{\partial t}\bigg|_{i+1/2,j+1/2}^{n} = {} & \mathcal{C}(\Delta y)\left(E_x\big|_{i+1/2,j+1}^{n} - E_x\big|_{i+1/2,j}^{n}\right) \\ & + \mathcal{C}(\Delta x)\left(E_y\big|_{i,j+1/2}^{n} - E_y\big|_{i+1,j+1/2}^{n}\right)\end{aligned} \quad \text{for } \mathcal{C}(\Delta h) = \frac{a - b(\Delta h)^2\nabla^2}{\Delta h}, \tag{2.79}$$

where a and b are supplementary degrees of freedom, whose proper evaluation appends artificial dispersion effects that compensate for the numerical dispersion errors and control phase correction. Moreover, operator ∇^2 converts second-order temporal derivatives to second-order spatial ones via the homogeneous Helmholtz equation. For illustration, the fully discretized FDTD form of (2.77) becomes

$$
\begin{aligned}
E_x\big|_{i+1/2,j}^{n+1} &= E_x\big|_{i+1/2,j}^{n} + \frac{\Delta t}{\Delta y}\left\{\left[a + b\left(3 + 2\frac{(\Delta y)^2}{(\Delta x)^2}\right)\right]\left(H_z\big|_{i+1/2,j+1/2}^{n+1/2} - H_z\big|_{i+1/2,j-1/2}^{n+1/2}\right)\right. \\
&+b\frac{(\Delta y)^2}{(\Delta x)^2}\left(H_z\big|_{i+3/2,j-1/2}^{n+1/2} - H_z\big|_{i+3/2,j+1/2}^{n+1/2} + H_z\big|_{i-1/2,j-1/2}^{n+1/2} - H_z\big|_{i-1/2,j+1/2}^{n+1/2}\right) \\
&\left.+b\left(H_z\big|_{i+1/2,j-3/2}^{n+1/2} - H_z\big|_{i+1/2,j+3/2}^{n+1/2}\right)\right\}. \qquad (2.80)
\end{aligned}
$$

Concerning the stability of the previous schemes, von Neumann analysis leads to the Courant criterion

$$
\upsilon\Delta t \leq \frac{\Delta h}{\sqrt{2}(a+8b)} \quad \text{for } \Delta h = \Delta x = \Delta y. \qquad (2.81)
$$

If a monochromatic wave is substituted in (2.80) and the other FDTD equations, the dispersion relation is

$$
\left[\frac{\sqrt{2}}{\gamma}\sin\left(\frac{\gamma k \Delta h}{2\sqrt{2}}\right)\right] = \mathcal{X}^2 + \mathcal{Y}^2, \qquad (2.82)
$$

in which $\gamma = \upsilon\Delta t\sqrt{2}/\Delta h$ is the Courant-Friedrichs-Lewy number (CFLN), k is the wavenumber with $k_x = k\cos\theta$, $k_y = k\sin\theta$ and $k = \sqrt{k_x^2 + k_y^2}$, θ is the angle of propagation, and

$$
\mathcal{X} = b\sin\left(\frac{3k_y\Delta h}{2}\right) - \left[a + 3b + 4b\sin^2\left(\frac{k_x\Delta h}{2}\right)\right]\sin\left(\frac{k_y\Delta h}{2}\right), \qquad (2.83a)
$$

$$
\mathcal{Y} = b\sin\left(\frac{3k_x\Delta h}{2}\right) - \left[a + 3b + 4b\sin^2\left(\frac{k_y\Delta h}{2}\right)\right]\sin\left(\frac{k_x\Delta h}{2}\right). \qquad (2.83b)
$$

It is apparent that for a given set of θ and γ, the phase-velocity error can be zeroed if a, b are appropriately selected. This implies that one may solve for a, b from (2.82) or define the dispersion error

$$
\nu = \left[\frac{\sqrt{2}}{\gamma}\sin\left(\frac{\gamma k \Delta h}{2\sqrt{2}}\right)\right] - \left[\mathcal{X}^2 + \mathcal{Y}^2\right], \qquad (2.84)
$$

and let $\nu = 0$. By expanding ν in terms of $p = \Delta h/\lambda$, for $k\Delta h = 2\pi\Delta h/\lambda$, it is derived that

$$
\begin{aligned}
\nu &= (1-a^2)\pi^2 p^2 + \frac{1}{12}\left\{a\left[3(a-32b) + a\cos(4\theta)\right] - 2\gamma^2\right\}\pi^4 p^4 \\
&+ \frac{1}{180}\left[720ab - 5a^2 - 2880b^2 - 3a(a-80b)\cos(4\theta)\right]\pi^6 p^6 + O(p^8). \qquad (2.85)
\end{aligned}
$$

In order to find a, b, the first two terms of (2.85) are set to zero. This yields

$$a = 1.0 \quad \text{and} \quad b = \frac{1}{96}\left[3 - 2\gamma^2 + \cos(4\theta)\right], \tag{2.86}$$

with an error of $O(p^6)$, while if $0 \le b \le 1/24$, the optimal γ value for a stable simulation is 0.75.

However, in practical applications, single-frequency studies are relatively rare. Thus, it would be instructive to optimize the dispersion error *around* a frequency or a propagation angle. For this aim, presuming that monomials p^m in (2.85) constitute a basis of an infinite-dimensional linear space, a different choice may be performed. According to [53], two processes are the most attractive: (a) the Butterworth filtering corresponding to basis $\{(p - p_0)^m\}$ and (b) the Chebyshev filtering corresponding to $\{T_m(p - p_0)/\Delta p\}$. In the previous, p_0 is the central frequency, Δp is the desired frequency range and $T_m(x)$ are the mth-order Chebyshev polynomials of the first kind. In this framework, (2.85) is expanded up to order six and written as

$$\nu = p^2[w_1, 0, w_3, 0, w_5]\mathbf{U}^T + O(p^8), \tag{2.87}$$

in which w_i, for $i = 1, 2, \ldots, 5$, are the coefficients assigned to each monomial.

The first alternative basis forms matrix $\mathbf{P} = [1, (p - p_0), (p - p_0)^2, (p - p_0)^3, (p - p_0)^4]$ and provides

$$\mathbf{P}^T = \mathbf{A}\mathbf{U}^T \quad \text{with} \quad \mathbf{A} = \begin{bmatrix} 1 & 0 & 0 & 0 & 0 \\ -p_0 & 1 & 0 & 0 & 0 \\ p_0^2 & -2p_0 & 1 & 0 & 0 \\ -p_0^3 & 3p_0^2 & -3p_0 & 1 & 0 \\ p_0^4 & -4p_0^3 & 6p_0^2 & -4p_0 & 1 \end{bmatrix}, \tag{2.88}$$

which modifies (2.85) as

$$\nu = p^2\left[w_1, 0, w_3, 0, w_5\right]\mathbf{A}^{-1}\mathbf{P}^T + O(p^8) = p^2\left[\bar{w}_1, \bar{w}_2, \bar{w}_3, \bar{w}_4, \bar{w}_5\right]\mathbf{P}^T + O(p^8). \tag{2.89}$$

To obtain the unknown a, b, elements $\bar{w}_1$ and $\bar{w}_2$ must be set to zero, namely

$$\begin{aligned} \bar{w}_1 = 0 \quad \Rightarrow \quad & \left[1 - \frac{1}{36}(36 - 9\pi^2 p_0^2 + \pi^4 p_0^4)\right]\pi^2 + \frac{1}{180}\left\{720(\pi^2 p_0^2 - 2)ab - 30\gamma^2\right. \\ & \left. + 2\pi^2 p_0^2(\gamma^4 - 1440b^2) - 3a\left[(\pi^2 p_0^2 - 5)a - 80\pi^2 p_0^2 b\right]\right\}\pi^4 p_0^2 = 0, \end{aligned} \tag{2.90a}$$

$$\begin{aligned} \bar{w}_2 = 0 \quad \Rightarrow \quad & \frac{1}{90}\left\{(45 - 10\pi^2 p_0^2)a^2 + 1440(\pi^2 p_0^2 - 1)ab - 30\gamma^2 + 4\pi^2 p_0^2(\gamma^2 - 1440b^2)\right. \\ & \left. - 3a\left[(5 - 2\pi^2 p_0^2)a + 160\pi^2 p_0^2 b\right]\cos(4\theta)\right\}\pi^4 p_0 = 0. \end{aligned} \tag{2.90b}$$

Consequently, the error around the central frequency is controlled by the term $p^2\tilde{w}_3(p - p_0)^2$.

For the next basis, the numerical dispersion around the desired frequency is governed by

$$\begin{aligned} \nu = p^2 \{&\tilde{w}_2 T_2 [(p - p_0)/\Delta p] + \tilde{w}_3 T_3 [(p - p_0)/\Delta p] \\ &+\tilde{w}_4 T_4 [(p - p_0)/\Delta p]\} + O(p^8), \end{aligned} \tag{2.91}$$

where $\tilde{w}_i$, for $i = 2, 3, 4$, are the respective Chebyshev coefficients. It is stated that the dispersion error near the central frequency varies very smoothly. Also, the angle-optimized algorithm is far more isotropic than the usual (2, 4) FDTD scheme. The latter may be greatly improved through the analysis of (2.82)–(2.90).

Likewise, the 3-D formulation expresses Maxwell's equations as

$$\left.\begin{aligned} (\boldsymbol{\mathcal{D}} \cdot \nabla) \times \mathbf{E} = -\partial \mathbf{B}/\partial t \\ (\boldsymbol{\mathcal{D}} \cdot \nabla) \times \mathbf{H} = \partial \mathbf{D}/\partial t \end{aligned}\right\} \text{ with } \boldsymbol{\mathcal{D}} = \operatorname{diag}\left\{a - b(\Delta x)^2\nabla^2, a - b(\Delta y)^2\nabla^2, a - b(\Delta z)^2\nabla^2\right\}. \tag{2.92}$$

Application of the above steps in (2.92) yields the dispersion error of

$$\begin{aligned} v = (1 - a^2)\pi^2 p^2 + \frac{1}{288}\{&63a^2 - 2304ab - 32\gamma^2 \\ &+3a^2\left[4\cos(2\theta) + 7\cos(4\theta) + 8\cos(4\phi)\sin^4\theta\right]\}\pi^4 p^4 + O(p^6), \end{aligned} \tag{2.93}$$

with θ and ϕ the polar angles of propagation that allow broadband simulations. By implementing the two filtering schemes, additional enhancement may be equivalently achieved.

For the numerical verification of the preceding schemes, a 2-D example implementing a nonfiltered higher order FDTD technique optimized at 22.5° is analyzed. The respective coefficients are $a = 1.0$ and $b = 0.0198417$ for a central frequency that yields $p_0 = 0.08$. Figure 2.10(a) provides the relative error of the normalized phase velocity versus wavelengths per cell for different angles. Notice the significantly small discrepancies around the optimization angle. Analogous results are obtained for the 3-D problem of Figure 2.10(b) that addresses the above error for a Chebyshev angle-optimized FDTD formulation at $\theta = 90^\circ$ and $\varphi = 0^\circ$ for different center frequencies, i.e., different p_c from the basis $\{(p - p_c)^m\}$. An important remark, herein, is that the local dispersion error is practically *smaller* for higher frequencies than for lower ones as opposed to ordinary FDTD simulations. Such a property is very helpful, since at the upper part of the spectrum, the domain is electrically larger and therefore the accumulated phase errors tend to increase [54].

2.5.3 Dispersion-Relation-Preserving FDTD Schemes

The design of optimal second-order and higher order FDTD algorithms with dispersion-relation-preserving (DRP) properties constitutes a promising tool for the drastic reduction of dispersion errors, as firstly presented and extensively investigated in [55, 56]. Actually, the key

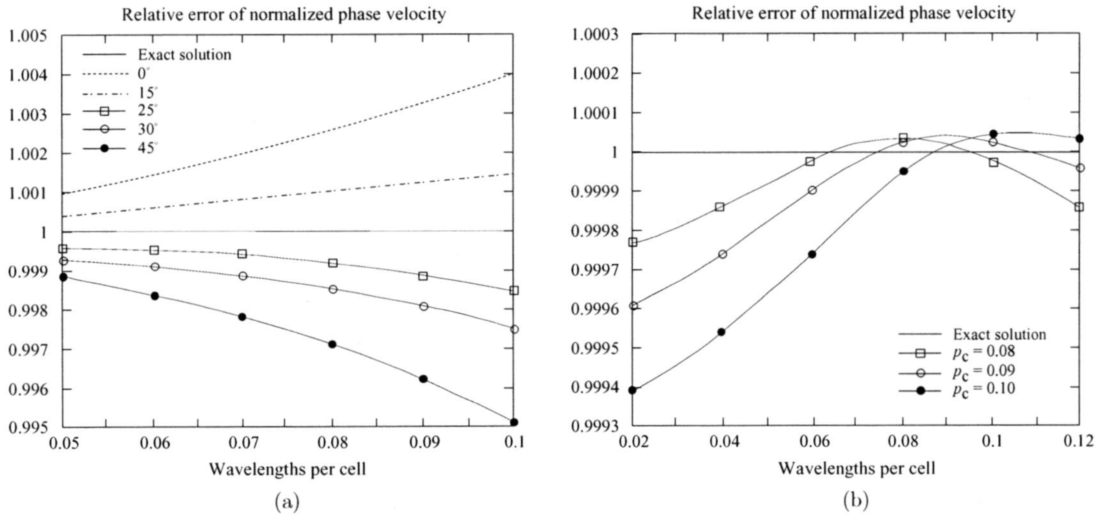

FIGURE 2.10: Relative error of the normalized phase velocity (a) for a 2-D angle-optimized FDTD scheme, optimized at 22.5° without filtering and (b) for a 3-D Chebyshev angle-optimized FDTD scheme at $\theta = 90°$, $\varphi = 0°$ with different p_c values

goal of a higher order DRP scheme, as compared to the traditional one, resides in the evaluation of a certain set of coefficients that *weigh* every spatial derivative in Maxwell's curl equations.

Let us focus on the 3-D (2, 4) FDTD equations

$$\begin{aligned} \mathbf{E}|^{n+1} &= \mathbf{E}|^{n} + \frac{\Delta t}{\varepsilon}\boldsymbol{\mathcal{D}} \times \mathbf{H}|^{n+1/2} \\ \mathbf{H}|^{n+1/2} &= \mathbf{H}|^{n-1/2} - \frac{\Delta t}{\mu}\boldsymbol{\mathcal{D}} \times \mathbf{E}|^{n} \end{aligned} \tag{2.94}$$

$$\text{with}\quad \boldsymbol{\mathcal{D}} = \frac{1}{\Delta x}\left(A_x\mathbf{d}_x + B_x\bar{\mathbf{d}}_x\right)\hat{\mathbf{x}} + \frac{1}{\Delta y}\left(A_y\mathbf{d}_y + B_y\bar{\mathbf{d}}_y\right)\hat{\mathbf{y}} + \frac{1}{\Delta z}\left(A_z\mathbf{d}_z + B_z\bar{\mathbf{d}}_z\right)\hat{\mathbf{z}}$$

A_ζ, B_ζ unknown coefficients, for $\zeta \in (x, y, z)$, $\mathbf{d}_\zeta\left[f\big|^{n}_{l+1/2}\right] = f\big|^{n}_{l+1} - f\big|^{n}_{l}$, and $\bar{\mathbf{d}}_\zeta\left[f\big|^{n}_{l+1/2}\right] = f\big|^{n}_{l+2} - f\big|^{n}_{l-1}$.

Expanding electric and magnetic fields into a discrete set of Fourier modes and substituting the outcomes into (2.94), along with $\mathbf{E}(t) = \boldsymbol{\mathcal{E}}e^{j\omega n\Delta t}$ and $\mathbf{H}(t) = \boldsymbol{\mathcal{H}}e^{j\omega(n+1/2)\Delta t}$, one acquires

$$\boldsymbol{\mathcal{E}}\sin\left(\frac{\omega\Delta t}{2}\right) = -\frac{\Delta t}{\varepsilon}\mathbf{G} \times \boldsymbol{\mathcal{H}}, \tag{2.95}$$

$$\boldsymbol{\mathcal{H}}\sin\left(\frac{\omega\Delta t}{2}\right) = \frac{\Delta t}{\mu}\mathbf{G} \times \boldsymbol{\mathcal{E}}, \tag{2.96}$$

for

$$\mathbf{G} = G_x\hat{\mathbf{x}} + G_y\hat{\mathbf{y}} + G_z\hat{\mathbf{z}} \quad\text{and}\quad G_\zeta = \frac{1}{\Delta\zeta}\left[A_\zeta\sin\left(\frac{k_x\Delta\zeta}{2}\right) + B_\zeta\sin\left(\frac{3k_x\Delta\zeta}{2}\right)\right].$$

The dispersion relation of (2.94) can be extracted if either $\mathcal{E}$ or $\mathcal{H}$ is eliminated from (2.95). To obtain coefficients A_ζ and B_ζ, the maximum dispersion error will be initially expanded in a rapidly convergent series as a function of propagation angles θ and φ and then its dominant terms will be set to zero. Hence, the coefficients, so computed, minimize the maximum dispersion error at all angles in completely adjustable way.

Consider a plane wave traveling in the (θ, φ) direction with $k_x = k\sin\theta\cos\varphi$, $k_y = k\sin\theta\sin\varphi$, and $k_z = k\cos\theta$. Recalling that such a wave can be decomposed into a TE and a TM polarization, the former part is selected for our analysis, yielding

$$\mathcal{E}_x = -\mathcal{E}\sin\varphi, \qquad \mathcal{E}_y = -\mathcal{E}\cos\varphi, \qquad \mathcal{E}_z = 0, \tag{2.97a}$$

$$\mathcal{H}_x = -\mathcal{H}\cos\theta\cos\varphi, \qquad \mathcal{H}_y = -\mathcal{H}\cos\theta\sin\varphi, \qquad \mathcal{H}_z = -\mathcal{H}\sin\theta. \tag{2.97a}$$

If (2.97) are plugged into (2.96),

$$\mathcal{H}\cos\theta\sin\left(\frac{\omega\Delta t}{2}\right) = \frac{\Delta t}{\mu\Delta z}\left[A_z\sin\left(\frac{k_z\Delta z}{2}\right) + B_z\sin\left(\frac{3k_z\Delta z}{2}\right)\right]\mathcal{E}, \tag{2.98}$$

$$\begin{aligned}\mathcal{H}\sin\theta\sin\left(\frac{\omega\Delta t}{2}\right) = \frac{\Delta t}{\mu}\Bigg\{&\frac{1}{\Delta x}\cos\varphi\left[A_x\sin\left(\frac{k_x\Delta x}{2}\right) + B_x\sin\left(\frac{3k_x\Delta x}{2}\right)\right]\\ &+\frac{1}{\Delta y}\sin\varphi\left[A_y\sin\left(\frac{k_y\Delta y}{2}\right) + B_y\sin\left(\frac{k_y\Delta y}{2}\right)\right]\Bigg\}\mathcal{E}.\end{aligned} \tag{2.99}$$

Due to the symmetry of the problem, (2.98) suffices for the determination of A_z and B_z, whereas the remaining coefficients will be simultaneously computed. In view of these observations, (2.98) becomes

$$\frac{\sqrt{3}}{\gamma_\zeta}\sin\left(\frac{\pi p_\zeta\gamma_\zeta}{\sqrt{3}}\right)\cos\theta = A_\zeta\sin(\pi p_s\cos\theta) + B_\zeta\sin(3\pi p_s\cos\theta), \tag{2.100}$$

where $\gamma_\zeta = \sqrt{3}\upsilon\Delta t/\Delta\zeta$ is the corresponding CFLN and $p_\zeta = \Delta\zeta/\lambda$ for $\zeta \in (x, y, z)$. From (2.100), the necessary error functional is defined as

$$\begin{aligned}\nu(A_\zeta, B_\zeta, \theta, \varphi) = &\frac{\sqrt{3}}{\gamma_\zeta}\sin\left(\frac{\pi p_\zeta\gamma_\zeta}{\sqrt{3}}\right)\cos\theta\\ &-\left[A_\zeta\sin(\pi p_s\cos\theta) + B_\zeta\sin(3\pi p_s\cos\theta)\right],\end{aligned} \tag{2.101}$$

and in order to evaluate A_ζ and B_ζ, it will be set to zero. This is performed if (2.100) is expanded in a series of spherical harmonics $Y_{l,m}(\theta, \varphi)$, weighted by coefficients $c_{l,m}$:

$$c_{l,m} = \int_0^{2\pi}\int_0^{\pi}\nu(A_\zeta, B_\zeta, \theta, \varphi)Y_{l,m}(\theta, \varphi)\sin\theta\,d\theta\,d\varphi. \tag{2.102}$$

Notice that since (2.101) is actually independent of φ, only $m = 0$ azimuthal harmonics are required. So,

$$Y_{l,0}(\theta, \varphi) = \frac{1}{\sqrt{2\pi}} \Theta_{l,0}(\theta) \quad \text{and} \quad \Theta_{l,0} = \sqrt{\frac{2l+1}{2}} P_l^0(\cos\theta),$$

with $P_l^0(\cos\theta)$ the associated Legendre functions, which if $m = 0$ are equal to $P_l(\cos\theta)$ for $l \geq 0$ and $-l \leq m \leq l$. After the relevant mathematical manipulations, (2.101) results in

$$\begin{aligned} v(A_\zeta, B_\zeta, \theta, \varphi) = & \sum_{l=0}^{\infty} \sqrt{(4l+3)\pi} \left[A_\zeta I_{2l+1}(\pi p_\zeta) + B_\zeta I_{2l+1}(3\pi p_\zeta) \right] Y_{2l+1,0}(\theta, \varphi) \\ & + \frac{2\sqrt{\pi}}{\gamma_\zeta} \sin\left(\frac{\pi p_\zeta \gamma_\zeta}{\sqrt{3}} \right) Y_{1,0} \end{aligned} \tag{2.103}$$

for

$$I_{l(\text{odd})}(x) = \left[\frac{2(2l-3)}{2l-5} - \frac{(2l-1)(2l-3)}{x^2} \right] I_{l-2}(x) - \frac{2l-1}{2l-5} I_{l-4}(x)$$

$$I_1(x) = \frac{2(\sin x - x\cos x)}{x^2} \quad \text{and} \quad I_2(x) = \frac{6(2x^2-5)\sin x - 2x(x^2-15)\cos x}{x^4}.$$

Enforcing the first two terms, $Y_{1,0}(\theta, \varphi)$ and $Y_{3,0}(\theta, \varphi)$, to be zero and expanding the outcomes in Taylor series, coefficients A_ζ and B_ζ are given by

$$A_\zeta = \frac{9}{8} - \frac{1}{64} 4\pi^2 p_\zeta^2 (\gamma_\zeta^2 - 1) \quad \text{and} \quad B_\zeta = -\frac{1}{24} - \frac{1}{1728} 4\pi^2 p_\zeta^2 (\gamma_\zeta^2 - 9), \tag{2.104}$$

which, in an effort to avoid third-order temporal differentiation, are written in the form of (2.77)–(2.79) as

$$A_\zeta = \frac{9}{8} - \frac{1}{64} (\gamma_\zeta^2 - 1)(\Delta\zeta)^2 \nabla^2 \quad \text{and} \quad B_\zeta = -\frac{1}{24} - \frac{1}{1728} (\gamma_\zeta^2 - 9)(\Delta\zeta)^2 \nabla^2. \tag{2.105}$$

Observe that the first terms on the right hand side of (2.105) are the coefficients of the traditional fourth-order spatial operator. The stability criterion of the aforementioned technique, for $\Delta h = \Delta x = \Delta y = \Delta z$, is

$$v\Delta t \leq \frac{\gamma \Delta h}{\sqrt{3}} = \frac{\Delta h}{\sqrt{3}\,|C|_{\max}}, \quad \text{where } |C|_{\max} = \max\left[\frac{7}{6} + \frac{36\gamma^2}{144} \right], \tag{2.106}$$

maximized for $\gamma = 0$ [55]. Obviously, the use of the filtering schemes, discussed in Section 2.4.2, will offer further improvements especially where strenuous expansion series have to be employed.

To investigate the efficiency of the DRP-FDTD method, its phase error per wavelength for a Chebyshev filtering process is examined in Figure 2.11(a). As can be observed,

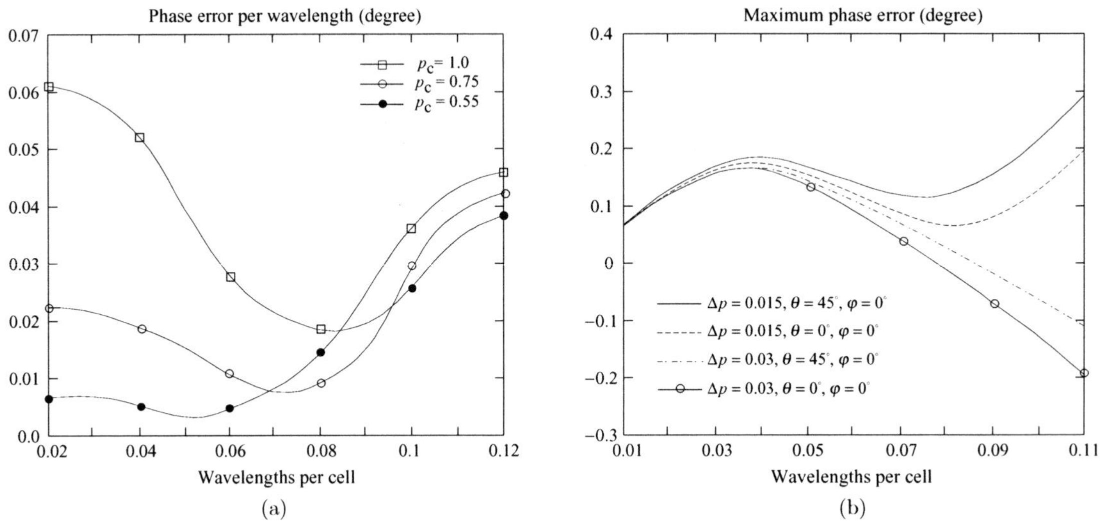

FIGURE 2.11: (a) Phase error per wavelength of Chebyshev filtering for the 3-D DRP schemes with different p_c. (b) Maximum phase error over the largest wavelength via diverse Δp and $p_c = 0.15$

for systematically chosen p_c values, the error may become very small. Moreover, Figure 2.11(b) presents the accumulated phase error at two angle sets with the largest error difference, namely $(\theta, \varphi) = (0°, 0°)$ and $(\theta, \varphi) = (45°, 0°)$ for a variety of Δp. Results indicate that the magnitude of the error can be diminished if Δp values vary according to a specific rate.

In conclusion, the (2, 4) DRP-FDTD method accomplishes a noteworthy reduction of dispersion errors leading to reliable, accurate as well as convergent numerical simulations and thus may be deemed a significant contribution in the advancement of conventional higher order time-domain modeling.

2.5.4 Design of Controllable Dispersion-Error Techniques

This section deals with the construction of optimal higher order FDTD schemes with adjustable dispersion error. Rather than implementing the ordinary approaches, based on Taylor series expansion, the modified finite-difference operators are designed via alternative procedures that enhance the wideband capabilities of the resulting numerical techniques. First, an algorithm founded on the separate optimization of spatial and temporal derivatives is developed. Additionally, a second method is derived that reliably reflects artificial lattice inaccuracies via the necessary algebraic expressions. Utilizing the same kind of differential operators as the typical fourth-order scheme, both approaches retain their reasonable computational complexity and memory requirements. Furthermore, analysis substantiates that important error compensation is achieved around a specific frequency, while reduced errors are acquired for higher frequencies as well.

2.5.4.1 Optimized Finite-Difference Operators

For the discretization of spatial derivatives, the particular method uses a family of central difference operators, while a parametric expression with an *extra* degree of freedom for the temporal derivatives is employed [57]. In two dimensions, these approximants have the general forms

$$\left.\frac{\partial f}{\partial \zeta}\right|_l^n \cong \frac{A_\zeta}{\Delta\zeta}\left(f\big|_{l+1/2}^n - f\big|_{l-1/2}^n\right) + \frac{B_\zeta}{\Delta\zeta}\left(f\big|_{l+3/2}^n - f\big|_{l-3/2}^n\right) \quad \text{and}$$
$$\left.\frac{\partial f}{\partial t}\right|_l^n \cong \frac{R}{\Delta t}\left(f\big|_l^{n+1/2} - f\big|_l^{n-1/2}\right), \tag{2.107}$$

with $\zeta \in (x, y)$. The optimization process uses the term that determines the phase variation of plane waves, i.e., $p = e^{j(\omega t - \mathbf{k}\cdot\mathbf{r})}$ as a test function in an effort to resolve relations for the errors that arise from the application of (2.107). Considering the corresponding exact differentiation

$$\frac{\partial p}{\partial t} = j\omega p \quad \text{and} \quad \frac{\partial p}{\partial \zeta} = -jk_\zeta p, \tag{2.108}$$

as well as the analogous outcomes of finite differencing

$$\left.\frac{\partial f}{\partial t}\right|_l^n = \frac{2j}{\Delta t} R \sin\left(\frac{\omega \Delta t}{2}\right) p^n \quad \text{and}$$
$$\left.\frac{\partial f}{\partial \zeta}\right|_l^n = -\frac{2j}{\Delta\zeta}\left[A_\zeta \sin\left(\frac{k_\zeta \Delta\zeta}{2}\right) + B_\zeta \sin\left(\frac{3k_\zeta \Delta\zeta}{2}\right)\right] p_l, \tag{2.109}$$

two distinct error indicators may be defined as the difference between the multiplicative factors of p in (2.108) and (2.109). In this manner, temporal and spatial errors are expressed as

$$\nu_{\mathrm{T}} = \left[\frac{2R}{\Delta t}\sin\left(\frac{\omega\Delta t}{2}\right) - \omega\right]^2 \quad \text{and}$$
$$\nu_\zeta^{2\mathrm{D}} = \left[\frac{2A_\zeta}{\Delta\zeta}\sin\left(\frac{k_\zeta\Delta\zeta}{2}\right) + \frac{2B_\zeta}{\Delta\zeta}\sin\left(\frac{3k_\zeta\Delta\zeta}{2}\right) - k_\zeta\right]^2, \tag{2.110}$$

where the inequality between the numerical and physical wavenumber has been enforced. The prior expressions can be viewed as polynomials of the unknown coefficients whose values minimize (2.110). Focusing on ν_{T} and bearing in mind the condition $R^2 \geq 0$, one deduces that the solution of $\partial\nu_{\mathrm{T}}/\partial R = 0$ provides a value of R that minimizes the temporal error. More specifically, this process yields

$$R = \frac{\omega\Delta t}{2\sin\left(\frac{\omega\Delta t}{2}\right)}. \tag{2.111}$$

To construct the optimized spatial operator, the complicated angular dependence of ν_ζ^{2D} must be somehow circumvented. Since an error reduction over all propagation angles is desired, the usage of the integrated error terms is proven to be fairly convenient. Therefore, it is denoted that

$$V_\zeta^{2D} = \int_0^{2\pi} \nu_\zeta^{2D} d\varphi. \tag{2.112}$$

The unknown coefficients can be computed via the minimization of V_ζ^{2D} at a preselected frequency, which is guaranteed by the vanishing of the V_ζ^{2D} gradient or

$$\frac{\partial V_\zeta^{2D}}{\partial A_\zeta} = \frac{\partial V_\zeta^{2D}}{\partial B_\zeta} = 0. \tag{2.113}$$

As a consequence, the spatial operators are evaluated by

$$\begin{bmatrix} 1 - J_0(k\Delta\zeta) & J_0(k\Delta\zeta) - J_0(2k\Delta\zeta) \\ J_0(k\Delta\zeta) - J_0(2k\Delta\zeta) & 1 - J_0(3k\Delta\zeta) \end{bmatrix} \begin{bmatrix} A_\zeta \\ B_\zeta \end{bmatrix} = \begin{bmatrix} J_1\left(\frac{k\Delta\zeta}{2}\right) \\ J_1\left(\frac{3k\Delta\zeta}{2}\right) \end{bmatrix} k\Delta\zeta, \tag{2.114}$$

with J_m the mth-order Bessel function of the first kind. Observe that the minimization of V_ζ^{2D} is assured by the positive definite character of the Hessian matrix of (2.114).

Extension to the 3-D case does not present any difficulties, as temporal derivatives receive exactly the same treatment, while for the spatial ones, the ν_ζ^{3D} error indicator for $\zeta \in (x, y, z)$ has to be minimized by the integrated error V_ζ^{3D}. The fulfillment of appropriate conditions like those in (2.113) gives

$$\begin{bmatrix} 1 - \frac{\sin(k\Delta\zeta)}{k\Delta\zeta} & \frac{2\sin(k\Delta\zeta) - \sin(2k\Delta\zeta)}{2k\Delta\zeta} \\ \frac{2\sin(k\Delta\zeta) - \sin(2k\Delta\zeta)}{2k\Delta\zeta} & 1 - \frac{\sin(3k\Delta\zeta)}{3k\Delta\zeta} \end{bmatrix} \begin{bmatrix} A_\zeta \\ B_\zeta \end{bmatrix} = \begin{bmatrix} \frac{4\sin\left(\frac{k\Delta\zeta}{2}\right)}{k\Delta\zeta} - 2\cos\left(\frac{k\Delta\zeta}{2}\right) \\ \frac{4\sin\left(\frac{3k\Delta\zeta}{2}\right)}{9k\Delta\zeta} - \frac{2}{3}\cos\left(\frac{3k\Delta\zeta}{2}\right) \end{bmatrix}. \tag{2.115}$$

It should be mentioned, herein, that the selection of the *correct design frequency* around which the above schemes exhibit their enhanced behavior comprises the most serious factor for their broadband utilization.

2.5.4.2 Controllable Error Estimators

According to the initial optimization concept, this technique launches a set of parametric expressions for the differential operators, whose evaluation is based on proper *error estimators* [59]. What is actually desirable – yet analytically not feasible – for these estimators is that their dependence on frequency and propagation angle should resemble that of the true dispersion error. Hence, an expansion of these expressions in terms of basis angular functions is performed, which enables the accuracy improvement regardless the direction of propagation. Depending on the policy, the foregoing parametric operators are computed, diverse higher order FDTD algorithms with controllable features may be devised.

For a 2-D domain with a lossless homogeneous medium, Maxwell's first equation $\varepsilon\,\partial E_x/\partial t = \partial H_z/\partial y$ for the TE case (similarly for TM and the other equations) is discretized by

$$\left.\frac{\partial f}{\partial \zeta}\right|_{i,j}^{n} = \frac{1}{\Delta\zeta}\left[\sum_{s=1}^{M} A_\zeta^s \left(f\big|_{i+(2s-1)/2,j}^{n} - f\big|_{i-(2s-1)/2,j}^{n}\right) + B_\zeta\left(f\big|_{i+1/2,j+1}^{n} + f\big|_{i+1/2,j-1}^{n}\right.\right.$$
$$\left.\left. - f\big|_{i-1/2,j+1}^{n} - f\big|_{i-1/2,j-1}^{n}\right)\right] \quad \text{for} \quad \zeta \in (x, y), \tag{2.116a}$$

and

$$\left.\frac{\partial f}{\partial t}\right|_{i,j}^{n} = \frac{f\big|_{i,j}^{n+1/2} - f\big|_{i,j}^{n-1/2}}{\Delta t}. \tag{2.116b}$$

Actually, the prior spatial operator comprises two parts. The first includes nodal points along the differentiation axis, while the second involves points symmetrically located around the node under study. Differently speaking, (2.116a) may be regarded as the linear combination of diverse second-order spatial operators. Although the value of parameter M could be arbitrarily high, it is proven that if $M = 1, 2, 3$, the resulting schemes will constitute the ideal choices from a computational overhead and accuracy perspective.

Assuming a plane-wave propagation toward a direction depicted by angle φ, one may write $E_x = -E\sin\varphi$, $E_y = E\cos\varphi$, and $H_z = H$ and obtain $E_x = -\sqrt{\mu/\varepsilon}\,H_z \sin\varphi$ through $E/H = \sqrt{\mu/\varepsilon}$. Application of the latter expression to the aforementioned Maxwell's equation yields

$$\sqrt{\mu\varepsilon}\,\sin\varphi\frac{\partial H_z}{\partial t} + \frac{\partial H_z}{\partial t} = 0. \tag{2.117}$$

Subsequently, (2.117) is discretized by the operators of (2.116) and H_z is replaced with plane-wave solutions $H_{z0}e^{-j(k_x^{\mathrm{num}}x + k_y^{\mathrm{num}}y)}$, where $\mathbf{k}^{\mathrm{num}} = k_x^{\mathrm{num}}\hat{\mathbf{x}} + k_y^{num}\hat{\mathbf{y}}$ is the numerical wavenumber. These steps lead to

$$\nu(w, k^{\mathrm{num}}, \varphi) = \frac{1}{\upsilon\Delta t}\sin\left(\frac{\omega\Delta t}{2}\right)\sin\varphi - \frac{1}{\Delta y}\sum_{s=1}^{M} A_y^s \sin\left[\frac{(2s-1)k^{\mathrm{num}}\sin\varphi\Delta y}{2}\right]$$
$$-\frac{2}{\Delta y}B_y\Delta^y\cos(k^{\mathrm{num}}\cos\varphi\Delta x)\sin\left(\frac{k^{\mathrm{num}}\sin\varphi\Delta y}{2}\right) = 0 \tag{2.118}$$

with $\sqrt{\mu\varepsilon} = 1/\upsilon$ and $k_x^{\mathrm{num}} = k^{\mathrm{num}}\cos\varphi$, $k_y^{\mathrm{num}} = k^{\mathrm{num}}\sin\varphi$. If the equality between the numerical and physical wavenumber is enforced ($k^{\mathrm{num}} = k = \omega/\upsilon$), it becomes apparent that (2.118) will be no longer valid, namely $\nu(\omega, k, \varphi) = \nu(\omega, \varphi) \neq 0$, because practically, $k^{\mathrm{num}} \neq k$.

Since the vanishing of ν for certain directions of propagation reveals that this FDTD scheme is exact, (2.118) can be selected as the desired discretization error estimator. Therefore, the parametric expressions of coefficients A_ζ and B_ζ are going to be determined with respect to the minimization of ν.

To estimate ν, irrespective of φ, and enhance the range of the resulting operators, (2.118) is analyzed in terms of $\{\sin(m\varphi), \cos(m\varphi)\}(m = 1, 2, \ldots)$ cylindrical harmonic basis functions. Moreover, for our mathematical manipulations, formula

$$\sin(\alpha \sin\varphi) = 2\sum_{m=0}^{\infty} J_{2m-1}(\alpha)\sin[(2m-1)\varphi] \tag{2.119}$$

is employed in conjunction with

$$\sin(\alpha \sin\varphi)\cos(\beta\cos\varphi) = 2\sum_{m=0}^{\infty} J_{2m-1}(\xi)\cos[(2m-1)\tau]\sin[(2m-1)\varphi], \tag{2.120}$$

for $\xi = \sqrt{\alpha^2+\beta^2}$ and $\tau = \tan^{-1}(\beta/\alpha)$. If (2.119) and (2.120) are plugged into (2.118), one obtains

$$\nu(\omega,\varphi) = \sum_{m=0}^{\infty} \mathcal{K}_m(\omega)\sin[(2m-1)\varphi], \tag{2.121}$$

where

$$\begin{aligned}\mathcal{K}_m(\omega) = {} & \frac{1}{\upsilon\Delta t}\sin\left(\frac{\omega\Delta t}{2}\right)\delta_m - \frac{2}{\Delta y}\sum_{s=1}^{\infty} A_y^s J_{2m-1}\left[\frac{(2s-1)k\Delta y}{2}\right] \\ & -\frac{4}{\Delta y}B_y J_{2m-1}(\kappa)\cos[(2m-1)\varphi_0],\end{aligned} \tag{2.122}$$

with

$$\kappa = k\sqrt{(\Delta x)^2 + \frac{1}{4}(\Delta y)^2}, \quad \varphi_0 = \tan^{-1}\left(\frac{2\Delta y}{\Delta x}\right), \quad \delta_m = \begin{cases}1, & \text{if } m = 0\\ 0, & \text{if } m \neq 0\end{cases}. \tag{2.123}$$

The ideal case $\nu = 0$ for $0 \leq \varphi \leq 2\pi$ and a fixed frequency ω is equivalent to $\mathcal{K}_m(\omega) = 0$ for $m = 0, 1, 2, \ldots$. As a matter of fact, the main objective of the investigation is a small, but nonzero, value for ν which is pursued from the vanishing of the leading terms in the corresponding expansion series.

In view of these observations, the optimized higher order FDTD schemes are promptly designed. The first one is probably the simplest, since it minimizes the dispersion error at one frequency ω_0. This is attained by zeroing the first three terms of (2.121) for all directions, namely $\mathcal{K}_0(\omega_0) = \mathcal{K}_1(\omega_0) = \mathcal{K}_2(\omega_0) = 0$, which provide the unknown coefficients A_y and B_y. For a more wideband improvement, two other schemes are constructed. Their common characteristic

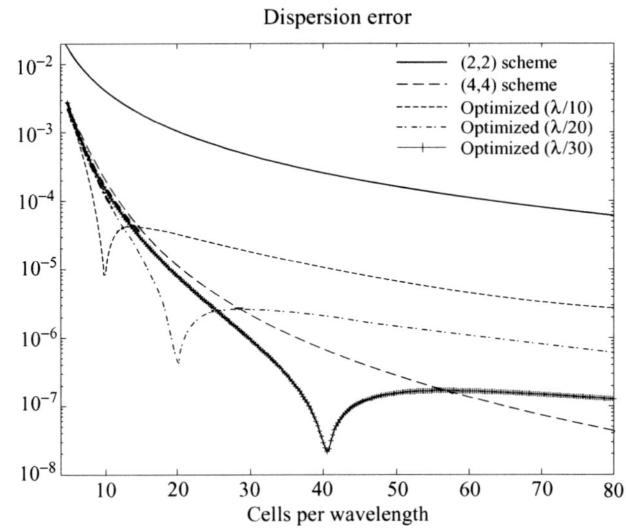

FIGURE 2.12: Dispersion error as a function of grid resolution for different FDTD configurations

is that they both require the vanishing of the first two terms in (2.121) at a specific ω_1, i.e., $\mathcal{K}_0(\omega_1) = \mathcal{K}_1(\omega_1) = 0$. Their difference lies on the way their broadband behavior is established. Hence in one case, second-order formal accuracy is imposed to spatial operators ($A_\zeta + 3B_\zeta = 1$), whereas in the other the vanishing of the first term of error ν is required at a second design frequency $\omega_2 \neq \omega_1$, i.e., $\mathcal{K}_0(\omega_2) = 0$. To mitigate the more negative influence of the dispersion error at higher than at lower frequencies, ω_2 is chosen lower than ω_1.

In Figure 2.12, the dispersion error versus cells per wavelength is examined. As anticipated, the controllable error estimators induce a minimum error at the design frequency points. Furthermore, they achieve a constantly reducing error toward lower frequencies and most importantly they remain more accurate than the conventional (4, 4) schemes in the high-frequency band.

Finally, concerning the extension of the optimization process to three dimensions, the necessary error estimators are analogously defined, while unknown coefficients are derived through spherical harmonic basis functions [55], as described in Section 2.4.3.

2.5.5 Coefficient-Modification Algorithms

This technique selects certain optimal parameters at *preassigned* grid densities and time increments, decreasing so the numerical dispersion along desired directions of propagation, such as along the axes, the diagonals, or anywhere in the mesh [58]. The basic idea stems from the use of a simple coefficient-modification approach which for a broadband signal can minimize the lattice reflection and the average-accumulated phase error. Hence, avoiding Taylor series, the procedure is based on the incorporation of weight parameters w for the optimization of all relative spatial contributions to the construction of the (2, 4) FDTD scheme. For instance, from the 2-D TE mode in an isotropic and inhomogeneous medium, E_x component becomes

$$E_x|_{i+1/2,j}^{n+1} = E_x|_{i+1/2j}^{n} + \omega\frac{\Delta t}{\varepsilon\Delta y}\left(H_z|_{i+1/2,j+1/2}^{n+1/2} - H_z|_{i+1/2,j-1/2}^{n+1/2}\right)$$
$$+(1-\omega)\frac{\Delta t}{3\varepsilon\Delta y}\left(H_z|_{i+1/2,j+3/2}^{n+1/2} - H_z|_{i+1/2,j-3/2}^{n+1/2}\right). \tag{2.124}$$

Extracting the amplification factor by the pertinent Fourier analysis, the numerical dispersion relation is

$$\frac{1}{(c\,\Delta t)^2}\sin^2\left(\frac{\omega\Delta t}{2}\right) = \sum_{\zeta=x,y}\left[\frac{w}{\Delta\zeta}\sin\left(\frac{k_\zeta\Delta\zeta}{2}\right) + \frac{1-w}{3\Delta\zeta}\sin\left(\frac{3k_\zeta\Delta\zeta}{2}\right)\right]^2, \tag{2.125}$$

with $k_x = k\cos\varphi$, $k_y = k\sin\varphi$, and k the respective wavenumber. Since (2.125) depends on w, its behavior may be improved in various ways. Thus, to eliminate the dispersion error toward the axes of a square mesh ($\Delta h = \Delta x = \Delta y$), (2.125) provides a weight parameter w_a:

$$w_a = \frac{\frac{1}{c\Delta t}\sin\left(\frac{\omega\Delta t}{2}\right) - \frac{1}{3\Delta h}\sin\left(\frac{3k_a\Delta h}{2}\right)}{\frac{1}{\Delta h}\sin\left(\frac{k_a\Delta h}{2}\right) - \frac{1}{3\Delta h}\sin\left(\frac{3k_a\Delta h}{2}\right)}, \tag{2.126}$$

for k_a the numerical phase constant along the axes. Expression (2.126) is solved if k_a is set equal to the theoretical constant $k_0 = 2\pi/\lambda$ at the design frequency. Moreover, when a grid diagonal is considered, where $\varphi = \pi/4$ and $k_d = k_0\sqrt{2}/2$, parameter w_d is equivalently derived. On the other hand, the combination of both cases may render the numerical velocity independent of the traveling direction. After the mutual elimination of Δt from w_a and w_d, the optimal value of w is given by

$$w = \frac{\frac{\sqrt{2}}{3\Delta h}\sin\left(\frac{3k_d\Delta h}{2}\right) - \frac{1}{3\Delta h}\sin\left(\frac{3k_a\Delta h}{2}\right)}{\frac{1}{\Delta h}\sin\left(\frac{k_a\Delta h}{2}\right) - \frac{1}{3\Delta h}\sin\left(\frac{3k_a\Delta h}{2}\right) - \sqrt{2}\left[\frac{1}{\Delta h}\sin\left(\frac{k_d\Delta h}{2}\right) - \frac{1}{3\Delta h}\sin\left(\frac{3k_d\Delta h}{2}\right)\right]}. \tag{2.127}$$

Finally, for the stability criterion of the algorithm, von Neumann's method yields

$$v\Delta t \le \left(\frac{3}{4w-1}\right)\left[\frac{1}{(\Delta x)^2} + \frac{1}{(\Delta y)^2}\right]^{-1/2}, \tag{2.128}$$

which results in a slightly smaller time-step than that of the usual (2, 4) FDTD scheme.

2.6 HIGHER ORDER FDTD SCHEMES IN CURVILINEAR COORDINATES

It has been long recognized that the extension of Yee's technique in curvilinear coordinates is a very important issue, since it enables the modeling of various strenuous problems. Belonging to the general family of FDTD schemes, the higher order versions could not be an exception to this convention. In fact, a poor discretization of a curved surface is likely to cancel the advantages of the increased temporal and spatial accuracy. For this aim, the specific section is devoted to the development of a fourth-order FDTD method in orthogonal curvilinear meshes, based on the aspects of [30], which conform to metallic boundaries or dielectric surfaces. The benefit from the use of such grids is the preservation of simplicity in the extraction of Maxwell's equations and the treatment of continuity conditions at the interfaces.

Consider the 2-D coordinate transformation $x = x(u, v)$, $y = y(u, v)$ that converts the curvilinear domain in the physical space, $P = \{P; x, y\}$, into a rectangular region in the transformed space, $\mathcal{P} = \{\mathcal{P}; u, v\}$. In this context, the TM mode of Maxwell's equations (2.1) and (2.2) in a homogeneous, isotropic, and lossy medium with $\mathbf{J}_s = \mathbf{M}_s = \mathbf{0}$ and $\rho_e = \rho_m = 0$ are transformed to

$$\mu \frac{\partial H^u}{\partial t} = -\frac{1}{\mathcal{V}} \frac{\partial E^z}{\partial v} - \sigma^* H^u, \tag{2.129}$$

$$\mu \frac{\partial H^v}{\partial t} = \frac{1}{\mathcal{V}} \frac{\partial E^z}{\partial u} - \sigma^* H^v, \tag{2.130}$$

$$\varepsilon \frac{\partial E^z}{\partial t} = \frac{1}{\mathcal{V}} \left[\gamma_v \frac{\partial H^v}{\partial u} - \gamma_u \frac{\partial H^u}{\partial v} - \vartheta \left(\frac{\partial H^v}{\partial v} - \frac{\partial H^u}{\partial u} \right) \right] + \frac{1}{\mathcal{V}} \left(\mathcal{M} H^u + \mathcal{N} H^v \right) - \sigma E^z, \tag{2.131}$$

where $\mathcal{V} = x_u y_v - x_v y_u$ is the Jacobian, H^u and H^v are the contravariant magnetic field components (see Section 3.3.1), $\gamma_u = x_u^2 + y_u^2$ and $\gamma_v = x_v^2 + y_v^2$ are system metrics, $\vartheta = x_u x_v + y_u y_v$ and

$$\mathcal{M} = y_{uu} y_v - y_{uv} y_u + x_{uu} x_v - x_{uv} x_u, \qquad \mathcal{N} = y_{vu} y_v - y_{vv} y_u + x_{vu} x_v - x_{vv} x_u \tag{2.132}$$

with the subscripts denoting differentiation with respect to u and v.

If the above transformation is also orthogonal, namely $\vartheta = 0$ and $V \neq 0$, then (2.131) becomes

$$\varepsilon \frac{\partial E^z}{\partial t} = \frac{1}{\mathcal{V}} \left(\gamma_v \frac{\partial H^v}{\partial u} - \gamma_u \frac{\partial H^u}{\partial v} \right) + \frac{1}{\mathcal{V}} \left(\mathcal{M} H^u + \mathcal{N} H^v \right) - \sigma E^z. \tag{2.133}$$

In this manner, the algorithm exploits the properties of lattice orthogonality, such as the zero off-diagonal elements in the metric tensors, which lessen the overhead of the computations.

To construct the fourth-order FDTD scheme, let us apply the Taylor expansion series to H^u (similarly for the other components) as in (2.14), and obtain

$$\left.\frac{\partial H^u}{\partial t}\right|^n = \frac{H^u|^{n+1/2} - H^u|^{n-1/2}}{\Delta t} - \frac{(\Delta t)^2}{24}\left.\frac{\partial^3 H^u}{\partial t^3}\right|^n + O[(\Delta t)^4]. \tag{2.134}$$

The next concern lies on the computation of the term $\partial^3/\partial t^3$ in (2.134). Taking the third-order temporal derivative of (2.129), one gets

$$\left.\frac{\partial^3 H^u}{\partial t^3}\right|^n = -\frac{\partial^2}{\partial t^2}\left(\frac{1}{\mu \mathcal{V}}\frac{\partial E^z}{\partial v} - \frac{\sigma^*}{\mu}H^u\right)\right|^n, \tag{2.135}$$

where for the resulting $\partial^2 H^u/\partial t^2$, the respective differentiation of (2.5) gives

$$\left.\frac{\partial^2 H^u}{\partial t^2}\right|^n = -\frac{1}{\mu \mathcal{V}}\left[\frac{\partial}{\partial t} - \frac{\sigma^*}{\mu}\right]\left.\frac{\partial E^z}{\partial v}\right|^n + \left(\frac{\sigma^*}{\mu}\right)^2 H^u|^n. \tag{2.136}$$

For the evaluation of $H^u|^n$ in (2.136) via its known values at $n+1/2$ and $n-1/2$, the following fourth-order time-averaging process is employed

$$H^u|^n = \frac{H^u|^{n+1/2} + H^u|^{n-1/2}}{2} - \frac{(\Delta t)^2}{8}\left.\frac{\partial^2 H^u}{\partial t^2}\right|^n + O\left[(\Delta t)^4\right]. \tag{2.137}$$

Thus, from (2.135)–(2.137) and a uniform 2-D grid $(u, v) = (i\Delta u, j\Delta v)$ with $\Delta h = \Delta u = \Delta v$, the update FDTD expression of (2.5) takes the form of

$$\begin{aligned}(1+\sigma_{\mathrm{B}}^*)\, H^u\big|_{i,j+1/2}^{n+1/2} = (1-\sigma_{\mathrm{B}}^*)\, H^u\big|_{i,j+1/2}^{n-1/2} - \frac{\Delta t}{\mu\, \mathcal{V}|_{i,j+1/2}}\Bigg\{1 + \frac{(\Delta t)^2}{24}\frac{\partial^2}{\partial t^2} + \frac{\sigma_{\mathrm{A}}^*}{12}\\ \left[1+\frac{(\sigma_{\mathrm{A}}^*)^2}{8}\right]^{-1}\left(\Delta t\frac{\partial}{\partial t} - \sigma_{\mathrm{A}}^*\right)\Bigg\}\, \Gamma_v\left[E^z\big|_{i,j+1/2}^n\right],\end{aligned} \tag{2.138}$$

in which

$$\sigma_{\mathrm{A}}^* = \frac{\sigma^*\Delta t}{\mu}, \qquad \sigma_{\mathrm{B}}^* = \sigma_{\mathrm{A}}^*\frac{1+(\sigma_{\mathrm{A}}^*)^2/24}{2\left[1+(\sigma_{\mathrm{A}}^*)^2/8\right]}, \tag{2.139}$$

and

$$\Gamma_v\left[f\big|_{i,j}^n\right] = \frac{9}{8}\left(\frac{f\big|_{i,j+1/2}^n - f\big|_{i,j-1/2}^t}{\Delta h}\right) - \frac{1}{24}\left(\frac{f\big|_{i,j+3/2}^n - f\big|_{i,j-3/2}^n}{\Delta h}\right), \tag{2.140}$$

is the curvilinear version of the fourth-order spatial operator (2.18). Notice that, for brevity, the stencils of constitutive parameters, indicating the actual location of a certain material inside a cell, have been omitted.

In a completely analogous way, (2.129) and (2.130) lead to

$$(1+\sigma_{\mathrm{B}}^{*})\, H^{v}\big|_{i+1/2,j}^{n+1/2} = (1-\sigma_{\mathrm{B}}^{*})\, H^{v}\big|_{i+1/2,j}^{n-1/2} - \frac{\Delta t}{\mu\, \mathcal{V}|_{i+1/2,j}} \left\{ 1 + \frac{(\Delta t)^2}{24}\frac{\partial^2}{\partial t^2} + \frac{\sigma_{\mathrm{A}}^{*}}{12} \right.$$

$$\left. \times \left[1+\frac{(\sigma_{\mathrm{A}}^{*})^2}{8}\right]^{-1} \left(\Delta t \frac{\partial}{\partial t} - \sigma_{\mathrm{A}}^{*}\right) \right\} \Gamma_u \left[E^{z}\big|_{i+1/2,j}^{n} \right], \tag{2.141}$$

$$(1+\sigma_{\mathrm{B}})\, E^{z}\big|_{i,j}^{n+1} = (1-\sigma_{\mathrm{B}})\, E^{z}\big|_{i,j}^{n} - \frac{\Delta t}{\varepsilon\, \mathcal{V}|_{i,j}} \left\{ 1 + \frac{(\Delta t)^2}{24}\frac{\partial^2}{\partial t^2} + \frac{\sigma_{\mathrm{A}}}{12}\left[1+\frac{(\sigma_{\mathrm{A}})^2}{8}\right]^{-1} \right.$$

$$\left. \times \left(\Delta t \frac{\partial}{\partial t} - \sigma_{\mathrm{A}}\right) \right\} \left(\gamma_v|_{i,j}\, \Gamma_u\left[H^{v}\big|_{i,j}^{n+1/2}\right] - \gamma_u|_{i,j}\, \Gamma_v\left[H^{u}\big|_{i,j}^{n+1/2}\right] \right.$$

$$\left. + \mathcal{M}|_{i,j}\, H^{u}\big|_{i,j}^{n+1/2} + \mathcal{N}|_{i,j}\, H^{v}\big|_{i,j}^{n+1/2} \right), \tag{2.142}$$

with σ_{A} and σ_{B} defined by (2.139) under the replacement of μ with ε and σ^* with σ, while $\Gamma_u[.]$ is given by (2.140) for an interchange of the $\pm 1/2$ shift from index j to i. Moreover, in the case of H^u or H^v at integer mesh points (i, j), the appropriate interpolations are conducted by the fourth-order relations

$$H^{u}\big|_{i,j}^{n+1/2} = \frac{9}{16}\left(H^{u}\big|_{i,j-1/2}^{n+1/2} + H^{u}\big|_{i,j+1/2}^{n+1/2} \right) - \frac{1}{16}\left(H^{u}\big|_{i,j+3/2}^{n+1/2} + H^{u}\big|_{i,j-3/2}^{n+1/2} \right), \tag{2.143a}$$

$$H^{v}\big|_{i,j}^{n+1/2} = \frac{9}{16}\left(H^{v}\big|_{i-1/2,j}^{n+1/2} + H^{v}\big|_{i+1/2,j}^{n+1/2} \right) - \frac{1}{16}\left(H^{v}\big|_{i+3/2,j}^{n+1/2} + H^{v}\big|_{i-3/2,j}^{n+1/2} \right). \tag{2.143b}$$

Finally, the discrete values of coefficients $\mathcal{M}, \mathcal{N}$ and metrics γ_u, γ_v require the knowledge of second-order grid derivatives with respect to u and v. These quantities, usually reduced to first-order ones, are computed by the five-point schemes

$$x_u|_{i,j} = \frac{1}{12\Delta h}\left(x|_{i-2,j} - x|_{i+2,j} \right) + \frac{2}{3\Delta h}\left(x|_{i+1,j} - x|_{i-1,j} \right), \tag{2.144a}$$

$$y_v|_{i,j} = \frac{1}{12\Delta h}\left(y|_{i,j-2} - y|_{i,j+2} \right) + \frac{2}{3\Delta h}\left(y|_{i,j+1} - y|_{i,j-1} \right). \tag{2.144b}$$

The preceding technique, notwithstanding its requirements for additional memory, does not influence the total computational burden. On the contrary, its ability to manipulate curvilinear grids and relatively smooth material interfaces is proven to be fairly convenient and instructive.

REFERENCES

[1] A. Taflove and S. C. Hagness, *Computational Electrodynamics: The Finite-Difference Time-Domain Method.* 3rd ed., Norwood, MA: Artech House, 2005.

[2] C. A. Balanis, *Advanced Engineering Electromagnetics.* New York: John Wiley & Sons, 1989.

[3] J. Fang, *Time Domain Finite Difference Computation for Maxwell's Equations.* Ph.D. thesis, Univ. California, Berkeley, 1989.

[4] T. Deveze, L. Beaulie, and W. Tabbara, "A fourth order scheme for the FDTD algorithm applied to Maxwell equations," in *Proc. IEEE Antennas Propag. Soc. Int. Symp.*, Chicago, IL, July 1992, vol. 1, pp. 346–349.

[5] S. K. Lele, "Compact finite difference schemes with spectral resolution," *J. Comput. Phys.*, vol. 101, pp. 16–42, July 1992.

[6] P. G. Petropoulos, "Phase error control for FD-TD methods of second and fourth order accuracy," *IEEE Trans. Antennas Propag.*, vol. 42, no. 6, pp. 859–862, June 1994.

[7] M. H. Carpenter, D. Gottlieb, and S. Abarbanel, "Time-stable boundary conditions for finite-difference schemes solving hyperbolic systems: Methodology and application to high-order compact schemes," *J. Comput. Phys.*, vol. 111, no. 2, pp. 220–236, Apr. 1994.

[8] C. W. Manry, S. L. Broschat, and J. B. Schneider, "Higher-order FDTD methods for large problems," *Appl. Comput. Electromagn. Soc. J.*, vol. 10, no. 2, pp. 17–29, 1995.

[9] B. Gustafsson and P. Olsson, "Fourth-order difference methods for hyperbolic IBVPs," *J. Comput. Phys.*, vol. 117, no. 2, pp. 300–317, Mar. 1995.

[10] D. W. Zingg, H. Lomax, and H. Jurgens, "High-accuracy finite-difference schemes for linear wave propagation," *SIAM J. Sci. Comput.*, vol. 17, pp. 328–346, 1996.

[11] G. Cohen and P. Joly, "Construction and analysis of fourth-order finite difference schemes for the acoustic wave equations in nonhomogeneous media," *SIAM J. Numer. Anal.*, vol. 33, pp. 1266–1302, 1996.

[12] Y. Liu, "Fourier analysis of numerical algorithms for Maxwell equations," *J. Comput. Phys.*, vol. 124, pp. 396–406, Mar. 1996.

[13] D. W. Zingg, "A review of high-order and optimized finite-difference methods for simulating linear wave phenomena," Tech. Rep. 97-2088, June 1997, AIAA.

[14] J. L. Young, D. Gaitonde, and J. S. Shang, "Toward the construction of a fourth-order difference scheme for transient EM wave simulation: Staggered grid approach," *IEEE Trans. Antennas Propag.*, vol. 45, no. 11, pp. 1573–1580, Nov. 1997.

[15] E. Turkel, "High-order methods," in *Advances in Computational Electrodynamics: The Finite-Difference Time-Domain Method*, A. Taflove, Ed. Norwood, MA: Artech House, 1998, ch. 2, pp. 63–110.

[16] K. Mahesh, "A family of high order finite difference schemes with good spectral resolution," *J. Comput. Phys.*, vol. 145, no. 1, pp. 332–358, Sep. 1998.

[17] T. A. Driscoll and B. Fornberg, "A block pseudospectral method for Maxwell's equations: I. One-dimensional case," *J. Comput. Phys.*, vol. 140, no. 1, pp. 39–80, Feb. 1998.

[18] T. A. Driscoll and B. Fornberg, "Block pseudospectral methods for Maxwell's equations: II. Two-dimensional, discontinuous-coefficient case," *SIAM J. Sci. Comput.*, vol. 21, pp. 1146–1167, 1999.

[19] E. Turkel and A. Yefet, "On the construction of a high order difference scheme for complex domains in a Cartesian grid," *Appl. Numer. Math.*, vol. 33, nos. 1–4, pp. 113–124, May 2000.

[20] A. Ditkowski, K. Dridi, and J. S. Hesthaven, "Convergent Cartesian grid methods for Maxwell's equations in complex geometries," *J. Comput. Phys.*, vol. 170, no. 1, pp. 39–80, June 2001.

[21] G. C. Cohen, *Higher-Order Numerical Methods for Transient Wave Equations*. Berlin, Germany: Springer-Verlag, 2002.

[22] J. F. Nystrom, "High-order time-stable numerical boundary scheme for the temporally dependent Maxwell equations in two dimensions," *J. Comput. Phys.*, vol. 178, no. 2, 290–306, May 2002.

[23] J. S. Hesthaven, "High-order accurate methods in time-domain computational electromagnetics: A review," in *Advances in Imaging and Electron Physics*, P. Hawkes, Ed. New York: Academic Press, 2003, vol. 127, pp. 59–123.

[24] I. Singer and E. Turkel, "High-order finite-difference methods for the Helmholtz equation," *Comp. Methods Appl. Mech. Engrg.*, vol. 163, nos. 1–4, pp. 343–358, Sep. 1998.

[25] H. M. Jurgens and D. W. Zingg, "Numerical solution of the time-domain Maxwell equations using high-accuracy finite-difference methods," *SIAM J. Sci. Comput.*, vol. 22, pp. 1675–1696, 2000.

[26] A. Yefet and P. G. Petropoulos, "A staggered fourth-order accurate explicit finite difference scheme for the time-domain Maxwell's equations," *J. Comput. Phys.*, vol. 168, no. 2, pp. 286–315, Apr. 2001.

[27] J. Nordström and R. Gustafsson, "High order finite difference approximations of electromagnetic wave propagation close to material discontinuities," *J. Sci. Comput.*, vol. 18, no. 2, pp. 215–234, Apr. 2003.

[28] S. Zhao and G. W. Wei, "High-order FDTD methods via derivative matching for Maxwell's equations with material interfaces," *J. Comput. Phys.*, vol. 200, no. 1, pp. 60–103, Oct. 2004.

[29] N. V. Kantartzis, T. I. Kosmanis, and T. D. Tsiboukis, "Fully nonorthogonal higher-order accurate FDTD schemes for the systematic development of 3-D reflectionless PMLs in

general curvilinear coordinate systems," *IEEE Trans. Magn.*, vol. 36, no 4, pp. 912–916, July 2000.

[30] Z. Xie, C.-H. Chan, and B. Zhang, "An explicit fourth-order orthogonal curvilinear staggered grid FDTD method for Maxwell's equations," *J. Comput. Phys.*, vol. 175, no. 2, pp. 739–763, Dec. 2002.

[31] N. V. Kantartzis, "A generalised higher-order FDTD-PML algorithm for the enhanced analysis of 3-D waveguiding EMC structures in curvilinear coordinates," *IEE Proc. Microw., Antennas Propag.*, vol. 150, no. 5, pp. 351–359, Oct. 2003.

[32] W. Yu and R. Mittra, "A new higher-order subgridding method for finite difference time domain (FDTD) algorithm," in *Proc. IEEE Antennas Propag. Soc. Int. Symp.*, Atlanta, GA, Jun. 1998, vol. 1, pp. 608–611.

[33] N. V. Kantartzis, T. I. Kosmanis, T. V. Yioultsis, and T. D. Tsiboukis, "A nonorthogonal higher-order wavelet-oriented FDTD technique for 3-D waveguide structures on generalised curvilinear grids," *IEEE Trans. Magn.*, vol. 37, pp. 3264–3268, 2001.

[34] S.-T. Chun and J. Y. Choe, "A higher order FDTD method in integral formulation," *IEEE Trans. Antennas Propag.*, vol. 53, no. 7, pp. 2237–2246, July 2005.

[35] M. Aidam and P. Russer, "New high order time-stepping schemes for finite differences," in *Proc. 15th Ann. Rev. Prog. Appl. Comput. Electromagn.*, Monterey, CA, Mar. 1999, pp. 578–585.

[36] J. L. Young, "High-order, leapfrog methodology for the temporally dependent Maxwell's equations," *Radio Sci.*, vol. 36, no. 1, pp. 9–17, Feb. 2001.

[37] H. Spachmann, R. Schuhmann, and T. Weiland, "Convergence, stability and dispersion analysis of higher order leapfrog schemes for Maxwell's equations," in *Proc. 17th Ann. Rev. Prog. Appl. Comput. Electromagn.*, Monterey, CA, Mar. 2001, pp. 655–662.

[38] S. Gottlieb, C.-W. Shu, and E. Tadmor, "Strong stability-preserving high-order time discretization methods," *SIAM Rev.*, vol. 43, no. 1, pp. 89–9112, 2001.

[39] T. Rylander and A. Bondeson, "Stability of explicit-implicit hybrid time-stepping schemes for Maxwell's equations," *J. Comput. Phys.*, vol. 179, no. 2, pp. 426–438, July 2002.

[40] H. Spachmann, R. Schuhmann, and T. Weiland, "Higher order explicit time integration schemes for Maxwell's equations," *Int. J. Numer. Model.*, vol. 15, nos. 5–6, pp. 419–437, Sep.–Dec. 2002.

[41] S. V. Georgakopoulos, *Higher-Order Finite Difference Methods for Electromagnetic Radiation and Penetration.* Ph.D. thesis, Arizona State Univ., Tempe, AZ, 2001.

[42] S. V. Georgakopoulos, R. A. Renaut, C. A. Balanis, and C. R. Birtcher, "A hybrid fourth-order FDTD utilizing a second-order FDTD subgrid," *IEEE Microw. Wireless Compon. Lett.*, vol. 11, no. 11, pp. 462–464, Nov. 2001.

[43] S. V. Georgakopoulos, C. R. Birtcher, C. A. Balanis, and R. A. Renaut. "HIRF penetration and PED coupling analysis for scaled fuslage models using a hybrid subgrid FDTD(2,2)/FDTD(2,4) method," *IEEE Trans. Electromagn. Compat.*, vol. 45, no. 2, pp. 293–305, May 2003.

[44] K. P. Prokopidis and T. D. Tsiboukis, "Higher-order FDTD (2,4) scheme for accurate simulations in lossy dielectrics," *IEE Electron. Lett.*, vol. 39, no. 11, pp. 835–836, May 2003.

[45] M. Fujii, M. Tahara, I. Sakagami, W. Freude, and P. Russer, "High-order FDTD and auxiliary differential equation formulation of optical pulse propagation in 2-D Kerr and Raman nonlinear dispersive media," *IEEE J. Quantum Electron.*, vol. 40, no. 2, pp. 175–182, Feb. 2004.

[46] K. P. Prokopidis, E. P. Kosmidou, and T. D. Tsiboukis, "An FDTD algorithm for wave propagation in dispersive media using higher-order schemes," *J. Electromagn. Waves Appl.*, vol. 18, no. 9, 1171–1194, 2004.

[47] M. F. Hadi and M. Piket-May, "A modified FDTD (2,4) scheme for modeling electrically large structures with high-phase accuracy, *IEEE Trans. Antennas Propag.*, vol. 45, no. 2, pp. 254–264, Feb. 1997.

[48] G. Haussmann and M. Piket-May, "FDTD M24 dispersion and stability in three dimensions," in *Proc. 14th Ann. Rev. Prog. Appl. Comput. Electromagn.*, Monterey, CA, Mar. 1998, vol. 1, pp. 82–89.

[49] N. V. Kantartzis and T. D. Tsiboukis, "A higher-order FDTD technique for the implementation of enhanced dispersionless perfectly matched layers combined with efficient absorbing boundary conditions," *IEEE Trans. Magn.*, vol. 34, no 5, pp. 2736–2739, Sep. 1998.

[50] K. Lan, Y. Liu, and W. Lin, "A higher order (2,4) scheme for reducing dispersion in FDTD algorithms," *IEEE Trans. Electromagn. Compat.*, vol. 41, no. 2, pp. 160–165, May 1999.

[51] H. E. Abd El-Raouf, E. A. El-Diwani, A. E.-H. Ammar, and F. El-Hefnawi, "A low-dispersion 3-D second-order in time fourth-order in space FDTD scheme ($M3d_{24}$)," *IEEE Trans. Antennas Propag.*, vol. 52, no. 7, pp. 1638–1646, July 2004.

[52] N. V. Kantartzis, T. D. Tsiboukis, and E. E. Kriezis, "A topologically consistent class of 3-D higher-order curvilinear FDTD schemes for dispersion-optimized EMC and material modeling," *J. Mat. Processing Technol.*, vol. 161, nos. 1–2, pp. 210–217, Apr. 2005.

[53] S. Wang and F. L. Teixeira, "A three-dimensional angle-optimized finite-difference time-domain algorithm," *IEEE Trans. Microw. Theory Tech.*, vol. 51, no. 3, pp. 811–817, Mar. 2003.

[54] S. Wang and F. L. Teixeira, "A finite-difference time-domain algorithm for arbitrary propagation angles," *IEEE Trans. Antennas Propag.*, vol. 51, no. 9, pp. 2456–2463, Sep. 2003.

[55] S. Wang and F. L. Teixeira, "Dispersion-relation-preserving FDTD algorithms for large-scale three-dimensional problems," *IEEE Trans. Antennas Propag.*, vol. 51, no. 8, pp. 1818–1828, Aug. 2003.

[56] S. Wang and F. L. Teixeira, "Grid-dispersion error reduction for broadband FDTD electromagnetic simulations," *IEEE Trans. Magn.*, vol. 40, no. 2, pp. 1440–1443, Mar. 2004.

[57] T. T. Zygiridis and T. D. Tsiboukis, "Low-dispersion algorithms based on the higher order (2,4) FDTD method," *IEEE Trans. Microw. Theory Tech.*, vol. 52, no. 4, pp. 1321–1327, Apr. 2004.

[58] G. Sun and C. W. Trueman, "Optimized finite-difference time-domain methods based on the (2,4) stencil," *IEEE Trans. Microw. Theory Tech.*, vol. 53, no. 3, pp. 832–842, Mar. 2005.

[59] T. T. Zygiridis and T. D. Tsiboukis, "Development of higher-order FDTD schemes with controllable dispersion error," *IEEE Trans. Antennas Propag.*, vol. 53, no. 9, pp. 2952–2960, Sep. 2005.

[60] K. L. Shlager and J. B. Schneider, "Comparison of the dispersion properties of higher order FDTD schemes and equivalent-sized MRTD schemes," *IEEE Trans. Antennas Propag.*, vol. 52, no. 4, pp. 1095–1104, Apr. 2004.

[61] S. V. Georgakopoulos, C. R. Birtcher, C. A. Balanis, and R. A. Renaut, "Higher-order finite-difference schemes for electromagnetic radiation, scattering, and penetration, part I: Theory," *IEEE Antennas Propag. Mag.*, vol. 44, 134–142, Feb. 2002.

[62] S. V. Georgakopoulos, C. R. Birtcher, C. A. Balanis, and R. A. Renaut, "Higher-order finite-difference schemes for electromagnetic radiation, scattering, and penetration, part II: Applications," *IEEE Antennas Propag. Mag.*, vol. 44, pp. 92–101, Apr. 2002.

[63] K.-P. Hwang and A. C. Cangellaris, "Computational efficiency of Fang's fourth-order FDTD schemes," *Electromagn.*, vol. 23, pp. 89–102, 2003.

CHAPTER 3
Higher Order Nonstandard FDTD Methodology

3.1 INTRODUCTION

The conventional higher order FDTD schemes, described in the previous chapter, exhibit some indisputably significant benefits regarding the accomplishment of correct models and credible numerical simulations. Actually, their structural robustness mitigates the undesired weaknesses of the classical Yee's technique and leads to the extraction of acceptable solutions for a variety of electromagnetic field problems. However, in several practical applications, the reliability of these powerful constituents is still influenced by dispersion and dissipation errors. Typical examples are the curvilinear waveguide or antenna configurations and the arbitrarily aligned – with respect to mesh axes – dissimilar media interfaces. Examining the mechanism of conventional higher order differentiation, a critical source of error should be pursued in the approximation of spatial and temporal derivatives appearing in the discrete counterparts of partial differential equations. In particular, the idea to employ a relatively restricted set of additional nodes in the computational lattice may yield inadequate sampling rates or improperly oversized time intervals. Furthermore, the elemental cells so produced tend to exponentially growing artificial oscillations and vector parasites which, owing to their accumulating nature, induce rather unstable results. These deficiencies constitute the principal motive for the development of an enhanced algorithm able to attain rigorous simulations for complicated devices and realistic arrangements. Of crucial importance toward this aim is the observation that lattice reflection errors present a *nontrivial sensitivity* on the direction of propagation due to the anisotropy of existing finite-difference approximants. As a consequence, research must concentrate on the derivation of fully isotropic Laplacian operators with controllable accuracy order in an ample frequency range and drastically decreased errors even for coarse grid resolutions. This is exactly where the discretization strategy of *nonstandard differencing* can be implemented. Since its mathematical background offers substantial profits from a theoretical interpretation point of view, it is anticipated that the preceding prerequisites will be more effectively fulfilled.

In this chapter, after a brief overview of the nonstandard finite-difference theory for the two-dimensional (2-D) and three-dimensional (3-D) modeling of the wave and Maxwell's equations, a higher order FDTD methodology in Cartesian coordinates is presented. The formulation is founded on the principles of algebraic topology and predominantly on the dual-cell notion that is utilized for the representation of computational domains. By launching a specific set of spatial and temporal operators, the 3-D algorithm achieves the immediate transition from the continuous to discrete state. Subsequently, the technique is extended in curvilinear coordinates through a parametric covariant and contravariant approach, intended for the manipulation of the resulting div-curl problem. The analysis then proceeds to the appropriate dispersion error and stability investigation that reveals the most prominent advantages. Finally, the chapter closes with several practical aspects intending to furnish useful hints and facilitate the realization of this systematic higher order FDTD method.

3.2 THE NONSTANDARD FINITE-DIFFERENCE ALGORITHM

Nonstandard finite-difference schemes are, primarily, a *generalization* of the customary models of differential equations [1–4]. While the latter cannot satisfactorily model cumbersome physical phenomena, the particular forms maintain the underlying properties of their continuous analogs and evade elementary instabilities. In this manner, they have gained a noteworthy popularity in several scientific fields [5], like fluid mechanics, electromagnetics, photoconductivity, nonlinear optics, and mathematical biology to name a few.

The efficient construction of a nonstandard finite-difference method arises from the concept of exact schemes in order to guarantee that consistency, stability, and convergence concerns do not arise. Nonetheless, the *a priori* design of an exact differencing model for a differential equation is basically not feasible without knowing its general solution. Moreover, there are several realistic situations, where the complication of the governing equations is such that even if exact solvers were derived their performance could hardly be validated. To reinforce these issues, it is stressed that a correctly established technique ensures the unperturbed mapping of every continuous physical component. As a result, it clarifies all elementary mechanisms and enables the smooth transition to the discrete state. This last assertion plays a decisive role in the competence of a potential differencing formula, since it permits the successful handling of numerous strenuous cases without the need of mandatory simplifications. It is, therefore, plainly understood that the determination of the key principles regarding the development of nonstandard operators constitutes a greatly instructive guide.

Based on elaborate analytical and numerical investigations, a set of *rules* that accumulate the focal characteristics of a prospective framework has been introduced in [1]. These may be summarized as

Rule 1: The order of the discrete derivatives must be equal to the order of the corresponding derivatives appearing in the differential equations.

Rule 2: Denominator functions for the discrete derivatives must be generally defined by means of more sophisticated expressions of spatial and temporal increments than those of the conventional ones.

Rule 3: Nonlinear terms must be represented in a nonlocal discrete sense.

Rule 4: Special conditions holding for the solutions of a differential equation should be equivalently valid for the solutions of the finite-difference scheme.

Rule 5: The resulting formulation must not produce spurious oscillations or lattice artificialities.

Rule 6: For differential equations having $N \geq 3$ terms, it is effective to develop finite-difference schemes for diverse subequations comprising $M < N$ terms, and then combine all schemes in a comprehensive finite-difference model.

Although these rules cannot ensure the presence of a unique representation for a certain differential equation, they do confine the possible available models. Overall, a nonstandard analog, acquired in this way, (while not fully exact) displays the aforementioned desirable properties that provide very accurate, stable, and convergent outcomes as compared to those of existing approaches.

Before proceeding to the main formulation, let us draw our attention to the definition of the nonstandard operator. It is well known that the conventional second-order finite-difference counterpart is

$$\mathbf{d}_x^{\text{conv}}\left[f\,\big|_x^t\right] = \tfrac{1}{\Delta x}\mathbf{d}_x\left[f\,\big|_x^t\right] \quad \text{with } \mathbf{d}_x\left[f\,\big|_x^t\right] = f\,\big|_{x+\Delta x/2}^t - f\,\big|_{x-\Delta x/2}^t. \tag{3.1}$$

The nonstandard finite-difference operator [1] is defined by

$$\mathbf{d}_x^{\text{nst}}\left[f\,\big|_x^t\right] = \frac{1}{\Psi(k,\,\Delta x)}\mathbf{d}_x\left[f\,\big|_x^t\right], \tag{3.2}$$

where $\Psi(k,\,\Delta x)$ is a *correction function* selected to minimize the difference $\left|\left(\partial_x - \mathbf{d}_x^{\text{nst}}\right) f(x)\right|$ with respect to a set of basis functions, customarily comprising sinusoidal and exponential terms with arguments toward each propagation direction in the mesh. A frequent choice for $\Psi(k,\,\Delta x)$ may be obtained as follows: assume the computation of $\mathbf{d}_x\left[e^{jkx}\right]$, which because of (3.1) is found to be $\mathbf{d}_x\left[e^{jkx}\right] = 2je^{jkx}\sin(k\Delta x/2)$. In order to fulfill the minimization criterion, mentioned above, $\Psi(k,\,\Delta x)$ is denoted as

$$\Psi(k,\,\Delta x) = 2\sin(k\Delta x/2)/k. \tag{3.3}$$

Hence, for an $f \in \{\sin(kx), \cos(kx), e^{\pm jkx}\}$, correction function (3.3) achieves $\partial_x e^{jkx} = \mathbf{d}_x^{\text{nst}}[e^{jkx}]$ and produces a zero-error finite-difference technique with regard to plane waves propagating in the prefixed direction of wavevector $\mathbf{k}$. Unfortunately, this directional anisotropy restrains the expediency of the ensuing scheme, thus calling for a more effective exploitation of the nonstandard finite-difference properties. In fact, this requirement is systematically satisfied in the subsequent sections of this chapter.

3.2.1 Discretization of the Wave Equation

Let us consider the general form of the wave equation that governs the evolution of any electric or magnetic field component f. Its homogeneous version in a Cartesian coordinate system (x, y, z) is given by

$$\frac{\partial^2 f(\mathbf{r}, t)}{\partial t^2} - \upsilon^2(\mathbf{r})\nabla^2 f(\mathbf{r}, t) = 0 \quad \text{with} \quad \nabla^2 \equiv \frac{\partial}{\partial x^2} + \frac{\partial}{\partial y^2} + \frac{\partial}{\partial z^2}, \tag{3.4}$$

$\mathbf{r}$ the position vector, and $\upsilon(\mathbf{r}, t)$ the propagation velocity. For the sake of clarity and notational convenience, the nonstandard finite-difference concepts are initially applied in a 2-D domain.

Two-Dimensional Analysis

In this case and according to the theoretical aspects of [2, 3], there are two Laplacian difference operators for the discretization of (3.4). The first one is expressed as

$$\mathbf{L}_1^2\left[f\,\big|_{x,y}^{t}\right] = \mathbf{d}_x^2\left[f\,\big|_{x,y}^{t}\right] + \mathbf{d}_y^2\left[f\,\big|_{x,y}^{t}\right], \tag{3.5}$$

$$\text{where} \quad \mathbf{d}_x^2\left[f\,\big|_{x,y}^{t}\right] = f\,\big|_{x+\Delta h,y}^{t} - 2f\,\big|_{x,y}^{t} + f\,\big|_{x-\Delta h,y}^{t},$$
$$\mathbf{d}_y^2\left[f\,\big|_{x,y}^{t}\right] = f\,\big|_{x,y+\Delta h}^{t} - 2f\,\big|_{x,y}^{t} + f\,\big|_{x,y-\Delta h}^{t},$$

and $\Delta h = \Delta x = \Delta y$ the spatial increment which, without loss of generality, refers to the cells of a uniform lattice. On the other hand, the second operator is defined by the more extensively arranged nodal structure of

$$\mathbf{L}_2^2\left[f\,\big|_{x,y}^{t}\right] = \frac{1}{2}\Big(f\,\big|_{x+\Delta h,y+\Delta h}^{t} + f\,\big|_{x-\Delta h,y+\Delta h}^{t} + f\,\big|_{x+\Delta h,y-\Delta h}^{t} + f\,\big|_{x-\Delta h,y-\Delta h}^{t}\Big) - 2f\,\big|_{x,y}^{t}, \tag{3.6}$$

offering so a supplementary degree of freedom for the approximation of $\nabla^2 f(x, y)$ via the use of a weighted superposition of $\mathbf{L}_1^2[.]$ and $\mathbf{L}_2^2[.]$. This process leads to

$$\nabla^2 f(x, y) \cong \frac{1}{[\Psi(k, \Delta h)]^2}\mathbf{L}_0^2\left[f\,\big|_{x,y}^{t}\right] = \mathbf{L}_0^{2,\mathrm{nst}}\left[f\,\big|_{x,y}^{t}\right] \tag{3.7}$$

with

$$\mathbf{L}_0^2\left[f\,\big|_{x,y}^{t}\right] = p\mathbf{L}_1^2\left[f\,\big|_{x,y}^{t}\right] + (1-p)\mathbf{L}_2^2\left[f\,\big|_{x,y}^{t}\right],$$

for a quantity $0 \leq p \leq 1$ principally dependent on position and the parameters of the partial differential equation under study. In (3.7) and through (3.2), $\mathbf{L}_0^{2,\mathrm{nst}}[.]$ is designated as the *generalized nonstandard operator* whose task is to attain the most precise evaluation of every spatial derivative. Herein, correction function $\Psi(k, \Delta h)$ is chosen to minimize the difference $\left|(\nabla^2 - \mathbf{L}_0^{2,\text{nst}}) f(x, y)\right|$. Note that, due to its advanced configuration, $\mathbf{L}_0^{2,\text{nst}}[.]$ can extend its zero-error aptitude in a far more wideband sense compared to (3.2).

Having determined the basic attributes of (3.7), analysis will now focus on the development of a representation for $\nabla^2 e^{j\mathbf{k}\cdot\mathbf{r}}$ with $\mathbf{r} = x\hat{\mathbf{x}} + y\hat{\mathbf{y}}$ and $\hat{\mathbf{x}}, \hat{\mathbf{y}}$ the respective unit vectors. Since in the one-dimensional case, application of (3.5) gives $\mathbf{d}_x^2\left[e^{jkx}\right] = 2\left[\cos(k\Delta h) - 1\right] e^{jkx}$, the intention is to find a value of p so that

$$\mathbf{L}_0^2\left[e^{j\mathbf{k}\cdot\mathbf{r}}\right] \cong 2\left[\cos(k\Delta h) - 1\right] e^{j\mathbf{k}\cdot\mathbf{r}}, \tag{3.8}$$

for all directions of $\mathbf{k}$ at a prefixed magnitude $k = |\mathbf{k}|$. Substituting (3.8) in (3.7), one obtains

$$p\left(\mathbf{L}_1^2\left[e^{j\mathbf{k}\cdot\mathbf{r}}\right] - \mathbf{L}_2^2\left[e^{j\mathbf{k}\cdot\mathbf{r}}\right]\right) + \mathbf{L}_2^2\left[e^{j\mathbf{k}\cdot\mathbf{r}}\right] = 2\left[\cos(k\Delta h) - 1\right] e^{j\mathbf{k}\cdot\mathbf{r}}, \tag{3.9}$$

which, if satisfied by a certain value p at an acceptable level of accuracy, enables the extraction of a correspondingly efficient computation of $\nabla^2 e^{j\mathbf{k}\cdot\mathbf{r}}$. In this context, let us define $\mathbf{L}_i^2\left[e^{j\mathbf{k}\cdot\mathbf{r}}\right] = 2e^{j\mathbf{k}\cdot\mathbf{r}}\Lambda_i(k\Delta h, \theta)$ for $i = 1, 2$ and after some algebraic manipulations in terms of $k_x = k\cos\theta$, $k_y = k\sin\theta$, find

$$\Lambda_1(k\Delta h, \theta) = \cos(k_x\Delta h) + \cos(k_y\Delta h) - 2, \tag{3.10a}$$

$$\Lambda_2(k\Delta h, \theta) = \cos(k_x\Delta h)\cos(k_y\Delta h) - 1, \tag{3.10b}$$

where θ is the angle formed by $\mathbf{k}$ and x-axis. Consequently, from (3.9), $p = p(k\Delta h, \theta)$ is provided by

$$p(k\Delta h, \theta) = \frac{\cos(k\Delta h) - \Lambda_2(k\Delta h, \theta) - 1}{\Lambda_1(k\Delta h, \theta) - \Lambda_2(k\Delta h, \theta)}. \tag{3.11}$$

Inspection of (3.11) leads to the direct conclusion that, despite our primary goal, p depends on angle θ. Such a relation is fairly weak, implying that a $p(k\Delta h, \theta)$, which exactly satisfies (3.11) at a specific value $\theta = \theta_{\text{ex}}$, still remains a sensible approximation for $\theta \neq \theta_{\text{ex}}$. An indicative θ_{ex} for minimizing the deviation of $\mathbf{L}_0^2[e^{j\mathbf{k}\cdot\mathbf{r}}]/e^{j\mathbf{k}\cdot\mathbf{r}}$ from $2\left[\cos(k\Delta h) - 1\right]$ may be derived via $\cos^4\theta_{\text{ex}} = 1/2$. Under these considerations and through the assistance of (3.10), expression (3.11) becomes

$$p(k\Delta h) = p(k\Delta h, \theta_{\text{ex}}) = \frac{\cos(k_x^{\text{ex}}\Delta h)\cos(k_y^{\text{ex}}\Delta h) - \cos(k\Delta h)}{1 + \cos(k_x^{\text{ex}}\Delta h)\cos(k_y^{\text{ex}}\Delta h) - \cos(k_x^{\text{ex}}\Delta h) - \cos(k_y^{\text{ex}}\Delta h)}, \tag{3.12}$$

with $k_x^{\text{ex}} = k\cos\theta_{\text{ex}}$ and $k_y^{\text{ex}} = k\sin\theta_{\text{ex}}$. Bearing in mind the prior procedure and denoting as $\mathbf{d}_t^2[.]$ the temporal counterpart of $\mathbf{d}_x^2[.]$ or $\mathbf{d}_y^2[.]$, wave equation (3.4) receives the following form

$$\begin{aligned} &\mathbf{d}_t^2\left[f\Big|_{x,y}^t\right] - \upsilon^2(\mathbf{r})\frac{1}{[\Psi(\Delta h)]^2}\mathbf{L}_0^2\left[f\Big|_{x,y}^t\right] = 0 \\ &\Rightarrow \mathbf{d}_t^2\left[f\Big|_{x,y}^t\right] - \upsilon^2(\mathbf{r})\mathbf{L}_0^{2,\text{nst}}\left[f\Big|_{x,y}^t\right] = 0, \end{aligned} \tag{3.13}$$

which, after the necessary replacements and the use of time increment Δt, yields

$$f\big|_{x,y}^{t+\Delta t} = 2f\big|_{x,y}^{t} - f\big|_{x,y}^{t-\Delta t} + \upsilon^2(\mathbf{r})\mathbf{L}_0^{2,\mathrm{nst}}\left[f\big|_{x,y}^{t}\right]. \tag{3.14}$$

The solution of (3.4) in a 3-D domain is straightforward, since the derivation process preserves the preceding abstractions and only operator $\mathbf{L}_0^2[.]$ has to be appropriately redefined.

Three-Dimensional Analysis

The extension of the nonstandard finite-difference method in three dimensions requires the definition of three Laplacian operators for the calculation of spatial derivatives in the wave equation. Keeping the notation of the 2-D case and assuming a uniform grid, $\mathbf{L}_1^2[.]$ is now depicted by

$$\mathbf{L}_1^2\left[f\big|_{x,y,z}^{t}\right] = \mathbf{d}_x^2\left[f\big|_{x,y,z}^{t}\right] + \mathbf{d}_y^2\left[f\big|_{x,y,z}^{t}\right] + \mathbf{d}_z^2\left[f\big|_{x,y,z}^{t}\right] \tag{3.15}$$

for

$$\begin{aligned}
\mathbf{d}_x^2\left[f\big|_{x,y,z}^{t}\right] &= f\big|_{x+\Delta h,y,z}^{t} - 2f\big|_{x,y,z}^{t} + f\big|_{x-\Delta h,y,z}^{t} \\
\mathbf{d}_y^2\left[f\big|_{x,y,z}^{t}\right] &= f\big|_{x,y+\Delta h,z}^{t} - 2f\big|_{x,y,z}^{t} + f\big|_{x,y-\Delta h,z}^{t} \\
\mathbf{d}_z^2\left[f\big|_{x,y,z}^{t}\right] &= f\big|_{x,y,z+\Delta h}^{t} - 2f\big|_{x,y,z}^{t} + f\big|_{x,y,z-\Delta h}^{t}.
\end{aligned}$$

The second operator $\mathbf{L}_2^2[.]$ extends its structural arrangement at two different z-planes as

$$\begin{aligned}
\mathbf{L}_2^2\left[f\big|_{x,y,z}^{t}\right] = \frac{1}{4}\Big(& f\big|_{x+\Delta h,y+\Delta h,z+\Delta h}^{t} + f\big|_{x-\Delta h,y+\Delta h,z+\Delta h}^{t} + f\big|_{x+\Delta h,y-\Delta h,z+\Delta h}^{t} \\
&+ f\big|_{x-\Delta h,y-\Delta h,z+\Delta h}^{t} + f\big|_{x-\Delta h,y-\Delta h,z-\Delta h}^{t} + f\big|_{x+\Delta h,y-\Delta h,z-\Delta h}^{t} \\
&+ f\big|_{x-\Delta h,y+\Delta h,z-\Delta h}^{t} + f\big|_{x+\Delta h,y+\Delta h,z-\Delta h}^{t}\Big) - 2f\big|_{x,y,z}^{t}.
\end{aligned} \tag{3.16}$$

Additionally, for the new operator $\mathbf{L}_3^2[.]$, one has to combine three 2-D Laplacians with the discretization logic of (3.6), for each of the three coordinate planes $x-y$, $x-z$, and $y-z$. This notion leads to

$$\begin{aligned}
\mathbf{L}_3^2\left[f\big|_{x,y,z}^{t}\right] &= \frac{1}{2}\left(\mathbf{L}_2^2\left[f\big|_{x,y,z}^{t}\right]_{xy} + \mathbf{L}_2^2\left[f\big|_{x,y,z}^{t}\right]_{xz} + \mathbf{L}_2^2\left[f\big|_{x,y,z}^{t}\right]_{yz}\right) \\
&= \frac{1}{4}\Big[f\big|_{x+\Delta h,y-\Delta h,z}^{t} + f\big|_{x-\Delta h,y+\Delta h,z}^{t} + f\big|_{x+\Delta h,y,z-\Delta h}^{t} + f\big|_{x-\Delta h,y,z+\Delta h}^{t} \\
&\quad + f\big|_{x,y+\Delta h,z-\Delta h}^{t} + f\big|_{x,y-\Delta h,z+\Delta h}^{t}\Big] - 3f\big|_{x,y,z}^{t}.
\end{aligned} \tag{3.17}$$

If $\mathbf{L}_i^2[.]$ $(i = 1, 2, 3)$ is applied to $e^{j\mathbf{k}\cdot\mathbf{r}}$ and denote $\mathbf{L}_i^2\left[e^{j\mathbf{k}\cdot\mathbf{r}}\right] = 2e^{j\mathbf{k}\cdot\mathbf{r}}\Lambda_i(k\Delta h, \theta, \varphi)$, functions Λ_i are found to be

$$\Lambda_1(k\Delta h, \theta, \varphi) = \cos(k_x\Delta h) + \cos(k_y\Delta h) + \cos(k_z\Delta h) - 3, \tag{3.18a}$$

$$\Lambda_2(k\Delta h, \theta, \varphi) = \cos(k_x\Delta h)\cos(k_y\Delta h)\cos(k_z\Delta h) - 1, \tag{3.18b}$$

$$\begin{aligned} 2\Lambda_3(k\Delta h, \theta, \varphi) = {} & \cos(k_x\Delta h)\cos(k_y\Delta h) + \cos(k_x\Delta h)\cos(k_z\Delta h) \\ & + \cos(k_y\Delta h)\cos(k_z\Delta h) - 3, \end{aligned} \tag{3.18c}$$

with $k_x = k\sin\theta\cos\varphi$, $k_y = k\sin\theta\sin\varphi$, and $k_z = k\cos\theta$ at the spherical coordinate system (r, θ, φ).

For the construction of the 3-D $\mathbf{L}_0^2[.]$ analog of (3.7), analysis looks for a combination like

$$\mathbf{L}_0^2\left[f\,\Big|_{x,y,z}^{t}\right] = \sum_{i=1}^{3} s_i\mathbf{L}_i^2\left[f\,\Big|_{x,y,z}^{t}\right] \quad \text{where} \quad s_1 + s_2 + s_3 = 1, \tag{3.19}$$

such that

$$\Lambda_0 = \sum_{i=1}^{3} s_i\Lambda_i \cong \cos(k\Delta h) - 1 \tag{3.20}$$

is isotropic for every θ and φ. Toward this objective, we first define $\mathbf{L}_{12}^2[.] = p_{12}\mathbf{L}_1^2[.] + (1 - p_{12})\mathbf{L}_2^2[.]$ and $\mathbf{L}_{13}^2[.] = p_{13}\mathbf{L}_1^2[.] + (1 - p_{13})\mathbf{L}_3^2[.]$ and attempt to minimize the variation of linear combinations

$$\Lambda_{12} = p_{12}\Lambda_1 + (1 - p_{12})\Lambda_2 \quad \text{and} \quad \Lambda_{13} = p_{13}\Lambda_1 + (1 - p_{13})\Lambda_3, \tag{3.21}$$

with respect to θ angle at $\varphi = 0°$. Substitution of (3.18) into (3.21) gives

$$p_{12}(k\Delta h, \theta, 0) = \frac{\cos(k\Delta h) - \Lambda_2(k\Delta h, \theta, 0) - 1}{\Lambda_1(k\Delta h, \theta, 0) - \Lambda_2(k\Delta h, \theta, 0)} \tag{3.22a}$$

$$p_{13}(k\Delta h, \theta, 0) = \frac{\cos(k\Delta h) - \Lambda_3(k\Delta h, \theta, 0) - 1}{\Lambda_1(k\Delta h, \theta, 0) - \Lambda_3(k\Delta h, \theta, 0)}. \tag{3.22b}$$

Obviously, from (3.11), (3.12), and (3.22) for the optimal value of $\theta = \theta_{\text{ex}}$ and the dependence of p on θ, it can be shown that $p_{12}(k\Delta h, \theta) = p(k\Delta h, \theta_{\text{ex}}) = p(k\Delta h)$ and $p_{13}(k\Delta h, \theta) = 2p(k\Delta h) - 1$. Thus,

$$\mathbf{L}_{12}^2\left[f\,\Big|_{x,y,z}^{t}\right] = p\mathbf{L}_1^2\left[f\,\Big|_{x,y,z}^{t}\right] + (1 - p)\mathbf{L}_2^2\left[f\,\Big|_{x,y,z}^{t}\right], \tag{3.23a}$$

$$\mathbf{L}_{13}^2\left[f\,\Big|_{x,y,z}^{t}\right] = (2p - 1)\mathbf{L}_1^2\left[f\,\Big|_{x,y,z}^{t}\right] + 2(1 - p)\mathbf{L}_3^2\left[f\,\Big|_{x,y,z}^{t}\right] \tag{3.23b}$$

may be deemed virtually isotropic at $\varphi = 0°$. The next step is to superimpose (3.23a) and (3.23b) as

$$\mathbf{L}_0^2\left[f\,\big|_{x,y,z}^{t}\right] = s\mathbf{L}_{12}^2\left[f\,\big|_{x,y,z}^{t}\right] + (1-s)\mathbf{L}_{13}^2\left[f\,\big|_{x,y,z}^{t}\right], \tag{3.24}$$

and suppress its φ dependence by redefining $\Lambda_{12} = p\Lambda_1 + (1-p)\Lambda_2$, $\Lambda_{13} = (2p-1)\Lambda_1 + 2(1-p)\Lambda_3$ and assuming that $\theta = \pi/4$. If the idea of (3.19)–(3.21) is repeated for (3.24), one gets

$$s(k\Delta h) = s(k\Delta h, \varphi_{\text{ex}}) = \frac{\Lambda_{13}(k\Delta h, \pi/4, \varphi_{\text{ex}}) - \cos(k\Delta h) + 1}{\Lambda_{13}(k\Delta h, \pi/4, \varphi_{\text{ex}}) - \Lambda_{12}(k\Delta h, \pi/4, \varphi_{\text{ex}})}. \tag{3.25}$$

It is stressed that φ_{ex} is the value that gives a weak dependence of s on φ; i.e., it minimizes φ-variation. After the suitable numerical calculations, $\varphi \cong 0.11811\pi$. Finally, $\mathbf{L}_0^{2,\text{nst}}[.]$ becomes

$$\begin{aligned}\mathbf{L}_0^{2,\text{nst}}\left[f\,\big|_{x,y,z}^{t}\right] &= \frac{1}{\Psi(\Delta h)}\mathbf{L}_0^2\left[f\,\big|_{x,y,z}^{t}\right]\\ &= \frac{1}{\Psi(\Delta h)}\left(s_1\mathbf{L}_1^2\left[f\,\big|_{x,y,z}^{t}\right] + s_2\mathbf{L}_2^2\left[f\,\big|_{x,y,z}^{t}\right] + s_3\mathbf{L}_3^2\left[f\,\big|_{x,y,z}^{t}\right]\right),\end{aligned} \tag{3.26}$$

with

$$s_1 = s(1-p) + (2p-1), \qquad s_2 = s(1-p), \qquad s_3 = 1 - s_1 - s_2, \tag{3.27}$$

yielding, again, (3.14) as the nonstandard finite-difference solution of the 3-D wave equation.

3.2.2 Discretization of Maxwell's Equations

When the time-dependent Maxwell's equations are to be discretized, the nature of vector difference operator requires the decomposition of the prospective $\mathbf{L}_0^2[.]$ to guarantee the pertinent consistency. Hence,

$$\mathbf{L}_0^2\left[f(\mathbf{r},t)\right] = \sum_{\zeta=1}^{\dim(\mathbf{r})} \mathbf{d}_\zeta^{\text{A}}\left[f(\mathbf{r},t)\right]\mathbf{d}_\zeta^{\text{B}}\left[f(\mathbf{r},t)\right], \tag{3.28}$$

where $\mathbf{d}_\zeta^{\text{B}}[.]$ signifies the central difference operator described in (3.1), $\mathbf{d}_\zeta^{\text{A}}[.]$ is a novel operator that must be determined, and $\dim(\mathbf{r})$ is the dimensionality of vector $\mathbf{r}$, depending on the problem [i.e. $2 \to (x, y)$ or $3 \to (x, y, z)$], which controls the size of the summation.

Next and via (3.2), $\mathbf{d}_t^{\text{nst}}\left[f(\mathbf{r},t)\right] = \mathbf{d}_t\left[f(\mathbf{r},t)\right]/\Psi(\omega,\Delta t)$ are defined, with $\Psi(\omega,\Delta t) = 2\sin(\omega\Delta t/2)/\omega$ the respective correction function for time differentiation and $\mathbf{d}_t\left[f(\mathbf{r},t)\right] = f(\mathbf{r}, t+\Delta t/2) - f(\mathbf{r}, t-\Delta t/2)$. In this occasion, the nonstandard expression of Maxwell's equations, (2.1) and (2.2), is written as

$$\mu(\mathbf{r}_{\rm fc})\frac{\partial \mathbf{H}(\mathbf{r}_{\rm fc},t)}{\partial t} = -\nabla\times\mathbf{E}(\mathbf{r}_{\rm ed},t) \Rightarrow \mu(\mathbf{r}_{\rm fc})\mathbf{d}_t^{\rm nst}\left[\mathbf{H}(\mathbf{r}_{\rm fc},t)\right] = -\overline{\mathbf{L}}_{\rm A}^{\rm nst}\left[\mathbf{E}(\mathbf{r}_{\rm ed},t)\right], \tag{3.29a}$$

$$\varepsilon(\mathbf{r}_{\rm ed})\frac{\partial \mathbf{E}(\mathbf{r}_{\rm ed},t+\Delta t/2)}{\partial t} = \nabla\times\mathbf{H}(\mathbf{r}_{\rm fc},t+\Delta t/2) \Rightarrow \varepsilon(\mathbf{r}_{\rm ed})\mathbf{d}_t^{\rm nst}[\mathbf{E}(\mathbf{r}_{\rm ed},t+\Delta t/2)] = \overline{\mathbf{L}}_{\rm B}^{\rm nst}[\mathbf{H}(\mathbf{r}_{\rm fc},t+\Delta t/2)], \tag{3.29b}$$

with $\mathbf{r}_{\rm fc}$ and $\mathbf{r}_{\rm ed}$ denoting – as in the FDTD method – that $\mathbf{H}$ and $\mathbf{E}$ field components are located at the faces and the edges of the elementary cell, respectively. Also, vector operators $\overline{\mathbf{L}}_{\rm A}^{\rm nst}[.]$ and $\overline{\mathbf{L}}_{\rm B}^{\rm nst}[.]$ are depicted by

$$\overline{\mathbf{L}}_i^{\rm nst}\left[f(\mathbf{r},t)\right] = \frac{1}{\Psi(k,\Delta h)}\overline{\mathbf{L}}_i\left[f(\mathbf{r},t)\right] = \frac{1}{\Psi(k,\Delta h)}\sum_{\zeta=1}^{\dim(\mathbf{r})}\mathbf{d}_\zeta^i\left[f(\mathbf{r},t)\right]\hat{\zeta} \quad \text{for } i = \mathrm{A, B} \tag{3.30}$$

$\Psi(k,\Delta x) = 2\sin(k\Delta x/2)/k$, and $\hat{\zeta}$ the appropriate unit vector depending on $\dim(\mathbf{r})$, as mentioned above. Expanding $\mathbf{d}_t^{\rm nst}[.]$ and $\overline{\mathbf{L}}_i^{\rm nst}[.]$, the system of (3.29) turns into

$$\mathbf{H}(\mathbf{r}_{\rm fc},t+\Delta t/2) = \mathbf{H}(\mathbf{r}_{\rm fc},t-\Delta t/2) - \frac{\Psi(\omega,\Delta t)}{\mu(\mathbf{r}_{\rm fc})\Psi(k,\Delta h)}\overline{\mathbf{L}}_{\rm A}\left[\mathbf{E}(\mathbf{r}_{\rm ed},t)\right], \tag{3.31a}$$

$$\mathbf{E}(\mathbf{r}_{\rm ed},t+\Delta t) = \mathbf{E}(\mathbf{r}_{\rm ed},t) + \frac{\Psi(\omega,\Delta t)}{\varepsilon(\mathbf{r}_{\rm ed})\Psi(k,\Delta h)}\overline{\mathbf{L}}_{\rm B}\left[\mathbf{H}(\mathbf{r}_{\rm fc},t+\Delta t/2)\right]. \tag{3.31b}$$

Let us now return to the extraction of $\mathbf{d}_\zeta^{\rm A}[.]$. For the 2-D case, namely $\zeta \in (x,y)$, an auxiliary operator $\mathbf{d}_\zeta^{\rm C}[.]$ is introduced as

$$\mathbf{d}_x^{\rm C}\left[f\big|_{x,y}^t\right] = \frac{1}{2}\Big(f\big|_{x+\Delta h/2,y+\Delta h}^t - f\big|_{x-\Delta h/2,y+\Delta h}^t + f\big|_{x+\Delta h/2,y-\Delta h}^t - f\big|_{x-\Delta h/2,y-\Delta h}^t\Big), \tag{3.32a}$$

$$\mathbf{d}_y^{\rm C}\left[f\big|_{x,y}^t\right] = \frac{1}{2}\Big(f\big|_{x+\Delta h,y+\Delta h/2}^t + f\big|_{x-\Delta h,y+\Delta h/2}^t - f\big|_{x+\Delta h,y-\Delta h/2}^t - f\big|_{x-\Delta h,y-\Delta h/2}^t\Big), \tag{3.32b}$$

and consider the weighted superposition

$$\mathbf{d}_\zeta^{\rm A}\left[f\big|_{x,y}^t\right] = q\mathbf{d}_\zeta^{\rm B}\left[f\big|_{x,y}^t\right] + (1-q)\mathbf{d}_\zeta^{\rm C}\left[f\big|_{x,y}^t\right] \quad \text{for } \zeta = (x,y). \tag{3.33}$$

It can be readily proven that

$$\sum_{\zeta=x,y}\mathbf{d}_\zeta^{\rm B}\left[f\big|_{x,y}^t\right]\mathbf{d}_\zeta^{\rm C}\left[f\big|_{x,y}^t\right] = 2\mathbf{L}_2^2\left[f\big|_{x,y}^t\right] - \mathbf{L}_1^2\left[f\big|_{x,y}^t\right], \tag{3.34}$$

where $\mathbf{L}_1^2[.]$ and $\mathbf{L}_2^2[.]$ are given from (3.5) and (3.6), respectively. Parameter q is extracted so as (3.28) is fully satisfied. Plugging (3.33) into (3.28) and utilizing (3.34), one reaches to

$$q = (1 + p)/2, \tag{3.35}$$

in which p is acquired by (3.12).

For the 3-D case, inspection of (3.28) reveals that the weighted superposition of $\mathbf{d}_\zeta^{\mathrm{A}}[.]$ for $\zeta \in (x, y, z)$ requires three independent difference operators, i.e., $\mathbf{d}_\zeta^{\mathrm{B}}[.]$ – provided, again, by (3.1) – as well as the two auxiliary ones $\mathbf{d}_\zeta^{\mathrm{C}}[.]$ and $\mathbf{d}_\zeta^{\mathrm{D}}[.]$. Hence, studying a certain partial derivative for illustration ∂_z, the unknown operators $\mathbf{d}_z^{\mathrm{C}}[.]$ and $\mathbf{d}_z^{\mathrm{D}}[.]$ receive the form of

$$\begin{aligned}\mathbf{d}_z^{\mathrm{C}}\left[f\left|_{x,y,z}^t\right.\right] = \frac{1}{4}\Big(& f\left|_{x+\Delta h,y+\Delta h,z+\Delta h/2}^t\right. - f\left|_{x+\Delta h,y+\Delta h,z-\Delta h/2}^t\right. + f\left|_{x+\Delta h,y-\Delta h,z+\Delta h/2}^t\right. \\ & - f\left|_{x+\Delta h,y-\Delta h,z-\Delta h/2}^t\right. + f\left|_{x-\Delta h,y+\Delta h,z+\Delta h/2}^t\right. - f\left|_{x-\Delta h,y+\Delta h,z-\Delta h/2}^t\right. \\ & + f\left|_{x-\Delta h,y-\Delta h,z+\Delta h/2}^t\right. - f\left|_{x-\Delta h,y-\Delta h,z-\Delta h/2}^t\right.\Big),\end{aligned} \tag{3.36}$$

$$\begin{aligned}\mathbf{d}_z^{\mathrm{D}}\left[f\left|_{x,y,z}^t\right.\right] = \frac{1}{4}\Big(& f\left|_{x,y+\Delta h,z+\Delta h/2}^t\right. - f\left|_{x,y+\Delta h,z-\Delta h/2}^t\right. + f\left|_{x+\Delta h,y,z+\Delta h/2}^t\right. \\ & - f\left|_{x+\Delta h,y,z-\Delta h/2}^t\right. + f\left|_{x,y-\Delta h,z+\Delta h/2}^t\right. - f\left|_{x,y-\Delta h,z-\Delta h/2}^t\right. \\ & + f\left|_{x-\Delta h,y,z+\Delta h/2}^t\right. - f\left|_{x-\Delta h,y,z-\Delta h/2}^t\right.\Big).\end{aligned} \tag{3.37}$$

Finally, the superposition for $\mathbf{d}_z^{\mathrm{A}}[.]$ may be written as

$$\mathbf{d}_z^{\mathrm{A}}\left[f\left|_{x,y,z}^t\right.\right] = \sum_{i=\mathrm{B,C,D}} q_i \mathbf{d}_z^i\left[f\left|_{x,y,z}^t\right.\right] \quad \text{with} \sum_{i=\mathrm{B,C,D}} q_i = 1 \tag{3.38}$$

and a similar process holding for ∂_x and ∂_y. To compute coefficients q_i, it is proven that

$$\sum_{\zeta=x,y,z} \mathbf{d}_\zeta^{\mathrm{B}}\left[f\left|_{x,y,z}^t\right.\right]\mathbf{d}_\zeta^{\mathrm{C}}\left[f\left|_{x,y,z}^t\right.\right] = 3\mathbf{L}_2^2\left[f\left|_{x,y,z}^t\right.\right] - \mathbf{L}_3^2\left[f\left|_{x,y,z}^t\right.\right], \tag{3.39a}$$

$$\sum_{\zeta=x,y,z} \mathbf{d}_\zeta^{\mathrm{B}}\left[f\left|_{x,y,z}^t\right.\right]\mathbf{d}_\zeta^{\mathrm{D}}\left[f\left|_{x,y,z}^t\right.\right] = 2\mathbf{L}_3^2\left[f\left|_{x,y,z}^t\right.\right] - \mathbf{L}_1^2\left[f\left|_{x,y,z}^t\right.\right], \tag{3.39b}$$

where operators $\mathbf{L}_1^2[.]$, $\mathbf{L}_2^2[.]$, and $\mathbf{L}_3^2[.]$ are given by (3.15), (3.16), and (3.17), respectively. Substitution of (3.38) and (3.39) into the 3-D analog of (3.28) yields

$$q_{\mathrm{B}} = s_1 + \frac{1}{3}s_2 + \frac{1}{2}s_3, \qquad q_{\mathrm{C}} = \frac{1}{3}s_2, \qquad q_{\mathrm{D}} = \frac{1}{3}s_2 + \frac{1}{2}s_3, \tag{3.40}$$

with s_i $(i = 1, 2, 3)$ retrieved from (3.27), whose involvement produces the alternative set of

$$q_{\mathrm{B}} = \frac{1}{3}s(1-p) + p, \qquad q_{\mathrm{C}} = \frac{1}{3}s(1-p), \qquad q_{\mathrm{D}} = 1 - q_{\mathrm{B}} - q_{\mathrm{C}}. \tag{3.41}$$

In this manner, difference operators $\mathbf{d}_{\zeta}^{A}[.]$ and $\mathbf{d}_{\zeta}^{B}[.]$ are fully determined, and so allowing the consistent design of (3.30) and the subsequent time update of the 3-D Maxwell's equations (3.31). Conclusively, a noteworthy feature of these operators is the use of *extra nodal points* for the approximation of partial derivatives. This implies that, unlike the limited stencil of the FDTD technique, the nonstandard concepts offer an enhanced manipulation of the elementary cells and through additional degrees of freedom permit the *significant suppression* of dispersion and anisotropy errors. These merits are much more prominent in higher order formulations, where the abruptly curved waveguide or antenna components, the arbitrary material discontinuities, and the dissimilar interfaces stipulate very robust simulations.

3.3 DEVELOPMENT OF THE HIGHER ORDER NONSTANDARD FORMS IN CARTESIAN COORDINATES

A serious difficulty in the realization of a second-order time-domain numerical technique is the appearance of the undesirable dispersion errors that can contaminate the overall solution of several problems. Having analyzed the theoretical background of the nonstandard operators in Section 3.1, it becomes apparent that a mitigation of these defects can be sought in the proper spatial sampling of the unknown field variables. Also, a key issue in their successful suppression is the explicit or implicit inclusion of the divergence conditions in the discrete model. For this aim, the present section describes a higher order nonstandard FDTD method in Cartesian coordinates with superior modeling characteristics. To manipulate the inevitably widened spatial stencils, the compact-operator process is adopted, whereas time update is conducted via a family of higher order leapfrog-type or Runge–Kutta integrators. Therefore, this concept enables the uncollocated placement of electric and magnetic field components in a spatially proportional manner.

3.3.1 Theory of the Precise Spatial/Temporal Operators

The most important feature of this dispersion-optimized FDTD method is the *higher order nonstandard finite-difference schemes* [6, 7] that substitute their conventional counterparts in the differentiation of Ampère's and Faraday's laws, as already described in (3.31). The proposed technique can be occasionally even 7 to 8 orders of magnitude more accurate than the fourth-order implementations of Chapter 2. Although the cost is slightly increased, the overall simulation benefits from the low resolutions and the reduced number of iterations. Thus, for spatial derivative approximation, the following two operators are defined:

$$\mathbf{DS}_x\left[f\big|_{x,y,z}^{t}\right] = \frac{9}{8}\mathbf{D}_{x,\Delta x}^{\mathrm{nst}}\left[f\big|_{x,y,z}^{t}\right] - \frac{1}{24}\mathbf{D}_{x,3\Delta x}^{\mathrm{nst}}\left[f\big|_{x,y,z}^{t}\right], \tag{3.42}$$

$$\mathbf{DS}_x\left[f\big|^t_{x,y,z}\right] = \frac{11}{12}\mathbf{D}^{\text{nst}}_{x,\Delta x}\left[f\big|^t_{x,y,z}\right] \\ -\frac{1}{24}\left(\mathbf{D}^{\text{nst}}_{x,3\Delta x}\left[f\big|^t_{x,y,z}\right] + \frac{f\big|^t_{x-\Delta x/2,y,z} - f\big|^t_{x-3\Delta x/2,y,z}}{\Delta x}\right), \quad (3.43)$$

where $\mathbf{D}^{\text{nst}}_{x,\Delta}[.]$ (for $\Delta = \Delta x,\ 3\Delta h$) is the 3-D nonstandard finite-difference operator which is defined as

$$\mathbf{D}^{\text{nst}}_{x,\Delta}\left[f\big|^t_{x,y,z}\right] \equiv \frac{1}{\Psi(k,\Delta)}\left(q_1\mathbf{d}^{\text{A}}_{x,\Delta}\left[f\big|^t_{x,y,z}\right] + q_2\mathbf{d}^{\text{B}}_{x,\Delta}\left[f\big|^t_{x,y,z}\right] \\ + q_3\mathbf{d}^{\text{C}}_{x,\Delta}\left[f\big|^t_{x,y,z}\right]\right). \quad (3.44)$$

Similar operators with respect to y and z can be simply obtained. Correction function $\Psi(k,\Delta)$ accepts multiple arguments and therefore it can drastically contribute to the minimization of the undesirable FDTD dispersion and dissipation errors. Among its various possible forms, some of the most convenient are

$$\Psi_1(k,\Delta) = \frac{2}{k}\sin\left(\frac{k\Delta}{2}\right), \qquad \Psi_2(k,\Delta) = \frac{5}{k}\cos\left(\frac{k\Delta}{4}\right), \\ \Psi_3(k,\Delta) = \frac{12}{k^2}\sin\left(\frac{k\Delta}{4}\right)\cos\left(\frac{k\Delta}{8}\right). \quad (3.45)$$

Let us now examine the choice of argument $k\Delta$ in (3.45) together with the possible treatments devised to handle broadband transient fields via the nonstandard higher order concepts. Since, $\Psi_i(k,\Delta)$, for $i = 1, 2, 3$, and $\mathbf{D}^{\text{nst}}_{x,\Delta}[.]$ depend on k, solution error appears for frequencies deviating from the exact value k^{ex} at which both quantities are defined. Nonetheless, in this algorithm, the structural profile of (3.42) and (3.43) alleviates these discrepancies. After some algebra and the inspection of (3.8), it is concluded that the constraint $\mathbf{D}^{\text{nst}}_{x,\Delta}[e^{jkx}]/\partial_x e^{jkx} = 1$ can be approximated by $\mathbf{D}^{\text{nst}}_{x,\Delta}[e^{jkx}]/\partial_x e^{jkx} \cong 2(\cos k\Delta - 1)$. To pick an acceptable $k\Delta$ value in the case of multifrequency excitations, two procedures are implemented. In the first one, the Fourier transform of the already computed transient field components (**E** or **H**) at predetermined grid locations is employed. After their frequency range is found, the maximum or mean value of this spectrum is selected through which the optimum $k\Delta$ argument is specified. The process not affecting the total CPU burden takes place at every time-step, while its accuracy increases as the number of the lattice points becomes larger. The second procedure lessens the difference $[\cos(k_{\text{min}}\Delta) - \cos(k_{\text{max}}\Delta)]$ by setting the minimum wavelength λ_{min} (in grid units), corresponding to the highest wavenumber k_{max}, where k_{min} is the lowest wavenumber. Its efficiency depends on the configuration of every problem, the frequency range of the incident field and the accumulated phase error. So, a broad frequency spectrum implies that the value of λ_{min} should be augmented, which in its turn requires a denser lattice relative to wavelengths.

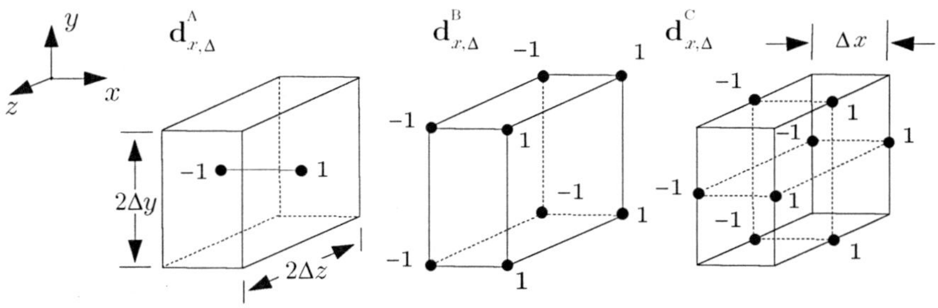

FIGURE 3.1: Graphical representation of difference operators. The numbers at the vertices indicate the sign of summation for the corresponding values of the approximated function f. The origin (x, y, z) is located at the center of the cell

Operators $\mathbf{d}^{i}_{x,\Delta}[.]$ ($i = $ A, B, C) are defined by means of (3.1) and (3.32)–(3.37), whereas their graphical depiction is shown in Figure 3.1. Hence, one can write

$$\mathbf{d}^{\mathrm{A}}_{x,\Delta}\left[f\,\big|^{t}_{x,y,z}\right] = f\,\big|^{t}_{x+\Delta/2,y,z} - f\,\big|^{t}_{x-\Delta/2,y,z}, \tag{3.46}$$

$$\begin{aligned}\mathbf{d}^{\mathrm{B}}_{x,\Delta}\left[f\,\big|^{t}_{x,y,z}\right] = \frac{1}{4}\Big(&f\,\big|^{t}_{x+\Delta/2,y+\Delta,z+\Delta} + f\,\big|^{t}_{x+\Delta/2,y+\Delta,z-\Delta} + f\,\big|^{t}_{x+\Delta/2,y-\Delta,z+\Delta}\\ &+f\,\big|^{t}_{x+\Delta/2,y-\Delta,z-\Delta} - f\,\big|^{t}_{x-\Delta/2,y+\Delta,z+\Delta} - f\,\big|^{t}_{x-\Delta/2,y+\Delta,z-\Delta}\\ &-f\,\big|^{t}_{x-\Delta/2,y-\Delta,z+\Delta} - f\,\big|^{t}_{x-\Delta/2,y-\Delta,z-\Delta}\Big),\end{aligned} \tag{3.47}$$

$$\begin{aligned}\mathbf{d}^{\mathrm{C}}_{x,\Delta}\left[f\,\big|^{t}_{x,y,z}\right] = \frac{1}{4}\Big(&f\,\big|^{t}_{x+\Delta/2,y+\Delta,z} + f\,\big|^{t}_{x+\Delta/2,y-\Delta,z} + f\,\big|^{t}_{x+\Delta/2,y,z+\Delta}\\ &+f\,\big|^{t}_{x+\Delta/2,y,z-\Delta} - f\,\big|^{t}_{x-\Delta/2,y+\Delta,z} - f\,\big|^{t}_{x-\Delta/2,y-\Delta,z}\\ &-f\,\big|^{t}_{x-\Delta/2,y,z+\Delta} - f\,\big|^{t}_{x-\Delta/2,y,z-\Delta}\Big),\end{aligned} \tag{3.48}$$

Parameters q_i, for $i = 1, 2, 3$, are given by (3.40) and (3.41) in terms of condition (3.28). Note that despite their different notation, they have exactly the same function. In fact, their values certify the stable profile of the algorithm as well as its enhanced levels of convergence.

On the other hand, temporal differentiation in Maxwell's equations is conducted via the ensuing nonstandard operator, which also involves third-order conventional counterparts as those discussed in Chapter 2. So,

$$\mathbf{DT}^{\mathrm{nst}}\left[f\,\big|^{t}_{x,y,z}\right] = \frac{f\,\big|^{t+\Delta t/2}_{x,y,z} - f\,\big|^{t-\Delta t/2}_{x,y,z}}{\Psi(\omega,\Delta t)} - \frac{[\Psi(\omega,\Delta t)]^2}{24}\,\frac{\partial^3 f}{\partial t^3}\bigg|^{t}_{x,y,z}, \tag{3.49}$$

with $\Psi(\omega, \Delta t)$ its respective correction function [8].

The foregoing conceptual study reveals the mathematical robustness of the nonstandard strategy, mainly attributed to the structural reliability of the discretization technique it embodies. These characteristics guarantee the consistent construction of the lattices and bridge the gap between the continuous and approximate state. Moreover, their closer assessment by specific quantitative tools is proven to be extremely useful, since it leads to the clarification of several mechanisms. This interpretation enables the definition of exact time-stepping relations which, with respect to traditional approaches, yield an alternative update intrinsically built into the fundamental laws of electromagnetism. Thus, numerical schemes complying with the above guidelines can be naturally conveyed in terms of field quantities associated with the propagation of both propagating and evanescent waves [7]. As a matter of fact, the importance of such a treatment will become more evident in nonorthogonal or unstructured meshes where the incorporation of lattice duality is necessary.

Recalling the structure of any higher order FDTD solver, a fairly substantial issue that may hinder its implementation is of the inevitably *widened spatial stencils* in the vicinity of PEC interfaces or absorbing walls. What seems apparent in the second-order Yee's technique, where such discontinuities are efficiently simulated, is not adequate when dealing with higher order formulae. As readily observed, their stencil extends at least two nodes on either side of a lattice point at which the unknown quantity is defined. Therefore, when one reaches at the preceding areas, problems arise. To circumvent these difficulties, various ideas have been proposed so far. Among them, the idea of *compact central operators* is recognized as an effective means. Since the topic of compact differencing is going to be more generally covered in Section 4.3, for the sake of convenience and completion, herein, only a brief description will be provided. In particular, extending the abstractions of [9, 10], a *higher order* version of the implicit compact central operators is presented.

Consider the computation of first- and second-order partial derivatives ∂_x and ∂_x^2, respectively, at a PEC interface, with an alike treatment holding for quantities with respect to y- and z-direction. For these quantities, the general expression of compact central operators is

$$a_1 \frac{\partial f}{\partial x}\bigg|_x^t + a_2 \left(\frac{\partial f}{\partial x}\bigg|_{x+\Delta x}^t + \frac{\partial f}{\partial x}\bigg|_{x-\Delta x}^t \right) + a_3 \Delta x \left(\frac{\partial^2 f}{\partial x^2}\bigg|_{x+\Delta x}^t - \frac{\partial^2 f}{\partial x^2}\bigg|_{x-\Delta x}^t \right) + \cdots = b \frac{f\big|_{x+\Delta x/2}^t - f\big|_{x-\Delta x/2}^t}{\Delta x}, \tag{3.50}$$

$$c_1 \frac{\partial^2 f}{\partial x^2}\bigg|_x^t + c_2 \left(\frac{\partial^2 f}{\partial x^2}\bigg|_{x+\Delta x}^t + \frac{\partial^2 f}{\partial x^2}\bigg|_{x-\Delta x}^t \right) + \frac{c_3}{2\Delta x} \left(\frac{\partial f}{\partial x}\bigg|_{x+\Delta x}^t - \frac{\partial f}{\partial x}\bigg|_{x-\Delta x}^t \right) + \cdots = d \frac{f\big|_{x+\Delta x}^t - 2f\big|_x^t + f\big|_{x-\Delta x}^t}{(\Delta x)^2}, \tag{3.51}$$

where only the x variable – along which all stencil modifications occur – is displayed, while coefficients a_i, c_i (for $i = 1, 2, \ldots$) and b, d are unknown and have to be specified. In this case, where just first-order spatial derivatives are approximated by higher order schemes, (3.50) and (3.51) reduce to

$$a_2 \left.\frac{\partial f}{\partial x}\right|_{x+\Delta x}^{t} + a_1 \left.\frac{\partial f}{\partial x}\right|_{x}^{t} + a_2 \left.\frac{\partial f}{\partial x}\right|_{x-\Delta x}^{t} = b \frac{f\left|_{x+\Delta x/2}^{t} - f\right|_{x-\Delta x/2}^{t}}{\Delta x}. \tag{3.52}$$

For the evaluation of coefficients a_1, a_2, and b, in (3.52), the Fourier analysis is selected. Assume that f represents a plane wave propagating toward the x-axis, namely f has the exponential form of $e^{j(kx-\omega t)}$. Application of (3.52) to this function gives the dispersion relation of

$$k^{\text{num}}(k\Delta x) = \frac{2b \sin(k\Delta x/2)}{[2a_1 \cos(k\Delta x) + a_2]\Delta x}, \tag{3.53}$$

with k^{num} the numerical wavenumber. Ideally, the preceding operator should produce a k^{num}/k ratio equal to 1 for every wavenumber. Nevertheless in real experimentations, this ratio satisfies such a condition for only one frequency and approaches 1 for a relatively narrow frequency range. Typically, the wavenumber that must be chosen in order to attain $k^{\text{num}}/k = 1$ is equal to 0. For example, if (3.53) is expanded around point $k\Delta x = 0$, by means of Taylor's series, and ignore the second-order terms, it is feasible to design the fourth-order compact operator; i.e., $a_1 = 1$, $a_2 = 22$, and $b = 24$. Alternative values of a_1, a_2, and b may also be obtained through other optimization criteria. Thus, (3.52) regardless of its implicit character handles hard-to-model walls without restricting the convergence of the procedure and concurrently satisfies the boundary conditions pointwise, exactly.

3.3.2 Higher Order Nonstandard Leapfrog-Type Integrators

This section investigates the essential issue of numerical integration for the nonstandard FDTD method by developing a family of *generalized leapfrog schemes* with adjustable order of accuracy and gradual evolution. Motivated by the necessity of seriously suppressing any oscillatory artifacts, due to the incompatible order between spatial and temporal approximants, the technique alleviates the ill-posed character of several simulations. Therefore, it guarantees a notable performance without imposing nonphysical conditions.

The approach is based on the leapfrog concept and can be implemented for any even-order FDTD and compact-operator formulation with reasonable CPU and memory requirements. Analysis starts from the following expansion for the magnetic field at time-step $t = n\Delta t$

$$\mathbf{H}|^{n+1/2} = \mathbf{H}|^{n-1/2} + 2\sum_{\substack{\kappa=1 \\ (\text{odd})}}^{K} \frac{1}{\kappa!}\left[\frac{\Psi(\omega, \Delta t)}{2}\right]^{\kappa} \mathbf{DT}^{\text{nst},\kappa}[\mathbf{H}|^{n}], \tag{3.54}$$

and the corresponding one for the electric field at $t = (n + 1/2)\Delta t$

$$\mathbf{E}|^{n+1} = \mathbf{E}|^{n} + 2\sum_{\substack{\kappa=1 \\ \text{(odd)}}}^{K} \frac{1}{\kappa!}\left[\frac{\Psi(\omega, \Delta t)}{2}\right]^{\kappa} \mathbf{DT}^{\mathrm{nst},\kappa}[\mathbf{E}|^{n+1/2}], \tag{3.55}$$

where $\mathbf{DT}^{\mathrm{nst},\kappa}[.]$ operator indicates the κth-order nonstandard temporal derivative. It is mentioned that in the conventional higher order FDTD case, the above integration process leads to the one of [11], while when $K = 1$, the well-known Yee's scheme is derived. On the contrary, when $K \neq 1$, the resulting κth-order temporal derivatives must be somehow calculated. For small values of K, this can be conducted by converting them to spatial $(\kappa - 1)$th-order ones via successive mutual substitutions in Ampère's and Faraday's laws. However, if $K > 4$, the above practice becomes prohibitively cumbersome, owing to the excessive complexity of the derivatives and their incorrect association with the topology of the cells.

An efficient way to circumvent this difficulty is to devise a gradually extending leapfrog algorithm, which can be reliably employed for any K value. For illustration, the fourth-order integrator is given by

$$\begin{aligned} &\Omega_1 = -[\Psi(\omega, \Delta t)/\mu]\,\overline{\mathbf{DS}}\left[\mathbf{E}|^{n}\right], && \Omega_2 = [\Psi(\omega, \Delta t)/\varepsilon]\,\overline{\mathbf{DS}}\,[\Omega_1], \\ &\Omega_3 = -[\Psi(\omega, \Delta t)/\mu]\,\overline{\mathbf{DS}}\,[\Omega_2], && \mathbf{H}|^{n+1/2} = \mathbf{H}|^{n-1/2} + \Omega_1 + \Omega_3/24, \end{aligned} \tag{3.56}$$

with $\overline{\mathbf{DS}}[.]$ the nonstandard finite-difference curl operator expressed in matrix form as

$$\overline{\mathbf{DS}}[.] = \nabla\times = \begin{bmatrix} 0 & \mathbf{DS}_z & \mathbf{DS}_y \\ \mathbf{DS}_z & 0 & -\mathbf{DS}_x \\ -\mathbf{DS}_y & \mathbf{DS}_x & 0 \end{bmatrix}. \tag{3.57}$$

Vectors Ω_i, for $i = 1, 2, 3$, in (3.56), are temporary storage quantities for the intermediate values of the progressive integration. In an analogous manner, the remaining part of the leapfrog scheme is completed as

$$\begin{aligned} &\Omega_1 = [\Psi(\omega, \Delta t)/\varepsilon]\,\overline{\mathbf{DS}}\left[\mathbf{H}|^{n+1/2}\right], && \Omega_2 = -[\Psi(\omega, \Delta t)/\mu]\,\overline{\mathbf{DS}}\,[\Omega_1], \\ &\Omega_3 = [\Psi(\omega, \Delta t)/\varepsilon]\,\overline{\mathbf{DS}}\,[\Omega_2], && \mathbf{E}|^{n+1} = \mathbf{E}|^{n} + \Omega_1 + \Omega_3/24. \end{aligned} \tag{3.58}$$

A graphical description of the generalized integrator for the case of $K = 4$ is illustrated in Figure 3.2. Apparently, (3.56) and (3.58) compensate the slight increase of the required memory with their low complexity. Also, their structure resembles the Runge–Kutta model (see Section 2.3.3), which can also be utilized for the time advancing of the higher order nonstandard FDTD method. Nonetheless, there is a significant difference between the two approaches: The former conducts a continuous temporal integration of *every* field component, while the latter advances *all* quantities in space but *not* in time at every time-step.

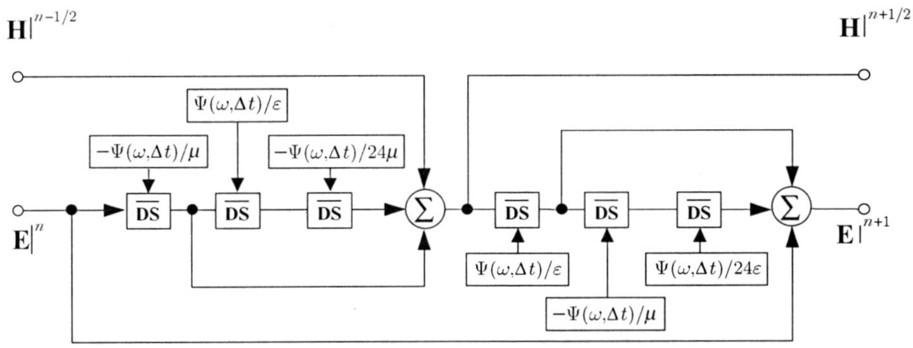

FIGURE 3.2: Schematic depiction of the fourth-order nonstandard leapfrog-type integrator

3.3.3 Alternative Approaches

The investigation of the aforementioned algorithm has shown that further accuracy improvements may be accomplished by the introduction of additional degrees of freedom. In view of this observation, an enhanced operator with optimal stability can be constructed.

Considering the integral form of Maxwell's equations and (3.44), spatial derivatives are evaluated by

$$\begin{aligned}\mathbf{DS}_x^{\text{nst}}\left[f\big|_{x,y,z}^t\right] = {} & \frac{A_1}{\Delta x}\left(\mathbf{D}_{x,3\Delta x/2}^{\text{nst}}\left[f\big|_{x,y,z}^t\right] - \mathbf{D}_{x,-3\Delta x/2}^{\text{nst}}\left[f\big|_{x,y,z}^t\right]\right)\\ & + \frac{1-A_1-A_2}{\Delta x}\left(\mathbf{D}_{x,\Delta x/2}^{\text{nst}}\left[f\big|_{x,y,z}^t\right] - \mathbf{D}_{x,-\Delta x/2}^{\text{nst}}\left[f\big|_{x,y,z}^t\right]\right)\\ & + \frac{A_2}{6\Delta x}\left(\mathbf{D}_{x,3\Delta x/2}^{\text{nst}}\left[f\big|_{x,y+\Delta x,z}^t\right] - \mathbf{D}_{x,-3\Delta x/2}^{\text{nst}}\left[f\big|_{x,y+\Delta x,z}^t\right]\right.\\ & \left. + \mathbf{D}_{x,3\Delta x/2}^{\text{nst}}\left[f\big|_{x,y-\Delta x,z}^t\right] - \mathbf{D}_{x,-3\Delta x/2}^{\text{nst}}\left[f\big|_{x,y-\Delta x,z}^t\right]\right),\end{aligned} \tag{3.59}$$

instead of (3.42) or (3.43) [12, 13]. Tuning parameters A_1 and A_2 satisfy constraints $A_1 < 0$ and $A_2 < 1.5 - 2A_1$. Specifying their range entails the solution of a convoluted system of nonlinear equations that yields complex expressions inappropriate for the extraction of any useful deductions. Nonetheless and without loss of generality, it is adequate to state that the above constraints can provide the correct A_1 and A_2 values for the majority of the simulations. To obtain the ideal A_1 and A_2, the respective dispersion relation (formed according to the selected coordinate metrics) is solved for the numerical value of k as a function of A_1 and A_2 for all propagation angles. This outcome is then used in the calculation of the arithmetic difference between the numerical and physical value of k, which is minimized in terms of A_1 and A_2. Evidently, the above process depends on grid resolution and material properties. Therefore, it must be conducted for every new mesh, despite the fact that some past values may suffice. Bearing in mind the variety of shapes in waveguide and antenna modeling, one can realize the reasons for this decision. It is worth mentioning that the precision of tuning parameters should be high because they regulate the phases of all electromagnetic vectors that are far more sensitive to perturbations than their magnitudes. A safe choice would be at least *ten significant digits* to accomplish rigorous outcomes. Similar conclusions are also deduced in [14], whose authors employ an analogous optimization for the construction of their conventional (2, 4) FDTD scheme.

3.4 GENERALIZED HIGHER ORDER CURVILINEAR FDTD METHOD

The optimal design of a realistic generally curved structure necessitates the meticulous modeling of every geometric peculiarity as well as the detailed characterization of any media interface. When these attributes are to be explored by a staircase time-domain method, the requirement of credible discretizations cannot be easily fulfilled because of its inherent restriction to orthogonal lattices. As a consequence, material boundaries not aligned to the mesh axes are poorly or even erroneously approximated. To evade these difficulties, a number of proficient techniques have been proposed [15–30], ranging from curvilinear or nonorthogonal FDTD variations to modified Cartesian and conformal algorithms. Since an in-depth analysis of this topic is not in the scope of this confined section, the reader is referred to the detailed study of [31].

A chief trait of these methods is that, in majority, they implement second-order approximations for spatial and temporal derivatives, which, under particular circumstances, are disposed to mesh reflection errors. So, the incorporation of higher order FDTD ideas in the prior approaches is a promising perspective for the improvement of their performance and their applicability to even more challenging situations. In this context, the conventional case would obviously constitute our first choice. However, when dual-cell or nonorthogonal tessellations are considered, the relatively rigid stencil manipulation of the operators, so derived, suffers from some nontrivial difficulties. These are summarized in a) the centers of primary cells do not always coincide with the centers of secondary edges, b) in the modified Maxwell's equations, temporal derivatives act on field components that are normal to the cell faces, while spatial derivatives treat the tangential quantities, and c) the algorithm exhibits loss of accuracy because of the gradually evolving time integration.

Bearing in mind these weaknesses, a higher order FDTD methodology for the systematic examination of complicated applications in 3-D *generalized curvilinear* coordinates, which has been developed in [32–34], is next presented. Through a *covariant/contravariant* vector field flux formulation, a parametric dual-cell classification of higher order nonstandard operators for the nonorthogonal div-curl problem is introduced. This tensorial character preserves the consistency and hyperbolic character of Maxwell's equations and includes the prerequisite of being fully explicit. In the laborious situation of dissimilar-material interfaces, convergence is preserved through efficient grid concepts accounting for the proper continuity or absorbing boundary conditions. For the temporal variable, the technique employs the leapfrog approach of Section 3.2.2 or the Runge–Kutta integrators that allow diverse excitations. Hence, the accumulating dispersion errors are radically diminished without the need of unmanageable meshes and small temporal increments.

3.4.1 Formulation of the Nonorthogonal Div-Curl Problem

The crucial point for numerical consistency regarding the FDTD solution of Maxwell's laws in curvilinear coordinates is the choice of the appropriate *basis* on which electric and magnetic quantities are expressed. Unfortunately, an improper differentiation of the basis' vectors may give rise to Cristoffel symbols [35], whose evaluation, via time-domain algorithms, is proven to be strenuous and usually inaccurate. Such erroneous instances are encountered in various problems involving multifrequency radiation, especially when the mesh lacks to track abrupt field singularities near highly curved surfaces. Since, Yee's method is not an exception, the above drawback may be overcome by developing a strategy for the formulation of the nonorthogonal div-curl problem that is a hyperset of Maxwell's equations. As it is already familiar, its main objective is the determination of a vector field $\mathbf{U}$, whose divergence q and curl $\mathbf{W}$ are assumed known. Thus, one has to extract dynamic expressions that will associate the inherent physical profile to a group of dual topological structures able to achieve an *optimal* grid phase velocity. The resulting scheme involves a higher order rendition of the covariant and contravariant vector component theory in which all metric terms are taken into consideration and are also fully conservative.

Preliminaries

Consider a 3-D domain that can be adequately described by the generalized curvilinear coordinate system (u, v, w) and that its mappings are adequately smooth to allow consistent definitions. Then, any vector $\mathbf{F}$ can be decomposed into three components with respect to the contravariant $\hat{\mathbf{a}}^u$, $\hat{\mathbf{a}}^v$, $\hat{\mathbf{a}}^w$ or the covariant $\hat{\mathbf{a}}_u$, $\hat{\mathbf{a}}_v$, $\hat{\mathbf{a}}_w$ linearly independent basis system as

$$\mathbf{F} = \sum_{\xi=u,v,w} (\hat{\mathbf{a}}_\xi \cdot \mathbf{F})\hat{\mathbf{a}}^\xi = \sum_{\xi=u,v,w} f_\xi \hat{\mathbf{a}}^\xi = \sum_{\zeta=u,v,w} (\hat{\mathbf{a}}^\zeta \cdot \mathbf{F})\hat{\mathbf{a}}_\zeta = \sum_{\zeta=u,v,w} f^\zeta \hat{\mathbf{a}}_\zeta . \tag{3.60}$$

The quantities f_ξ and f^ζ are, respectively, the *covariant* and *contravariant* components of $\mathbf{F}$ which, due to their reciprocity, satisfy the relation $\hat{\mathbf{a}}_\xi \cdot \hat{\mathbf{a}}^\zeta = \delta_{\xi\zeta}$ ($\delta_{\xi\zeta}$ is the Kronecker's delta). The system's metrical coefficients $g_{\zeta\xi}$ and $g^{\zeta\xi}$ are given by $g_{\zeta\xi} = g_{\xi\zeta} = \hat{\mathbf{a}}_\zeta \cdot \hat{\mathbf{a}}_\xi$ and $g^{\zeta\xi} = g^{\xi\zeta} = \hat{\mathbf{a}}^\zeta \cdot \hat{\mathbf{a}}^\xi$, and the determinant of the covariant metrical tensor by $\sqrt{g} = \hat{\mathbf{a}}_{\xi+1} \cdot (\hat{\mathbf{a}}_{\xi+2} \times \hat{\mathbf{a}}_{\xi+3})$, where $\xi + 1$, $\xi + 2$, and $\xi + 3$ denote a consecutive cyclic permutation of coordinates u, v, w [35]. The last expression reveals that $\sqrt{g}$ is the Jacobian of the transformation that maps curvilinear coordinates to Cartesian ones. Finally, the components of vector field **F** are related to each other in terms of the following rules:

$$f_\xi = g_{\xi\zeta} f^\zeta \quad \text{and} \quad f^\zeta = g^{\zeta\xi} f_\xi. \tag{3.61}$$

According to this theoretical framework, one can easily construct the three fundamental vector operators. So, the *gradient* of a scalar quantity ϕ is represented by the covariant components as

$$\nabla\phi = \frac{1}{\sqrt{g}} \sum_{\xi=u,v,w} \frac{\partial\phi}{\partial\xi}\hat{\mathbf{a}}^\xi. \tag{3.62}$$

Conversely, the *divergence* of **F** uses the contravariant components and becomes

$$\nabla\cdot\mathbf{F} = \frac{1}{\sqrt{g}} \sum_{\zeta=u,v,w} \frac{\partial}{\partial\zeta}\left(\sqrt{g}\,\mathbf{F}\cdot\hat{\mathbf{a}}^\zeta\right), \tag{3.63}$$

while its *curl* is expressed by

$$\nabla\times\mathbf{F} = \frac{1}{\sqrt{g}} \sum_{\xi=u,v,w} \frac{\partial}{\partial\xi}\left(\sqrt{g}\,\hat{\mathbf{a}}^\xi\times\mathbf{F}\right). \tag{3.64}$$

The Div-Curl Problem

In its general form the div-curl problem reads: Let us assume an open bounded domain V, with boundary S. Inside V, there exists a unique vector **U**, which satisfies

$$\nabla\times\mathbf{U} = \mathbf{W} \quad \text{in } V, \qquad \nabla\cdot\mathbf{U} = q \quad \text{in } V, \qquad \mathbf{U}\cdot\hat{\mathbf{n}} = \psi \quad \text{on } S, \tag{3.65}$$

where $\hat{\mathbf{n}}$ is the outward, normal to S unit vector, **W** is the solenoidal vector field, and q a scalar function. Moreover, ψ, defined on S, must fulfill the following compatibility constraint:

$$\iint_S \psi\, dS = \iiint_V q\, dV. \tag{3.66}$$

If the space V is m_p-multiply connected, then c_k circulations of vector **U** should be imposed to the m_p independent loops l_k as

$$\int_{l_k} \mathbf{U}\cdot\hat{\mathbf{t}}\, dl = c_k \quad \text{for } k = 1, 2, \ldots, m_p, \tag{3.67}$$

where $\hat{\mathbf{t}}$ is tangential to loop l_k unit vector and $c_k \in \mathbb{R}$.

To select the correct basis for (3.65), covariant (contravariant) components should be expressed as a function of their contravariant (covariant) counterparts, via the pertinent system metrics. Unfortunately, the differentiation of these metrics produces the demanding Cristoffel symbols that obstruct the solution of (3.65). This difficulty can be circumvented by classical

Helmholtz analysis. So, a vector field with the desired curl is evaluated and projected in the space of divergence-free vectors. This concept leads to the next assertion:

Let us presume that there is a vector $\overline{\mathbf{U}}$, which fulfills

$$\nabla \times \overline{\mathbf{U}} = \mathbf{W} \quad \text{in } V, \qquad \overline{\mathbf{U}} \cdot \hat{\mathbf{n}} = \psi \quad \text{on } S, \qquad \int_{l_k} \overline{\mathbf{U}} \cdot \hat{\mathbf{t}}\, dl = c_k \quad \text{for } k = 1, 2, \ldots, m_p, \tag{3.68}$$

and let ϕ be a solution of

$$\nabla^2 \phi = q - \nabla \cdot \overline{\mathbf{U}} \quad \text{in } V, \qquad \nabla \phi \cdot \hat{\mathbf{n}} = 0 \quad \text{on } S, \tag{3.69}$$

then vector field $\mathbf{U}$, given by $\mathbf{U} = \overline{\mathbf{U}} + \nabla \phi$, is the solution of (3.6).

Actually, the handling of the nonorthogonal div-curl problem comprises two stages: a) The stage of *prediction*, which involves the computation of a vector field with the desired curl $\mathbf{W}$ and b) the stage of *correction/projection* that attempts to solve the corresponding Poisson problem with the suitable Neumann boundary conditions. In this manner, the divergence operator, (3.63), acts on the contravariant components, while the gradient operator, (3.62), can be expressed in relation with the covariant ones, thus enabling transformation (3.61). Differently speaking, (3.68) and (3.69) permit the calculation of two fields for vector $\mathbf{U}$: one contravariant with divergence q and one covariant with curl $\mathbf{W}$.

3.4.2 The Topologically Consistent Higher Order Nonstandard Framework

Comprehending the role of reflection errors in nonorthogonal domains and the defects of the conventional higher order forms, a class of higher order nonstandard curvilinear schemes for the approximation of space and time derivatives in Maxwell's equations has been introduced. The principal attribute focuses on the representation of fields by a family of robust models that involve a significantly increased amount of lattice points in the vicinity of the central node [33, 36] as opposed to existing techniques.

Consider a general coordinate system (u, v, w) defined by its $g(u, v, w)$ metrics which typify all implementation issues and the appropriate grid selection. Assuming that all derivatives throughout the formulation will be evaluated with an Mth-order accuracy, the parametric expressions for the spatial and temporal approximants, respectively, are given by

$$\mathcal{S}_\zeta \left[f \big|^t_{u,v,w} \right] = \frac{r_A}{4\Delta\zeta} \left\{ \mathcal{Q}^M_{\zeta,L} \left[f \big|^t_{u,v,w} \right] + \sum_{s=1}^{3} f \big|^t_{\zeta \pm s\Delta\zeta/2} \right\} \quad \text{for } \zeta \in (u, v, w), \tag{3.70}$$

$$\mathcal{T} \left[f \big|^t_{u,v,w} \right] = \frac{f \big|^{t+\Delta t/2}_{u,v,w} - f \big|^{t-\Delta t/2}_{u,v,w}}{\Psi(\omega, \Delta t)} - 2 r_B \sum_{\substack{\kappa=3 \\ (\text{odd})}}^{K} \frac{1}{\kappa!} \left[\frac{\Psi(\omega, \Delta t)}{2} \right]^{\kappa-1} \frac{\partial^\kappa f}{\partial t^\kappa} \bigg|^t_{u,v,w}, \tag{3.71}$$

where $\Delta\zeta \in (\Delta u, \Delta v, \Delta w)$ is the space increment prescribed by $\mathcal{S}_\zeta[.]$ toward the u, v, w axes and Δt the time increment of the entire simulation. Parameter κ, in (3.71), runs from 3 to K, where K is a predecided upper bound that controls the order of temporal approximation according to the desired accuracy level. Note that κ must be *odd.* This constraint along with the limits of the summation term is obtained from Taylor's series expansion in which all even time derivatives are mutually canceled. Theoretically, one can employ a very large K value for a secure evaluation. However, this is not feasible, since the burden is increased. Through a detailed study, it has been found that among other choices, $K = 5$ provides optimal performance. Quantities r_{A} and r_{B} are polynomial functions of $\Delta\zeta$ and Δt which enhance the reliability of derivative approximation in Maxwell's laws. Their key role is to ensure the correct relation between different mesh mappings in (3.70) and (3.71) as well as to control the impact of any additional node contributions to algorithmic efficiency. Particularly, after the necessary mathematical manipulations regarding well posedness and stability conditions of finite-difference arrangements, one obtains

$$r_{\mathrm{A}}(\Delta\zeta) = (\alpha + 1)(\Delta\zeta)^M - \alpha(\Delta\zeta)^{M-1} \quad \text{with } \alpha \in [0, 1/2], \tag{3.72}$$

$$r_{\mathrm{B}}(\Delta t) = -(\gamma + 1)(\Delta t)^{M-1} + (\gamma - 1)(\Delta t)^M \quad \text{with } \gamma \in [0, 1]. \tag{3.73}$$

Proceeding to (3.70), special attention must be drawn to the multidirectional operator $\mathcal{Q}^M_{\zeta,L}[.]$, whose extra degree of freedom L denotes the suitable stencil size, $l\Delta\zeta$ along each lattice direction with a regular value of $L = 3$. Its generalized definition in terms of $g(u, v, w)$ metrics is

$$\mathcal{Q}^M_{\zeta,L}\left[f\,\Big|^t_{u,v,w}\right] = \frac{g(u, v, w)}{\Psi(k_\zeta, L\Delta\zeta)} \sum_{m=1}^{M} Y^\zeta_m \left\{\sum_{l=1}^{L} R^\zeta_{m,l} \mathcal{V}^{(m)}_{\zeta,l\Delta\zeta}\left[f\,\Big|^t_{u,v,w}\right]\right\}. \tag{3.74}$$

Parameters Y_m and R_m improve total robustness in the manipulation of the fine structural details, encountered at a waveguide or an antenna, by the fulfillment of the subsequent Mth-order gauges

$$\sum_{m=1}^{M} Y^\zeta_m = 1/2, \qquad \sum_{l=1}^{L} R^\zeta_{m,l} = 1 \quad \forall m. \tag{3.75}$$

Their selection should be carefully conducted, since they function as pure weighting factors and therefore they can *strongly* subdue dispersion and dissipation errors. Also, correction functions $\Psi(\omega, \Delta t)$ and $\Psi(k_\zeta, L\Delta\zeta)$ of (3.71) and (3.74) are selected to lessen grid discrepancies and certify the proper transition from the continuous physical space to the discretized domain. In fact, their arguments have a substantial contribution in the method's wideband profile and hence involve an in-depth examination. More specifically, by considering the excitation frequency content and duration, they subdue oscillatory or spurious modes that corrupt the final waveform envelope. Probing $\Psi(k_\zeta, L\Delta\zeta)$ analysis indicates that its argument should opt for values that furnish $\mathcal{S}_\zeta[e^{jk\zeta}]/\partial_\zeta e^{jk\zeta} \to 1$. In higher order simulations, this constraint has been approximated by $\mathcal{S}_\zeta[e^{jk\zeta}]/\partial_\zeta e^{jk\zeta} \approx 2[\cos(kL\Delta\zeta) - 1]$. On the contrary, conventional fourth-order approaches can not easily comply with this prerequisite in curvilinear applications and sometimes lead to severe instabilities. To decide on a satisfactory argument for distinct wavenumbers k, one may simply follow the process described in Section 3.2.1. As an additional remark, it is mentioned that the locations of the preset points for the Fourier scheme are not arbitrary, while their specification is carefully conducted at the beginning of the numerical realization. Normally, they are mesh points in the vicinity of the object – i.e., near acute corners, inclined slots, or discontinuities – since in these areas electromagnetic waves vary more abruptly due to multiple interactions. Furthermore, although the choice of k is mostly global, one might pick different local k – by determining the corresponding mesh locations – which optimally describe the problem. The procedure, which does not affect the total overhead, occurs at every time-step, whereas its convergence increases with the number of preset points despite the geometry, incident-wave angles, or time intervals. Herein, $\Psi(\omega, \Delta t)$ and $\Psi(k_\zeta, L\Delta\zeta)$ are described by the general combination of exponential and hyperbolic terms as

$$\Psi(\omega, \Delta t) = \frac{e^{j\omega\Delta t/2}}{\sqrt{1-(\Delta t/2)^2}} \sinh\left[\Delta t\sqrt{1-(\Delta t/2)^2}\right], \tag{3.76}$$

$$\Psi(k_\zeta, L\Delta\zeta) = -\frac{e^b\,\Delta\zeta}{2\sqrt{b^2-1}} \sinh\left(\Delta\zeta\sqrt{b^2-1}\right) + e^b\cosh\left(\Delta\zeta\sqrt{b^2-1}\right), \tag{3.77}$$

with k_ζ denoting the relevant k_u, k_v, or k_w component of the 3-D wavenumber vector $\mathbf{k}$ and $b = k_\zeta L\Delta\zeta/2$.

To complete the formulation, scalable order m operators $\mathcal{V}^{(m)}_{\zeta,l\Delta\zeta}[.]$, in (3.75), are defined. Their primary feature is the coverage of *all possible* nodal arrangements according to the requirements of the interaction. Thus, unlike Yee's approach that employs only two mesh points for derivative approximation, these concepts involve a *full* set of nodes which results in coarse but very convergent grid topologies. A typical u-directed $\mathcal{V}^{(m)}_{u,l\Delta u}[.]$ (for $L = 3$), depicted in Figure 3.3, receives the compact expression of

$$\mathcal{V}^{(m)}_{u,l\Delta u}\left[f\big|^t_{u,v,w}\right] = \frac{(\Delta u)^m}{3m-2}\sum_{r=-1}^{+1}\left[f\big|^t_{u+rl\Delta u/2,v-\Delta v,w+r\Delta w} - f\big|^t_{u-rl\Delta u/2,v+\Delta v,w-r\Delta w}\right]. \tag{3.78}$$

The aim of parameter r is the suitable handling of areas near PEC walls or composite absorber interfaces, namely cases where stencils extend *at least* two nodes on each side of a grid point. Therefore, instead of compact central operators, widened stencils may be competently treated by (3.78) as well.

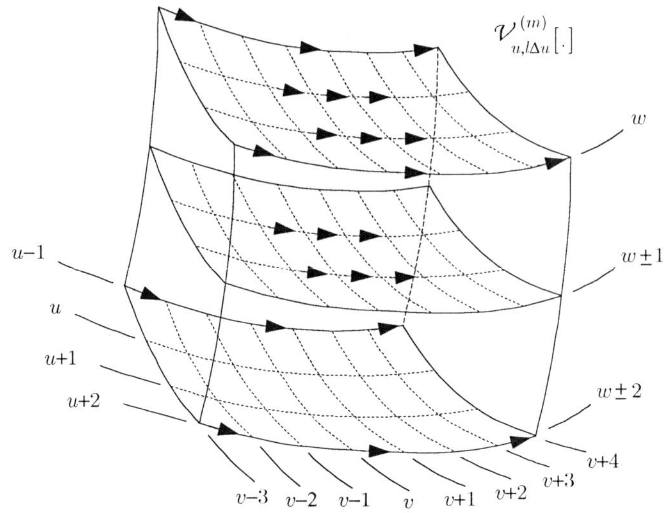

FIGURE 3.3: Spatial samples and summation directions established by a typical $\mathcal{V}^{(m)}_{u,l\Delta u}[.]$ operator in a curvilinear lattice

Alternatively and for smoother geometric characteristics, operator $\mathcal{Q}^{M}_{\zeta,L}[.]$ in (3.70) may be replaced by the curvilinear version of (3.44) which is simpler in its implementation. In this case, operators $\mathbf{d}^{i}_{u,\Delta}[.]$, for $i = \mathrm{A, B, C}$, are acquired by means of the graphical representation of Figure 3.4. Hence,

$$\mathbf{d}^{\mathrm{A}}_{u,\Delta u}\left[f\,\Big|^{t}_{u,v,w}\right] = f\,\Big|^{t}_{b_2^{\mathrm{A}}} - f\,\Big|^{t}_{b_1^{\mathrm{A}}}, \tag{3.79}$$

$$\begin{aligned}\mathbf{d}^{\mathrm{B}}_{u,\Delta u}\left[f\,\Big|^{t}_{u,v,w}\right] = \frac{1}{4}\Big(& f\,\Big|^{t}_{b_8^{\mathrm{B}}} + f\,\Big|^{t}_{b_4^{\mathrm{B}}} + f\,\Big|^{t}_{b_3^{\mathrm{B}}} + f\,\Big|^{t}_{b_7^{\mathrm{B}}} \\ & - f\,\Big|^{t}_{b_5^{\mathrm{B}}} - f\,\Big|^{t}_{b_1^{\mathrm{B}}} - f\,\Big|^{t}_{b_2^{\mathrm{B}}} - f\,\Big|^{t}_{b_6^{\mathrm{B}}}\Big),\end{aligned} \tag{3.80}$$

$$\begin{aligned}\mathbf{d}^{\mathrm{C}}_{u,\Delta u}\left[f\,\Big|^{t}_{u,v,w}\right] = \frac{1}{4}\Big(& f\,\Big|^{t}_{b_3^{\mathrm{C}}} + f\,\Big|^{t}_{b_8^{\mathrm{C}}} + f\,\Big|^{t}_{b_4^{\mathrm{C}}} + f\,\Big|^{t}_{b_7^{\mathrm{C}}} \\ & - f\,\Big|^{t}_{b_2^{\mathrm{C}}} - f\,\Big|^{t}_{b_5^{\mathrm{C}}} - f\,\Big|^{t}_{b_1^{\mathrm{C}}} - f\,\Big|^{t}_{b_6^{\mathrm{C}}}\Big),\end{aligned} \tag{3.81}$$

where notation b^{i} provides only the stencil variations along u, v, w. For example, $b_1^{\mathrm{B}} = (-\Delta u/2, \Delta v, -\Delta w)$ means that the specific field component is computed at point $(u - \Delta u/2, v + \Delta v, w - \Delta w)$.

3.4.3 Systematic Construction of Multidirectional Tessellations

Having developed the appropriate numerical approximants and before describing the modified Ampère's and Faraday's laws, it is significant to discuss the key points of lattice construction, especially for irregular shapes. The consistency of such FDTD solutions is accomplished by

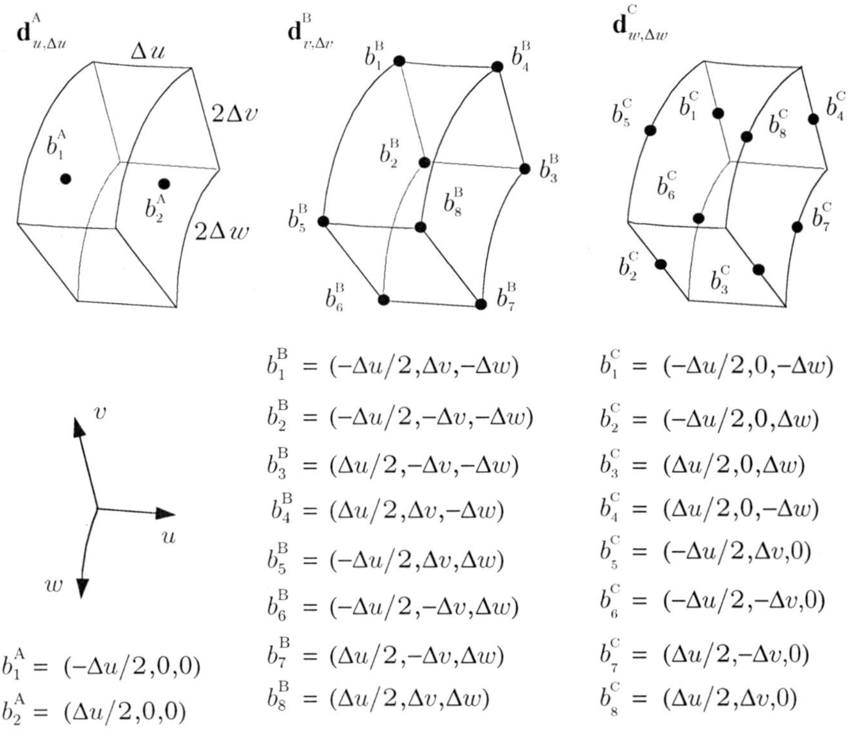

FIGURE 3.4: Graphical analysis of the alternative difference operators and the lattice points used for their computation

robust nonorthogonal grids [17, 31], which must be cautiously built, as difficult tessellations generate unexpected vector parasites cumbersome to confront. These challenging situations are often experienced in large applications involving multifrequency radiation or scattering, mainly when mesh resolution lacks to track field singularities near prominent discontinuities. As long as the regular Yee's technique is excluded from this rule, this difficulty is overcome by presenting an algorithm adjusted to meet the needs of higher order time-domain analysis. Our goal is to construct two *dual-nodal ensembles* and extract reliable expressions that correlate the problem's profile to a classification of adaptive topological structures able to support optimal phase velocities. The solver embodies the previously derived nonstandard formulas to maintain duality without any artificial restrains.

Let us assume a 3-D domain, adequately defined by the general system of curvilinear coordinates ($u = i\Delta u$, $v = j\Delta v$, $w = k\Delta w$). The space is divided into two nonuniform dual meshes. The center of each primary cell is placed at node (i, j, k), while secondary ones are mutually centered at the vertices of the primary grid $(i - \frac{1}{2}, j - \frac{1}{2}, k - \frac{1}{2})$. Following an equivalent process applicable to both lattices, the center of every cell face is obtained by shifting one of its

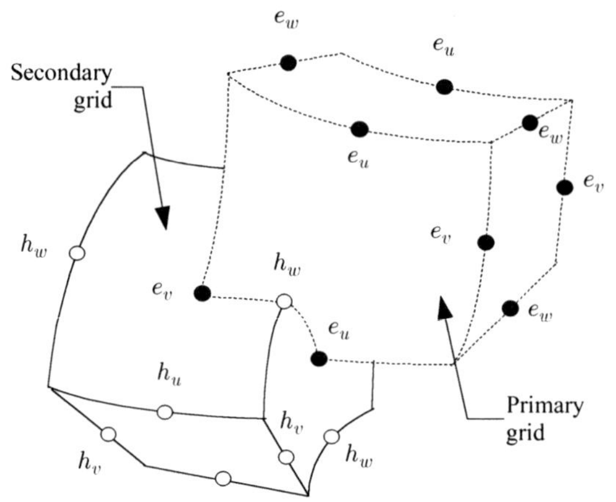

FIGURE 3.5: A nonorthogonal dual mesh indicating the locations of covariant electric and magnetic field components on the primary and dual grid, respectively

indices along half a stencil. The graphical representation of Figure 3.5 displays that covariant, h_ξ, and contravariant, h^ζ, magnetic **H**-field components are located at primary face centers remaining so in absolute staggering with e_ξ and e^ζ electric **E**-field quantities placed at edge centers. The curvilinear scheme uses the idea of *fluxes* for the vectors across the faces defined by $f^{(\zeta)} = \sqrt{g} f^\zeta$ given that $f = e, h$. Next, through (3.61), it expresses f_ξ values as a function of f^ζ ones, without any grid interchanges. The whole configuration for a nonorthogonal FDTD complex is presented in Figure 3.6. To complete the treatment of the div-curl problem (3.65), *linear operator* $\mathcal{P}^{(\zeta)}[.]$ is launched so that $f^{(\zeta)} = \mathcal{P}^{(\zeta)}\left[f_\xi\right]$, which involves local components f_ξ in conjunction with the neighboring ones $f_{\xi+1}$ and $f_{\xi+2}$, multiplied by the discrete metrics

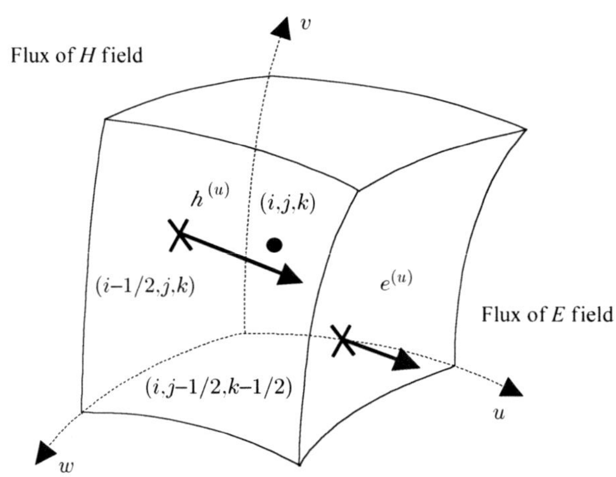

FIGURE 3.6: The generalized Yee's element cell for the evaluation of field fluxes

$\tilde{g}^{\zeta\xi} = \sqrt{g}g^{\zeta\xi}$ [33, 34]. For instance, the $\mathcal{P}^{(\zeta)}[h_u]$ becomes

$$\begin{aligned}\mathcal{P}^{(u)}\left[h_u\big|^{n+1/2}_{i-1/2,j,k}\right] &= \tilde{g}^{uu}_{i-1/2,j,k} h_u\big|^{n+1/2}_{i-1/2,j,k} + \frac{1}{4}\left[\tilde{g}^{uv}_{i,j,k}\left(h_v\big|^{n+1/2}_{i,j-1/2,k} + h_v\big|^{n+1/2}_{i,j+1/2,k}\right)\right.\\ &+ \tilde{g}^{uv}_{i-1,j,k}\left(h_v\big|^{n+1/2}_{i-1,j-1/2,k} + h_v\big|^{n+1/2}_{i-1,j+1/2,k}\right) + \tilde{g}^{uw}_{i,j,k}\left(h_w\big|^{n+1/2}_{i,j,k-1/2} + h_w\big|^{n+1/2}_{i,j,k+1/2}\right)\\ &\left.+ \tilde{g}^{uw}_{i-1,j,k}\left(h_w\big|^{n+1/2}_{i-1,j,k-1/2} + h_w\big|^{n+1/2}_{i-1,j,k+1/2}\right)\right], \qquad (3.82)\end{aligned}$$

which employs notation $f(u, v, w, t) = f(i\Delta u, j\Delta v, k\Delta w, n\Delta t) = f\big|^n_{i,j,k}$ and the same averaging technique as the one in [17]. Recalling the concepts of (3.70), the gradient of scalar field ϕ, the curl, and the divergence of vectors $\mathbf{F} = \mathbf{E}, \mathbf{H}$ – for the transverse plane in Figure 3.7 – are given by

$$\operatorname{grad}[\phi]\big|^n_{i,j,k} = \mathcal{S}_u\left[\varphi|^n_{i,j,k}\right]\hat{\mathbf{u}} + \mathcal{S}_v\left[\varphi|^n_{i,j,k}\right]\hat{\mathbf{v}} + \mathcal{S}_w\left[\varphi|^n_{i,j,k}\right]\hat{\mathbf{w}}, \qquad (3.83)$$

$$\operatorname{div}[\mathbf{F}]\big|^n_{i,j,k} = \mathcal{S}_u\left[f^{(u)}\Big|^n_{i,j,k}\right] + \mathcal{S}_v\left[f^{(v)}\Big|^n_{i,j,k}\right] + \mathcal{S}_w\left[f^{(w)}\Big|^n_{i,j,k}\right], \qquad (3.84)$$

$$\operatorname{curl}[\mathbf{F}]\big|^n_{i,j,k} = \mathcal{P}^{(u)}\left[G_u|^n_{i,j,k}\right]\hat{\mathbf{u}} + \mathcal{P}^{(v)}\left[G_v|^n_{i,j,k}\right]\hat{\mathbf{v}} + \mathcal{P}^{(w)}\left[G_w|^n_{i,j,k}\right]\hat{\mathbf{w}}, \qquad (3.85)$$

where the intermediate scalar quantities C_ζ correspond to

$$G_u|^n_{i,j,k} = \mathcal{S}_v\left[f_w|^n_{i,j,k}\right] - \mathcal{S}_w\left[f_v|^n_{i,j,k}\right], \qquad G_v|^n_{i,j,k} = \mathcal{S}_w\left[f_u|^n_{i,j,k}\right] - \mathcal{S}_u\left[f_w|^n_{i,j,k}\right],$$
$$G_w|^n_{i,j,k} = \mathcal{S}_u\left[f_v|^n_{i,j,k}\right] - \mathcal{S}_v\left[f_u|^n_{i,j,k}\right].$$

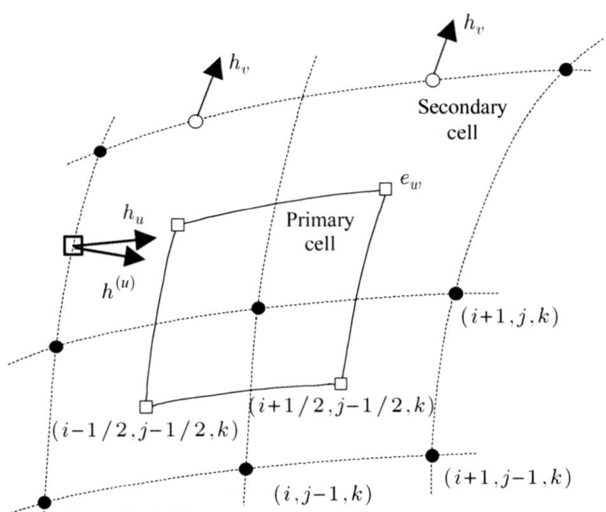

FIGURE 3.7: A primary–secondary cell complex for the evaluation of vector operators

To ensure the consistent termination of infinite space, again observe the boundary-adjacent cell of Figure 3.7. The contravariant component is evaluated by linear interpolation over the covariant ones at the white-dotted points, while the gradient is calculated through a linear extrapolation. A significant feature of (3.83)–(3.85) is their *fully conservative* nature that has a serious impact on the profile of the FDTD analysis. If the prior aspects are inserted into the complete form of Maxwell's equations (see Chapter 2), one acquires

$$\left(\mathbf{I}+\frac{1}{2}\mathbf{Y}_t\right)\mathbf{E}_{\mathrm{cv}}^{n+1}=\left(\mathbf{I}-\frac{1}{2}\mathbf{Y}_t\right)\mathbf{E}_{\mathrm{cv}}^{n}-\Psi(\omega,\,\Delta t)\varepsilon^{-1}\mathbf{J}_{\mathrm{cv}}^{n+1/2}$$
$$+\mathbf{G}^{\mathrm{H}}\overline{\mathcal{S}}\left[\mathbf{H}_{\mathrm{cv}}^{n+1/2}\right]+\mathbf{T}_{\mathrm{E}}^{n+1/2}, \tag{3.86a}$$

$$\left(\mathbf{I}+\frac{1}{2}\mathbf{R}_t\right)\mathbf{H}_{\mathrm{cv}}^{n+1/2}=\left(\mathbf{I}-\frac{1}{2}\mathbf{R}_t\right)\mathbf{H}_{\mathrm{cv}}^{n-1/2}-\Psi(\omega,\,\Delta t)\mu^{-1}\mathbf{M}_{\mathrm{cv}}^{n}$$
$$-\mathbf{G}^{\mathrm{E}}\overline{\mathcal{S}}\left[\mathbf{E}_{\mathrm{cv}}^{n}\right]+\mathbf{T}_{\mathrm{H}}^{n}, \tag{3.86b}$$

in which $\overline{\mathcal{S}}$ is the nonstandard curl counterpart of (3.57) that approximates the appropriate derivatives with respect to u, v, w. Furthermore, $\mathbf{E}_{\mathrm{cv}}=[e_u\ e_v\ e_w]^T$ and $\mathbf{H}_{\mathrm{cv}}=[h_u\ h_v\ h_w]^T$ are matrices of the unknown covariant electric and magnetic components, $\mathbf{J}=\sigma\mathbf{E}$ and $\mathbf{M}=\sigma^*\mathbf{H}$ the corresponding conduction current densities, $\mathbf{Y}_t$ and $\mathbf{R}_t$ constitutive matrices defining every material or interface of the problem, and $\mathbf{G}^{\mathrm{H}}$ and $\mathbf{G}^{\mathrm{E}}$ the suitable matrix metric tensors, respectively. Besides, matrices $\mathbf{T}_{\mathrm{E}}$ and $\mathbf{T}_{\mathrm{H}}$ congregate all higher order temporal derivatives, due to $\mathcal{S}[.]$ operators, and support their time marching. By substituting the appropriate metric coefficients in (3.86), after some algebra, one derives the final FDTD update expressions:

$$\begin{bmatrix}e_1\\e_2\\e_3\end{bmatrix}_{\mathrm{pos}}^{n+1}=\begin{bmatrix}e_1\\e_2\\e_3\end{bmatrix}_{\mathrm{pos}}^{n}-\Psi(\omega,\,\Delta t)\left\{\frac{\sigma}{\varepsilon}-\frac{[\Psi(\omega,\Delta t)]^2}{24}\frac{\partial^3}{\partial t^3}\right\}\begin{bmatrix}e_1\\e_2\\e_3\end{bmatrix}_{\mathrm{pos}}^{n+1/2}$$
$$+\frac{\Psi(\omega,\Delta t)}{\varepsilon\sqrt{g}}\begin{bmatrix}g_{11}&g_{12}&g_{13}\\g_{21}&g_{22}&g_{23}\\g_{31}&g_{32}&g_{33}\end{bmatrix}\begin{bmatrix}\mathcal{S}_v[h_3]-\mathcal{S}_w[h_2]\\\mathcal{S}_w[h_1]-\mathcal{S}_u[h_3]\\\mathcal{S}_u[h_2]-\mathcal{S}_v[h_1]\end{bmatrix}_{\mathrm{pos}}^{n+1/2}, \tag{3.87a}$$

$$\begin{bmatrix}h_1\\h_2\\h_3\end{bmatrix}_{\mathrm{pos}}^{n+1/2}=\begin{bmatrix}h_1\\h_2\\h_3\end{bmatrix}_{\mathrm{pos}}^{n-1/2}-\Psi(\omega,\,\Delta t)\left\{\frac{\sigma^*}{\mu}-\frac{[\Psi(\omega,\,\Delta t)]^2}{24}\frac{\partial^3}{\partial t^3}\right\}\begin{bmatrix}h_1\\h_2\\h_3\end{bmatrix}_{\mathrm{pos}}^{n}$$
$$-\frac{\Psi(\omega,\,\Delta t)}{\mu\sqrt{g}}\begin{bmatrix}g_{11}&g_{12}&g_{13}\\g_{21}&g_{22}&g_{23}\\g_{31}&g_{32}&g_{33}\end{bmatrix}\begin{bmatrix}\mathcal{S}_v[e_3]-\mathcal{S}_w[e_2]\\\mathcal{S}_w[e_1]-\mathcal{S}_u[e_3]\\\mathcal{S}_u[e_2]-\mathcal{S}_v[e_1]\end{bmatrix}_{\mathrm{pos}}^{n}. \tag{3.87b}$$

where pos signifies the exact position. Finally, the temporal integration of (3.87) can be conducted via Runge–Kutta or the leapfrog-type algorithm of Section 3.2.2, which must be now extended to curvilinear coordinates. This simply requires that operator $\overline{\mathbf{DS}}[.]$ is replaced by the $\bar{\mathcal{S}}[.]$ one in (3.56) and (3.58).

3.4.4 Extension to Non-Cartesian and Unstructured Grids

A *non-Cartesian* mesh focuses its competence on the approximation of spatial derivatives not only through data obtained from nodes along the coordinate axes, but also in several other positions. In this section, the discretization of a 2-D domain into *hexagonal* cells is described. Let us consider a hexagonal lattice with 7 points per cell. The primary mesh consists of hexagons divided in six equilateral triangles at the edge centers of which an equal number of normal magnetic field components, H_i ($i = 1, 2, \ldots, 6$), are defined, as depicted in Figure 3.8(a). The secondary mesh (shaded area) is a canonical hexagon as well, with the E_z quantities defined at its barycenter. Combining the six nearest neighbors with the higher order nonstandard FDTD operators (3.42)–(3.49), the ensuing set of update equations may be derived:

$$\varepsilon_0 \mathbf{DT}^{\mathrm{nst}}\left[E_z\big|_{i,j}^{n+1/2}\right] = \frac{2}{3\Delta h}\left[H_1\big|_{+1/4,j-\sqrt{3}/4}^{n+1/2} + H_2\big|_{i+1/2,j}^{n+1/2} + H_3\big|_{i+1/4,j+\sqrt{3}/4}^{n+1/2}\right.$$

$$\left. - H_4\big|_{i-1/4,j+\sqrt{3}/4}^{n+1/2} - H_5\big|_{i-1/2,j}^{n+1/2} - H_6\big|_{i+1/4,j-\sqrt{3}/4}^{n+1/2}\right], \tag{3.88a}$$

$$\mu_0 \mathbf{DT}^{\mathrm{nst}}\left[H_1\big|_{i+1/4,j-\sqrt{3}/4}^{n}\right] = \mathbf{DS}_x\left[E_z\big|_{i+1/4,j-\sqrt{3}/2}^{n}\right], \tag{3.88b}$$

$$\mu_0 \mathbf{DT}^{\mathrm{nst}}\left[H_2\big|_{i+1/2,j}^{n}\right] = \mathbf{DS}_x\left[E_z\big|_{i,j}^{n}\right], \tag{3.88c}$$

$$\mu_0 \mathbf{DT}^{\mathrm{nst}}\left[H_3\big|_{i+1/4,j+\sqrt{3}/4}^{n}\right] = \mathbf{DS}_x\left[E_z\big|_{i+1/4,j+\sqrt{3}/2}^{n}\right]. \tag{3.88d}$$

Apart from hexagonal grids, the analysis may be evenly applied to *octagonal* ones [Figure 3.8(b)]. As far as the extension of (3.88) to three dimensions is concerned, it should be stated that, despite its advanced discretization characteristics, total complexity is greatly increased. This is the reason for the relatively limited use of such lattices, especially in waveguide problems where their merits would be indeed fairly instructive.

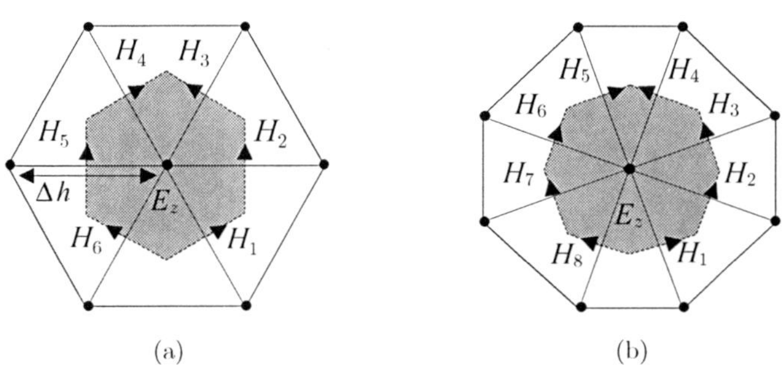

FIGURE 3.8: Location of field components in a (a) hexagonal and (b) octagonal cell

On the other hand, the use of *unstructured lattices* for the modeling of arbitrary curvatures by varying the size and position of the elementary cells constitutes another source of research. This kind of grids, composed of unstructured arrays of generally fitted polyhedral elements, can conform to very complex shapes and surfaces, while preserving sufficient cell aspect ratios and global uniformity. Conversely, due to the deformation of their cells near a geometric discontinuity, some inherent attributes of time-domain modeling are likely to be distorted, thus affecting the consistency of the simulation. Actually, the notion of *grid duality* is one of the most crucial factors for the correct discretization of several contemporary designs. Hence, unstructured meshes should receive a fairly careful treatment, especially in conventional higher order FDTD implementations, where the widened spatial stencils are likely to deteriorate the quality of the tessellation.

However, nonstandard finite differencing is proven to be far more robust, since it fully exploits the idea of dual-cell complexes. Duality in higher order lattice construction is strongly related to the issue of oriented differential forms [13, 36–39]. Essentially, one can discern two kinds of orientations for 3-D discretizations, as shown in Figure 3.9: the *externally oriented* arrangement (external or outer orientation) that circulates the object and involves additional dimensions and the *internal* one (internal or inner orientation) specifying a certain direction along the object without increasing its dimensionality. Obviously, from a geometrical point of view, there are two types of forms associated with these orientations. The primary forms that are internally oriented and the secondary forms share an outer orientation. As anticipated, the meaning of orientation should be applicable to cell complexes, establishing so the duality

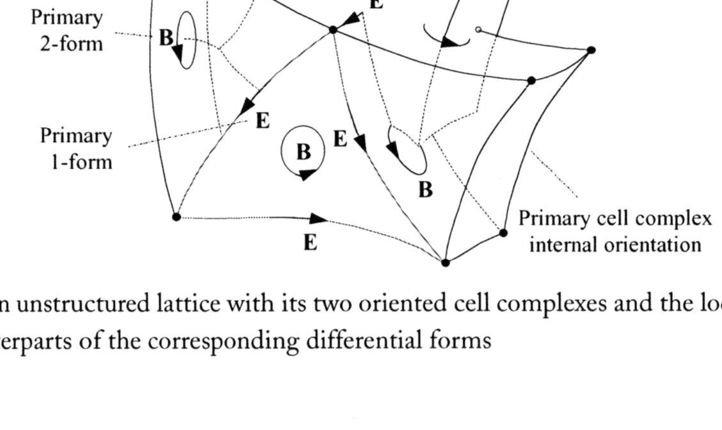

FIGURE 3.9: An unstructured lattice with its two oriented cell complexes and the locations of fields as the discrete counterparts of the corresponding differential forms

premise in the discrete state. Therefore, a *primary* cell complex consists of internally oriented cells that represent primary forms, whereas the *secondary* cell complex contains cells with external orientation associated to secondary forms. Then, for every p cell of the primary complex there exists a corresponding $(D-p)$ cell of the dual one. This assignment consists of the fact that a primary p cell contains or crosses a dual $(D-p)$ cell (Figure 3.9). To conduct a one-to-one mapping between electromagnetic fields and the prior classification, analysis denotes the electric field $\mathbf{E}$ as a primary 1-form (internally oriented lines) and the magnetic field $\overline{\mathbf{H}}$ as a secondary 1-form (externally oriented lines). Likewise, magnetic flux $\mathbf{B}$ is defined as a primary 2-form (internally oriented surface) and electric flux $\overline{\mathbf{D}}$ as a secondary 2-form (externally oriented surface). Adopting this framework, the higher order nonstandard FDTD method with the formulation of (3.82)–(3.85) develops update equations that use modified system metrics for field component projection. In this way, the properties of unstructured discretizations can constitute a powerful tool for intricate waveguide and antenna problems.

3.5 DISPERSION ERROR AND STABILITY ANALYSIS

3.5.1 The Enhanced Dispersion Relation

As expected, the dispersion relation of higher order nonstandard FDTD algorithms and especially that of (3.70)–(3.81) is greatly improved due to the scalable accuracy parameters M and L. The optimization level may be directly experienced by comparing the resulting dispersion relation with the corresponding one, $\mathcal{F}_{\mathrm{FDTD}}(.)$, of Yee's scheme. So, through the plane-wave substitution approach [40–45] and some algebra

$$\sin(\omega\Delta t) = \frac{19(\Delta t)^{(M-1)/4}}{3^{M+L}}\mathcal{F}_{\mathrm{FDTD}}(S_u, S_v, S_w) \quad \text{for } S_\zeta = \frac{\Delta t}{\Delta\zeta}\sin\left(\frac{k_\zeta\Delta\zeta}{2}\right), \tag{3.89}$$

and $\zeta \in (u, v, w)$. As an illustration, the dispersion relations of the two methods, for $M = 4$, $L = 3$, υ^{num} the numerical phase velocity, $\beta = 2\pi/(k\Delta\zeta)$, and θ the incident angle are provided. Hence,

$$\left(\frac{\upsilon}{\upsilon^{\mathrm{num}}}\right)^{\mathrm{Yee}} \cong 1 - \frac{\pi^2}{\beta^2}\left[\frac{1}{8} + \frac{1}{24}\cos(4\theta)\right], \tag{3.90}$$

$$\left(\frac{\upsilon}{\upsilon^{\mathrm{num}}}\right)^{\mathrm{nst}} \cong 1 - \frac{5\pi^7}{670\beta^7}\left[\frac{1}{19} + 12\cos(4\theta)\right]. \tag{3.91}$$

In fact, it is the careful choice of parameters Y_m^ζ and $\mathcal{R}_{m,l}^\zeta$ in (3.75) that leads to this serious improvement. To illustrate these issues, consider Figures 3.10(a) and 3.10(b) that present the variation of the maximum L_2 error norm versus $k\Delta h$ (Δh denotes a uniform $\Delta u = \Delta v = \Delta w$ mesh) and the relative error of the temporal evolution normalized phase velocity, respectively.

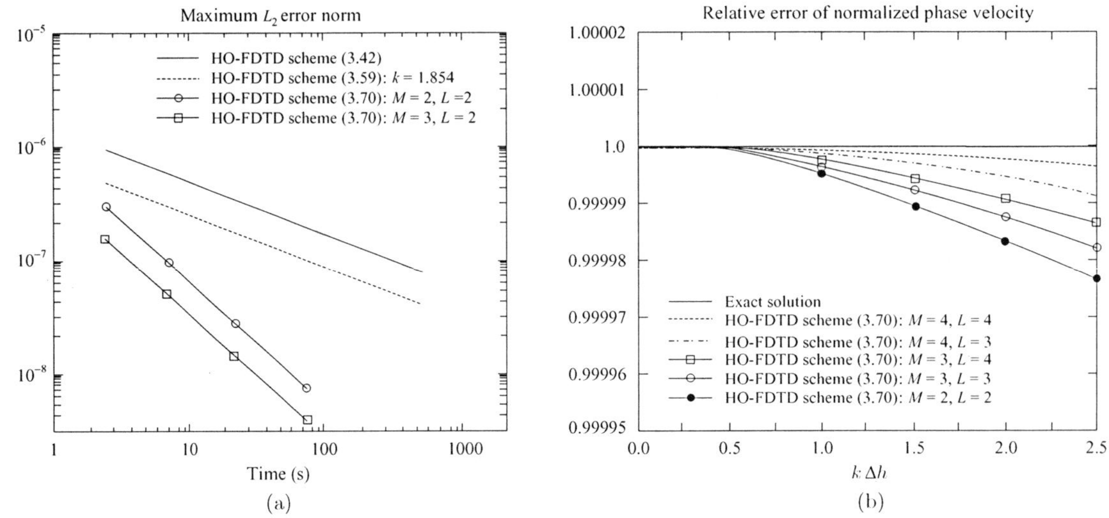

FIGURE 3.10: (a) Maximum L_2 error norm versus spatial frequency $k\Delta h$ and (b) relative error of the normalized phase velocity versus time for different higher order nonstandard FDTD arrangements

Evidently, from the former it is deduced that the strategy of (3.70)–(3.81) yields rapidly convergent schemes with remarkably small lattice reflection errors. In addition, the latter figure demonstrates the highly accurate nature of the algorithm, since, despite the large values of $k\Delta h$, it offers advanced levels of accuracy and negligible discrepancies. These promising observations can be confirmed through the comparative analysis of Table 3.1 as well.

TABLE 3.1: Maximum Dispersion Error

METHOD	GRID RESOLUTION	MAXIMUM DISPERSION ERROR	CONVERGENCE RATE
(2, 2) FDTD	1/15	0.00980	
(4, 4) FDTD	1/15	3.6521×10^{-4}	3.9826
Scheme (3.42)	1/10	2.1054×10^{-8}	4.0019
Scheme (3.59): $k = 1.432$	1/10	4.0418×10^{-8}	3.9954
Scheme (3.59): $k = 2.865$	1/9	3.9441×10^{-9}	5.8732
Scheme (3.70): $M = 2$, $L = 2$	1/9	6.9525×10^{-10}	2.0027
Scheme (3.70): $M = 3$, $L = 2$	1/8	4.8213×10^{-11}	2.9634
Scheme (3.70): $M = 4$, $L = 3$	1/7	5.2677×10^{-12}	3.9712
Scheme (3.70): $M = 4$, $L = 4$	1/7	8.6243×10^{-14}	4.0035

Overall, the higher order nonstandard technique creates a set of topologically robust forms which, by constructing a more consistent cell, diminish late-time instabilities, eliminate dispersion errors, and avoid conformal implementations. Their cooperation with the absorbing boundary condition is straightforward and as it will be shown in the next chapter, the resulting forms retain the original field variation and properties.

3.5.2 The Modified Stability Criterion

The integration process of (3.42), (3.59), and (3.70)–(3.71) *remains* stable for all types of numerical simulations. In the light of this remark and by means of the practical von Neumann's method, the Courant criteria for the first two and the third class of operators are expressed as

$$\upsilon\Delta t \leq \frac{3\sqrt{g}\,\min[\Delta u,\,\Delta v,\,\Delta w]}{\sqrt{2(1-5A_1)(3-4A_1-7A_2)}} \quad \text{and} \quad \upsilon\Delta t \leq \frac{3}{\pi}\sin^{-1}(0.7)\left(\sum_{l=u}^{w}\sum_{m=u}^{w}\frac{g^{lm}}{\Delta l\Delta m}\right)^{-1/2}, \tag{3.92}$$

with the double summation denoting a consecutive cyclic permutation of $(u,\, v,\, w)$. Notice the less strict character of (3.92) than their second-order counterparts. This implies that temporal increments can be selected fairly larger without inducing any oscillatory exponential modes that spoil the final results. It should be stressed that after a thorough application of the method to many realistic problems – provided in Chapters 7 and 8 – the prior criteria have been proven very satisfactory without any unpredictable situations.

3.6 ISSUES OF PRACTICAL IMPLEMENTATION

The realization of the nonstandard FDTD concepts requires some additional computer instructions apart from the common ones. In fact, all correction functions are evaluated only once during each time-step, setting around 45 per cell operations. Also, the matrices of (3.86) and (3.87) are sparse and, despite their 10–20% increased algorithmic complexity, do not augment storage or CPU needs. Conversely, and as will be deduced from numerical results, the nonstandard method exhibits notable system savings due to its enhanced accuracy achieved by the substantially fewer points per wavelength (even 1/120 of the Yee's algorithm). This simply calls for larger cell sizes, given a preset error level, and is translated to almost 90% lattice reduction and up to 80% less computational time, as compared to the FDTD method or other time-domain formulations.

REFERENCES

[1] R. E. Mickens, *Nonstandard Finite Difference Models of Differential Equations*. Singapore: World Scientific, 1994.

[2] J. B. Cole, "A high accuracy FDTD algorithm to solve microwave propagation and scattering problems on a coarse grid," *IEEE Trans. Microw. Theory Tech.*, vol. 43, no. 9, pp. 2053–2058, Sep. 1995.

[3] J. B. Cole, "A high-accuracy realization of the Yee algorithm using non-standard finite differences," *IEEE Trans. Microw. Theory Tech.*, vol. 45, no. 6, pp. 991–996, June 1997.

[4] R. E. Mickens, Ed., *Applications of Nonstandard Finite Difference Schemes*. Singapore: World Scientific, 2000.

[5] K. C. Patidar, "On the use of nonstandard finite difference methods," *J. Differ. Equ. Appl.*, vol. 11, no. 8, pp. 735–758, July 2005.

[6] N. V. Kantartzis and T. D. Tsiboukis, "A higher-order FDTD technique for the implementation of enhanced dispersionless perfectly matched layers combined with efficient absorbing boundary conditions," *IEEE Trans. Magn.*, vol. 34, no. 5, pp. 2736–2739, Sep. 1998.

[7] N. V. Kantartzis and T. D. Tsiboukis, "A higher-order nonstandard FDTD-PML method for the advanced modeling of complex EMC problems in generalized 3-D curvilinear coordinates," *IEEE Trans. Electromagn. Compat.*, vol. 46, no. 1, pp. 2–11, Feb. 2004.

[8] N. V. Kantartzis, J. S. Juntunen, and T. D. Tsiboukis, "An enhanced higher-order FDTD technique for the construction of efficient reflectionless PMLs in 3-D generalized curvilinear coordinates," in *Proc. IEEE Antennas Propag. Soc. Int. Symp.*, Orlando, FL, July 1999, vol. 3, pp. 1894–1897.

[9] S. K. Lele, "Compact finite difference schemes with spectral resolution," *J. Comput. Phys.*, vol. 101, no. 1, pp. 16–42, July 1992.

[10] J. S. Shang, "High-order compact-difference schemes for time-dependent Maxwell equations," *J. Comput. Phys.*, vol. 153, no. 1, pp. 312–333, July 1999.

[11] J. L. Young, "High-order, leapfrog methodology for the temporally dependent Maxwell's equations," *Radio Sci.*, vol. 36, no. 1, pp. 9–17, Feb. 2001.

[12] T. Kashiwa, H. Kudo, Y. Sendo, T. Ohtani, and Y. Kanai, "The phase velocity error and stability condition of the three-dimensional nonstandard FDTD method," *IEEE Trans. Magn.*, vol. 38, no. 2, pp. 661–664, Mar. 2002.

[13] N. V. Kantartzis and T. D. Tsiboukis, "Higher-order non-standard schemes in generalised curvilinear coordinates: A systematic strategy for advanced numerical modeling and consistent topological perspectives," *ICS Newslett.*, vol. 9, no. 3, pp. 5–14, 2002.

[14] M. F. Hadi and M. Piket-May, "A modified FDTD (2, 4) scheme for modeling electrically large structures with high-phase accuracy," *IEEE Trans. Antennas Propag.*, vol. 45, no. 2, pp. 254–264, Feb. 1997.

[15] R. Holland, "Finite-difference solution of Maxwell's equations in generalized nonorthogonal coordinates," *IEEE Trans. Nucl. Sci.*, vol. NS–30, no. 6, pp. 4589–4591, Dec. 1983.

[16] M. Fusco, M. Smith, and L. Gordon, "A three-dimensional FDTD algorithm in curvilinear coordinates," *IEEE Trans. Antennas Propag.*, vol. 39, no. 10, pp. 1463–1471, Oct. 1991.

[17] J.-F. Lee, R. Palendech, and R. Mittra, "Modeling three-dimensional discontinuities in waveguides using the nonorthogonal FDTD algorithm," *IEEE Trans. Microw. Theory Tech.*, vol. 40, no. 2, pp. 346–352, Feb. 1992.

[18] S. L. Ray, "Numerical dispersion and stability characteristics of time-domain methods on nonorthogonal meshes," *IEEE Trans. Antennas Propag.*, vol. 41, no. 2, pp. 233–235, Feb. 1993.

[19] N. Madsen, "Divergence preserving discrete surface integral methods for Maxwell's equations using nonorthogonal unstructured grids," *J. Comput. Phys.*, vol. 119, no. 1, pp. 34–45, June 1995.

[20] E. A. Navarro, C. Wu, P. Y. Chung, and J. Litva, "Some considerations about the finite difference time domain method in general curvilinear coordinates," *IEEE Microw. Guided Wave Lett.*, vol. 6, pp. 193–195, June 1996.

[21] J. B. Schneider and K. L. Shlager, "FDTD simulations of TEM horns and the implicatons for staircased representations," *IEEE Trans. Antennas Propag.*, vol. 45, no. 12, pp. 1830–1838, Dec. 1997.

[22] S. Dey, R. Mittra, and S. Chebolu, "A technique for implementing the FDTD algorithm on a nonorthogonal grid," *Microw. Opt. Technol. Lett.*, vol. 14, no. 4, pp. 213–215, Mar. 1997.

[23] R. Schuhmann and T. Weiland, "A stable interpolation technique for FDTD on nonorthogonal grids," *Int. J. Numer. Model.*, vol. 11, no. 6, pp. 299–306, Nov.–Dec. 1998.

[24] A. Bossavit and L. Kettunen, "Yee-like schemes on a tetrahedral mesh, with diagonal lumping," *Int. J. Numer. Model.*, vol. 12, nos. 1–2, pp. 129–142, Jan.–Apr. 1999.

[25] R. Schuhmann and T. Weiland, "FDTD on non-orthogonal grids with triangular fillings," *IEEE Trans. Magn.*, vol. 35, no. 3, pp. 1470–1473, May 1999.

[26] S. D. Gedney and J. A. Roden, "Numerical stability of nonorthogonal FDTD methods," *IEEE Trans. Antennas Propag.*, vol. 48, no. 2, pp. 231–239, Feb. 2000.

[27] A. J. Ward and J. B. Pendry, "A program for calculating photonic band structures, Green's functions and transmission/reflection coefficients using a non-orthogonal FDTD method," *Comput. Phys. Commun.*, vol. 128, no. 3, pp. 590–621, June 2000.

[28] W. Yu and R. Mittra, "A conformal FDTD software package for modeling antennas and microstrip circuit components," *IEEE Trans. Antennas Propag. Mag.*, vol. 42, no. 5, pp. 28–39, Oct. 2000.

[29] R. Nilavalan, I. J. Craddock, and C. J. Railton, "Quantifying numerical dispersion in non-orthogonal FDTD meshes," *IEE Proc. Microw., Antennas Propag.*, vol. 149, no. 1, pp. 23–27, Feb. 2002.

[30] M. R. Visbal and D. V. Gaitonde, "On the use of higher-order finite-difference schemes on curvilinear and deforming meshes," *J. Comput. Phys.*, vol. 181, no. 1, pp. 155–185, Sep. 2002.

[31] S. D. Gedney, J. A. Roden, N. K. Madsen. A. H. Mohammadian, W. F. Hall, V. Sankar, and C. Rowell, "Explicit time-domain solutions of Maxwell's equations via generalized grids," in *Advances in Computational Electrodynamics: The Finite-Difference Time-Domain Method*, A. Taflove, Ed. Norwood, MA: Artech House, 1998, ch. 4, pp. 163–262.

[32] N. V. Kantartzis, T. I. Kosmanis, T. V. Yioultsis, and T. D. Tsiboukis, "A nonorthogonal higher-order wavelet-oriented FDTD technique for 3-D waveguide structures on generalised curvilinear grids," *IEEE Trans. Magn.*, vol. 37, no. 5, pp. 3264–3268, Sep. 2001.

[33] N. V. Kantartzis, T. T. Zygiridis, and T. D. Tsiboukis, "A nonstandard higher-order FDTD algorithm for 3-D arbitrarily and fractal-shaped antenna structures on general curvilinear lattices," *IEEE Trans. Magn.*, vol. 38, no. 2, pp. 737–740, Mar. 2002.

[34] N. V. Kantartzis, T. D. Tsiboukis, and E. E. Kriezis, "A topologically consistent class of 3-D higher-order curvilinear FDTD schemes for dispersion-optimized EMC and material modeling," *J. Mat. Processing Technol.*, vol. 161, nos. 1–2, pp. 210–217, Apr. 2005.

[35] J. A. Stratton, *Electromagnetic Theory*. New York: McGraw-Hill, 1941.

[36] F. L. Teixeira and W. C. Chew, "Lattice electromagnetic theory from a topological viewpoint," *J. Math. Phys.*, vol. 40, no. 1, pp. 169–187, Jan. 1999.

[37] C. Mattiussi, "The finite volume, finite element and finite difference methods as numerical methods for physical field problems," in *Advances in Imaging and Electron Physics*, P. Hawkes, Ed. New York: Academic Press, 2000, vol. 113, pp. 1–146.

[38] E. Tonti, "Finite formulation of electromagnetic field," *ICS Newslett.*, vol. 8, no. 1, pp. 5–11, 2001.

[39] P. W. Gross and P. R. Kotiuga, *Electromagnetic Theory and Computation: A Topological Approach*. London, UK: Cambridge University Press, 2004.

[40] K. L. Shlager and J. B. Schneider, "Comparison of the dispersion properties of several low-dispersion finite-difference time-domain algorithms," *IEEE Trans. Antennas Propag.*, vol. 51, no. 3, pp. 642–653, Mar. 2003.

[41] T. Kashiwa, Y. Sendo, K. Taguchi, T. Ohtani, and Y. Kanai, "Phase velocity errors of the nonstandard FDTD method and comparison with high-accuracy FDTD methods," *IEEE Trans. Magn.*, vol. 39, no. 4, pp. 2125–2128, July 2003.

[42] J. B. Cole, "High-accuracy Yee algorithm based on nonstandard finite differences: New developments and verifications," *IEEE Trans. Antennas Propag.*, vol. 50, no. 9, pp. 1185–1191, Sep. 2002.

[43] H. Kudo, T. Kashiwa, and T. Ohtani, "Numerical dispersion and stability condition of the three-dimensional nonstandard FDTD method," *Electron. Commun. Japan (Part II: Electronics)*, vol. 86, no. 5, pp. 66–76, May 2003.

[44] N. V. Kantartzis, "A generalised higher-order FDTD-PML algorithm for the enhanced analysis of 3-D waveguiding EMC structures in curvilinear coordinates," *IEE Proc. Microw. Antennas Propag.*, vol. 150, no. 5, pp. 351–359, Oct. 2003.

[45] C. L. Wagner, "Theoretical basis for numerically exact three-dimensional time-domain algorithms," *J. Comput. Phys.*, vol. 205, no. 1, pp. 343–356, Nov. 2005.

CHAPTER 4

Absorbing Boundary Conditions and Widened Spatial Stencils

4.1 INTRODUCTION

The overall efficiency of higher order FDTD schemes – either conventional or nonstandard – is strongly associated with the appropriate manipulation of unbounded domains. As long as their discretization procedures presume a linear mapping to limitless lattices in terms of more involved nodal configurations and elementary cells, any attempt regarding space minimization is of crucial importance. This imposes additional demands on the precision and performance of any prospective truncation technique well beyond that of customary schemes. Hence, the design and implementation of consistent absorbing boundary conditions (ABCs) for higher order time-domain simulations should be very careful and systematic.

Equivalent attention, however, should be drawn to the nontrivial shortcoming of the inevitably widened spatial stencils near perfect electric conducting (PEC) or absorbing walls. Actually, the straightforward treatment of such areas via the second-order FDTD method is not applicable to higher order configurations, since the resulting stencils may extend two or more nodes on every side of a specific mesh point. Mitigation of this inherent weakness can be trailed in alternative discretization strategies that approximate the unknown field values at the problematic boundary nodes by means of calculated data from the interior of the domain. In this manner, numerical outcomes are neither contaminated by severe instabilities, nor do they require cumbersome conventions to provide credible outcomes for scattering or radiation electromagnetic applications.

The purpose of this chapter is to investigate the possibilities of constructing advanced ABCs pertinent for higher orders of spatial and temporal accuracy. Initiating from existing local or global techniques, basically formulated for the second-order case, analysis proceeds to more advanced algorithms, such as the powerful perfectly matched layer (PML) absorber. Various realizations are discussed, while a nonstandard differencing process copes with the intricate case of the near-grazing incidence effect. Next, the treatment of the enlarged spatial stencils is considered through diverse approaches and the chapter closes with an accurate scheme for the modeling of arbitrarily curved material interfaces not aligned with the gird axes.

4.2 HIGHER ORDER FDTD FORMULATION OF ANALYTICAL ABCs

Analytical ABCs have been the first techniques for the termination of infinite regions and, depending upon their theoretical derivation, they are basically classified into two major categories. The methods of the former category, also known as *global* ABCs, are nonlocal in space and time and usually provide high accuracy and stability. Since their implementation involves integral transforms along the boundary, they are not suited for domains of irregular shape (poor universality). Also, their nonlocal profile often implies mathematical complexity and excessively large computational cost, thus creating a problem of practical efficacy.

Treatment to this shortcoming can be obtained from the second group of ABCs, designated as *local*. The majority of these schemes are realized by means of a single partial differential equation imposed on the fictitious boundary. Their construction is founded on diverse principles, such as the one way wave equation, the asymptotic field expansion, and the pseudodifferential operator theory. Local ABCs are proven to be numerically inexpensive, algorithmically elegant, routinely feasible, and geometrically universal. As a consequence, they are simple in their combination with various numerical algorithms and particularly with the FDTD method. Some of the most representative members in the category are the Engquist–Majda [1], the Bayliss–Turkel [2], and Lindman [3] families of ABCs, the popular Mur [4], Trefethen and Halpern [5], Higdon [6], and Fang [7] differential annihilators as well as the Liao [8] multitransmitting absorbing operators and the Grote–Keller algorithm [9]. However, these methods exhibit lower levels of accuracy as compared to the global ones, a fact that can be critical in many up-to-date applications, as elaborately investigated in [10, 11]. Although various improvements have been presented, none of the previous schemes became entirely faultless.

Apart from these approaches, a set of techniques which are not ABCs by themselves but rather numerical procedures for the enhancement of existing formulations has been additionally presented. Two indicative examples of such algorithms are the superabsorption [12] and the complementary operators (COM) method [13]. The former embodies a mutually error-canceling practice according to which the same ABC is applied to both electric and magnetic field components on and near the outer boundaries, relative to the polarization examined. In contrast, the COM is based on a different framework. Its major strength lies on the fact that it can effectively suppress first-order reflections regarding both propagating and evanescent waves independent of their wave number and angle of incidence.

Obviously, the combination of the aforementioned ABCs with higher order FDTD methods requires systematic modifications. This is mainly attributed to the dissimilar nodal configurations of the lattice in the connecting regions which, under certain directions of wave propagation, may behave like artificial "material discontinuities" and generate partial reflections, as indicated in [14]. Therefore, in the following, the possibility of realizing some indicative ABCs according to higher order operators is explored.

Let us consider Lindman annihilators [3], which are constructed through the use of projection operators incorporating past data at the boundary. Primarily, they involve the suitable field approximations by solving a system of partial differential equations in terms of certain correction functions. Focusing on the absorption of E_x electric component at the outer boundary, $z = L\Delta z$, the higher order nonstandard FDTD form of its update expression for a lossy medium, in conjunction with (3.70) and (3.71), becomes

$$\mathcal{T}\left[E_x\big|_{i,j,k}^{n}\right] + \upsilon \mathcal{S}_z\left[E_x\big|_{i,j,k}^{n}\right] + \sigma\big|_{i,j,L} E_x\big|_{i,j,k}^{n} = -\sum_{m=1}^{M} \psi_m\big|_{i,j,k}^{n}, \tag{4.1}$$

$$\mathcal{T}\left[\psi_m\big|_{i,j,k}^{n}\right] - b_m \upsilon^2 \mathcal{S}_{yy}\left[\psi_m\big|_{i,j,k}^{n}\right] = a_m \upsilon^2 \mathcal{S}_{yz}\left[\psi_m\big|_{i,j,k}^{n}\right], \tag{4.2}$$

where the double subscript appearing in $\mathcal{S}$ operators of (4.2) denote double or cross differentiation with respect to the indicated directions. In (4.1), M represents the order of the ABC, υ is the propagation velocity, σ accounts for the losses, and the intermediate quantities ψ_m are obtained by (4.2). Correction terms a_m and b_m are derived by the minimization of a specific function so that the reflection coefficient of (4.1) approaches zero. By increasing their number, one can broaden the range of incident angles within which reflection errors are sufficiently small. It is noteworthy to state that the above formulation avoids the use of nodal interpolations and hence it hardly affects the overall burden. A similar treatment holds for Fang differential ABCs [7]. In fact, higher order differencing is applied to their adjustable degrees of freedom that may be properly chosen to accommodate different levels of suppression of nonphysical reflections. Apparently, the conventional higher order discretization is also applicable without any additional changes.

Complementary operators, on the other hand, comprise an efficient boundary procedure that improves the absorption of numerical reflections by using previously developed ABCs [13]. Their competence is based on the implementation of two boundary operators that are complementary in their action. In this manner, new ABCs that produce prespecified reflection coefficients are derived. By solving the problem with each of the two operators and then averaging the two solutions, the technique annihilates the first-order artificial reflections of both obliquely propagating and evanescent waves, irrespective of their wave number. More specifically, absorption of the latter occurs even if the original ABC reflects them totally. The modified higher order development of COM starts from a well-posed and stable ABC that can be expressed by a single differential equation, defined as $\mathcal{F}$. If nonstandard operator $S_x[.]$ is applied to $\mathcal{F}$ (similarly for $\mathcal{S}_y[.]$ and $\mathcal{S}_z[.]$), the complementary boundary condition $\mathcal{F}^c$ is given by $\mathcal{F}^c = \mathcal{S}_x[\mathcal{F}]$. After some quantitative manipulations the reflection coefficient R is found to be $R[\mathcal{F}^c] = (-1)R[\mathcal{F}]$, which confirms the complementary action of $\mathcal{F}^c$ over $\mathcal{F}$. Furthermore, this process guarantees that the accuracy of operator $\mathcal{F}^c$ will be built upon that of the original $\mathcal{F}$. Normally, the most applicable ABCs to the concepts of COM are the Higdon-type or the previously described Fang annihilators. Thus, the nonstandard form of these concepts is

$$\mathcal{S}_z\left[\prod_{m=1}^{M-1}\left(\mathcal{S}_x\left[E_x|^n_{i,j,k}\right]+\frac{\vartheta_m}{\upsilon}\mathcal{T}\left[E_x|^n_{i,j,k}\right]+\sigma|_{i,j,L}\,E_x|^n_{i,j,k}\right)\right]=0, \tag{4.3}$$

$$\mathcal{T}\left[\prod_{m=1}^{M-1}\left(\mathcal{S}_x\left[E_x|^n_{i,j,k}\right]+\frac{\vartheta_m}{\upsilon}\mathcal{T}\left[E_x|^n_{i,j,k}\right]+\sigma|_{i,j,L}\,E_x|^n_{i,j,k}\right)\right]=0, \tag{4.4}$$

where coefficient ϑ_m is responsible for the fine tuning of absorption angle and phase velocity. Evidently, the COM realization requires two independent runs of the computer code. This may initially lead us to the deduction that these operators increase the cost of the overall simulation. Nevertheless, this is a deceptive impression. Due to their absorptive capabilities, COM enables the imposition of the outer boundaries closer to the scatterer, compensating so for the two runs and at the same time sufficiently annihilating waves of grazing incidence. In this manner, a much smaller computational domain – which induces significantly lower dispersion errors as well – is constructed and fewer time-steps for the solution of the problem are required.

4.3 HIGHER ORDER PML ABSORBERS

The introduction of the *PML* [15] for the effectual absorption of outgoing electromagnetic waves has been proven to have a profound impact on the overall quality and computational performance of numerical algorithms and especially the FDTD method. Founded on the use of some fictitious variables through an arbitrary field-splitting, this absorber enlarges the FDTD space via a finite lossy medium, thus, annihilating all outward plane waves. Then, by suitably adjusting the set of constitutive parameters in the layer, the vacuum–PML interface becomes – at least in a theoretical sense – completely transparent for any frequency and angle of incidence. Soon after its initial presentation, a considerable effort has been invested in its general enhancement and alleviation of its possible deficiencies. Hence, several contributions may be perceived such as complex coordinate stretching [16–19] and unsplit-field [20–23] formulations, extensions to curvilinear coordinate systems [24–28] or nonorthogonal meshes [29–31], material-based implementations [32–38], and complex-shifted low-frequency techniques [39–42]. For a comprehensive analysis on the evolution and performance of the PML technique, the reader is referred to the extensive investigations of [29, 31] and comparative studies of [43–46]. As far as the higher order configuration of the PML is concerned, relative studies [47–51] indicate that the resulting absorbers due to their optimal design attain large annihilation rates even for complicated domains or broadband frequency spectra. In the ensuing paragraphs, conventional and nonstandard higher order realizations in Cartesian and generalized coordinate systems, which can be combined with any existing PML, are described and verified by some typical applications.

4.3.1 Conventional Realization in Cartesian Coordinates

Theory

Let us assume a 3-D domain, Ω, in a linear, homogeneous, and isotropic medium. To absorb outgoing waves, Ω is surrounded by a set of PMLs that dissipate the waves propagating through their interior, as shown in Figure 4.1. If adequate field attenuation is conducted by these absorbers, zero field values may be presumed at their outer border, thus permitting for simple Dirichlet conditions to be imposed at the ends of the domain without creating spurious reflections. However, other local ABCs may also be utilized [27].

The split-field forms of time-dependent Maxwell's equations along the x-direction are

$$a_x\mu\,\partial_t H_x + \Upsilon_x\mu H_x = \partial_z E_y - \partial_y E_z \quad \text{and} \quad a_x\varepsilon\,\partial_t E_x + \Upsilon_x\varepsilon E_x = \partial_y H_z - \partial_z H_y, \tag{4.5}$$

with similar equations holding toward y, z and a_x, Υ_x quantities that define the intrinsic characteristics of the coordinate system, such as its metrical coefficients. Proceeding to the discretization of (4.5), one gets

$$(\overleftarrow{a}_x\overleftarrow{\mu})\,\overrightarrow{\partial}_t\,\overleftarrow{H}_x\big|_{\text{ps2}}^{n-1/2} + (\overleftarrow{\Upsilon}_x\overleftarrow{\mu})\,\overleftarrow{H}_x\big|_{\text{ps2}}^{n+1/2} = \left(\overrightarrow{\partial}_z\overrightarrow{E}_y - \overrightarrow{\partial}_y\overrightarrow{E}_z\right)\big|_{\text{ps1}}^{n}, \tag{4.6a}$$

$$(\overrightarrow{a}_x\overrightarrow{\varepsilon})\,\overleftarrow{\partial}_t\overrightarrow{E}_x\big|_{\text{ps1}}^{n-1} + (\overrightarrow{\Upsilon}_x\overrightarrow{\varepsilon})\,\overrightarrow{E}_x\big|_{\text{ps1}}^{n} = \left(\overleftarrow{\partial}_y\overleftarrow{H}_z - \overleftarrow{\partial}_z\overleftarrow{H}_y\right)\big|_{\text{ps2}}^{n-1/2}, \tag{4.6b}$$

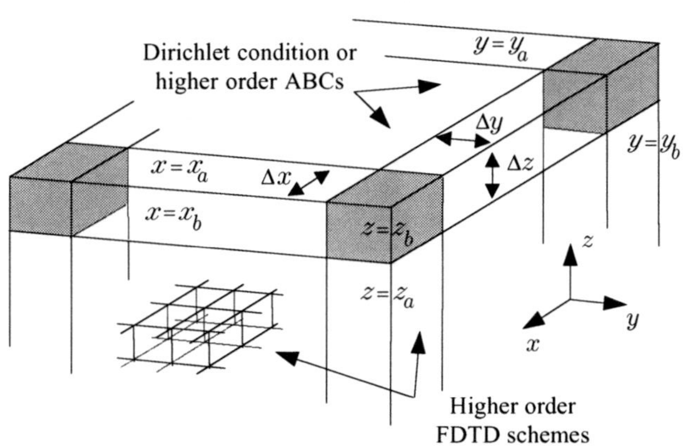

FIGURE 4.1: Geometry of the conventional higher order PML

where ps1 $= (i, j, k)$ and ps2 $= (i + \frac{1}{2}, j + \frac{1}{2}, k + \frac{1}{2})$, while $\vec{\partial}_y, \vec{\partial}_z, \vec{\partial}_t$ and $\overleftarrow{\partial}_y, \overleftarrow{\partial}_z, \overleftarrow{\partial}_t$ represent the forward and backward finite-difference approximation of the continuous analogs. A fully comparable treatment is used for the electric and magnetic field intensity, which are denoted as forward, $\vec{\mathbf{E}}$, and backward, $\overleftarrow{\mathbf{H}}$, vectors, respectively. Therefore, when a forward vector resides on a ps1 node, its field components correspond to the spatial vectors located in front of the specific node. For example, the x component of $\vec{\mathbf{E}}$ at ps1 is associated with node $(i + \frac{1}{2}, j, k)$, whereas that of $\overleftarrow{\mathbf{H}}$ at ps2 is related to $(i, j + \frac{1}{2}, k + \frac{1}{2})$. The arrows above the quantities a_x, ε, μ, and Y_x point out that their calculation will be conducted at the position of the forward or backward vector, respectively. These conventions are next applied to a plane wave:

$$\vec{\mathbf{E}}\Big|_{\text{ps1}}^{n} = \vec{\mathbf{E}}_0\, e^{-j\omega n\Delta t} \quad \text{and} \quad \overleftarrow{\mathbf{H}}\Big|_{\text{ps2}}^{n-1/2} = \overleftarrow{\mathbf{H}}_0 e^{-j\omega(n-1/2)\Delta t} \quad \text{for} \quad j^2 = -1, \tag{4.7}$$

where $\vec{\partial}_t \Rightarrow j\,\vec{\mathrm{Y}}_t$ and $\overleftarrow{\partial}_t \Rightarrow -j\,\overleftarrow{\mathrm{Y}}_t$ with $\vec{\mathrm{Y}}_t = \omega\, \sin(\omega\Delta t/2)e^{-j\omega\Delta t/2}$ and $\overleftarrow{\mathrm{Y}}_t = \omega\, \sin(\omega\Delta t/2)e^{j\omega\Delta t/2}$. So,

$$\begin{aligned} j\,\vec{\mathrm{Y}}_t\,(\overleftarrow{S}_x\overleftarrow{\mu})\overleftarrow{H}_x\Big|_{\text{ps2}} &= \left(\vec{\partial}_y \vec{E}_z - \vec{\partial}_z \vec{E}_y\right)\Big|_{\text{ps1}} \\ j\overleftarrow{\mathrm{Y}}_t\,(\vec{S}_x\vec{\varepsilon})\,\vec{E}_x\Big|_{\text{ps1}} &= \left(\overleftarrow{\partial}_z \overleftarrow{H}_y - \overleftarrow{\partial}_y \overleftarrow{H}_z\right)\Big|_{\text{ps2}}, \end{aligned} \tag{4.8}$$

for $S_x = a_x + j\mathrm{Y}_x/\mathrm{Y}_t$. Note that ratio $\mathrm{Y}_x/(\mathrm{Y}_t\, a_x)$ is equivalent to the loss tangent of the PML. Repeating the analysis for the other components and combining the results, the following vector equations are acquired:

$$j\,\vec{\mathrm{Y}}_t\,\overleftarrow{\mu}\overleftarrow{\mathbf{H}}\Big|_{\text{ps2}} = \vec{\nabla}_e \times \vec{\mathbf{E}}\Big|_{\text{ps1}} \quad \text{and} \quad j\overleftarrow{\mathrm{Y}}_t\,\vec{\varepsilon}\vec{\mathbf{E}}\Big|_{\text{ps1}} = -\overleftarrow{\nabla}_e \times \overleftarrow{\mathbf{H}}\Big|_{\text{ps2}}, \tag{4.9}$$

with

$$\vec{\nabla}_e = \hat{\mathbf{x}}\frac{1}{\overleftarrow{S}_x}\,\vec{\partial}_x + \hat{\mathbf{y}}\frac{1}{\overleftarrow{S}_y}\,\vec{\partial}_y + \hat{\mathbf{z}}\frac{1}{\overleftarrow{S}_z}\,\vec{\partial}_z \quad \text{and} \quad \overleftarrow{\nabla}_e = \hat{\mathbf{x}}\frac{1}{\vec{S}_x}\,\overleftarrow{\partial}_x + \hat{\mathbf{y}}\frac{1}{\vec{S}_y}\,\overleftarrow{\partial}_y + \hat{\mathbf{z}}\frac{1}{\vec{S}_z}\,\overleftarrow{\partial}_z. \tag{4.10}$$

If (4.7) is written in the form of $\vec{\mathbf{E}}\,\Big|_{\text{ps1}} = \vec{\mathbf{E}}_0\, e^{j\mathbf{k}\cdot\mathbf{r}}$ and $\overleftarrow{\mathbf{H}}\Big|_{\text{ps2}} = \overleftarrow{\mathbf{H}}_0 e^{j\mathbf{k}\cdot\mathbf{r}}$, its substitution in (4.9) yields

$$\vec{\mathrm{Y}}_t\,\overleftarrow{\mu}\overleftarrow{\mathbf{H}}_0 = \vec{\mathbf{K}}_e \times \vec{\mathbf{E}}_0 \quad \text{and} \quad \overleftarrow{\mathrm{Y}}_t\,\vec{\varepsilon}\vec{\mathbf{E}}_0 = -\overleftarrow{\mathbf{K}}_e \times \overleftarrow{\mathbf{H}}_0, \tag{4.11}$$

with

$$\vec{\mathbf{K}}_e = \frac{\vec{K}_x}{\overleftarrow{S}_x}\hat{\mathbf{x}} + \frac{\vec{K}_y}{\overleftarrow{S}_y}\hat{\mathbf{y}} + \frac{\vec{K}_z}{\overleftarrow{S}_z}\hat{\mathbf{z}} \quad \text{and} \quad \overleftarrow{\mathbf{K}}_e = \frac{\overleftarrow{K}_x}{\vec{S}_x}\hat{\mathbf{x}} + \frac{\overleftarrow{K}_y}{\vec{S}_y}\hat{\mathbf{y}} + \frac{\overleftarrow{K}_z}{\vec{S}_z}\hat{\mathbf{z}}, \tag{4.12}$$

where $\vec{K}_\zeta = k_p\, \mathrm{sinc}\,(k_\zeta\Delta\zeta/2)e^{jk_\zeta\Delta\zeta/2}$ and $\overleftarrow{K}_\zeta = k_p\, \mathrm{sinc}\,(k_\zeta\Delta\zeta/2)e^{-jk_\zeta\Delta\zeta/2}$ for $\zeta \in (x, y, z)$. It is important to mention that S_ζ and K_ζ are the only complex quantities. Differently speaking, the losses of the absorber can be controlled independently along x-, y-, and z-direction by means of diverse choices of stretched coordinates S_ς. The most prominent advantage of this generalized PML formulation is the introduction of the necessary *degrees of freedom* that enable its higher order FDTD implementation.

As an example, consider the original PML absorber in a 2-D computational space. Regarding the propagation of the TE case, the fourth-order electric field expression for E_x in the interior of the layer, becomes

$$\left[\varepsilon - \frac{\left(\sigma_y^* \Delta t\right)^2}{24}\right] \frac{\partial E_x}{\partial t} = \frac{\partial (H_{zx} + H_{zy})}{\partial y} - \sigma_y E_x + \frac{\varepsilon (\Delta t)^2}{24\mu} \left\{ \frac{1}{\varepsilon} \frac{\partial^2}{\partial y\, \partial x} \left[\frac{\partial (H_{zx} + H_{zy})}{\partial x} + \sigma_x E_y \right] \right.$$
$$+ \frac{1}{\varepsilon} \frac{\partial^2}{\partial y^2} \left[\frac{\partial (H_{zx} + H_{zy})}{\partial y} - \sigma_y E_x \right] + \left(\frac{\sigma_x^*}{\mu} + \frac{\sigma_y}{\varepsilon} \right) \frac{\partial}{\partial y} \left(\frac{\partial E_y}{\partial x} + \sigma_x^* H_{zx} \right)$$
$$\left. - \left(\frac{\sigma_y^*}{\mu} + \frac{\sigma_y}{\varepsilon} \right) \frac{\partial}{\partial y} \left(\frac{\partial E_x}{\partial y} - \sigma_y^* H_{zy} \right) \right\}, \tag{4.13}$$

where the split-field components H_{zx} and H_{zy} are analogously updated.

For the 3-D case, the study yields more complicated expressions. Thus, for reasons of brevity, only one of the twelve split-field PML subcomponents will be provided here. The remaining update equations can be then likewise extracted. For illustrative purposes, the electric field E_{xy} subcomponent is given by

$$\left[\varepsilon - \frac{(\sigma_y^* \Delta t)^2}{24}\right] \frac{\partial E_{xy}}{\partial t} = \frac{\partial (H_{zx} + H_{zy})}{\partial y} - \sigma_y E_{xy} + \frac{\varepsilon (\Delta t)^2}{24\mu}$$
$$\cdot \left\{ \frac{1}{\varepsilon} \frac{\partial^2}{\partial y\, \partial x} \left[\frac{\partial (H_{zx} + H_{zy})}{\partial x} + \sigma_x E_{yx} \right] + \frac{1}{\varepsilon} \frac{\partial^2}{\partial y^2} \left[\frac{\partial (H_{zx} + H_{zy})}{\partial y} - \sigma_y E_{xy} \right] + \left(\frac{\sigma_x^*}{\mu} + \frac{\sigma_y}{\varepsilon} \right) \right.$$
$$\cdot \frac{\partial}{\partial y} \left[\frac{\partial (E_{yx} + E_{yz})}{\partial x} + \sigma_x^* H_{zx} \right] - \frac{1}{\varepsilon} \frac{\partial^2}{\partial y\, \partial x} \left[\frac{\partial (H_{xy} + H_{xz})}{\partial z} - \sigma_z E_{yz} \right] + \frac{1}{\varepsilon} \frac{\partial^2}{\partial y^2}$$
$$\left. \cdot \left[\frac{\partial (H_{yx} + H_{yz})}{\partial z} - \sigma_z E_{xz} \right] - \left(\frac{\sigma_y^*}{\mu} + \frac{\sigma_y}{\varepsilon} \right) \frac{\partial}{\partial y} \left[\frac{\partial (E_{xy} + E_{xz})}{\partial y} - \sigma_y^* H_{zy} \right] \right\}. \tag{4.14}$$

In the above, the first spatial derivative on the right-hand side is approximated by (2.18), and the remaining terms by the well-known second-order formulas. Moreover, electric and magnetic conductivities in the layer are proportional to $\sigma(s) = \sigma_{\max} s^m$, where s is the depth coordinate, $m = 0, 1, 2$, and $\sigma_{\max}$ a tuning parameter for the achievement of the maximum absorption of outgoing waves for given layer thickness. The constants of proportionality in the above configuration are, of course, the constitutive parameters.

It is to be emphasized that this methodology appoints a new characteristic to the absorbers, designated as the *multidirectional absorption property* due to its ability to introduce additional attenuation terms along new directions in the layer as compared to the original PML [49]. These terms entail considerable suppression of the nonphysical FDTD lattice errors and contribute to the accuracy of the entire simulation. For instance, let us take into account the update equation (4.13). If the second-order PML had been employed, electric component E_x would have been attenuated only toward y-axis, as indicated by the $\sigma_y E_x$ term. Nonetheless, when the complete higher order version is implemented, the same component receives supplementary annihilation attributed to the presence of σ_x, σ_x^*, and σ_y^* losses. This enhanced performance is more prominent in the 3-D case (4.14), where losses toward all directions are involved. As a concluding observation, it can be stated that the multidirectional absorption property in conjunction with the improved treatment of the FDTD deficiencies gives higher levels of efficiency to the PML rendering it more competitive for the solution of difficult radiation and scattering applications.

Another point of notification is that both electric and magnetic field quantities are required on the right-hand side of the previous expressions at a certain time-step. However, most of them are not assumed to be stored in the computer's memory at this particular moment. This problem is overcome in two different ways. The first and the simpler one uses the *most recent* temporal value of the component, available via the time-advancing procedure, while the second way applies the semi-implicit approximation, assuming that the unknown quantity can be interpolated by averaging two of its already calculated values.

Due to the generalized structure of (4.5)–(4.12), the previous technique can be profitably applied to the unsplit-field PML as well. For example, a potential scheme for further enhancement is the robust reflectionless sponge layer, developed in [48]. As expected, the use of fourth-order spatial and temporal operators increases the absorber's versatility, while maintaining its well-posed and stable profile.

Indicative Applications

The merits of the conventional higher order PMLs are validated by a set of indicative scattering and waveguide problems. For the excitation, a compact smooth electric pulse is launched either as a hard source or via the total/scattered field technique [11]. For notational compatibility, the original representation of the PML is utilized, namely PML(d, P, R) denotes an absorber d- cell thick with a parabolic P loss variation, and a prefixed reflection coefficient R. In most simulations, PML is terminated by local ABCs, whose higher order derivation has been discussed in Section 4.2.

The first example examines the behavior of the PML global error for various conventional higher order PMLs. The 2-D TM domain is initially discretized in a $2.4\upsilon\tau \times 2.4\upsilon\tau$ ($\upsilon = 1/\sqrt{\mu\varepsilon}$ is the wavefront speed and $\tau = 10^{-9}$s the relaxation time of the smooth pulse) with

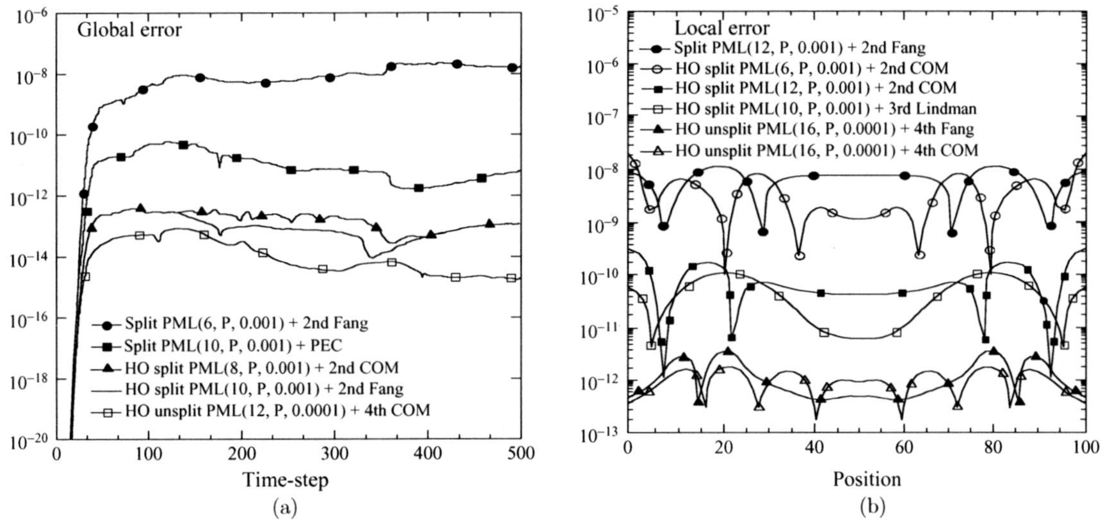

FIGURE 4.2: (a) Global error versus time-step and (b) local error versus position along a line of 100 cells generated by diverse higher order (HO) PML implementations

$\Delta x = \Delta y = 0.0072$ m. The space contains a centered cylindrical lossy anisotropic scatterer with a radius of $7\upsilon\tau/11$ and $\varepsilon_r = 2.0$. For the simulation length the time interval of $[0,\ T = 4\tau]$ is selected, while the dimensions of the reference domain are $9.5\upsilon\tau \times 9.5\upsilon\tau$. Observing Figure 4.2(a), it is deduced that the higher order schemes combined with the fourth-order COM [13] create significantly less error, almost in the order of 10^{-4}, than the original PML. Moreover, the performance of a 10-cell higher order PML is found to be equivalent to the one of a 16-cell second-order absorber, thus illustrating the considerable CPU and memory savings.

The local error for the previous 2-D case in the presence of a centered $0.8\upsilon\tau \times 0.8\upsilon\tau$ PEC scatterer with the rest of the problem's specifications remaining identical is shown in Figure 4.2(b). Again, the superiority of higher order schemes is prominent. More specifically, the enhanced formulations provide the best absorption rates, requiring almost half the thickness of the regular PML for the same order of local error.

In the next example, the ability of higher order PMLs to absorb evanescent waves is investigated. It is well known that the original PML formulation lacks to annihilate them. For this goal, a 50-mm parallel-plate metallic waveguide is considered, the TM_1 mode of which has a cutoff frequency at 3GHz. The open ends of this structure are terminated by PMLs, terminated by lossy ABCs, and results are given in Figure 4.3(a). It is apparent that when frequencies belong to the propagating spectrum, all techniques provide reasonable absorptions. But, as they move below the cutoff frequency, things deteriorate for the second-order schemes and only the higher order PMLs can sufficiently attenuate evanescent waves.

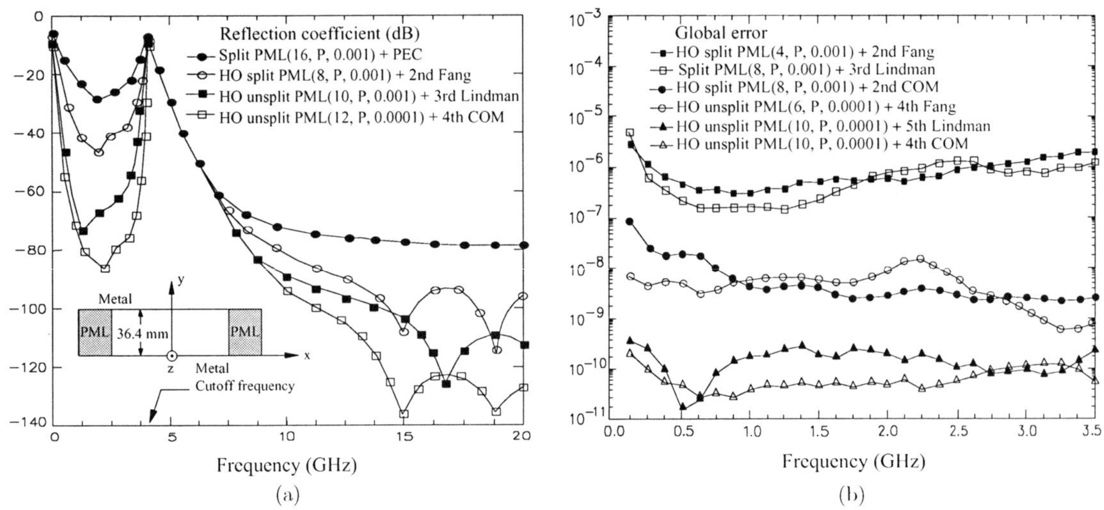

FIGURE 4.3: (a) Reflection coefficients of the TM_1 mode and (b) frequency spectrum of the global error induced by various higher order (HO) PML arrangements

Finally, the global error of the enhanced PMLs is explored in the frequency domain for a 3-D FDTD $2.4\upsilon\tau \times 1.8\upsilon\tau \times 2.0\upsilon\tau$ lattice in the presence of a cubic $0.5\upsilon\tau \times 0.5\upsilon\tau \times 0.5\upsilon\tau$ lossy scatterer ($\varepsilon_r = 2.5$). The structure is excited by a ramped sinusoidal excitation $E_z|^n = [1 - e^{-(n/30)^2}]\sin(2\pi f n \Delta t)$ with the exponential term reducing any form of transients, while the reference domain is $4.0\upsilon\tau \times 2.5\upsilon\tau \times 3.0\upsilon\tau$. Figure 4.3(b) proves that the errors induced by the higher order absorbers are very low for a broad frequency range.

4.3.2 Nonstandard Curvilinear Implementation

The construction of the higher order nonstandard unsplit-field PMLs in *curvilinear meshes* initiates from the division of a 3-D space, Ω, into two areas (separated with a nonplanar interface Γ) such that $\Omega = \Omega_{CS} \cup \Omega_{PML}$, where Ω_{CS} refers to the computational domain and Ω_{PML} is the area occupied by the PML under research. Considerable in the procedure is the extension of the stretched-coordinate theory to nonstandard models [30]. This is performed through the following steps:

- Incorporation of the field profile in the layer via stretched coordinate metrics $s_p = \upsilon_p + \frac{\sigma_p}{\chi_p + j\omega\kappa}$, with $\upsilon_p, \sigma_p, \chi_p \in \mathbb{R}^+$, $p \in (u, v, w)$, and κ the constitutive parameter to avoid field-splitting.
- Development of the elementary equations whose treatment provides the required attenuation rate. For $\chi_p > 0$, a high absorption of the laborious evanescent waves is attained.

- Analysis of the resulting expressions in the form of higher order time-dependent laws.
- Deduction of the constitutive relations in Ω_{PML} to completely identify the absorber.

Construction Procedure

To exhibit the advantages of the above conventions and without loss of generality, the most frequently encountered case of spherical coordinates (r, θ, φ) is analyzed. In this particular system, the usual PML arrangements suffer from several weaknesses, especially from an accuracy and convergence outlook. It is mentioned that the proposed method has also been applied – in an analogous way – to other coordinate systems with a very satisfactory performance.

Let us study the derivation of the FDTD-PML expressions under a Maxwellian manner. Unfortunately, such complicated curvilinear environments create highly dispersive reflections that form bands of transmitted modes growing spatially instead of being damped in the layer. For the goals of the approach, an inhomogeneous, isotropic, and lossless dielectric medium is examined in the frequency domain. Thus,

$$j\omega\varepsilon\mathbf{E}' = \nabla' \times \mathbf{H}', \quad \nabla' \cdot (\varepsilon\mathbf{E}') = 0, \quad -j\omega\mu\mathbf{H}' = \nabla' \times \mathbf{E}', \quad \nabla' \cdot (\mu\mathbf{H}') = 0. \qquad (4.15)$$

Here, Ω_{CS} is a sphere of radius r_0, and Ω_{PML} – extending from r_0 to infinity $(r' \geq r_0)$ – is the area to be backed up to a radius r_1 by a PEC wall, as depicted in Figure 4.4. The advanced PML is built via the right scaling between the independent (primed) and dependent variables in (4.15) that maintain the original field propagation in Ω_{CS} and host the proper coordinate stretching in Ω_{PML}. This treatment may be deemed as a mapping of the homogeneous isotropic

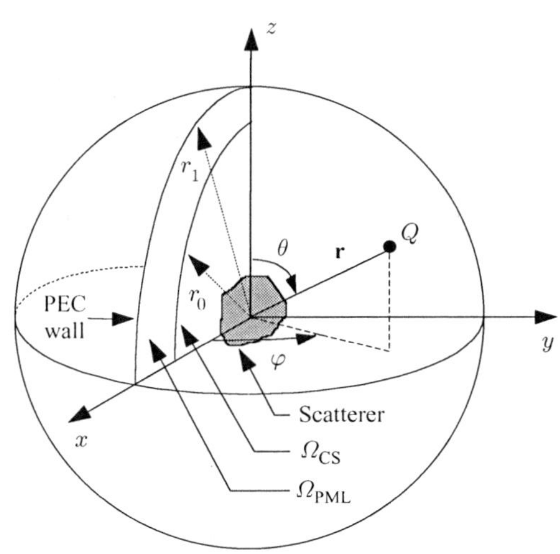

FIGURE 4.4: The higher order nonstandard PML in spherical coordinates

dielectric to an inhomogeneous uniaxial one. For our medium,

$$\mathbf{F}' = \begin{cases} \mathbf{F}, & \text{for } \mathbf{r}' \in \Omega_{\mathrm{CS}} \\ \operatorname{diag}\{\zeta_r, \xi_r^{-1}, \xi_r^{-1}\}\mathbf{F}, & \text{for } \mathbf{r}' \in \Omega_{\mathrm{PML}}, \end{cases} \tag{4.16}$$

with $\mathbf{r}' = \operatorname{diag}\{\xi_r, 1, 1\}\mathbf{r}$ and $\mathbf{r} = (r, \theta, \varphi)^T$. Substitution of (4.16) in the curl laws in (4.15) gives

$$j\omega\varepsilon\mathcal{R}\cdot\mathbf{E} = \nabla\times\mathbf{H}, \qquad -j\omega\mu\mathcal{R}\cdot\mathbf{H} = \nabla\times\mathbf{E}, \tag{4.17}$$

where the material tensor $\mathcal{R}$ is the diagonal matrix $\mathcal{R} = \operatorname{diag}\left[\xi_r^2\zeta_r, \zeta_r^{-1}, \zeta_r^{-1}\right]$ for a homogeneous dielectric background with constant constitutive parameters. The Gauss law for the electric and the magnetic field yields

$$\nabla\cdot(\varepsilon\mathcal{R}\cdot\mathbf{E}) = 0, \qquad \nabla\cdot(\mu\mathcal{R}\cdot\mathbf{H}) = 0. \tag{4.18}$$

Equations (4.17) and (4.18) constitute a causal hyperbolic system, according to the mathematical analysis of [26], which allows the physical implementation of the unsplit PML. Hence, these curvilinear absorbers *do not* generally create wave modes that grow linearly in time and for the majority of the problems are not vulnerable to perturbations. In the expressions written above,

$$\zeta_r(r,\omega) = \left[1 + \frac{\sigma_r(r)}{j\omega}\right]^{-1}, \qquad \xi_r(r,\omega) = 1 + \frac{\sigma_r'(r)}{j\omega}, \tag{4.19a}$$

for

$$\sigma_r(r) = \sigma_r^{\max}[(r - r_0)/\delta]^{m_d}; \quad m_d \geq 0, \qquad \sigma_r'(r) = r^{-1}\int_{r_0}^{r} \sigma_r(s)\,ds, \tag{4.19b}$$

with Ω_{PML} restricted to a sphere of radius $r_0 + \delta$ and $\delta = d_{\mathrm{PML}}\Delta r$ the layer's depth spanning to d_{PML} cells. Because of its ability to bend outgoing waves toward the normal to the PML and augment its absorption rate, $\sigma_r^{\max} \in \mathbb{R}^+$ must be carefully tuned.

The extraction of the higher order nonstandard FDTD-PML expressions is based on (4.17). Owing to the duality of Maxwell's equations, substitution of (4.19) in Faraday's law leads to

$$-j\omega\mu\xi_r^2\zeta_r H_r = (\nabla\times\mathbf{E})_r, \qquad -j\omega\mu\zeta_r^{-1}H_\theta = (\nabla\times\mathbf{E})_\theta, \qquad -j\omega\mu\zeta_r^{-1}H_\varphi = (\nabla\times\mathbf{E})_\varphi. \tag{4.20}$$

The update of H_r field components in (4.20) is conducted in conjunction with the magnetic flux density $B_r = \mu\xi_r\zeta_r H_r$. Thus, via (4.19a), both give

$$-j\omega B_r - \sigma_r'(r)B_r = (\nabla\times\mathbf{E})_r, \qquad j\omega B_r + \sigma_r(r)B_r = j\omega H_r + \mu\sigma_r'(r)H_r. \tag{4.21a}$$

Likewise, one derives

$$-j\omega B_\theta = (\nabla \times \mathbf{E})_\theta, \qquad j\omega B_\theta = j\omega H_\theta + \mu\sigma_r'(r)H_\theta, \tag{4.21b}$$

$$-j\omega B_\varphi = (\nabla \times \mathbf{E})_\varphi, \qquad j\omega B_\varphi = j\omega H_\varphi + \mu\sigma_r'(r)H_\varphi. \tag{4.21c}$$

It is stated that (4.21) are robust and the resulting update relations (after the appropriate transform in the time domain) are stable and adequately nondissipative, even when the usual Yee's scheme is implemented for their discretization. By approximating the spatial and temporal derivatives in the time-domain versions of (4.21), through the operators of (3.70)–(3.78), the final time-advancing formulas are

$$\mathcal{T}[B_r] + \sigma_r'(r)B_r = g_\theta \mathcal{S}_\theta[E_\varphi] - g_\varphi \mathcal{S}_\varphi[E_\theta], \quad \mathcal{T}[B_r] + \sigma_r(r)B_r = \mu\mathcal{T}[H_r] + \mu\sigma_r'(r)H_r, \tag{4.22a}$$

$$\mathcal{T}[B_\theta] = g_\varphi \mathcal{S}_\varphi[E_r] - g_r \mathcal{S}_r[E_\varphi], \quad \mathcal{T}[B_\theta] = \mu\mathcal{T}[H_\theta] + \mu\sigma_r(r)B_\theta, \tag{4.22b}$$

$$\mathcal{T}[B_\varphi] = g_r \mathcal{S}_r[E_\theta] - g_\theta \mathcal{S}_\theta[E_r], \quad \mathcal{T}[B_\varphi] = \mu\mathcal{T}[H_\varphi] + \mu\sigma_r(r)H_\varphi, \tag{4.22c}$$

in which g_p for $p \in (r, \theta, \varphi)$ are the spherical system metrics [28]. Apart from the previous merits, the proposed method assigns a basic characteristic to higher order PMLs: the ability to introduce *supplementary attenuation terms* along more directions in the layer, accomplishing a considerable suppression of the anisotropy discrepancies, especially in multifrequency environments.

The number of layers can be adequately decreased through a spherical mesh expansion technique, which gradually enlarges the FDTD cells in the PML region, as explained in Figure 4.5. Consequently, if τ is the expansion factor, the modified spatial increment $\overline{\Delta r}$ grows via the

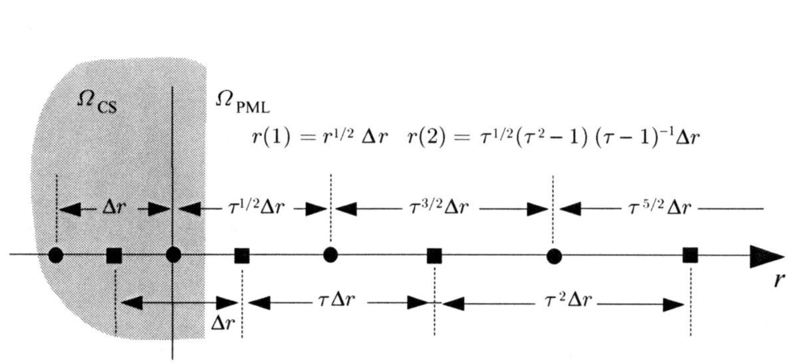

FIGURE 4.5: The first stages of the curvilinear mesh expansion approach describing the gradual cell enlargement

relations

$$\overline{\Delta}r(0) = \left(\frac{1}{2} + \frac{\tau - \sqrt{\tau}}{\tau - 1}\right)\Delta r, \qquad \overline{\Delta}r(L > 0) = \tau^{\rho}\Delta r, \tag{4.23}$$

for $\rho = \frac{1}{2}, 1, 1\frac{1}{2}, 2, \ldots, P$. Equation (4.23) introduces an additional stretching of the physical space that does not influence the effective action of s_m. A carefully selected τ enables the highest frequencies of a broadband excitation to propagate in the higher order elementary cells as well as the association of steeper sine or exponential conductivity profiles to the distinct layers of the absorber. Such choices decrease the inherent dispersion errors to a large extent and improve the overall convergence without affecting the grid.

Indicative Applications

To verify the higher order nonstandard PMLs, numerical analysis addresses several 3-D open-boundary applications expressed in spherical and elliptical coordinate systems.

The first is a scattering problem with a dielectric sphere ($\varepsilon = 4.8\varepsilon_0$) of radius $r_s = 0.5$ m. Due to the higher order stencils of (4.21), Ω_{CS} is divided into the coarse mesh of $7 \times 16 \times 32$ cells. So, grid resolution reaches the promising value of $\Delta r/\lambda_{\text{min}} = 1/5$, where λ_{min} is the minimum wavelength of a Gaussian pulse $P(t) = P_0 e^{-(t-t_0)^2/\tau^2}$ supported in $t \in [0, 4t_0]$. Its parameters are selected as $P_0 = 10^{-6}$, $t_0 = 10.482$ ns, and $\tau = 31.446$ ns to avoid the loss of any energy content. For completeness, a second-order spherical PML with $d_{\text{PML}} = 16$ cells and $\sigma_r^{\text{max}} = 4.147 \times 10^{10}$ is also studied. In this case, Ω_{CS} has $42 \times 68 \times 220$ cells, since common finite differences lack to trail the larger stencils. The global error is given in Figure 4.6(a) and as observed the optimized PMLs overwhelm the ordinary absorber without creating any oscillations. Furthermore, the ideal structure balancing absorption rate and d_{PML}, is the higher order eight-cell PML that suppresses the global error at 10^{-15} (four orders of magnitude lower than the second-order 12-cell PML).

In order to realize the influence of the grid size on the computational burden, a spherical domain $\Omega_{\text{CS}} = 11 \times 25 \times 38$, with a PEC scatterer of $r_s = 0.8$ m is examined. Given that the radius of Ω_{CS} is $r_0 = 1.2$ m, the higher order PML ($d_{\text{PML}} = 6$ cells) is expanded by $\tau = 1.68$, whereas the steeper Ricker wavelet,

$$P(t) = \gamma\sqrt{P_0/2}\,\frac{d}{dt} e^{-[(t-t_0)/\gamma]^2}\mathcal{H}(t), \tag{4.24}$$

excites the problem with $\gamma = 0.4391$, $t_0 = 1.5\mu$s, and $\mathcal{H}(t)$ the Heaviside pulse. Figure 4.6(b) gives the convergence rate of the global error versus grid resolution, $\Delta r/\lambda_{\text{min}}$, for different angles (even near-grazing ones) of wave impingement on the normal to the vacuum–PML interface. For comparison, a second-order PML ($d_{\text{PML}} = 10$ cells) is also imposed one cell away from the

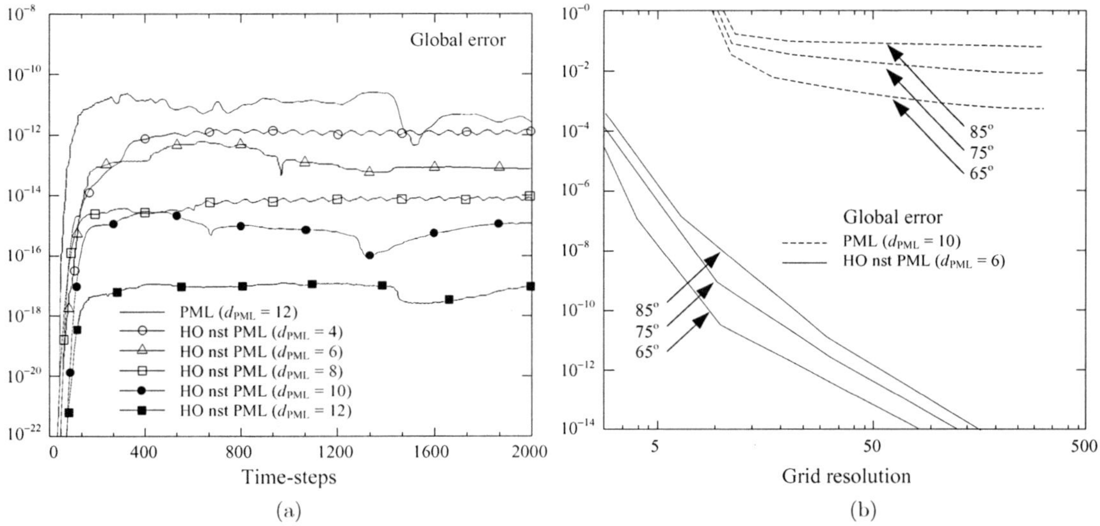

FIGURE 4.6: (a) Global error versus time-steps and (b) convergence rate of global error as a function of grid resolution at various angles of impingement for a second-order and several higher order (HO) nonstandard (nst) spherical PMLs

source. Results present that such a short distance is inadequate for the latter absorber, since its error is very high. Note that when $\Delta r/\lambda_{\min}$ has a low value, e.g., 1/5, this PML has a global error of 10^0, namely it reflects all outgoing waves. In opposition, its higher order nonstandard counterpart converges very fast, regardless of its small depth.

Next, a group of two lossy dielectric elliptical obstacles is embedded in a 3-D region of free space. The larger ($\varepsilon_1 = 3.6\varepsilon_0$, $\sigma_1 = 0.05$ S/m) is centered on the grid and contains the smaller one ($\varepsilon_2 = 6.9\varepsilon_0$, $\sigma_2 = 0.75$ S/m). The domain is discretised in $14 \times 29 \times 41$ cells and illuminated by a plane wave. A critical issue, herein, is that the boundary of the two scatterers does not align with any of the grid axes, thus making their modeling a cumbersome task. This shortcoming is attributed to the violation of the continuity conditions which, in its turn, arouses serious dispersion errors. In Figure 4.7(a), the maximum global error versus distance from the scattering group for different higher order PMLs is presented. Results are indeed very satisfactory and exhibit a smooth evolution for reasonable values of d_{PML}. For example, at the frequently used distance of three cells from the scatterer, the global error varies between 10^{-7} and 10^{-10}, enabling so large savings. Finally, Figure 4.7(b) displays the convergence of the reflection property, calculated for diverse orders of the global error norm $\|e(.)\|_l$, for $l = 1, 2, \infty$ (inf), on a sequence of progressively refined grids. As can be deduced, the higher order PML is again far more effective, especially at coarser lattice resolutions.

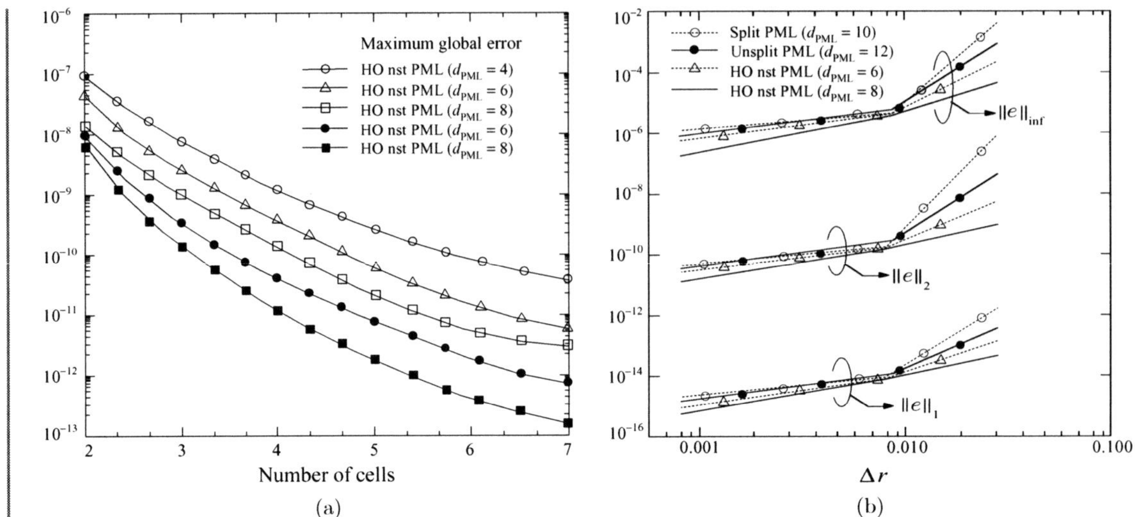

FIGURE 4.7: (a) Maximum global error versus distance from two lossy dielectric elliptical scatterers. (b) Convergence of reflection at the rate of the interior scheme

4.4 WIDENED SPATIAL STENCILS AND DISSIMILAR MEDIA INTERFACES

The attempt to alleviate the negative impact of enlarged spatial stencils in the modeling of material interfaces escapes the obvious concepts, e.g., the use of just a higher order approximation. Such practices may improve the situation quantitatively but do not bring a qualitative difference; i.e., the convergence rate essentially remains at a lower order. In addition, their extension to applications with increased dimensionality, where more complicated geometric details are involved, remains obscure. Hence, different methods should be explored. This is exactly the goal of this section which presents various contemporary algorithms that do not sacrifice the advantages of the general time-domain approach.

4.4.1 The Ghost-Node Process

A simple and rigorous way to cope with the widened stencils of higher order FDTD schemes is the design of an appropriate radiation boundary condition to be imposed at the outer ends of the lattice. For instance, let us take into account the 2-D (2, 4) algorithm that allows the stencils to reach up to the distance of $3\Delta h/2$ away from the tangential to the boundary field variable. This needs the use of two additional tangential nodes – one electric and one magnetic – interior to the tangential electric field boundary node. Treatment to this difficulty comes from Higdon's

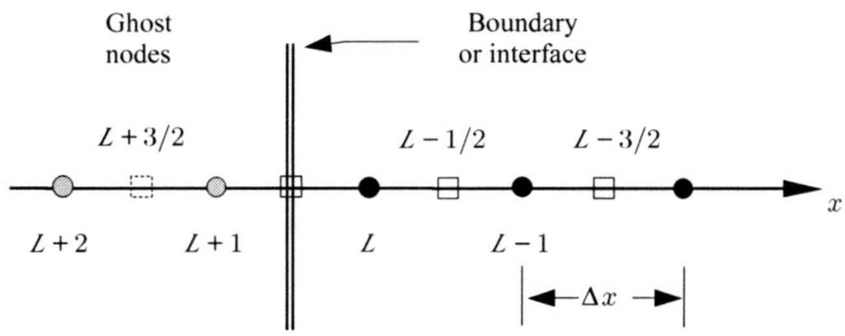

FIGURE 4.8: A typical (2, 4) lattice along the normal direction to the domain boundary. The ghost nodes on the left-hand side of the interface are depicted in grey scale

ABC [6]:

$$\mathcal{B}_M\left[\frac{\partial}{\partial n}, \frac{\partial}{\partial t}\right]F = \prod_{i=1}^{M}\left(\cos\theta_i \frac{\partial}{\partial t} - \upsilon \frac{\partial}{\partial n}\right)F = 0, \tag{4.25}$$

with M the order of accuracy, θ_i the angles of perfect wave annihilation, F a tangential field quantity, and $\partial/\partial n$ the spatial derivative in the outward normal direction to the boundary. A graphical depiction of the problem is given in Figure 4.8, where the points on the left-hand side of the boundary are designated as *"ghost" nodes* [48]. Presuming that all electric and magnetic components at time-step n and $n - \frac{1}{2}$, respectively, are already known, the approach that updates every quantity at both sides of the domain comprises the following stages: First, one has to advance H_z up to node $(L + \frac{1}{2}, j)$, via the (2, 4) scheme and enforce $\mathcal{B}_M\left[H_z\big|_{L+3/2,j}^{n+1/2}\right] = 0$ from (4.25) to acquire H_z at $(L + \frac{3}{2}, j)$. The next step is the update of H_x everywhere in the domain and the update of E_y up to node (L, j). Similarly, (4.24) is again applied as $\mathcal{B}_M\left[E_y\big|_{L+1,j}^{n+1}\right] = 0$ for E_y at $(L + 1, j)$, while $\mathcal{B}_M\left[E_y\big|_{L+2,j}^{n+1}\right] = 0$ is also employed to obtain E_y at $(L + 2, j)$. The practicality of this technique lies on the fact that operator $\mathcal{B}_M[.]$, for $M = 2, 3$, involves point (L, j) at time-step $n + 1$, the $(i \leq L, j)$ ones at previous time-steps, and points $(i \leq L - 1, j)$ at time-step $n + 1$. An equivalent procedure holds for the evaluation of the magnetic field with the suitable modifications in the spatial indices. The attractive feature of the ghost-node method is its 1-D structure that facilitates the manipulation of corners. In this case, the process is applied up to $j = j_{\max} - 1$ in the vertical direction next to the corner node, and then the same idea is implemented toward the horizontal direction including all i points at $j = j_{\max}$.

4.4.2 Self-Adaptive Compact Operators

Self-adaptive compact operators constitute a powerful tool for the unobstructed realization of higher order methods at boundary and material interfaces [52–58]. Having determined the basic form of their central version in Chapter 3, namely (3.50) and (3.51), the particular section provides a more detailed study regarding their generalized establishment and performance evaluation.

The most significant attribute of this classification is its implicit, nonsymmetric nature that satisfies the subsequent conditions: a) its implicit part comprises *tridiagonal* schemes and b) the orientation of its discretization strategy may be modified in the *same* way as in conventional finite differences. From a mathematical viewpoint, this simply implies that the approximating operator can change the sign of its self-adjoint part from positive to negative. In this context, our analysis will be based on the convention that nonsymmetric compact operators behave like additive corrections in the general first-order difference operators:

$$D_{\mathrm{co}}\left[f\big|_i^n\right] = \tfrac{1}{2}\left[f\big|_{i+1}^n - f\big|_{i-1}^n - s\left(f\big|_{i+1}^n - 2f\big|_i^n + f\big|_{i-1}^n\right)\right], \tag{4.26}$$

where s is a weighting parameter. Considering the difference between the continuous operator ∂_x and its discrete counterpart (4.26), one can write the following Taylor series:

$$\left.\frac{\partial f}{\partial x}\right|_i^n - \frac{1}{\Delta h}D_{\mathrm{co}}\left[f\big|_i^n\right] = \frac{s\,\Delta h}{2}\left.\frac{\partial^2 f}{\partial x^2}\right|_i^n - \frac{(\Delta h)^2}{6}\left.\frac{\partial^3 f}{\partial x^3}\right|_i^n + \frac{s(\Delta h)^3}{24}\left.\frac{\partial^4 f}{\partial x^4}\right|_i^n - \frac{(\Delta h)^4}{120}\left.\frac{\partial^5 f}{\partial x^5}\right|_i^n$$

$$+ O\left[(\Delta h)^5\right] = \left(I - \frac{\Delta h}{3s}\left.\frac{\partial f}{\partial x}\right|_i^n + \frac{(\Delta h)^2}{12}\left.\frac{\partial^2 f}{\partial x^2}\right|_i^n - \frac{(\Delta h)^3}{60s}\left.\frac{\partial^3 f}{\partial x^3}\right|_i^n\right)\frac{s\,\Delta h}{2}\left.\frac{\partial^2 f}{\partial x^2}\right|_i^n + O\left[(\Delta h)^5\right], \tag{4.27}$$

in which $s \neq 0$ and I the unit operator. Next, the power series

$$1 - \tfrac{1}{3s}z + \tfrac{1}{12}z^2 - \tfrac{1}{60s}z^3 \tag{4.28}$$

is substituted by the suitable Padé approximation, with the term $z = \Delta h\,\partial_x$ playing the role of its variable. Assuming that third-order accuracy is sufficient for the calculation of the left-hand side of (4.27), expression (4.28) becomes $(1 + z/3s)^{-1}$. The only requirement now is the discretization of $\Delta h\,\partial_x$ and $(\Delta h)^2\,\partial_{xx}$ through first- and second-order approximants. If the possible choices are limited to tridiagonal operators, these approximants are defined as $D_{\mathrm{co}}^{(1)}[.]$ and $D_{\mathrm{co}}^{(2)}[.]$. In this manner, one obtains the third-order operator:

$$\mathrm{LD}_3\left[f\big|_i^n\right] = \tfrac{1}{\Delta h}\left\{D_{\mathrm{co}}\left[f\big|_i^n\right] + \tfrac{s}{2}\left(I + \tfrac{1}{3s}D_{\mathrm{co}}^{(1)}\left[f\big|_i^n\right]\right)^{-1} D_{\mathrm{co}}^{(2)}\left[f\big|_i^n\right]\right\}. \tag{4.29}$$

Obviously, the preceding algorithm can be generalized to the construction of Mth-order compact operators LD_M. For example, using more terms in the Padé approximation (4.28), the fifth-order member is given by

$$LD_5\left[f\big|_i^n\right] = \tfrac{1}{\Delta h}\left\{D_{\mathrm{co}}\left[f\big|_i^n\right] + \tfrac{s}{2}\Gamma_{\mathrm{A}}^{-1}\Gamma_{\mathrm{B}}\left(I + \tfrac{1}{12}D_{\mathrm{co}}^{(2)}\left[f\big|_i^n\right]\right)^{-1} D_{\mathrm{co}}^{(1)}\left[f\big|_i^n\right]\right\}, \tag{4.30}$$

where

$$\Gamma_{\mathrm{A}} = I - \left(\tfrac{1}{15s} + \tfrac{s'}{4}\right)\left(f|_{i+1}^{n} - f|_{i-1}^{n}\right) + \left(\tfrac{1}{6} + \tfrac{s'}{15s}\right) D_{\mathrm{co}}^{(2)}\left[f|_{i}^{n}\right],$$
$$\Gamma_{\mathrm{B}} = I + \left(\tfrac{1}{10s} - \tfrac{s'}{4}\right)\left(f|_{i+1}^{n} - f|_{i-1}^{n}\right) + \left(\tfrac{1}{6} - \tfrac{s'}{10s}\right) D_{\mathrm{co}}^{(2)}\left[f|_{i}^{n}\right],$$

with a new independent variable s' and $s^2 = \frac{4}{5}$. Based on these abstractions and the respective Taylor series expansion, compact operators LD3 and LD5 yield

$$\mathrm{LD3}\left[f|\right]_{i}^{n} = \left.\frac{\partial f}{\partial x}\right|_{x=x_i}^{t} + s\left(\frac{s'}{12s} + \frac{1}{18s^2}\right)(\Delta h)^3 \left.\frac{\partial^4 f}{\partial x^4}\right|_{x=\bar{x}}^{t}, \tag{4.31}$$

$$\mathrm{LD5}\left[f|\right]_{i}^{n} = \left.\frac{\partial f}{\partial x}\right|_{x=x_i}^{t} - s\left(\frac{s'}{144s} + \frac{1}{900s^2}\right)(\Delta h)^5 \left.\frac{\partial^6 f}{\partial x^6}\right|_{x=\bar{x}}^{t}, \tag{4.32}$$

in which $\bar{x} \in [x_i - \Delta h, x_i + \Delta h]$. It is stressed that the relatively smaller coefficients in (4.31) and (4.32), as compared to those of the conventional operators, guarantee a more precise representation of field components and therefore a better modeling of the underlying physical wave interactions.

4.4.3 Treatment of Arbitrarily Curved Material Interfaces

The effects of staircasing and the lack of properly enforced jump conditions on both sides of arbitrarily embedded material interfaces – which do not coincide with the axes of the coordinate system – have notable consequences on the stability of the FDTD method. In fact, for cases where a field component is discontinuous along a curvilinear grid line, the scheme presents local divergence and loss of global convergence [59, 60]. Focusing on modern waveguide and antenna structures, one can come up with a more subtle problem, namely the use of coarse modeling around a nonmetallic surface. Although the relations associating the field components near such media are quite familiar, Yee's algorithm cannot perform a sufficient enforcement whenever required. To overcome these drawbacks, a higher order convergent process is developed. The technique modifies the spatial stencils around 3-D interfaces to correctly weight the contribution of their properties and impose the proper continuity conditions.

Assume the transverse cut of the curvilinear boundary of Figure 4.9, appearing in the secondary mesh, with $\hat{\mathbf{n}} = [\hat{n}_u, \hat{n}_v, \hat{n}_w]^T$ a normal unit vector in the (u, v, w) general coordinate system. The covariant components of $\mathbf{E}_{\mathrm{cv}}^{\mathrm{mt}} = \left[e_u^{\mathrm{mt}}, e_v^{\mathrm{mt}}, e_w^{\mathrm{mt}}\right]^T$ and $\mathbf{H}_{\mathrm{cv}}^{\mathrm{mt}} = \left[h_u^{\mathrm{mt}}, h_v^{\mathrm{mt}}, h_w^{\mathrm{mt}}\right]$ fields in the two regions of different materials, $\varepsilon^{\mathrm{mt}}$ and μ^{mt}, for mt = A, B, are connected by the suitable tangential/normal continuity rules, written as

$$\hat{\mathbf{n}} \times \mathbf{E}_{\mathrm{cv}}^{\mathrm{A}} = \hat{\mathbf{n}} \times \mathbf{E}_{\mathrm{cv}}^{\mathrm{B}}, \qquad \varepsilon^{\mathrm{A}}\hat{\mathbf{n}} \cdot \mathbf{E}_{\mathrm{cv}}^{\mathrm{A}} = \varepsilon^{\mathrm{B}}\hat{\mathbf{n}} \cdot \mathbf{E}_{\mathrm{cv}}^{\mathrm{B}}, \tag{4.33a}$$

$$\hat{\mathbf{n}} \times \mathbf{H}_{\mathrm{cv}}^{\mathrm{A}} = \hat{\mathbf{n}} \times \mathbf{H}_{\mathrm{cv}}^{\mathrm{B}}, \qquad \mu^{\mathrm{A}}\hat{\mathbf{n}} \cdot \mathbf{H}_{\mathrm{cv}}^{\mathrm{A}} = \mu^{\mathrm{B}}\hat{\mathbf{n}} \cdot \mathbf{H}_{\mathrm{cv}}^{\mathrm{B}}. \tag{4.33b}$$

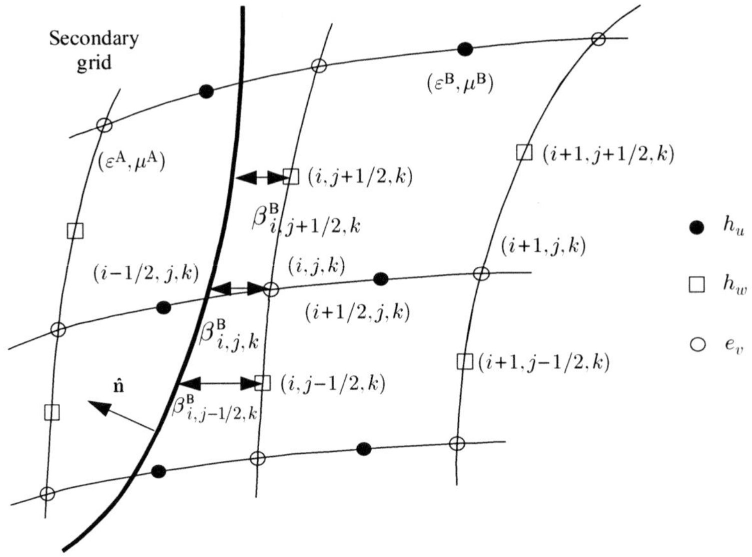

FIGURE 4.9: Treatment of a curvilinear arbitrary aligned material interface that does not align with the grid axes

To study the situation where the interface does not align with the axes of the grid, two parameters $\beta^{\mathrm{A}}_{i,j,k}$ and $\beta^{\mathrm{B}}_{i,j,k}$ are appointed to the separate regions as guidelines of the distance from the first/last cell to the physical position of the wall relative to the cell dimensions. Evidently, $\beta^{\mathrm{mt}}_{i,j,k}$ belongs to $[0, 1/2]$ and satisfies the global constraint $\beta^{\mathrm{A}}_{i,j,k} = \frac{1}{2} - \beta^{\mathrm{B}}_{i,j,k}$. Since these coefficients depend only on the boundary's geometrical shape, they may be evaluated and stored in a preprocessing stage once the lattice has been constructed. So,

$$h^{\mathrm{A}}_u\Big|^{n+1/2}_{i,j,k} = \left(1 + \beta^{\mathrm{A}}_{i,j,k}\right) h_u\Big|^{n+1/2}_{i-1/2,j,k} + \beta^{\mathrm{A}}_{i,j,k}\, h_u\Big|^{n+1/2}_{i-3/2,j,k} \tag{4.34}$$

is immediately derived. For the general case, one obtains an approximation of h^{B}_w through

$$\tilde{h}_w\Big|^{n+1/2}_{i,j\pm1/2,k} = h_w\Big|^{n+1/2}_{i,j\pm1/2,k} + \beta^{\mathrm{B}}_{i,j\pm1/2,k}\left(h_w\Big|^{n+1/2}_{i,j\pm1/2,k} - h_w\Big|^{n+1/2}_{i+1,j\pm1/2,k}\right), \tag{4.35}$$

$$h^{\mathrm{B}}_w\Big|^{n+1/2}_{i,j,k} = \frac{1}{2}\left(\tilde{h}_w\Big|^{n+1/2}_{i,j+1/2,k} + \tilde{h}_w\Big|^{n+1/2}_{i,j-1/2,k}\right), \tag{4.36}$$

where the tilded notation stands for some auxiliary magnetic variables. Retaining the basic idea that extrapolations do not span beyond half a cell, the h^{A}_v can be computed by means of four quantities that are found via the following collective equation, along the w-axis, as

$$\tilde{h}_v\Big|_{i,j\pm1/2,k\pm1/2}^{n+1/2} = h_v\Big|_{i-1/2,j\pm1/2,k\pm1/2}^{n+1/2} + \beta^{A}_{i,j\pm1/2,k\pm1/2} \cdot \left(h_v\Big|_{i-1/2,j\pm1/2,k\pm1/2}^{n+1/2} - h_v\Big|_{i-3/2,j\pm1/2,k\pm1/2}^{n+1/2}\right). \tag{4.37}$$

From these extrapolated values, h_v^{A} becomes

$$h_v^{A}\Big|_{i,j,k}^{n+1/2} = \frac{1}{4}\left(\tilde{h}_v\Big|_{i,j-1/2,k-1/2}^{n+1/2} + \tilde{h}_v\Big|_{i,j+1/2,k-1/2}^{n+1/2} + \tilde{h}_v\Big|_{i,j+1/2,k+1/2}^{n+1/2} + \tilde{h}_v\Big|_{i,j-1/2,k+1/2}^{n+1/2}\right), \tag{4.38}$$

and the unknown h_u^{B}, for the update of e_v, is then recovered by enforcing (4.33). Thus, one resolves

$$h_u^{B}\Big|_{i,j,k}^{n+1/2} = h_u^{A}\Big|_{i,j,k}^{n+1/2} + \hat{n}_u(\mu^{A} - \mu^{B})\frac{\hat{n}_u h_u^{A}\Big|_{i,j,k}^{n+1/2} + \hat{n}_v h_v^{A}\Big|_{i,j,k}^{n+1/2} + \hat{n}_w h_w^{B}\Big|_{i,j,k}^{n+1/2}}{\mu^{B}\hat{n}_u^2 + \mu^{B}\hat{n}_v^2 + \mu^{A}\hat{n}_w^2}. \tag{4.39}$$

Finally, the nonstandard approximation of h_u spatial derivative along w-direction is given by

$$S_w\left[h_u\Big|_{i,j,k}^{n+1/2}\right] = \frac{2}{(2\beta^{B}_{i,j,k} + 1)\Delta u}\left(h_u\Big|_{i+1/2,j,k}^{n+1/2} - h_u^{B}\Big|_{i,j,k}^{n+1/2}\right). \tag{4.40}$$

It must be emphasized that the remaining electric components are time marched in terms of a similar sequence, while the magnetic ones – placed next to the interface – may merely exploit the duality of the grid. To apply (4.34)–(4.40), a predictor-corrector stage is employed in which the higher order nonstandard FDTD algorithm, serving as the predictor, solves Maxwell's equations in the entire domain, while the above technique, acting as the corrector, alters the solutions locally. The slightly increased mathematical complexity is abundantly compensated by the substantial accuracy and the efficient handling of curvilinear peculiarities.

REFERENCES

[1] B. Engquist and A. Majda, "Absorbing boundary conditions for the numerical simulation of waves," *Math. Comput.*, vol. 31, pp. 629–651, 1977.

[2] A. Bayliss and E. Turkel, "Radiation boundary conditions for wave-like equations," *Comm. Pure Appl. Math.*, vol. 23, pp. 707–725, 1980.

[3] E. L. Lindman, "Free space' boundary conditions of the time dependent wave equation," *J. Comput. Phys.*, vol. 18, no. 1, pp. 66–78, May 1975.

[4] G. Mur, "Absorbing boundary conditions for the finite-difference approximation of the time-domain electromagnetic field equations," *IEEE Trans. Electromagn. Compat.*, vol. 23, no. 4, pp. 377–382, Nov. 1981.

[5] L. N. Trefethen and L. Halpern, "Well-posedness of one way wave equations and absorbing boundary conditions," *Math. Comput.*, vol. 47, pp. 421–435, 1986.

[6] R. L. Higdon, "Absorbing boundary conditions for difference approximations to the multi-dimensional wave equation," *Math. Comput.*, vol. 47, pp. 437–459, 1986.

[7] J. Fang, "Absorbing boundary conditions applied to model wave propagation in microwave integrated-circuits," *IEEE Trans. Microw. Theory Tech.*, vol. 42, no. 8, pp. 1506–1513, Aug. 1994.

[8] Z. P. Liao, H. L. Wong, B. P. Yang, and Y. F. Yuan, "A transmitting boundary for transient wave analyses," *Scientia Sinica (series A)*, vol. 27, pp. 1036–1076, 1984.

[9] M. J. Grote and J. B. Keller, "Nonreflecting boundary conditions for Maxwell's equations," *J. Comput. Phys.*, vol. 139, no. 2, pp. 327–342, Jan. 1998.

[10] S. V. Tsynkov, "Numerical solution of problems on unbounded domains. A review," *Appl. Numer. Math.*, vol. 27, no. 4, pp. 465–532, Aug. 1998.

[11] A. Taflove and S. C. Hagness, *Computational Electrodynamics: The Finite-Difference Time-Domain Method*, 3rd ed. Norwood, MA: Artech House, 2005.

[12] K. K. Mei and J. Fang, "Superabsorption – A method to improve absorbing boundary conditions," *IEEE Trans. Antennas Propag.*, vol. 40, no. 9, pp. 1001–1010, Sep. 1992.

[13] O. M. Ramahi, "Complementary operators: A method to annihilate artificial reflections arising from the truncation of the computational domain in the solution of partial differential equations," *IEEE Trans. Antennas Propag.*, vol. 43, no. 7, pp. 697–704, July 1995.

[14] C. W. Manry, S. L. Broschat, and J. B. Schneider, "Higher-order FDTD methods for large problems," *Appl. Comput. Electromagn. Soc. J.*, vol. 10, no. 2, pp. 17–29, 1995.

[15] J.-P. Bérenger, "A perfectly matched layer for the absorption of electromagnetic waves," *J. Comput. Phys.*, vol. 114, no. 2, pp. 185–200, Oct. 1994.

[16] W. C. Chew and W. H. Weedon, "A 3D perfectly matched medium from modified Maxwell's equations with stretched coordinates," *Microw. Opt. Technol. Lett.*, vol. 7, no. 13, pp. 599–604, 1994.

[17] C. M. Rappaport, "Perfectly matched absorbing boundary conditions based on anisotropic lossy mapping of space," *IEEE Microw. Guided Wave Lett.*, vol. 5, no. 3, pp. 90–92, Mar. 1995.

[18] R. Mittra and U. Pekel, "A new look at the perfectly matched layer (PML) concept for the reflectionless absorption of electromagnetic waves," *IEEE Microw. Guided Wave Lett.*, vol. 5, no. 3, pp. 84–86, Mar. 1995.

[19] Z. S. Sacks, D. M. Kingsland, R. Lee, and J.-F. Lee, "A perfectly matched anisotropic absorber for use as an absorbing boundary condition," *IEEE Trans. Antennas Propag.*, vol. 44, no. 12, pp. 1630–1639, Dec. 1995.

[20] S. D. Gedney, "An anisotropic perfectly matched layer-absorbing medium for the truncation of FDTD lattices," *IEEE Trans. Antennas Propag.*, vol. 44, no. 12, pp. 1630–1639, Dec. 1996.

[21] L. Zhao and A. C. Cangellaris, "GT-PML: Generalized theory of perfectly matched layers and its application to the reflectionless truncation of finite-difference time-domain grids," *IEEE Trans. Microw. Theory Tech.*, vol. 44, no. 12, pp. 2555–2563, Dec. 1996.

[22] A. C. Polycarpou and C. A. Balanis, "An optimized anisotropic PML for the analysis of microwave circuits," *IEEE Microw. Guided Wave Lett.*, vol. 8, no. 1, pp. 30–32, Jan. 1998.

[23] Y. S. Rickard and N. K. Nikolova, "Enhancing the PML absorbing boundary conditions for the wave equation," *IEEE Trans. Antennas Propag.*, vol. 53, no. 3, pp. 1242–1246, Mar. 2005.

[24] F. L. Teixeira and W. C. Chew, "PML-FDTD in cylindrical and spherical grids," *IEEE Microw. Guided Wave Lett.*, vol. 7, no. 9, pp. 285–287, Sep. 1997.

[25] F. Collino and P. Monk, "The perfectly matched layer in curvilinear coordinates," *SIAM J. Sci. Comp.*, vol. 19, no. 6, pp. 2061–2090, 1998.

[26] P. G. Petropoulos, "Reflectionless sponge layers as absorbing boundary conditions for the numerical solution of Maxwell's equations in rectangular, cylindrical and spherical coordinates," *SIAM J. Appl. Math.*, vol. 60, no. 3, pp. 1037–1058, 2000.

[27] N. V. Kantartzis, P. G. Petropoulos, and T. D. Tsiboukis, "Performance evaluation and absorption enhancement of the Grote–Keller and unsplit PML boundary conditions for the 3-D FDTD method in spherical coordinates," *IEEE Trans. Magn.*, vol. 35, no. 3, pp. 1418–1421, May 1999.

[28] K. P. Prokopidis, N. V. Kantartzis, and T. D. Tsiboukis, "A higher-order unsplit-feld perfectly matched layer for the reflectionless truncation of 3-D spherical FDTD lattices," *Electr. Engr.*, vol. 84, pp. 173–187, 2002.

[29] S. D. Gedney, "The perfectly matched layer absorbing medium," in *Advances in Computational Electrodynamics: The Finite-Difference Time-Domain Method*, A. Taflove, Ed. Norwood, MA: Artech House, 1998, ch. 5, pp. 263–344.

[30] N. V. Kantartzis, T. I. Kosmanis, and T. D. Tsiboukis, "Fully nonorthogonal higher-order accurate FDTD schemes for the systematic development of 3-D reflectionless PMLs in general curvilinear coordinate systems," *IEEE Trans. Magn.*, vol. 36, no 4, pp. 912–916, July 2000.

[31] F. L. Teixeira and W. C. Chew, "Advances in the theory of perfectly matched layers," in *Fast and Efficient Algorithms in Computational Electromagnetics*, W. C. Chew, J.-M. Jin, E. Michielssen, and J. M. Song, Eds. Norwood, MA: Artech House, 2001, ch. 8, pp. 283–346.

[32] J. Fang and Z. Wu, "Generalized perfectly matched layer for the absorption of propagating and evanescent waves in lossless and lossy media," *IEEE Trans. Microw. Theory Tech.*, vol. 44, no. 12, pp. 2216–2222, Dec. 1996.

[33] R. W. Ziolkowski, "The design of Maxwellian absorbers for numerical boundary conditions and for practical applications using engineered artificial materials," *IEEE Trans. Antennas Propag.*, vol. 45, no. 4, pp. 656–671, Apr. 1997.

[34] G.-X. Fan and Q. H. Liu, "An FDTD algorithm with perfectly matched layers for general dispersive media," *IEEE Trans. Antennas Propag.*, vol. 48, no. 5, pp. 637–646, May 2000.

[35] S. G. García, J. S. Juntunen, R. G. Martín, A. P. Zhao, B. G. Olmedo, and A. Räisänen, "A unified look at Bérenger's PML for general anisotropic media," *Microw. Opt. Technol. Lett.*, vol. 28, no. 6, pp. 414–416, Mar. 2001.

[36] M. Fujii, N. Omaki, M. Tahara, I. Sakagami, C. Poulton, W. Freude, and P. Russer, "Optimization of nonlinear dispersive APML ABC for the FDTD analysis of optical solitons," *IEEE J. Quantum Electron.*, vol. 41, no. 3, pp. 448–454, Mar. 2005.

[37] O. Ramadan and A. Y. Oztoprak, "Z-transform implementation of the perfectly matched layer for truncating FDTD domains," *IEEE Microw. Wireless Compon. Lett.*, vol. 13, no. 9, pp. 402–404, Sep. 2003.

[38] S. A. Cummer, "Perfectly matched layer behavior in negative refractive index materials," *IEEE Antennas Wireless Propag. Lett.*, vol. 3, no. 1, pp. 172–175, 2004.

[39] M. Kuzuoglou and R. Mittra, "Frequency dependence of the constitutive parameters of causal perfectly matched anisotropic absorbers," *IEEE Trans. Microw. Guided Wave Lett.*, vol. 6, no. 12, pp. 447–449, Dec. 1996.

[40] J. A. Roden and S. D. Gedney, "Convolutional PML (CPML): An efficient FDTD implementation of the CFS-PML for arbitrary media," *Microw. Optical Technol. Lett.*, vol. 27, no. 5, pp. 334–339, Dec. 2000.

[41] J.-P. Bérenger, "Application of the CFS PML to the absorption of evanescent waves in waveguides," *IEEE Microw. Wireless Compon. Lett.*, vol. 12, no. 6, pp. 218–220, June 2002.

[42] D. Correia and J.-M. Jin, "A simple and efficient implementation of CFS-PML in the FDTD analysis of periodic structures," *IEEE Microw. Wireless Compon. Lett.*, vol. 15, no. 7, pp. 487–489, July 2005.

[43] D. S. Katz, E. T. Thiele, and Taflove, "Validation and extension to three dimensions of the Bérenger PML absorbing boundary conditions," *IEEE Microw. Guided Wave Lett.*, vol. 4, no. 8, pp. 268–270, Aug. 1994.

[44] W. V. Andrew, C. A. Balanis, and P. A. Tirkas, "A comparison of the Bérenger perfectly matched layer and the Lindman higher-order ABC's for the FDTD method," *IEEE Microw. Guided Wave Lett.*, vol. 5, no. 6, pp. 192–194, June 1995.

[45] N. V. Kantartzis and T. D. Tsiboukis, "A comparative study of the Berenger perfectly matched layer, the superabsorption technique and several higher-order ABC's for the FDTD algorithm in two and three dimensional problems," *IEEE Trans. Magn.*, vol. 33, no. 2, pp. 1460–1463, Mar. 1997.

[46] O. M. Ramahi and J. B. Schneider, "Comparative study of the PML and C-COM mesh-truncation techniques," *IEEE Microw. Guided Wave Lett.*, vol. 8, no. 2, pp. 55–57, Feb. 1998.

[47] A. R. Roberts and J. Joubert, "PML absorbing boundary condition for higher-order FDTD schemes," *Electron. Lett.*, vol. 33, no. 1, pp. 32–34, Jan. 1997.

[48] P. G. Petropoulos, L. Zhao, and A. C. Cangellaris, "A reflectionless sponge layer absorbing boundary condition for the solution of Maxwell's equations with high-order staggered finite difference schemes," *J. Comput. Phys.*, vol. 139, no. 1, pp. 184–208, Jan. 1998.

[49] N. V. Kantartzis and T. D. Tsiboukis, "A generalised methodology based on higher-order conventional and nonstandard FDTD concepts for the systematic development of enhanced dispersionless wide-angle absorbing perfectly matched layers," *Int. J. Numer. Model.*, vol. 13, no 5, pp. 417–440, Sep.–Oct. 2000.

[50] M. Fujii, M. M. Tentzeris, and P. Russer, "Performance of nonlinear dispersive APML in high-order FDTD schemes," *in Proc. IEEE Microw. Theory Tech. Int. Symp.*, Philadelphia, PA, June 2003, vol. 2, pp. 1129–1132.

[51] H. A. Jamid, "Enhanced PML performance using higher-order approximation," *IEEE Trans. Microw. Theory Tech.*, vol. 52, no. 4, pp. 1166–1174, Apr. 2004.

[52] M. H. Carpenter, D. Gottlieb, and S. Abarbanel, "Time-stable boundary conditions for finite-difference schemes solving hyperbolic systems: Methodology and application to high-order compact schemes," *J. Comput. Phys.*, vol. 111, no. 2, pp. 220–236, Apr. 1994.

[53] Z. Zhang, "An explicit fourth-order compact finite difference scheme for the three-dimensional convection-diffusion equation," *Commun. Numer. Meth. Engng.*, vol. 14, pp. 219–218, 1998.

[54] D. V. Gaitonde, J. S. Shang, and J. L. Young, "Practical aspects of higher-order numerical schemes for wave propagation phenomena," *Int. J. Numer. Meth. Engng.*, vol. 45, no. 12, pp. 1849–1869, Aug. 1999.

[55] E. Turkel and A. Yefet, "Fourth-order compact implicit method for the Maxwell equations with discontinuous coefficients," *Appl. Numer. Math.*, vol. 33, nos. 1–4, pp. 125–134, May 2000.

[56] S. S. Abarbanel and A. E. Chertock, "Strict stability of high-order compact implicit finite-difference schemes: The role of boundary conditions for hyperbolic PDEs, I," *J. Comput. Phys.*, vol. 160, no. 1, pp. 42–66, May 2000.

[57] S. S. Abarbanel, A. E. Chertock, and A. Yefet, "Strict stability of high-order compact implicit finite-difference schemes: The role of boundary conditions for hyperbolic PDEs, II," *J. Comput. Phys.*, vol. 160, no. 1, pp. 67–87, May 2000.

[58] S. E. Sherer and J. N. Scott, "High-order compact finite-difference methods on general overset girds," *J. Comput. Phys.*, vol. 210, no. 2, pp. 459–496, Dec. 2005.

[59] K. H. Dridi, J. S. Hesthaven, and A. Ditkowski, "Staircase-free finite-difference time-domain formulation for general materials in complex geometries," *IEEE Trans. Antennas Propag.*, vol. 49, no. 5, pp. 749–758, May 2001.

[60] H. Spachmann, R. Schuhmann, and T. Weiland, "Higher-order spatial operators for the finite integration techniue," *ACES J.*, vol. 17, no. 1, pp. 11–22, Mar. 2002.

CHAPTER 5

Structural Extensions and Temporal Integration

5.1 INTRODUCTION

This chapter deals with the extension of higher order FDTD schemes to lossy and dispersive media along with their structural enhancement, from both a spatial- and temporal-approximation point of view. As competent as existing formulations are, there still exist several types of realistic problems that necessitate additional robustness. Therefore, it seems natural to look for alternatives, permitting one to preserve global accuracy and convergence in such situations. The main observation, herein, is that the efficiency of a higher order method is closely related to the smoothness of the solution. However, when the aforementioned applications are to be numerically solved, the global smoothness is generally decreased and the anticipated profits are degraded. Techniques to overcome these difficulties and establish a consistent modeling process remain active research areas. Actually, diverse approaches have been, so far, reported toward this direction with some promising outcomes. In the following sections, the lossy and frequency-dependent implementation of higher order forms is discussed and verified by some indicative examples. Subsequently, the improvement of spatial operators is conducted by means of material-parameter correction concepts, systematic stencil management and algorithms based on the use of dispersion relation. Finally, the popular leapfrog and Runge–Kutta temporal integrators are generalized under a unified theoretical framework in order to attain compatible orders of accuracy with their spatial counterparts, without discrepancies or instabilities.

5.2 MODELING OF LOSSY AND DISPERSIVE MEDIA WITH HIGHER ORDER FDTD SCHEMES

The incorporation of lossy and frequency-dependent materials in the production of modern devices, such as patch antennas, waveguides, or integrated circuits, has become a topic of intensive studies due to several attractive properties. Toward this direction, Yee's algorithm has already received the suitable modifications to cope with lossy [1–4] and dispersive configurations [5–19]. Given the competence of higher order schemes in material simulation, a possible extension would be fairly useful. Therefore, this section presents a set of techniques that embody the current trends of conventional and nonstandard finite differencing.

5.2.1 Lossy Formulations

Consider a conducting medium of electric conductivity σ and relative permittivity ε_r. Implementing the conventional (2, 4) FDTD schemes, described in Section 2.3, for the discretization of Maxwell's equations, (2.1)–(2.4), the three-dimensional (3-D) expression of E_x component [20] is given by

$$\begin{aligned} E_x\big|_{i+1/2,j,k}^{n+1} &= K E_x\big|_{i+1/2,j,k}^{n} \\ &-M_z\left(H_y\big|_{i+1/2,j,k-3/2}^{n+1/2} - 27H_y\big|_{i+1/2,j,k-1/2}^{n+1/2} + 27H_y\big|_{i+1/2,j,k+1/2}^{n+1/2} - H_y\big|_{i+1/2,j,k+3/2}^{n+1/2}\right) \\ &+M_y\left(H_z\big|_{i+1/2,j-3/2,k}^{n+1/2} - 27H_z\big|_{i+1/2,j-1/2,k}^{n+1/2} + 27H_z\big|_{i+1/2,j+1/2,k}^{n+1/2} - H_z\big|_{i+1/2,j+3/2,k}^{n+1/2}\right), \end{aligned} \tag{5.1}$$

where

$$K = \frac{\varepsilon_r\varepsilon_0 - \frac{1}{2}\sigma\Delta t}{\varepsilon_r\varepsilon_0 + \frac{1}{2}\sigma\Delta t}, \qquad M_\zeta = \frac{\Delta t}{24\Delta\zeta(\varepsilon_r\varepsilon_0 + \frac{1}{2}\sigma\Delta t)}, \qquad \text{for } \zeta \in (x, y, z).$$

Obviously the higher order nonstandard operators could also be applied in a similar sense. For the treatment of nodes located near boundaries or interfaces, the idea of one-sided approximations explained in Section 2.4.3 is employed. The stable profile of the algorithm is guaranteed by

$$\Delta t \le \frac{6}{7}\sqrt{\mu_0\varepsilon_0\varepsilon_r}\left[\sum_{\zeta=x,y,z}\frac{1}{(\Delta\zeta)^2}\right]^{-1/2} \tag{5.2}$$

while the dispersion relation, along the θ and φ propagation directions, has the form of

$$\begin{aligned} &\frac{\sin^2(\omega\Delta t/2)}{(\upsilon_0\Delta t)^2}\left[\varepsilon_r - \frac{j\sigma\Delta t}{2\varepsilon_0\tan(\omega\Delta t/2)}\right] \\ &\quad = \sum_{\zeta=x,y,z}\frac{1}{(\Delta\zeta)^2}\left[\frac{9}{8}\sin\left(\frac{k_\zeta^{\text{num}}\Delta\zeta}{2}\right) - \frac{1}{24}\sin\left(\frac{3k_\zeta^{\text{num}}\Delta\zeta}{2}\right)\right]^2, \end{aligned} \tag{5.3}$$

where $j^2 = -1$, $\upsilon_0 = 1/\sqrt{\mu_0\varepsilon_0}$, and k^{num} signifies the numerical wavenumber.

To examine the accuracy of the technique, the ratio of numerical to physical phase velocity, $\upsilon^{\text{num}}/\upsilon$, is examined for a medium of $\sigma = 0.1$ S/m and a uniform lattice with $\Delta\zeta = 0.1$ m. Solving (5.3) for k^{num} via Newton's iterative method, the results for three different grid resolutions, $N_\lambda = \lambda/\Delta\zeta$ and $\theta = \pi/2$, are shown in Figure 5.1(a), where the efficacy of the higher order schemes is obvious. The next example takes into account a parallel-plate waveguide

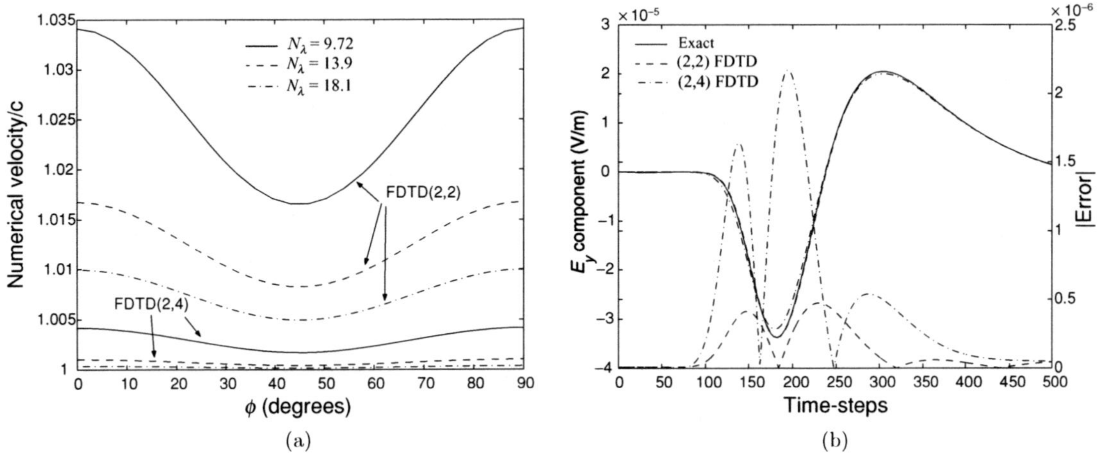

FIGURE 5.1: (a) Ratio of numerical to physical phase velocity, $\upsilon^{\mathrm{num}}/\upsilon$, versus propagation angle φ for second- and fourth-order FDTD implementations. (b) Electric field component (left axis) of analytic and FDTD solutions as well as their absolute errors (right axis) at an observation point $20\Delta x$ from the source

of thickness w occupied by a lossy dielectric material of conductivity $\sigma = 0.05$ S/m and $\varepsilon_{\mathrm{r}} = 1$. Assuming that E_y is determined over the structure's cross section at $x = 0$, the study is limited to the TM_1 mode. The two-dimensional (2-D) FDTD space is truncated by a four-cell perfectly matched layer, while the excitation is given by $E_y(0, y, t) = \cos(\pi y/w)p(t)$ with $p(t) = \left[1 - 2\pi^2 f_{\mathrm{R}}^2 (t - t_{\mathrm{R}})^2\right] \exp\left[-\pi^2 f_{\mathrm{R}}^2 (t - t_{\mathrm{R}})^2\right]$ in which f_{R} is the peak frequency and t_{R} the time delay. The rest of its parameters are $\Delta\zeta = 0.2$ m, $w = 20\Delta y$, $f_{\mathrm{R}} = 1/t_{\mathrm{R}} = f_{\mathrm{c}}$, and $f_{\mathrm{c}} = \upsilon_0/(2w\sqrt{\varepsilon_{\mathrm{r}}})$ for the cutoff frequency. Analytical solution is calculated by means of discrete Fourier series. Figure 5.1(b) presents the E_y component (left axis) and the absolute errors (right axis) of the (2, 2) and (2, 4) lossy FDTD solutions. The agreement between the latter and the exact one as well as its significantly reduced error are indeed promising. Finally, it is stressed that the prior formulation does not increase the total overhead, since all additional calculations are performed at a preprocessing stage.

5.2.2 Dispersive Media

In this case, the discretization process initiates from the fourth-order spatial operators:

$$\frac{1}{\Delta\zeta}\,\delta_\zeta\left[f\,|_l^n\right] = \frac{1}{24\Delta\zeta}\left(f\,|_{l-3/2}^n - 27\,f\,|_{l-1/2}^n + 27\,f\,|_{l+1/2}^n - f\,|_{l+3/2}^n\right). \tag{5.4}$$

Temporal differentiation, on the other hand, is approximated by central finite-difference (δ_t, δ_{2t}, δ_t^2) and central average (μ_t, μ_{2t}) operators defined in Table 5.1 [21].

TABLE 5.1: Temporal Approximations

OPERATOR TYPE	TIME DOMAIN	Z DOMAIN	FREQUENCY DOMAIN
$\delta_t[f\|^n]$	$f\|^{n+1/2} - f\|^{n-1/2}$	$Z^{1/2} - Z^{-1/2}$	$2j\sin(\omega\Delta t/2)$
$\delta_{2t}[f\|^n]$	$\frac{9}{8}\left(f\|^{n+1} - f\|^{n-1}\right)$	$\frac{1}{2}\left(Z - Z^{-1}\right)$	$j\sin(\omega\Delta t)$
$\mu_t[f\|^n]$	$\frac{1}{2}\left(f\|^{n+1/2} + f\|^{n-1/2}\right)$	$\frac{1}{2}\left(Z^{1/2} + Z^{-1/2}\right)$	$\cos(\omega\Delta t/2)$
$\mu_{2t}[f\|^n]$	$\frac{1}{2}\left(f\|^{n+1} + f\|^{n-1}\right)$	$\frac{1}{2}\left(Z + Z^{-1}\right)$	$\cos(\omega\Delta t)$
$\delta_t^2[f\|^n]$	$f\|^{n+1} - 2f\|^n + f\|^{n-1}$	$Z + Z^{-1} - 2$	$-4\sin^2(\omega\Delta t/2)$

The simulation of dispersive media is based on the auxiliary differential equation technique [6], while the frequency-dependent constitutive relation $\mathbf{D}(\mathbf{r},\omega) = \varepsilon(\omega)\mathbf{E}(\mathbf{r},\omega)$ governs wave propagation in their interior. Let us now investigate three different cases of such materials characterized by diverse $\varepsilon(\omega)$.

First, assume the general *M*th-order *Debye* medium, the complex permittivity of which is expressed as

$$\varepsilon(\omega) = \varepsilon_\infty + \sum_{m=1}^{M} \frac{\varepsilon_{sm} - \varepsilon_\infty}{1 + j\omega\tau_m}, \tag{5.5}$$

where $\varepsilon_s = \varepsilon(0)$, $\varepsilon_\infty = \varepsilon(\infty)$, and τ_m is the relaxation time constant corresponding to the mth pole. Introducing M supplementary electric field polarization terms $\mathbf{P}_\eta$ as

$$\mathbf{D} = \varepsilon_\infty \mathbf{E} + \sum_{\eta=1}^{M} \mathbf{P}_\eta, \tag{5.6}$$

$$P_\eta = \frac{\varepsilon_{sm} - \varepsilon_\infty}{1 + j\omega\tau_m}\mathbf{E} \tag{5.7}$$

leads to the subsequent differential equation for the ηth pole:

$$\mathbf{P}_\eta + \tau_\eta \frac{\partial \mathbf{P}_\eta}{\partial t} = (\varepsilon_{s\eta} - \varepsilon_\infty)\mathbf{E} \xrightarrow[\text{form}]{\text{discrete}} (\mu_t + \tau_\eta\delta_t)\mathbf{P}_\eta^n = (\varepsilon_{s\eta} - \varepsilon_\infty)\mu_t\mathbf{E}^n, \tag{5.8}$$

for $\eta = 1, 2, \ldots, M$. If (5.8) is combined with (5.6), the final update expression for the electric field becomes

$$\mathbf{E}^{n+1} = \frac{1}{\vartheta_4}\left[\mathbf{D}^{n+1} - \sum_{\eta=1}^{M} \frac{1}{\vartheta_1^\eta}\left(\vartheta_2^\eta \mathbf{P}_\eta^n + \vartheta_3^\eta \mathbf{E}^n\right)\right], \tag{5.9}$$

with

$$\vartheta_{1,2}^{\eta} = 2\tau_{\eta} \pm \Delta t, \qquad \vartheta_3^{\eta} = (\varepsilon_{s\eta} - \varepsilon_{\infty})\Delta t, \qquad \vartheta_4 = \varepsilon_{\infty} + \sum_{\eta=1}^{M} \frac{\varepsilon_{s\eta} - \varepsilon_{\infty}}{2\tau_{\eta} + \Delta t}\Delta t,$$

whereas the summation quantities $\mathbf{P}_{\eta}$ at $n+1$ time-step are computed via

$$\mathbf{P}_{\eta}^{n+1} = \frac{1}{\vartheta_1^{\eta}}\left[\vartheta_2^{\eta}\mathbf{P}_{\eta}^{n} + \vartheta_3^{\eta}\left(\mathbf{E}^{n+1} + \mathbf{E}^{n}\right)\right]. \tag{5.10}$$

Proceeding to an *M*th-order *Lorentz* medium with a resonant frequency ω_m and a damping coefficient α_m, the macroscopic permittivity function is described by

$$\varepsilon(\omega) = \varepsilon_{\infty} + (\varepsilon_{sm} - \varepsilon_{\infty})\sum_{m=1}^{M} \frac{G_m\omega_m^2}{\omega_m^2 + 2j\omega\alpha_m - \omega^2} \quad \text{with} \sum_{m=1}^{M} G_m = 1. \tag{5.11}$$

In similarity with (5.6)–(5.8), the respective difference equation for the ηth pole is

$$\left[\omega_{\eta}^2\mu_{2t} + \frac{2\alpha_{\eta}}{\Delta t}\delta_t\mu_t + \frac{\delta_t^2}{(\Delta t)^2}\mu_{2t}\right]\mathbf{P}_{\eta}^{n} = G_{\eta}(\varepsilon_s - \varepsilon_{\infty})\omega_{\eta}^2\mu_{2t}\mathbf{E}^{n}, \tag{5.12}$$

yielding

$$\mathbf{E}^{n+1} = \frac{1}{\vartheta_4}\left[\mathbf{D}^{n+1} - \sum_{\eta=1}^{M} \frac{1}{\vartheta_1^{\eta}}\left(4\mathbf{P}_{\eta}^{n} - \vartheta_2^{\eta}\mathbf{P}_{\eta}^{n-1} + \vartheta_3^{\eta}\mathbf{E}^{n-1}\right)\right], \tag{5.13}$$

with

$$\vartheta_{1,2}^{\eta} = \omega_{\eta}^2(\Delta t)^2 \pm 2\alpha_{\eta}\Delta t + 2, \qquad \vartheta_3^{\eta} = G_{\eta}(\varepsilon_s - \varepsilon_{\infty})\omega_{\eta}^2(\Delta t)^2, \qquad \vartheta_4 = \varepsilon_{\infty} + \sum_{\eta=1}^{M}\frac{\vartheta_3^{\eta}}{\vartheta_1^{\eta}}.$$

After updating $\mathbf{E}^{n+1}$, through Ampère's law for $\mathbf{D}^{n+1}$, the values of $\mathbf{P}_{\eta}^{n+1}$ are obtained by

$$\mathbf{P}_{\eta}^{n+1} = \frac{1}{\vartheta_1^{\eta}}\left[4\mathbf{P}_{\eta}^{n} - \vartheta_2^{\eta}\mathbf{P}_{\eta}^{n-1} + \vartheta_3^{\eta}\left(\mathbf{E}^{n+1} + \mathbf{E}^{n-1}\right)\right]. \tag{5.14}$$

When a *Drude* material is considered, its model related to cold plasma [10, 11], is described by

$$\varepsilon(\omega) = \varepsilon_0\left[1 + \frac{\omega_{\mathrm{p}}^2}{\omega(j\nu_{\mathrm{c}} - \omega)}\right], \tag{5.15}$$

with ω_{p} the radian plasma and ν_{c} the collision frequency. The governing differential equation is

$$\nu_c\frac{\partial\mathbf{D}}{\partial t} + \frac{\partial^2\mathbf{D}}{\partial t^2} = \varepsilon_0\left(\omega_{\mathrm{p}}^2\mathbf{E} + \nu_{\mathrm{c}}\frac{\partial\mathbf{E}}{\partial t} + \frac{\partial^2\mathbf{E}}{\partial t^2}\right), \tag{5.16}$$

which if discretized and solved for $\mathbf{E}^{n+1}$ gives

$$\mathbf{E}^{n+1} = \tfrac{1}{\varepsilon_0}\left\{4\varepsilon_\infty \mathbf{E}^n - \varepsilon_0\left[(\omega_\mathrm{p}\Delta t)^2 - \nu_\mathrm{c}\Delta t + 2\right]\mathbf{E}^{n-1} + (2+\nu_\mathrm{c}\Delta t)\mathbf{D}^{n+1} \right.$$
$$\left. - 4\mathbf{D}^n + (2-\nu_\mathrm{c}\Delta t)\mathbf{D}^{n-1}\right\} / \left[(\omega_\mathrm{p}\Delta t)^2 + \nu_\mathrm{c}\Delta t + 2\right]. \tag{5.17}$$

The stability attributes of the aforementioned method are explored via the combined von Neumann and Routh–Hurwitz technique [2]. According to this approach, the error that appears during the computation of any field component is described by a single term of a Fourier series expansion:

$$f|^n = f_0 Z^n e^{j\sum_{\zeta=x,y,z} k_\zeta^{\mathrm{num}} n\Delta\zeta}, \tag{5.18}$$

where the complex variable Z is associated to the growth factor of the error. Conducting the necessary mathematical analysis, the stability criterion for an Mth-order Debye and Lorentz medium are

$$\Delta t_{\mathrm{Debye}} \le \frac{6}{7}\sqrt{\mu\varepsilon_\infty}\left[\sum_{\zeta=x,y,z}\frac{1}{(\Delta\zeta)^2}\right]^{-1/2}$$
$$\Delta t_{\mathrm{Lorentz}} \le \frac{6}{7\sqrt{2}}\sqrt{\mu\varepsilon_\infty}\left[\sum_{\zeta=x,y,z}\frac{1}{(\Delta\zeta)^2}\right]^{-1/2}. \tag{5.19}$$

To facilitate the derivation of the dispersion relation, the discrete counterpart of the continuous permittivity, i.e., the *numerical permittivity*, is defined as the ratio of the discrete values of $\mathbf{D}$ and $\mathbf{E}$. So, in analogy to the continuous case $\mu_0\varepsilon^{\mathrm{num}}(\omega^{\mathrm{num}})(\omega^{\mathrm{num}})^2 = \mathbf{k}^{\mathrm{num}}\cdot\mathbf{k}^{\mathrm{num}}$. Employing the conventions of the third column of Table 5.1, acquired if $Z = \exp(j\omega\Delta t)$, the complex permittivity function for the Lorentz medium is

$$\varepsilon^{\mathrm{num}}_{\mathrm{Lorentz}} = \varepsilon_\infty + (\varepsilon_s - \varepsilon_\infty)\frac{(\omega_0^{\mathrm{num}})^2}{(\omega_0^{\mathrm{num}})^2 + 2j\omega^{\mathrm{num}}\alpha_0^{\mathrm{num}} - (\omega^{\mathrm{num}})^2}, \tag{5.20}$$

with $\omega^{\mathrm{num}} = \frac{2}{\Delta t}\sin(\omega\Delta t/2)$, $\alpha_0^{\mathrm{num}} = \cos(\omega\Delta t/2)$, and $\omega_0^{\mathrm{num}} = \omega_0\sqrt{\cos(\omega\Delta t)}$. Likewise, for the Debye and Drude materials, one gets

$$\varepsilon^{\mathrm{num}}_{\mathrm{Debye}} = \varepsilon_\infty + \frac{\varepsilon_\mathrm{s} - \varepsilon_\infty}{1 + j\omega^{\mathrm{num}}\tau^{\mathrm{num}}} \quad \text{and} \quad \varepsilon^{\mathrm{num}}_{\mathrm{Drude}} = \varepsilon_0\left[1 + \frac{(\omega_\mathrm{p}^{\mathrm{num}})^2}{\omega^{\mathrm{num}}(j\nu_\mathrm{c}^{\mathrm{num}} - \omega^{\mathrm{num}})}\right], \tag{5.21}$$

for $\tau^{\mathrm{num}} = \tau/\cos(\omega\Delta t/2)$, $\omega_\mathrm{p}^{\mathrm{num}} = \omega_\mathrm{p}\sqrt{\cos(\omega\Delta t)}$, and $\nu_\mathrm{c}^{\mathrm{num}} = \nu_\mathrm{c}\cos(\omega\Delta t/2)$. In this context, and without loss of generality, the one-dimensional dispersion relation for the (2, 4) and (2, 6) dispersive FDTD scheme is

$$\tfrac{2}{\Delta h}\left[\tfrac{9}{8}\sin(S) - \tfrac{1}{24}\sin(3S)\right] = \omega^{\mathrm{num}}\sqrt{\mu_0\varepsilon^{\mathrm{num}}}, \tag{5.22}$$

$$\tfrac{2}{\Delta h}\left[\tfrac{75}{64}\sin(S) - \tfrac{35}{384}\sin(3S) + \tfrac{3}{640}\sin(5S)\right] = \omega^{\mathrm{num}}\sqrt{\mu_0\varepsilon^{\mathrm{num}}}, \tag{5.23}$$

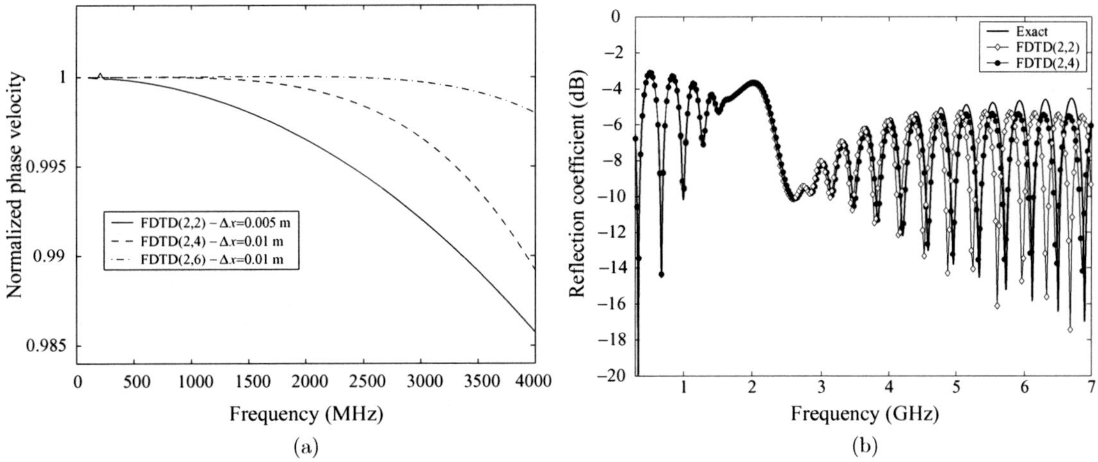

FIGURE 5.2: (a) Normalized phase velocity and (b) reflection coefficient versus frequency for a second-order Lorentz medium

where $S = k^{\text{num}}\Delta h/2$. Note that k^{num} is assumed to be real and if the right-hand side of (5.22) or (5.23) is complex, only the real part is considered. Moreover, since these expressions are not in a closed form, to obtain the numerical wavenumber, the iterative Newton's method is usually implemented.

As an indication of the algorithm's efficiency, compared to the second-order one, an air slab placed inside a second-order Lorentz-type medium is studied. The material's resonant frequency is $\omega_0 = 2 \times 10^9$rad/s, its damping coefficient is $\alpha_0 = 0.1\omega_0$, while $\varepsilon_\infty = 2.25\varepsilon_0$ and $\varepsilon_s = 3\varepsilon_0$. The domain has 2000 cells and the slab occupies the region from the 700th to the 750th cell. Figure 5.2(a) illustrates the normalized phase velocity $\upsilon^{\text{num}}/\upsilon$ for the $(2,4)$ and $(2,6)$ FDTD case, specifying their superiority over the usual approach. Also, Figure 5.2(b) shows the slight deviations of the $(2,4)$ reflection coefficient from the exact one, even for frequencies well above the resonant one, where the second-order scheme lacks to provide adequate results.

Next, the same configuration is modeled for a third-order Debye material in which the characteristic parameters of its three poles are $\varepsilon_{s1} = 3\varepsilon_0$, $\tau_1 = 9.4 \times 10^{-9}\ \text{s}^{-1}$, $\varepsilon_{s2} = 2\varepsilon_0$, $\tau_2 = 10^{-10}\ \text{s}^{-1}$, and $\varepsilon_{s3} = \varepsilon_0$, $\tau_3 = 10^{-6}\ \text{s}^{-1}$. The electric field in the time domain is presented in Figure 5.3(a), confirming again the improved accuracy of the higher order technique. Finally, an infinite-height cylinder made of cold plasma is placed in a 200×200-cell air region. The structure is excited by a modulated Gaussian pulse centered at 20 GHz, while the plasma frequency of the Drude-type medium is $\omega_\text{p} = 5.74$ GHz. Figure 5.3(b) provides a snapshot of the electric field at a time instant when the incident wave has already been scattered. Observing this outcome, one can easily discern the smoothness of the simulation without any undesired artificialities.

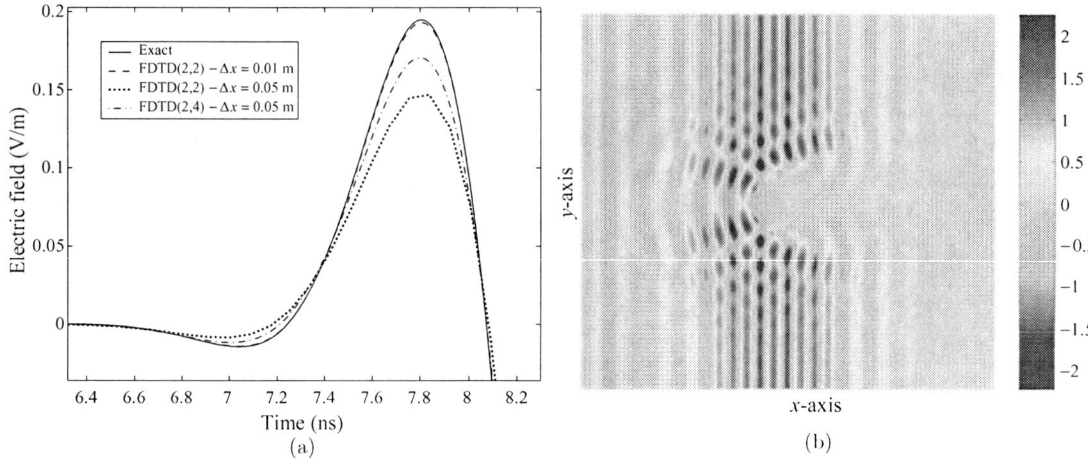

FIGURE 5.3: (a) Evolution of the electric field versus time for a third-order Debye medium. (b) Snapshot of the electric field for a Drude-type material

5.3 IMPROVEMENT VIA THE CORRECTION OF MATERIAL PARAMETERS

As already mentioned, the conventional (2, 4) FDTD scheme exhibits some nontrivial phase discrepancies due to the inaccurate value of numerical velocity. This shortcoming may be assessed as a lack of finite differencing to sufficiently resolve the profile of isotropic materials, especially in an FDTD grid. Hence, it is reasonable to consider an anisotropic medium, whose presence is likely to counterbalance the artificial anisotropy and subsequently alleviate field distortion. Rather than attempting to reformulate spatial operators or widen their stencil (see Chapter 2), the particular approach opts for diagonally anisotropic materials, the manipulation of which does not mandate cumbersome modifications in existing computer implementations [22, 23]. Their use offers a significant precision around a certain frequency, yet the total error remains small at a large bandwidth. Thus, the algorithm can be proven practical for broadband applications as well.

Let us focus on the 2-D TE case and a computational domain filled with a diagonally anisotropic material ($[\varepsilon] = \varepsilon_0 \operatorname{diag}[\varepsilon_{rx}, \varepsilon_{ry}]$, $\mu = \mu_0$). The respective dispersion relation for $S_\zeta = k_\zeta^{\mathrm{num}} \Delta\zeta/2$, $\zeta \in (x, y)$, is

$$\left[\frac{\sin(\omega\Delta t/2)}{\upsilon_0 \Delta t}\right]^2 = \frac{1}{\varepsilon_{ry}}\left[\frac{A_x \sin(S_x) + B_x \sin(3S_x)}{\Delta x}\right]^2 + \frac{1}{\varepsilon_{rx}}\left[\frac{A_y \sin(S_y) + B_y \sin(3S_y)}{\Delta y}\right]^2, \tag{5.24}$$

where spatial differentiation has been conducted by the general operator of (2.107). To determine the features of the material that optimally approximates free space, numerical wavenumber is enforced to be equal to its exact value for two different angles of propagation, φ_1 and φ_2. In view of this requirement and via (5.24), a system of two equations is constructed, whose solution permits the evaluation of ε_{rx} and ε_{ry}. So,

$$\varepsilon_{rx} = \left[\frac{2}{\omega \overline{\Delta t}} \sin^{-1} \left(\upsilon_0 \overline{\Delta t} \sqrt{\frac{\mathcal{X}_1^2 + b\mathcal{Y}_1^2}{b}} \right) \right]^2 \quad \text{and} \quad \varepsilon_{ry} = b\varepsilon_{rx}, \tag{5.25}$$

with

$$\overline{\Delta t} = Q \left[\upsilon_0 \sqrt{\frac{(|A_x| + |B_x|)^2}{b(\Delta x)^2} + \frac{(|A_y| + |B_y|)^2}{(\Delta y)^2}} \right]^{-1} \quad \text{for } Q \leq 1,\ b = \frac{\mathcal{X}_2^2 - \mathcal{X}_1^2}{\mathcal{Y}_1^2 - \mathcal{Y}_2^2}, \tag{5.26}$$

and

$$\begin{aligned} \mathcal{X}_m &= \frac{1}{\Delta x} \left[A_x \sin\left(S_x^{2\mathrm{D}} \right) + B_x \sin\left(3 S_x^{2\mathrm{D}} \right) \right], \\ \mathcal{Y}_m &= \frac{1}{\Delta y} \left[A_y \sin\left(S_y^{2\mathrm{D}} \right) + B_y \sin\left(3 S_y^{2\mathrm{D}} \right) \right], \end{aligned} \tag{5.27}$$

for $S_x^{2\mathrm{D}} = k \cos \varphi_m \Delta x/2$, $S_y^{2\mathrm{D}} = k \sin \varphi_m \Delta y/2$, and $m = 1, 2$. The calculation of the unknown parameters involves two steps. First, through the values of (5.25) for the material constants, a modified numerical velocity is obtained with a mean deviation from υ_0, denoted as $\Delta_{\mathrm{mean}} = \upsilon_{\mathrm{mean}}^{\mathrm{num}} - \upsilon_0$. During the second step, the material parameters are again determined from (5.25) by incorporating the slightly changed value of $\upsilon_0 - \Delta_{\mathrm{mean}}$ in an effort to derive a phase velocity with the *minimum mean error*. Consequently, the initial medium is substituted by a novel one that provides a numerical phase velocity much closer to its physical value at a preselected frequency. Evidently, in the case of a material different from free space ($\varepsilon,\ \mu \neq \varepsilon_0,\ \mu_0$), the preceding procedure is the same with the sole exception of replacing υ_0 with $\upsilon = 1/\sqrt{\mu\varepsilon}$.

In the above, coefficients A_ζ and B_ζ are acquired from the prerequisite that the corresponding operator (2.107) produces exact results when applied to plane-wave expressions along two prefixed directions of propagation. To fully optimize their wideband action, the most convenient directions are those along the differentiation axis ($\varphi_1 = 0^\circ,\ 90^\circ$) and the grid's diagonal ($\varphi_2 = 45^\circ$). These choices form a system of equations that gives the values of A_ζ and B_ζ for a certain mesh resolution.

Extension to 3-D problems is uncomplicated, with the dispersion error seriously decreased through the replacement of isotropic materials with others that can be both electrically and magnetically anisotropic. Thus, if $[\varepsilon_{\mathrm{r}}] = \mathrm{diag}[\varepsilon_{\mathrm{r}x},\ \varepsilon_{\mathrm{r}y},\ \varepsilon_{\mathrm{r}z}] = [\mu_{\mathrm{r}}]$, the inherent impedance of the physical medium is not modified, while the imposition of relation $\mathbf{k}^{\text{num}} = \mathbf{k}$ for three directions of propagation leads to

$$\varepsilon_{rx} = \frac{2}{\omega\overline{\Delta t}} \sin^{-1}\left(\upsilon_0 \overline{\Delta t}\sqrt{\frac{\mathcal{X}_1^2 + b_1\mathcal{Y}_1^2 + b_2\mathcal{Z}_1^2}{b_1 b_2}}\right), \qquad \varepsilon_{ry} = b_1\varepsilon_{rx}, \qquad \varepsilon_{rz} = b_2\varepsilon_{rx}. \quad (5.28)$$

with

$$\overline{\Delta t} = Q\left[\upsilon_0\sqrt{\frac{(|A_x| + |B_x|)^2}{b_1 b_2(\Delta x)^2} + \frac{(|A_y| + |B_y|)^2}{b_2(\Delta y)^2} + \frac{(|A_z| + |B_z|)^2}{b_1(\Delta z)^2}}\right]^{-1} \quad \text{for } Q \leq 1, \quad (5.29)$$

$$b_1 = -\left(\frac{\mathcal{X}_1^2 - \mathcal{X}_2^2}{\mathcal{Z}_1^2 - \mathcal{Z}_2^2} - \frac{\mathcal{X}_1^2 - \mathcal{X}_3^2}{\mathcal{Z}_1^2 - \mathcal{Z}_3^2}\right)\left(\frac{\mathcal{Y}_1^2 - \mathcal{Y}_2^2}{\mathcal{Z}_1^2 - \mathcal{Z}_2^2} - \frac{\mathcal{Y}_1^2 - \mathcal{Y}_3^2}{\mathcal{Z}_1^2 - \mathcal{Z}_3^2}\right)^{-1},$$

$$b_2 = -\frac{\mathcal{X}_1^2 - \mathcal{X}_2^2}{\mathcal{Z}_1^2 - \mathcal{Z}_2^2} - \frac{\mathcal{Y}_1^2 - \mathcal{Y}_2^2}{\mathcal{Z}_1^2 - \mathcal{Z}_2^2} b_1, \quad (5.30)$$

$$\text{and } \mathcal{W}_m = \tfrac{1}{\Delta w}\left[A_w \sin\left(S_w^{3\text{D}}\right) + B_w \sin\left(3S_w^{3\text{D}}\right)\right], \text{ for } \mathcal{W} = \mathcal{X},\ \mathcal{Y},\ \mathcal{Z},\ w \in (x,\ y,\ z),$$
$$S_x^{3\text{D}} = \sin\theta_m S_x^{2\text{D}},\ S_y^{3\text{D}} = \sin\theta_m S_y^{2\text{D}},\ S_z^{3\text{D}} = k\cos\theta_m \Delta z/2, \text{ and } m = 1,\ 2,\ 3. \quad (5.31)$$

The performance of the material-correction schemes is examined via various 2- and 3-D examples. Figure 5.4(a) presents the phase-velocity error of the (2, 4) FDTD, before and after the material modification, for a cubic lattice with a size of $\lambda/8$. The selected angles of propagation (θ_m, φ_m) are (90°, 0°), (90°, 90°), and (0°, 90°). As can be observed from the different axis scaling of the two figures, the improvement is quite satisfactory and most importantly, all

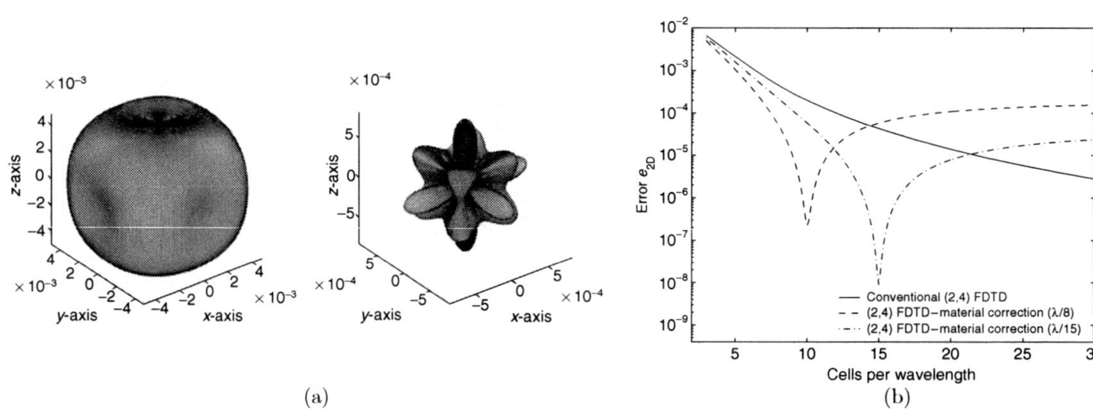

FIGURE 5.4: (a) Phase-velocity error of the conventional (2, 4) and the material-correction FDTD method, respectively. (b) Wideband performance of various (2, 4) FDTD implementations

optimal material parameters have been evaluated for the *maximum allowable time-step*, despite the fact that the latter is unknown in advance.

Although the nature of the material-correction method is basically single-frequency, the example of Figure 5.4(b) reveals its wideband capabilities as well. In particular, the error quantity of (2.75a), for all angles of propagation, is plotted for various mesh resolutions. As anticipated, the improvement around the preselected frequencies is indeed notable. Nonetheless, this behavior is preserved for coarser discretizations, proving that the technique is competent of confronting broadband phenomena with sufficient accuracy.

5.4 ENHANCED SPATIAL APPROXIMATIONS

An essential factor in the construction of accurate higher order FDTD techniques is the *correct stencil manipulation* provided by the respective spatial operators. Thus, apart from the most frequently encountered schemes, discussed in the previous chapters, a variety of rigorous approaches have also been developed [24–30]. As an indication of these interesting trends, this section presents three algorithms which, based on different principles, attempt to improve the behavior of higher order spatial sampling and approximation.

5.4.1 Alternative Stencil Management

Starting from spatial derivative $\partial^2/\partial\zeta^2$, for $\zeta \in (x, y, z)$, its higher order approximant, in accordance with the analysis of [27], is expressed as

$$\left.\frac{\partial^2 f}{\partial \zeta^2}\right|_l^n = \frac{1}{(\Delta\zeta)^2} \sum_{s=-(2M-1)}^{2M-1} c_{|s|}\, f|_{s+l}^n = \mathbf{D}_{\mathrm{A}}^{(2M)}\left[f|_l^n\right] \mathbf{D}_{\mathrm{B}}^{(2M)}\left[f|_l^n\right], \tag{5.32}$$

where the two $2M$th-order operators are given by

$$\mathbf{D}_{\mathrm{A}}^{(2M)}\left[f|_{l+1/2}^n\right] = \frac{1}{\Delta\zeta} \sum_{s=1}^{M} c_{\mathrm{A},s}\left(f|_{l+s}^n - f|_{l-s+1}^n\right), \tag{5.33a}$$

$$\mathbf{D}_{\mathrm{B}}^{(2M)}\left[f|_l^n\right] = \frac{1}{\Delta\zeta} \sum_{s=1}^{M} c_{\mathrm{B},s}\left(f|_{l+s-1/2}^n - f|_{l-s+1/2}^n\right). \tag{5.33b}$$

Coefficients $c_{\mathrm{A},s}$ and $c_{\mathrm{B},s}$ are related to $c_{|s|}$ in order to yield optimal weighting values or certain combinations that follow problem-dependent criteria. The prior operators may also receive the more convenient form of

$$\mathbf{D}_{\mathrm{A}}^{(2M)}\left[f|_l^n\right] = \sum_{p=1}^{M} \kappa_p \mathbf{D}_{\mathrm{A},p\Delta\zeta}^{(2)}\left[f|_l^n\right] \quad \text{and} \quad \mathbf{D}_{\mathrm{B}}^{(2M)}\left[f|_l^n\right] = \sum_{p=1}^{M} \kappa_p \mathbf{D}_{\mathrm{B},p\Delta\zeta}^{(2)}\left[f|_l^n\right], \tag{5.34}$$

with $$\mathbf{D}^{(2)}_{\mathrm{A},p\Delta\zeta}\left[f|^n_{l+1/2}\right] = \frac{1}{(2p-1)\Delta\zeta}\left(f|^n_{l+p} - f|^n_{l-p+1}\right), \tag{5.35a}$$

$$\mathbf{D}^{(2)}_{\mathrm{B},p\Delta\zeta}\left[f|^n_{l}\right] = \frac{1}{(2p-1)\Delta\zeta}\left(f|^n_{l+p-1/2} - f|^n_{l-p+1/2}\right). \tag{5.35b}$$

If (5.35a) is expanded in Taylor series at $p\Delta\zeta$, one obtains

$$\mathbf{D}^{(2)}_{\mathrm{A}p\Delta\zeta}\left[f|^n_{l+1/2}\right] = \sum_{s=0}^{M} \frac{(\Delta\zeta)^{2s}}{(2s+1)!2^{2s}} \left.\frac{\partial^{2s+1} f}{\partial\zeta^{2s+1}}\right|^n_l + O\left[(\Delta\zeta)^{2M+2}\right]. \tag{5.36}$$

The relevant relation for (5.35b) is extracted via the substitution of l by $l+\frac{1}{2}$ in (5.36). Moreover, if $\Delta\zeta/2$ is replaced with $(2p-1)\Delta\zeta/2$ in the same relation and the outcome is plugged into (5.34), it is derived that

$$\mathbf{D}^{(2)}_{\mathrm{A},\Delta\zeta}\left[f|^n_{l}\right] = \sum_{s=0}^{M}\left[\frac{(\Delta\zeta)^{2s}}{(2s+1)!2^{2s}} \left.\frac{\partial^{2s+1} f}{\partial\zeta^{2s+1}}\right|^n_l \sum_{p=1}^{M+1} \kappa_p (2p-1)^{2s}\right] + O\left[(\Delta\zeta)^{2M+2}\right], \tag{5.37}$$

which leads to the ensuing system of equation

$$\begin{aligned} \kappa_1 + \kappa_2 + \cdots + \kappa_{M+1} &= 1 \\ \kappa_1 + 4\kappa_2 + \cdots + (M+1)^2\kappa_{M+1} &= 0 \\ \kappa_1 + 16\kappa_2 + \cdots + (M+1)^4\kappa_{M+1} &= 0 \\ &\vdots \\ \kappa_1 + 2^{2M}\kappa_2 + \cdots + (M+1)^{2M}\kappa_{M+1} &= 0. \end{aligned} \tag{5.38}$$

Some representative solutions for $M = 1, 2, \ldots, 6$ are summarized in Table 5.2. Observe that coefficients κ_p are different from those presented in Table 2.1 for the ordinary higher order

TABLE 5.2: Coefficients κ_p for Higher Order Spatial Approximations

ORDER $2M$	κ_1	κ_2	κ_3	κ_4	κ_5	κ_6
2	1	0	0	0	0	0
4	$\frac{9}{8}$	$-\frac{1}{8}$	0	0	0	0
6	$\frac{75}{64}$	$-\frac{25}{128}$	$\frac{3}{128}$	0	0	0
8	$\frac{1225}{1024}$	$-\frac{245}{1024}$	$\frac{49}{1024}$	$-\frac{5}{1024}$	0	0
10	$\frac{19845}{16384}$	$-\frac{2205}{8192}$	$\frac{567}{8192}$	$-\frac{405}{32768}$	$\frac{35}{32768}$	0
12	$\frac{160083}{131072}$	$-\frac{38115}{131072}$	$\frac{22869}{262144}$	$-\frac{5445}{262144}$	$\frac{847}{262144}$	$-\frac{63}{262144}$

FDTD schemes. Actually, these values constitute the basic contribution of (5.32) in the systematic design of robust spatial approximations.

5.4.2 Improvement Through the Dispersion Relation

This paragraph complements the analysis of Section 2.5 by presenting a technique for the construction of improved fourth-order spatial operators through the use of the *discrete* dispersion relation. Principally, the algorithm considers the ordinary leapfrog scheme for time marching, while it involves the parametric expression of (2.107) for spatial differentiation. By substituting plane-wave constituents in Maxwell's equations, the 2-D dispersion relation for an isotropic medium is

$$\left[\frac{\sin(\omega\Delta t/2)}{\upsilon\Delta t}\right]^2 = \sum_{\zeta=x,y}\frac{1}{(\Delta\zeta)^2}\left[A_\zeta\,\sin\left(\frac{k_\zeta^{\text{num}}\Delta\zeta}{2}\right) + B_\zeta\,\sin\left(\frac{3k_\zeta^{\text{num}}\Delta\zeta}{2}\right)\right]^2. \tag{5.39}$$

The basic idea of the algorithm stems from the observation that the requirement $\mathbf{k}^{\text{num}} = \mathbf{k}$ is *too strict* to fulfill for all frequencies and angles of propagation. To circumvent this problem, an approximate version of (5.39), based on Taylor expansion, is utilized and then $\mathbf{k}^{\text{num}} = \mathbf{k}$ is applied in the modified equation [28]. The difference between the two sides of the expression, so extracted, is defined as the error function $e_{2\text{D}}$:

$$\begin{aligned} e_{2\text{D}} = {} & \frac{\omega^2}{4\upsilon^2}\left[1 - \sum_{\zeta=x,y}(A_\zeta + 3B_\zeta)^2\eta_\zeta^2\right] \\ & - \frac{\omega^4}{48\upsilon^4}\left[\upsilon^2\Delta t^2 - \sum_{\zeta=x,y}(A_\zeta + 3B_\zeta)(A_\zeta + 27B_\zeta)\eta_\zeta^4(\Delta\zeta)^2\right], \end{aligned} \tag{5.40}$$

where terms with order larger than two have been neglected and $\eta_x = \cos\varphi$ and $\eta_y = \sin\varphi$. It is straightforward to comprehend that the zeroth-order term of (5.40) can be readily eliminated if

$$A_\zeta + 3B_\zeta = 1 \quad \text{for} \quad \zeta \in (x,\ y). \tag{5.41}$$

The next step is the treatment of the second-order quantities in (5.40). Nonetheless, the presence of φ entails the selection of certain propagation directions along which the extraction of the respective conditions would be conducted. Rather than proceeding to this restrictive approach, from an applicability viewpoint, a more general procedure for error reduction over all angles is developed. In this context, the remaining term

$$\nu_{2\text{D}} = (\upsilon\Delta t)^2 - \sum_{\zeta=x,y}(A_\zeta + 3B_\zeta)(A_\zeta + 27B_\zeta)\eta_\zeta^4(\Delta\zeta)^2, \tag{5.42}$$

in (5.40) can be expressed as a sum of harmonic functions of the propagation angle $\{\sin(m\varphi), \cos(n\varphi)\}$:

$$\begin{aligned}\nu_{2\text{D}} = (\upsilon\Delta t)^2 &- \tfrac{3}{8}\left[(A_x + 27B_x)(\Delta x)^2 + (A_y + 27B_y)(\Delta y)^2\right] \\ &- \tfrac{1}{2}\left[(A_x + 27B_x)(\Delta x)^2 - (A_y + 27B_y)(\Delta y)^2\right]\cos(2\varphi) \\ &- \tfrac{1}{8}\left[(A_x + 27B_x)(\Delta x)^2 + (A_y + 27B_y)(\Delta y)^2\right]\cos(4\varphi).\end{aligned} \tag{5.43}$$

Then, the requirement $\nu_{2\text{D}} = 0$ is equivalent to the nullification of the three harmonic terms in (5.43), where (5.41) has been partially utilized for the sake of clarity. To derive additional equations for the calculation of A_ς and B_ς, the first two terms of error indicator (5.43) are set to zero. In particular, if the constant term (i.e., the mean value of $\nu_{2\text{D}}$) is zeroed, it is obtained that

$$A_x + 27B_x + r^2(A_y + 27B_y) = \tfrac{8}{3}Q_{2\text{D}}^2, \tag{5.44}$$

while the zeroing of the coefficient of $\cos(2\varphi)$ gives

$$A_x + 27B_x - r^2(A_y + 27B_y) = 0. \tag{5.45}$$

For the simplification of the previous expressions, it has been presumed that cell dimensions satisfy $r = \Delta y/\Delta x$, with the time-step fulfilling

$$\Delta t = \frac{\tau_{2\text{D}}\Delta x}{\upsilon\sqrt{1 + r^{-2}}} = \frac{Q_{2\text{D}}\Delta x}{\upsilon}, \tag{5.46}$$

for $Q_{2\text{D}} = \tau_{2\text{D}}/\sqrt{1 + r^{-2}}$ and $\tau_{2\text{D}}$ determined through the classical von Neumann stability analysis. Finally, the unknown coefficients of spatial operators (2.107) and (5.40) can be promptly computed by solving the system of (5.41), (5.44), and (5.45), without the need of additional constraints. Such a process leads to

$$A_x = \frac{9}{8} - \frac{Q_{2\text{D}}^2}{6}, \qquad A_y = \frac{9}{8} - \frac{Q_{2\text{D}}^2}{6r^2}, \tag{5.47a}$$

$$B_x = -\frac{1}{24} + \frac{Q_{2\text{D}}^2}{18}, \qquad B_y = -\frac{1}{24} + \frac{Q_{2\text{D}}^2}{18r^2}. \tag{5.47b}$$

Some indicative values for two different mesh configurations are summarized in Table 5.3.

It becomes apparent that such spatial operators take into account both the time-step size and the shape of the elementary cells, namely they depend on the *anisotropic features* of the FDTD simulation. Besides, as compared to the schemes of Section 2.5.4, they exhibit a more narrow-band performance.

TABLE 5.3: Coefficients A_ζ and B_ζ for the Enhanced 2-D Higher-Order Spatial Operators

CELL DIMENSIONS	τ_{2D}	A_x	A_y	B_x	B_y
$\Delta y = \Delta x$	0.65	1.089791667	1.089791667	−0.029930556	−0.029930556
	0.75	1.078125000	1.078125000	−0.026041667	−0.026041667
	0.85	1.064791667	1.064791667	−0.021597222	−0.021597222
$\Delta y = 2\Delta x$	0.65	1.068666667	1.110916667	−0.022888889	−0.036972222
	0.75	1.050000000	1.106250000	−0.016666667	−0.035416667
	0.85	1.028666667	1.100916667	−0.009555556	−0.033638889

The modified stability criterion by means of A_ζ and B_ζ is written as

$$\upsilon \Delta t \leq \frac{\Delta x}{\sqrt{\left(\frac{7}{6} - \frac{2Q_{2D}^2}{9}\right)^2 + \frac{1}{r^2}\left(\frac{7}{6} - \frac{2Q_{2D}^2}{9r^2}\right)^2}}, \tag{5.48}$$

on condition that $A_\zeta > 0$ and $B_\zeta < 0$ for specific values of Q_{2D}. Relation (5.48) is proven to be less strict than (2.25), where for the maximum temporal increment it holds $\tau_{2D} = 6/7 \simeq 0.857$. Indeed for a uniform grid, parameter τ_{2D} can reach up to the value of 0.935 (i.e., a 9% increase of the stability area). This limit is further augmented in the case of a general nonuniform lattice.

For illustration, let us estimate the dispersion error of the aforementioned narrow-band technique and compare it with the one induced by the wideband method of (2.107)–(2.115). The cell dimensions are chosen as $\Delta y = 2\Delta x$ and $\tau_{2D} = 0.85$ in (5.46). Figure 5.5 gives the results for two mesh resolutions with respect to Δx. In contrast with the performance of the latter scheme, the narrow-band approach achieves a remarkable reduction around the design frequency, whereas its accuracy deteriorates at finer resolutions. This is, however, not a serious shortcoming, since a given computational domain appears to have a smaller electrical size at lower frequencies and, consequently, the dispersion error is not as considerable as in the high-frequency band. Moreover, it is noteworthy to observe that the narrow-band scheme generates smaller errors for coarser lattices and thus, its application to broadband simulations should not be ruled out.

The extension of the preceding algorithm in 3-D meshes with $r_1 = \Delta y/\Delta x$ and $r_2 = \Delta z/\Delta x$ pursues similar lines. The dispersion relation is again provided by (5.39) with the exception of $\zeta \in (x, y, z)$ and $\eta_x = \sin\theta \cos\varphi$, $\eta_y = \sin\theta \sin\varphi$, $\eta_z = \cos\theta$ in spherical

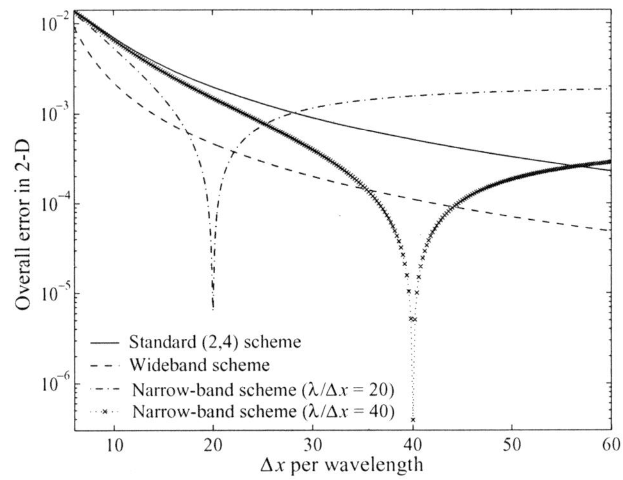

FIGURE 5.5: Dispersion error versus grid resolution for diverse (2, 4) FDTD configurations

coordinates. For the construction of the analogous error function e_{3D}, (5.39) is expanded in Taylor series where $\mathbf{k}^{\text{num}} = \mathbf{k}$ is applied. Function e_{3D} is obtained via the difference of the two sides of the resulting expression in which the time-step is denoted as

$$\Delta t = \frac{\tau_{3D}\Delta x}{\upsilon\sqrt{1 + r_1^{-2} + r_2^{-2}}} = \frac{Q_{3D}\Delta x}{\upsilon}, \tag{5.49}$$

with $Q_{3D} = \tau_{3D}/\sqrt{1 + r_1^{-2} + r_2^{-2}}$. Selecting the unknown coefficients to satisfy (5.41) for $\zeta \in (x,\ y,\ z)$, the nullification of the zeroth-order term is attained. Contrarily, for the second-order term, it is required that

$$\nu_{3D} = \sum_{\zeta=x,y,z} (A_\zeta + 3B_\zeta)(A_\zeta + 27B_\zeta)\tau_u^4(\Delta\zeta)^2 - Q_{3D}^2(\Delta x)^2, \tag{5.50}$$

which, owing to the θ, φ dependence, is additionally reduced through the use of even spherical harmonics (see Section 2.5.3). Hence, apart from (5.41), the system of equations for A_ζ and B_ζ is completed by

$$A_x + 27B_x + r_1^2\left(A_y + 27B_y\right) + r_2^2\left(A_z + 27B_z\right) = 5Q_{3D}^2, \tag{5.51}$$

$$A_x + 27B_x + r_1^2\left(A_y + 27B_y\right) - 2r_2^2\left(A_z + 27B_z\right) = 0, \tag{5.52}$$

$$A_x + 27B_x - r_1^2\left(A_y + 27B_y\right) = 0, \tag{5.53}$$

TABLE 5.4: Coefficients A_ζ and B_ζ for the Enhanced 3-D Higher-Order Spatial Operators

τ_{3D}	A_x	A_y	A_z	B_x	B_y	B_z
0.65	1.11871280	1.09985119	1.06841518	−0.03957093	−0.03328373	−0.02280506
0.75	1.11662946	1.09151786	1.04966518	−0.03887649	−0.03050595	−0.01655506
0.85	1.11424851	1.08199405	1.02823661	−0.03808284	−0.02733135	−0.00941220

yielding as a solution

$$A_x = \frac{9}{8} - \frac{5Q_{3D}^2}{24}, \qquad A_y = \frac{9}{8} - \frac{5Q_{3D}^2}{24r_1^2}, \qquad A_z = \frac{9}{8} - \frac{5Q_{3D}^2}{24r_2^2}, \tag{5.54a}$$

$$B_x = -\frac{1}{24} + \frac{5Q_{3D}^2}{72}, \qquad B_x = -\frac{1}{24} + \frac{5Q_{3D}^2}{72r_1^2}, \qquad B_z = -\frac{1}{24} + \frac{5Q_{3D}^2}{72r_2^2}. \tag{5.54b}$$

In the following, Table 5.4 supplies a set of typical A_ζ and B_ζ values for $r_1 = 2$ and $r_2 = 3$.

The respective 3-D stability condition becomes

$$\upsilon\Delta t \leq \frac{\Delta x}{\sqrt{\left(\frac{7}{6} - \frac{5Q_{3D}^2}{18}\right)^2 + \frac{1}{r_1^2}\left(\frac{7}{6} - \frac{5Q_{3D}^2}{18r_1^2}\right)^2 + \frac{1}{r_2^2}\left(\frac{7}{6} - \frac{5Q_{3D}^2}{18r_2^2}\right)^2}}, \tag{5.55}$$

which is also lighter than its usual (2, 4) FDTD counterpart as occurs with other approaches [29, 30].

Finally, to investigate the merits of the 3-D formulation, the numerical phase velocity in a uniform grid with $\tau_{3D} = 0.85$ is computed. Table 5.5 shows the outcomes which prove that the narrow-band technique, described herein, accomplishes a more substantial improvement

TABLE 5.5: Numerical Phase-Velocity Error for a 3-D Uniform Lattice

RESOLUTION	CONVENTIONAL (2, 4)	WIDEBAND	NARROW BAND
$\lambda/10$	3.668×10^{-3}	1.039×10^{-3}	2.986×10^{-5}
$\lambda/20$	9.717×10^{-4}	2.460×10^{-4}	1.825×10^{-6}
$\lambda/30$	4.365×10^{-4}	1.083×10^{-4}	3.591×10^{-7}
$\lambda/40$	2.464×10^{-4}	6.072×10^{-5}	1.134×10^{-7}
$\lambda/50$	1.580×10^{-4}	3.880×10^{-5}	2.240×10^{-8}

than the wideband one with its error decreasing at a fourth-order rate, unlike the second-order convergence of the common (2, 4) scheme.

5.5 GENERALIZING TEMPORAL INTEGRATION

Proceeding to the time update of electric and magnetic fields and having studied the application of several leapfrog and Runge–Kutta variations in Chapters 2 and 3, our attention will now focus on the generalization of these algorithms. For this goal, consider the semidiscrete version of Maxwell's equations (2.13) for a lossless medium and define the equivalents of curl operators as $\mathbf{D}_{\mathrm{F}}[.] = \nabla \times \mathbf{F}$ with $\mathbf{F} = \mathbf{E}, \mathbf{H}$. Then,

$$\frac{\partial}{\partial t}\begin{bmatrix}\mathbf{E}(t)\\ \mathbf{H}(t)\end{bmatrix} = \begin{bmatrix}\mathbf{0} & \varepsilon^{-1}\mathbf{D}_{\mathrm{H}}\\ -\mu^{-1}\mathbf{D}_{\mathrm{E}} & \mathbf{0}\end{bmatrix}\begin{bmatrix}\mathbf{E}(t)\\ \mathbf{H}(t)\end{bmatrix} = \mathbf{A}\begin{bmatrix}\mathbf{E}(t)\\ \mathbf{H}(t)\end{bmatrix}. \tag{5.56}$$

The solution of (2.56) with the concurrent substitution of time marching $t = n\Delta t$ leads to

$$\begin{bmatrix}\mathbf{E}\\ \mathbf{H}\end{bmatrix}^{n} = \begin{bmatrix}\mathbf{E}\\ \mathbf{H}\end{bmatrix}^{0} e^{\mathbf{A}n\Delta t} = \begin{bmatrix}\mathbf{E}\\ \mathbf{H}\end{bmatrix}^{0} e^{\mathbf{A}(n-1)\Delta t} e^{\mathbf{A}\Delta t} = \begin{bmatrix}\mathbf{E}\\ \mathbf{H}\end{bmatrix}^{n-1} e^{\mathbf{A}\Delta t}. \tag{5.57}$$

The above relation implies that temporal integration does not arise any errors as long as there exists an exact means of evaluating the sparse exponential matrix $e^{\mathbf{A}\Delta t}$. Nevertheless, such a procedure is practically unfeasible due to the size of **A** even for medium simulations, requesting so the use of approximate calculations. Actually, the leapfrog and Runge–Kutta techniques belong to this particular category [31–34].

5.5.1 Higher Order Leapfrog Schemes

According to the leapfrog concept, the $e^{\mathbf{A}\Delta t}$ quantity is approximated via Taylor expansion as

$$e^{\mathbf{A}\Delta t} = \sum_{\kappa=0}^{\infty}\frac{(\Delta t)^{\kappa}}{\kappa!}(\mathbf{A})^{\kappa} = \mathbf{I} + \mathbf{A}\Delta t + O\left[(\Delta t)^2\right], \tag{5.58}$$

where **I** is the appropriate unit matrix. Assuming a temporal shift equal to the duration of half time-step between the electric and magnetic field vectors, the time update of the algorithm is given by

$$\begin{aligned}\mathbf{E}|^{n+1} &= \mathbf{E}|^{n} + \varepsilon^{-1}\Delta t\mathbf{D}_{\mathrm{H}}\left[\mathbf{H}|^{n+1/2}\right]\\ \mathbf{H}|^{n+1/2} &= \mathbf{H}|^{n-1/2} - \mu^{-1}\Delta t\mathbf{D}_{\mathrm{E}}\left[\mathbf{E}|^{n}\right]^{n}.\end{aligned} \tag{5.59}$$

If higher order accuracy is required, the suitable number of additional terms in (5.58) should be considered. Hence, the general time-marching equations for $(2\kappa,\ M)$ FDTD schemes are acquired by

$$\mathbf{E}|^{n+1} = \mathbf{E}|^{n} + \left\{ \sum_{\kappa=1}^{K/2} \frac{(-1)^{\kappa+1}}{2^{2\kappa-2}(2\kappa-1)!} \left[(\Delta t)^2 \varepsilon^{-1} \mathbf{D}_{\mathrm{H}} \mu^{-1} \mathbf{D}_{\mathrm{E}} \right]^{\kappa-1} \cdot \Delta t \varepsilon^{-1} \mathbf{D}_{\mathrm{H}} \right\} \mathbf{H}|^{n+1/2}, \tag{5.60a}$$

$$\mathbf{H}|^{n+1/2} = \mathbf{H}|^{n-1/2} + \left\{ \sum_{\kappa=1}^{K/2} \frac{(-1)^{\kappa}}{2^{2\kappa-2}(2\kappa-1)!} \left[(\Delta t)^2 \mu^{-1} \mathbf{D}_{\mathrm{E}} \varepsilon^{-1} \mathbf{D}_{\mathrm{H}} \right]^{\kappa-1} \cdot \Delta t \mu^{-1} \mathbf{D}_{\mathrm{E}} \right\} \mathbf{E}|^{n}. \tag{5.60b}$$

In other words, the only difference between (5.60) and the second-order leapfrog scheme is of the supplementary terms – depending on the value of κ – which increase the computational overhead with the consecutive use of spatial operators at additional nodes. It is stressed that the presence of lossy materials receives an analogous treatment with the exception of more complicated expressions beyond the fourth order.

5.5.2 Higher Order Multistage Runge–Kutta Integrators

The temporal update through the single-stage Kth-order Runge–Kutta algorithms [33] is performed by

$$\mathbf{E}|^{n+1} = \sum_{\kappa=0}^{K/2} \frac{(-1)^{\kappa} (\upsilon \Delta t)^{2\kappa}}{(2\kappa)!} (\mathbf{D}_{\mathrm{H}} \mathbf{D}_{\mathrm{E}})^{\kappa} \, \mathbf{E}|^{n} + \frac{\Delta t}{\varepsilon} \sum_{\kappa=0}^{K/2-1} \frac{(-1)^{\kappa} (\upsilon \Delta t)^{2\kappa}}{(2\kappa+1)!} (\mathbf{D}_{\mathrm{H}} \mathbf{D}_{\mathrm{E}})^{\kappa} \, \mathbf{D}_{\mathrm{H}} \left[\mathbf{H}|^{n} \right], \tag{5.61a}$$

$$\mathbf{H}|^{n+1} = \sum_{\kappa=0}^{K/2} \frac{(-1)^{\kappa} (\upsilon \Delta t)^{2\kappa}}{(2\kappa)!} (\mathbf{D}_{\mathrm{E}} \mathbf{D}_{\mathrm{H}})^{\kappa} \, \mathbf{H}|^{n+1} - \frac{\Delta t}{\mu} \sum_{\kappa=0}^{K/2-1} \frac{(-1)^{\kappa} (\upsilon \Delta t)^{2\kappa}}{(2\kappa+1)!} (\mathbf{D}_{\mathrm{E}} \mathbf{D}_{\mathrm{H}})^{\kappa} \, \mathbf{D}_{\mathrm{E}} [\mathbf{E}|^{n}]. \tag{5.61b}$$

The most common case, however, is the multistage implementation in successive steps. Recalling the Taylor expansion of matrix $e^{\mathbf{A}\Delta t}$ up to pth-order terms, one obtains

$$e^{\mathbf{A}\Delta t} = \mathbf{I} + \Delta t \mathbf{A} \left\{ \mathbf{I} + \tfrac{\Delta t}{2} \mathbf{A} \left[\mathbf{I} + \tfrac{\Delta t}{3} \mathbf{A} \ldots \left(\mathbf{I} + \tfrac{\Delta t}{p} \mathbf{A} \right) \right] \right\} + O\left[(\Delta t)^{p+1} \right]. \tag{5.62}$$

So, the time advancement of field components can be achieved in p stages as

$$\begin{aligned} \mathbf{F}_1 &= \mathbf{F}|^{n} + \tfrac{\Delta t}{p} \mathbf{A} \, \mathbf{F}|^{n} & \mathbf{F}_1 &= \mathbf{F}|^{n} + \tfrac{\Delta t}{p-1} \mathbf{A} \mathbf{F}_1 \\ & & &\vdots \\ \mathbf{F}_{p-1} &= \mathbf{F}|^{n} + \tfrac{\Delta t}{2} \mathbf{A} \mathbf{F}_{p-2} & \mathbf{F}|^{n+1} &= \mathbf{F}|^{n} + \Delta t \mathbf{A} \mathbf{F}_{p-1} \end{aligned} \tag{5.63}$$

with a typical example of the fourth-order integrators of (2.41). Finally, it is significant to mention that the application of Runge–Kutta schemes introduces artificial magnitude attenuation, yielding thus a gradual decrease of the overall energy, even if the discretized problem does not predict a behavior of this kind.

REFERENCES

[1] K. S. Kunz and R. J. Luebbers, *The Finite Difference Time Domain Method for Electromagnetics*. Boca Raton, FL: CRC Press, 1993.

[2] J. A. Pereda, O. García, Á. Vegas, and A. Prieto, "Numerical dispersion and stability analysis of the FDTD technique in lossy dielectrics," *IEEE Trans. Microw. Guided Wave Lett.*, vol. 8, no. 7, pp. 245–247, July 1998.

[3] A. Grande, I. Barba, A. C. L. Cabeceira, J. Represa, P. P. M. So, and W. J. R. Hoefer, "FDTD modeling of transient microwave signals in dispersive and lossy bi-isotropic media," *IEEE Trans Microw. Theory Tech.*, vol. 52, no. 3, pp. 773–784, Mar. 2004.

[4] P. Kosmas and C. M. Rappaport, "A simple absorbing boundary condition for FDTD modeling of lossy, dispersive media based on the one-way wave equation," *IEEE Trans. Antennas Propag.*, vol. 52, no. 9, pp. 2476–2478, Sep. 2004.

[5] R. J. Luebbers, F. Hunsberger, and K. S. Kunz, "A frequency-dependent finite-difference time-domain formulation for transient propagation in plasma," *IEEE Trans. Antennas Propag.*, vol. 39, no. 1, pp. 29–34, Jan. 1991.

[6] R. M. Joseph, S. C. Hagness, and A. Taflove, "Direct time integration of Maxwell's equations in linear dispersive media with absorption for scattering and propagation of femtosecond electromagnetic pulse," *Opt. Lett.*, vol. 16, pp. 1412–1414, Sep. 1991.

[7] T. Kashiwa, Y. Ohtomo, and I. Fukai, "Formulation of dispersive characteristics associated with orientation polarization using the FD-TD method," *Electron. Commun. Japan (Part I: Communications)*, vol. 75, no. 6, pp. 87–96, 1992.

[8] D. M. Sullivan, "Frequency-dependent FDTD methods using Z-transforms," *IEEE Trans. Antennas Propag.*, vol. 40, no. 10, pp. 1223–1230, Oct. 1992.

[9] P. G. Petropoulos, "Stability and phase error analysis of FD-TD in dispersive dielectrics," *IEEE Trans. Antennas Propag.*, vol. 42, no. 1, pp. 62–69, Jan. 1994.

[10] J. L. Young, "A higher order FDTD method for EM propagation in a collisionless cold plasma," *IEEE Trans. Antennas Propag.*, vol. 44, no. 9, pp. 1283–1289, Sep. 1996.

[11] S. A. Cummer, "An analysis of new and existing FDTD methods for isotropic cold plasma and a method for improving their accuracy," *IEEE Trans. Antennas Propag.*, vol. 45, no. 3, pp. 392–400, Mar. 1997.

[12] M. Mrozowski and M. A. Stuchly, "Parameterization of media dispersive properties for FDTD," *IEEE Trans. Antennas Propag.*, vol. 45, no. 9, pp. 1438–1439, Sep. 1997.

[13] F. L. Teixeira, W. C. Chew, M. Straka, M. L. Oristaglio, and T. Wang, "Finite-difference time-domain simulation of ground penetrating radar on dispersive, inhomogeneous, and conductive soils," *IEEE Trans. Geosci. Remote Sens.*, vol. 36, no. 6, pp. 1928–1937, Nov. 1998.

[14] R. W. Ziolkowski and M. Tanaka, "Finite-difference time-domain modeling of dispersive-material photonic bandgap structures," *J. Opt. Soc. Amer. A Opt. Image Sci. Vision*, vol. 16, no. 4, pp. 930–940, Apr. 1999.

[15] J. H. Beggs, "Validation and demonstration of frequency approximation methods for modeling dispersive media in FDTD," *Appl. Comput. Electromagn. Soc. J.*, vol. 14, no. 2, pp. 52–58, July 1999.

[16] J. L. Young and R. O. Nelson, "A summary and systematic analysis of FDTD algorithms for linearly dispersive media," *IEEE Antennas Propag. Mag.*, vol. 43, no. 1, pp. 61–77, Feb. 2001.

[17] J. A. Pereda, L. A. Vielva, Á. Vegas, and A. Prieto, "Analyzing the stability of the FDTD technique by combining the von Neumann method with the Routh–Hurwitz criterion," *IEEE Trans. Microw. Theory Tech.*, vol. 49, no. 2, pp. 377–381, Feb. 2001.

[18] D. Popovic and M. Okoniewski, "Effective permittivity at the interface of dispersive dielectrics in FDTD," *IEEE Microw. Wireless Compon. Lett.*, vol. 13, no. 7, pp. 265–267, July 2003.

[19] M. K. Kärkkäinen, "FDTD model of electrically thick frequency-dispersive coatings on metals and semiconductors based on surface impedance boundary conditions," *IEEE Trans. Antennas Propag.*, vol. 53, no. 3, pp. 1174–1186, Mar. 2005.

[20] K. P. Prokopidis and T. D. Tsiboukis, "Higher-order FDTD (2,4) scheme for accurate simulations in lossy dielectrics," *IEE Electron. Lett.*, vol. 39, no. 11, pp. 835–836, May 2003.

[21] K. P. Prokopidis, E. P. Kosmidou, and T. D. Tsiboukis, "An FDTD algorithm for wave propagation in dispersive media using higher-order schemes," *J. Electromagn. Waves Appl.*, vol. 18, no. 9, 1171–1194, 2004.

[22] J. S. Juntunen and T. D. Tsiboukis, "Reduction of numerical dispersion in FDTD method through artificial anisotropy," *IEEE Trans. Microw. Theory Tech.*, vol. 48, no. 4, pp. 582–588, Apr. 2000.

[23] T. T. Zygiridis and T. D. Tsiboukis, "Higher order finite difference schemes with reduced dispersion errors for accurate time domain electromagnetic simulations," *Int. J. Numer. Model.*, vol. 17, no. 5, pp. 461–486, Sep.–Oct. 2004.

[24] M. Feliziani, F. Maradei, and G. Tribellini, "Field analysis of penetrable conductive shields by the finite-difference time-domain method with impedance network boundary

conditions (INBC's)," *IEEE Trans. Electromagn. Compat.*, vol. 41, pp. 307–319, Nov. 1999.

[25] M. Ghrist, B. Fornberg, and T. A. Driscoll, "Staggered time integrators for wave equations," *SIAM J. Numer. Anal.*, vol. 38, no. 3, pp. 718–741, 2000.

[26] J. E. Castillo, J. M. Hyman, M. Shashkov, and S. Steinberg, "Fourth- and sixth-order conservative finite difference approximations of the divergence and gradient," *Appl. Numer. Math.*, vol. 37, nos. 1–2, pp. 171–187, Apr. 2001.

[27] G. C. Cohen, *Higher-Order Numerical Methods for Transient Wave Equations.* Berlin, Germany: Springer-Verlag, 2002.

[28] T. T. Zygiridis and T. D. Tsiboukis, "Low-dispersion algorithms based on the higher order (2,4) FDTD method," *IEEE Trans. Microwave Theory Tech.*, vol. 52, no. 4, pp. 1321–1327, Apr. 2004.

[29] N. V. Kantartzis, T. K. Katsibas, C. S. Antonopoulos, and T. D. Tsiboukis, "Unified higher-order FDTD-PMLs for 3-D electromagnetics and advective acoustics," *COMPEL*, vol. 21, no. 3, pp. 451–471, 2002.

[30] F. Xiao, X. Tang, and H. Ma, "High-order US-FDTD based on the weighted finite-difference method," *Microw. Opt. Technol. Lett.*, vol. 45, no. 2, pp. 142–144, Apr. 2005.

[31] S. Gutschling, H. Krüger, T. Weiland, "Time domain simulation of dispersive media with the finite integration technique," *Int. J. Numer. Model.*, vol. 13, no. 4, pp. 329–348, July–Aug. 2000.

[32] E. Abenius, U. Andersson, F. Edelvik, L. Eriksson, and G. Ledfelt, "Hybrid time domain solvers for the Maxwell equations in 2D," *Int. J. Numer. Meth. Engng.*, vol. 53, no. 9, pp. 2185–2199, Mar. 2002.

[33] H. Spachmann, R. Schuhmann, and T. Weiland, "Higher order explicit time integration schemes for Maxwell's equations," *Int. J. Numer. Model.*, vol. 15, nos. 5–6, pp. 419–437, Sep.–Dec. 2002.

[34] F. Edelvik and B. Strand, "Frequency dispersive materials for 3-D hybrid solvers in time domain," *IEEE Trans. Antennas Propag.*, vol. 51, no. 6, pp. 1199–1205, June 2003.

CHAPTER 6

Hybrid and Alternative Higher Order FDTD Schemes

6.1 INTRODUCTION

An attractive characteristic of higher order FDTD operators is the *versatility* and *modular* nature of their finite-difference kernel, which enables them to profitably cooperate with other numerical techniques. Therefore, taking into account the individual parameters of every problem, these algorithms can offer structural improvements, additional means of enhanced modeling, and significant computational savings. Hybridization is performed for the approximation of spatial or temporal derivatives, while several features of the initial formulation may be preserved. Obviously, due to the widened stencils of higher order FDTD forms, the above combinations should be carefully designed in order to obstruct the appearance of incompatible lattice arrangements or artificial spurious oscillations. On the other hand, alternative approaches are also possible and sometimes their use is proven to be very instructive. In this case, of critical importance is the correct assignment of electric and magnetic field components to the nodes of the resulting tessellations.

This chapter is devoted to the presentation of such methods for both conventional and nonstandard concepts. More specifically, a hybrid (2, 2)/(2, 4) FDTD technique, founded on a robust subgridding framework, is first analyzed. Next, the potential of introducing the discrete singular convolution (DSC) along with the idea of symplectic operators, as an alternative way to conduct temporal integration, is examined. Furthermore, our interest will focus on a dispersion-optimized alternating-direction implicit (ADI) FDTD algorithm that allows the selection of time-steps significantly beyond the Courant limit. The chapter ends with an explicit formulation based on the properties of weighted essentially nonoscillatory (WENO) forms and higher order curvilinear nonstandard operators for the handling of abrupt discontinuities.

6.2 HYBRID SECOND-ORDER AND HIGHER ORDER FDTD TECHNIQUES

The constantly growing requirements for the certification of modern waveguide and antenna systems have lately stipulated challenging standards for the FDTD method and its higher order formulations. In particular, the latter are likely to demand rather increased, and sometimes unaffordable, computational overheads when structures with fine geometric details or large electrical sizes are to be modeled. To avoid these defects and diminish the overall burden, a viable solution can be pursued in the development of hybrid techniques. Merging the attractive properties of various schemes, dependent on the problem under study, with the merits of higher order FDTD concepts, these approaches exhibit significant features at a reasonable cost.

Toward this trend, this section presents a hybrid algorithm that combines a subridding strategy with the conventional (2, 4) FDTD technique. Actually, subgridding implementations constitute an efficient tool for the inexpensive discretization of relatively complex applications and therefore they have been intensely investigated [1–10]. The specific method, originally developed in [11, 12], applies the common second-order Yee's scheme on a fine grid to manipulate geometric peculiarities, while on a coarse lattice it uses the (2, 4) FDTD formulation. Hence, existing successful approaches for the incorporation of discontinuities, excitation patterns and absorbing boundary conditions can be easily realized on the fine grid. Conversely, the role of the higher order approach is the consistent simulation of propagating waves in the homogeneous domain, yielding so accurate outcomes without undesired vector parasites.

To guarantee the stability of the resulting algorithm, the ratio between the coarse and the fine lattice is set to 1:3, whereas their interface is collocated with electric instead of magnetic field components, as illustrated in the two-dimensional depiction of Figure 6.1(a). Furthermore, two types of weighting processes for the smooth transition and update of the unknown quantities are utilized in the vicinity of the coarse/fine-mesh boundary. According to the first one, the electric components near the boundary are given by

$$\tilde{E}_{\mathrm{B}} = 0.95\tilde{E}_{\mathrm{B}}^{\mathrm{fine}} + 0.025\left(E_{\mathrm{A}} + \tilde{E}_{\mathrm{C}}\right), \tag{6.1}$$

where $\tilde{E}_{\mathrm{B}}$ is the weighted electric intensity value of the fine lattice adjacent to the boundary, and $\tilde{E}_{\mathrm{B}}^{\mathrm{fine}}$ and $\tilde{E}_{\mathrm{C}}$ are the fields calculated by the second-order FDTD scheme on the fine grid. Analogously, E_{A} represents the field evaluated by spatial and temporal interpolation of the (2, 4) algorithm on the coarse/fine-mesh boundary [Figure 6.1(b)]. The second kind of weighting assigns to the coarse lattice magnetic components that are neighboring to the boundary and are located inside the area of the fine grid, in the following way:

$$H_{\mathrm{B}} = 0.7H_{\mathrm{B}}^{\mathrm{coarse}} + 0.3\tilde{H}_{\mathrm{B}}^{\mathrm{fine}} \quad \text{and} \quad \tilde{H}_{\mathrm{B}} = 0.3H_{\mathrm{B}}^{\mathrm{coarse}} + 0.7\tilde{H}_{\mathrm{B}}^{\mathrm{fine}}, \tag{6.2}$$

with the superscripts coarse and fine signifying the quantities on the coarse and fine mesh computed via the (2, 4) and the second-order FDTD method, respectively, as also shown in Figure 6.1(a).

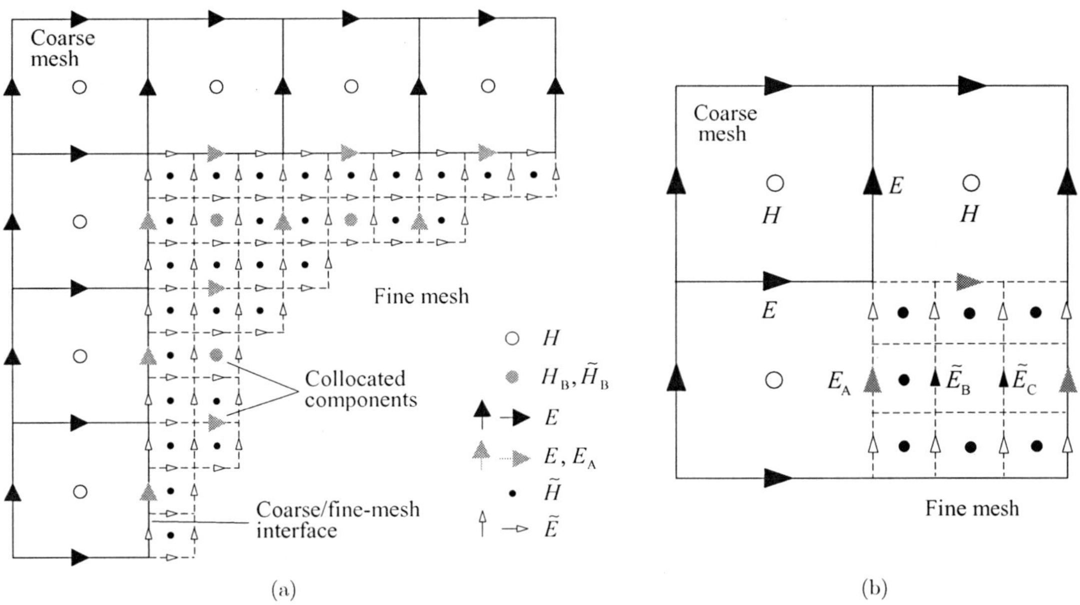

FIGURE 6.1: (a) Graphical depiction of the coarse/fine grids for the hybrid higher order FDTD method. (b) Detail of the previous lattice for the first weighting procedure. The tilde over the electric and magnetic components represents fine-grid field values

Another point of concern is the temporal interpolation of the fine-lattice electric field values on the interface. This is conducted through the third-order expressions of

$$\begin{aligned} E|^{n+1+1/3} &= \tfrac{2}{9}\ E|^{n} - \tfrac{7}{9}\ E|^{n+1} + \tfrac{14}{9}\ E|^{n+2} \\ E|^{n+2+2/3} &= -\tfrac{1}{9}\ E|^{n+1} + \tfrac{5}{9}\ E|^{n+2} + \tfrac{5}{9}\ E|^{n+3}, \end{aligned} \tag{6.3}$$

for the time increments given in Figure 6.2. Finally, the stability of the entire scheme is monitored by means of the Courant criterion of the (2, 4) technique (2.25) reduced by the safety percentage of 20%.

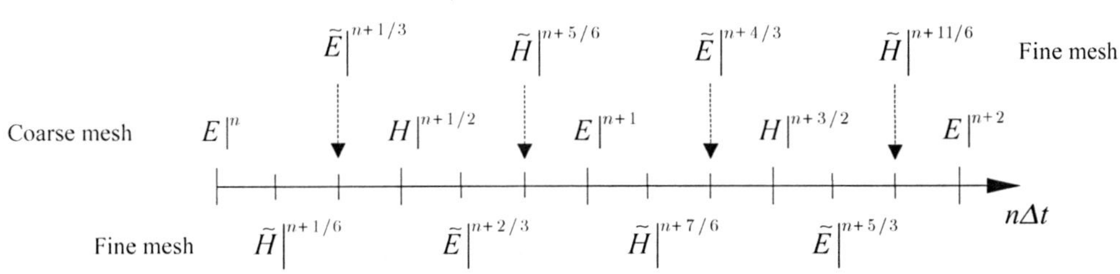

FIGURE 6.2: Temporal evolution of electric and magnetic field components for the hybrid higher order FDTD technique. The tilde over the components represents fine-grid field values

In summary, the basic steps of the hybrid FDTD formulation are described as

Step 1: Application of the (2, 4) FDTD scheme to every main mesh node, including those inside the fine grid, to compute H at $n + \frac{1}{2}$.
Step 2: Use of the second-order Yee's algorithm on the fine mesh to acquire $\tilde{H}$ at $n + \frac{1}{6}$.
Step 3: Application of the second-order Yee's scheme in the fine grid to derive $\tilde{E}$ at $n + \frac{1}{3}$. Time marching of $\tilde{E}$ at $n + \frac{1}{3}$ on the coarse/fine-lattice interface via spatial and temporal interpolations. Utilization of (6.1) to weigh $\tilde{E}$ at $n + \frac{1}{3}$ one cell inside the fine mesh.
Step 4: Use of the second-order Yee's scheme on the fine grid to calculate $\tilde{H}$ at $n + \frac{3}{6}$. Weighting of H at $n + \frac{1}{2}$ via (6.2), collocated one coarse-grid cell into the region of the fine lattice.
Step 5: Update E at $n + 1$ on the coarse/fine-mesh boundary through the known values of H at $n + \frac{1}{2}$.
Step 6: Repetition of a) Step 3 to obtain $\tilde{E}$ at $n + \frac{2}{3}$, b) Step 2 to derive $\tilde{H}$ at $n + \frac{5}{6}$, and c) Step 3 to evaluate $\tilde{E}$ at $n + 1$. Modify the values of $\tilde{E}$ at $n + 1$ on the interface in terms of spatial interpolation of E at $n + 1$.
Step 7: Transfer all fine lattice $\tilde{E}$ components to their collocated coarse-lattice E counterparts.

The aforementioned technique assumes that all subgrid second-order FDTD regions are internal to the coarse-mesh higher order domain. This is fairly convenient in the case of scattering or radiation problems, where all objects or excitation structures are positioned inside the computational space. Nonetheless, there exist some applications, such as electrically large shielding enclosures, whose analysis needs the opposite hybrid configuration. Herein, the largest portion of the domain is the interior of the enclosure, where the conventional (2, 4) FDTD scheme must be applied, while the second-order Yee's method for the subgridding is employed near the perfect electric conducting (PEC) walls. The resulting algorithm discussed in [12] is similarly developed, yielding remarkable savings in CPU and memory requirements by a factor of 2.5 to 8 times, as compared to the pure second-order FDTD configuration. For the stability of the opposite technique, the condition of (2.25) now reduced by a safety percentage of 50% is selected. This choice offers stable simulations for several tens of thousands of time-steps for diverse problems. It is stressed that spatial interpolation of the coarse-grid values on the interface is attained by ordinary second-order finite-difference schemes.

6.3 DISCRETE SINGULAR CONVOLUTION AND SYMPLECTIC OPERATORS

A nontrivial weakness of the (2, 4) and (4, 4) schemes is the incomplete modeling of PEC structures or dissimilar media configurations, whenever the stencils of the discrete derivatives have to surpass the interface. Typically, such realizations are prone to instabilities, especially in three dimensions, where wave interactions near these objects are more complicated. Although, this critical topic has been analyzed in Chapter 4, herein, an alternative approach that may be deemed as a generalization of the higher order FDTD concept is presented. Its main idea, stemming from [13, 14], combines the DSC technique with a symplectic integrator propagator [15–17] to design efficient schemes.

Let us first concentrate on the DSC method and assume that T is a distribution and $f(x)$ an element of the test-function space. Then, the singular convolution of $f(x)$ is denoted as

$$F(t) = (T * f)(t) = \int_{-\infty}^{+\infty} T(t-x) f(x)\, dx. \tag{6.4}$$

Its DSC counterpart, used for spatial differentiation, may be evaluated by

$$F_p(t) = \sum_i T_p(t - x_i) f\left(x|_i\right), \tag{6.5}$$

where T_p represents a set of weighting functions for T at points $x|_i$ and F_p approximates F of (6.4). Considering a Cartesian grid, the Mth-order discretized spatial derivative of f is expressed as

$$\frac{\partial^M f}{\partial x^M} = \sum_{i=-\Xi}^{-1} \delta_\sigma^M \left(x - x|_{i+1/2}\right) f(x|_{i+1/2}) + \sum_{i=1}^{\Xi} \delta_\sigma^M \left(x - x|_{i-1/2}\right) f(x|_{i-1/2}), \tag{6.6}$$

with

$$\delta_\sigma^M(x - x|_i) = \frac{d^M}{dx^M} \delta_\sigma (x - x|_i),$$

$\delta_\sigma \left(x - x|_i\right)$ the DSC delta kernel, and Ξ the bandwidth of possible points. There is a variety of delta kernels available in mathematical analysis. According to [13], the best choice is the LK kernel due to its optimal bandwidth for a preset level of accuracy, defined by

$$\mathrm{LK}_{\Xi,i}(x) = \prod_{\substack{j=i-\Xi \\ j\neq i}}^{i+\Xi} \frac{x - x|_j}{x|_i - x|_j}, \tag{6.7}$$

which if regularized receives the form of

$$\delta_\sigma \left(x - x|_j\right) = \left[\prod_{\substack{j=i-\Xi \\ j\neq i}}^{i+\Xi} \frac{x - x|_j}{x|_i - x|_j} \right] e^{-\frac{\left(x - x|_j\right)^2}{2\sigma^2}}. \tag{6.8}$$

The product operator of (6.8) is usually obtained by a recurrence method, like those depicted in [15]. As DSC kernels can be either symmetric or antisymmetric, they need f values that are located at points outside the computational domain. Therefore, specific boundaries must be utilized. These are:

- For a perfect electric wall, the tangential electric and the normal magnetic fields are acquired by antisymmetric extensions, while the normal electric and the tangential magnetic fields by symmetric ones.
- For a perfect magnetic wall, the tangential magnetic and normal electric fields are obtained by antisymmetric extensions, while the normal magnetic and the tangential electric fields by symmetric ones.
- For a periodic boundary, all electric and magnetic fields are calculated via periodic extensions.
- For open boundaries, specific absorbing boundary conditions are used for the simulations.

The next part of the method refers to temporal integration that is conducted by a symplectic integrator approach. To start with, one has to express the lossy version of Maxwell's equations in matrix form as

$$\frac{\partial}{\partial t}\begin{bmatrix}\mathbf{E}\\ \mathbf{H}\end{bmatrix} = \mathbf{A}\begin{bmatrix}\mathbf{E}\\ \mathbf{H}\end{bmatrix} \tag{6.9}$$

with

$$\mathbf{A} = \begin{bmatrix} -\varepsilon^{-1}\sigma\mathbf{I} & \varepsilon^{-1}\mathbf{D}_{\mathrm{H}} \\ -\mu^{-1}\mathbf{D}_{\mathrm{E}} & -\mu^{-1}\sigma^{*}\mathbf{I}\end{bmatrix} = \frac{1}{\mu}\begin{bmatrix} 0 & 0 \\ -\mathbf{D}_{\mathrm{E}} & -\sigma^{*}\mathbf{I}\end{bmatrix} + \frac{1}{\varepsilon}\begin{bmatrix} -\sigma\mathbf{I} & \mathbf{D}_{\mathrm{H}} \\ 0 & 0\end{bmatrix} = \mathcal{X} + \mathcal{Y},$$

which follows the analysis of Section 5.5. In the above, $\mathbf{I}$ is a 3×3 unity matrix and $\mathbf{D}_{\mathrm{F}}[.] = \nabla \times \mathbf{F}$ with $\mathbf{F} = \mathbf{E}, \mathbf{H}$ is the vector of curl operators. The solution of (6.9), in terms of (5.56) and (5.57), after a Δt time increment is derived through the well-known exponential operator

$$\begin{bmatrix}\mathbf{E}(\Delta t)\\ \mathbf{H}(\Delta t)\end{bmatrix} = \mathrm{e}^{\mathbf{A}\Delta t}\begin{bmatrix}\mathbf{E}(0)\\ \mathbf{H}(0)\end{bmatrix} \tag{6.10}$$

that is approximated by the symplectic integrator propagator, given by the multiproduct of exponential operator of $\mathcal{X}$ and $\mathcal{Y}$. More specifically,

$$e^{\mathbf{A}\Delta t} = \prod_{k=1}^{K} e^{b_k \mathcal{Y}\Delta t} e^{a_k \mathcal{X}\Delta t} + O\left[(\Delta t)^{K+1}\right], \tag{6.11}$$

with a_k and b_k real parameters used for the design of the propagator and K is the stage number of quantities

$$e^{\mathcal{X}\Delta t} = \begin{bmatrix} \mathbf{I} & \mathbf{0} \\ -\frac{1-e^{-\sigma^{*}\Delta t/\mu}}{\sigma^{*}}\mathbf{D}_{\mathrm{E}} & e^{-\sigma^{*}\Delta t/\mu}\mathbf{I}\end{bmatrix} \quad \text{and} \quad e^{\mathcal{Y}\Delta t} = \begin{bmatrix} e^{-\sigma\Delta t/\varepsilon}\mathbf{I} & \frac{1-e^{-\sigma\Delta t/\varepsilon}}{\sigma}\mathbf{D}_{\mathrm{H}} \\ \mathbf{0} & \mathbf{I}\end{bmatrix}. \tag{6.12}$$

A sufficient means for computing (6.12) are the Padé schemes of

$$e^{-x} \simeq \frac{1 - \frac{x}{2} + \frac{x^2}{12}}{1 + \frac{x}{2} + \frac{x^2}{12}} \quad \text{and} \quad \frac{1 - e^{-x}}{x} \simeq \frac{1 - \frac{x}{10} + \frac{x^2}{60}}{1 + \frac{2x}{5} + \frac{x^2}{20}}, \tag{6.13}$$

discussed in [16, 18], where $x = \sigma \Delta t/\varepsilon$ or $x = \sigma^* \Delta t/\mu$. If temporal and spatial differentiations are performed via the well-known fourth-order operators (2.14) and (2.18), a higher order symplectic FDTD algorithm is derived. For instance, the three-dimensional (3-D) expressions for the third-stage H_z and E_z components at the interface between two dielectric media (with ε^{A} and ε^{B} as their dielectric permittivities) are given by

$$\begin{aligned} H_z\big|^{n+3/5}_{i+1/2,j+1/2,k} = {} & H_z\big|^{n+2/5}_{i+1/2,j+1/2,k} \\ & - \tfrac{a_3 \Delta t}{\mu \Delta x}\left[\tfrac{9}{8}\left(E_y\big|^{n+2/5}_{i+1,j+1/2,k} - E_y\big|^{n+2/5}_{i,j+1/2,k}\right)\right. \\ & \left. - \tfrac{1}{24}\left(E_y\big|^{n+2/5}_{i+2,j+1/2,k} - E_y\big|^{n+2/5}_{i-1,j+1/2,k}\right)\right] \\ & + \tfrac{a_3 \Delta t}{\mu \Delta y}\left[\tfrac{9}{8}\left(E_x\big|^{n+2/5}_{i+1/2,j+1,k} - E_x\big|^{n+2/5}_{i+1/2,j,k}\right)\right. \\ & \left. - \tfrac{1}{24}\left(E_x\big|^{n+2/5}_{i+1/2,j+2,k} - E_x\big|^{n+2/5}_{i+1/2,j-1,k}\right)\right], \end{aligned} \tag{6.14}$$

$$\begin{aligned} E_z\big|^{n+3/5}_{i,j,k+1/2} = {} & E_z\big|^{n+2/5}_{i,j,k+1/2} \\ & + \tfrac{b_3 \Delta t}{\Delta x}\left[C_{\mathrm{A},x}\left(H_y\big|^{n+3/5}_{i+1/2,j,k+1/2} - H_y\big|^{n+3/5}_{i-1/2,j,k+1/2}\right)\right. \\ & \left. - C_{\mathrm{B},x}\left(H_y\big|^{n+3/5}_{i+3/2,j,k+1/2} - H_y\big|^{n+3/5}_{i-3/2,j,k+1/2}\right)\right] \\ & - \tfrac{b_3 \Delta t}{\Delta y}\left[C_{\mathrm{A},y}\left(H_x\big|^{n+3/5}_{i,j+1/2,k+1/2} - H_x\big|^{n+3/5}_{i,j-1/2,k+1/2}\right)\right. \\ & \left. - C_{\mathrm{B},y}\left(H_x\big|^{n+3/5}_{i,j+3/2,k+1/2} - H_x\big|^{n+3/5}_{i,j-3/2,k+1/2}\right)\right], \end{aligned} \tag{6.15}$$

in which

$$C_{\mathrm{A},\zeta} = \frac{9}{8\, \varepsilon^{\mathrm{A}}_{\zeta}\big|_{i,j,k+1/2}} \quad \text{and} \quad C_{\mathrm{B},\zeta} = \frac{1}{24\, \varepsilon^{\mathrm{B}}_{\zeta}\big|_{i,j,k+1/2}} \quad \text{for } \zeta \in (x, y),$$

and $\varepsilon^{\mathrm{mt}}_x$, $\varepsilon^{\mathrm{mt}}_y$($\mathrm{mt} = \mathrm{A},\ \mathrm{B}$) correspond to the appropriate field components. Equations (6.14) and (6.15) demonstrate that the successive update of internal data during the stages of one time-step does not need supplementary memory for their temporary storage. In this manner, the method manages to merge the accuracy of the higher order FDTD notions with its capability to model large numbers of conducting interfaces and most importantly, it demands 75% less memory than the Runge–Kutta process and 65% coarser grids along every direction as compared to the original Yee's technique.

6.4 THE HIGHER ORDER ADI-FDTD METHOD

6.4.1 Main Motives and Objectives

The procedure of time advancing plays a serious role in the performance of any time-domain numerical technique. It has been long recognized that a methodically founded temporal integrator assures algorithmic stability and can, under pertinent conditions, minimize the length of the entire simulation by employing several acceleration perspectives. Indeed, research activity on this particular subject has evolved alongside the developments regarding accuracy and convergence aspects. Unfortunately, the use of Yee's approach is often restricted by strict criteria that impose an upper threshold to time-step selection to avoid the uncontrollable exponential growth of field quantities. So, when electrically large problems comprising frequency-dependent materials, which allow multipath wave propagation, are to be treated by the ordinary FDTD method over a wide frequency range, the Courant criterion requires very small temporal increments, and so yielding elongated simulations. Possible solution to this problem may be obtained by the ADI-FDTD technique which, through an exact time-step subdivision, becomes *unconditionally stable* [19, 20]. However, extensive studies [21–41] indicated that the stability limit cannot be adequately surpassed because dispersion errors are constantly increasing. These artificial mechanisms depend closely on grid resolution and sampling rate, while their negative influence is amplified by the dispersive nature of modern media, especially in wideband evaluations.

To correct the above shortcomings, a variety of competent algorithms have been proposed so far [24, 28, 30, 35–44]. In order to exploit the merits of higher order nonstandard schemes, this section presents a 3-D curvilinear ADI-FDTD technique whose time intervals can greatly exceed the Courant condition without prohibitive dispersion errors [42–44]. Providing that the constitutive parameters $\varepsilon = \varepsilon_0\varepsilon_\mathrm{r}$ and $\mu = \mu_0\mu_\mathrm{r}$ of most realistic materials depend on frequency, broadband performance predictions entail that the specific method should be developed in a frequency-dependent framework. For the simulation of this dependence, an *N*th-order scheme that incorporates the first-order Debye differential model [29, 31, 33] is launched. Consequently, the excessive difficulty of convolution integral approaches is circumvented, while geometries that would otherwise require diverse nonuniform grids are now represented by the proper discretization.

6.4.2 Generalized Theoretical Establishment

Consider a dispersive material (with a homogeneous, linear, isotropic, and lossy profile), whose electrical frequency dependence is described by the variation of complex permittivity $\hat{\varepsilon}(\omega)$ as

$$\hat{\varepsilon}(\omega) = \varepsilon_0\varepsilon_\mathrm{r}(\omega) = \varepsilon_0\left(1 + \sum_{\eta=1}^{N} \frac{\chi_\eta}{1 + j\omega\tau_\eta}\right), \tag{6.16}$$

where χ_η and τ_η are fitting coefficients referring, respectively, to the maximum attenuation and relaxation time of the medium. Their values, for a given N, may be acquired from $\varepsilon_r(\omega)$ measurement data at prefixed frequencies through any fast optimization process or explicitly computed by a certain quantitative relation. Actually, ordinary applications need two or three terms for their precise representation. Hence, for a general coordinate system (u, v, w), the electric flux density $\mathbf{D} = [D_u\ D_v\ D_w]^T$ is written as

$$\mathbf{D} = \hat{\varepsilon}(\omega)\mathbf{E} = \varepsilon_0\mathbf{E} + \varepsilon_0 \sum_{\eta=1}^{N} \frac{\chi_\eta}{1 + j\omega\tau_\eta}\mathbf{E} = \varepsilon_0\mathbf{E} + \varepsilon_0 \sum_{\eta=1}^{N} \mathbf{P}_\eta, \tag{6.17}$$

with $\mathbf{E} = [E_u\ E_v\ E_w]^T$ the electric field intensity and $\mathbf{P}_\eta = \chi_\eta\mathbf{E}/(1 + j\omega\tau_\eta)$ a set of N intermediate electric polarization vectors for the efficient treatment of the summation. Under these assumptions and the use of the higher order nonstandard methodology of (3.70)–(3.81), Maxwell's time-dependent equations become

$$\textit{Ampere's law:} \quad \mathbf{G}^{\mathrm{H}}\overline{\mathcal{S}}[\mathbf{H}] = \mathcal{T}[\mathbf{D}] + \sigma\mathbf{E} + \mathbf{J}, \tag{6.18a}$$

$$\textit{Faraday's law:} \quad \mathbf{G}^{\mathrm{E}}\overline{\mathcal{S}}[\mathbf{E}] = -\mu\mathcal{T}[\mathbf{H}], \tag{6.18b}$$

$$\mathbf{D} = \varepsilon_0\mathbf{E} + \sum_{\eta=1}^{N}\mathbf{P}_\eta, \qquad \mathbf{P}_\eta + \tau_\eta\mathcal{T}\left[\mathbf{P}_\eta\right] = \chi_\eta\mathbf{E} \quad \text{for } \eta = 1, 2, \ldots, N. \tag{6.18c}$$

In (6.18), σ denotes the electric losses of the medium under study, $\mathbf{H} = [H_u\ H_v\ H_w]^T$ is the magnetic intensity, and $\mathbf{J} = [J_u\ J_v\ J_w]^T$ is a prearranged electric current density source used for the external excitation of the structure. Observe that (6.18c) constitutes the auxiliary differential equation form that provides the mathematical background of the frequency relationship between vectors $\mathbf{D}$ and $\mathbf{E}$. Specifically, it is derived via the inverse Fourier transform of the $\mathbf{P}_\eta$ definition considering an $e^{j\omega t}$ variation.

The dispersion-optimized ADI-FDTD algorithm preserves the simplicity of the common approach and circumvents the defects of conventional higher order finite-difference schemes. So, it splits the original iteration of a component into two subiterations, namely for time forwarding from the nth to the $(n + 1)$th time-step, one obtains, the first subiteration from n to $n + \frac{1}{2}$ and the second one from $n + \frac{1}{2}$ to $n + 1$. It is stressed that for the sake of symmetry during the ADI splitting process, the electric current density terms in (6.18a) are replaced with judiciously adjusted temporal averages. For instance, at time-step $n + \frac{1}{2}$,

$$\begin{aligned} \sigma\mathbf{E}^{n+1/2} &= \tfrac{\sigma}{4}\left[\left(\mathbf{E}^{n+1/2} + \mathbf{E}^{n}\right) + \left(\mathbf{E}^{n+1} + \mathbf{E}^{n+1/2}\right)\right] \\ \mathbf{J}^{n+1/2} &= \tfrac{1}{2}\left(\mathbf{J}^{n+1/4} + \mathbf{J}^{n+3/4}\right), \end{aligned} \tag{6.19}$$

from which the terms $\frac{\sigma}{4}(\mathbf{E}^{n+1/2} + \mathbf{E}^{n})$ and $\frac{1}{2}\mathbf{J}^{n+1/4}$ are employed in the first, and the $\frac{\sigma}{4}(\mathbf{E}^{n+1} + \mathbf{E}^{n+1/2})$ and $\frac{1}{2}\mathbf{J}^{n+3/4}$ ones in the second subiteration of the method.

Now, assume the dual-cell FDTD structure of Figure 3.5 and focus on the update of electric component E_u. In the first subiteration, the u-directed part of the nonstandard Ampere's law, (6.18a), yields

$$\mathcal{T}\left[D_u\big|_{\text{ps1}}^{n+1/2} \right] + \frac{\sigma}{4}\left(E_u\big|_{\text{ps1}}^{n+1/2} + E_u\big|_{\text{ps1}}^{n} \right) + \frac{1}{2} J_u\big|_{\text{ps1}}^{n+1/4}$$
$$= g_{uw} S_v \left[H_w\big|_{\text{ps1}}^{n+1/2} \right] - g_{uv} S_w \left[H_v\big|_{\text{ps1}}^{n} \right], \tag{6.20}$$

with the subscript $\text{ps1} = (i + \frac{1}{2}, j, k)$. Expanding operator $\mathcal{T}[.]$ and after some algebra, it is derived that

$$D_u\Big|_{\text{ps1}}^{n+1/2} = D_u\Big|_{\text{ps1}}^{n} + \Psi(\omega, \Delta t)\Big\{ g_{uw} S_v \left[H_w\big|_{\text{ps1}}^{n+1/2} \right] - g_{uv} S_w \left[H_v\big|_{\text{ps1}}^{n} \right]$$
$$- \frac{\sigma}{4}\left(E_u\big|_{\text{ps1}}^{n+1/2} + E_u\big|_{\text{ps1}}^{n} \right) - \frac{1}{2} J_u\big|_{\text{ps1}}^{n+1/4} + T_D^{\text{HO}} \Big\}, \tag{6.21}$$

where T_D^{HO} is a weighting function that contains all higher order time differentiations of D_u performed at n or earlier time-steps (known values) and offers extra field data for the update, as in (3.86). Combined with $\Psi(\omega, \Delta t)$ of (3.76) and the enhanced spatial operators, this degree of freedom leads to a large suppression of the dispersion error, even when the Courant stability criterion has been significantly exceeded [42].

As promptly detected from (6.21), partial derivative $S_v[H_w]$ must be implicitly evaluated provided that it involves only unknown H_w pivotal $(i \pm \frac{1}{2}, j + \frac{1}{2}, k)$ values at time-step $n + \frac{1}{2}$, while its $S_w[H_v]$ counterpart is explicitly represented by the already computed H_v quantities at the nth time-step. To eliminate H_w, the same ADI concept is implemented in the w-directed part of Faraday's law, (6.18b), described by

$$\mu \mathcal{T}\left[H_w\big|_{\text{ps2}}^{n+1/2} \right] = g_{wu} S_v \left[E_u\big|_{\text{ps2}}^{n+1/2} \right] - g_{wv} S_u \left[E_v\big|_{\text{ps2}}^{n} \right], \tag{6.22}$$

with a stencil value of $\text{ps2} = (i \pm \frac{1}{2}, j + \frac{1}{2}, k)$. After the application of $\mathcal{T}[.]$, (6.22) becomes

$$H_w\big|_{\text{ps2}}^{n+1/2} = H_w\big|_{\text{ps2}}^{n} + \frac{\Psi(\omega, \Delta t)}{\mu}\Big\{ g_{wu} S_v \left[E_u\big|_{\text{ps2}}^{n+1/2} \right] - g_{wv} S_u \left[E_v\big|_{\text{ps2}}^{n} \right] + T_H^{\text{HO}} \Big\}, \tag{6.23}$$

in which T_H^{HO} is the corresponding function for the higher order temporal derivatives of H_w based on known data. However, before the replacement of (6.23), our attention will be drawn to the final unknown data of (6.21), i.e. the D_u at $n + \frac{1}{2}$, whose presence is attributed to the material's frequency-dependent nature. The treatment of this component may be attained through an unconditionally stable Crank–Nicolson scheme applied to (6.18c) that takes into account the action of $\mathcal{T}[.]$ operator. Thus, for $\eta = 1, 2, \ldots, N$ terms in (6.17), the ηth differential equation is discretized as

$$P_{u\eta}\Big|_{\mathrm{ps1}}^{n+1/2} = \left[\frac{\chi_\eta \Psi(\omega, \Delta t)}{4\tau_\eta + \Psi(\omega, \Delta t)}\right]\left(E_u\Big|_{\mathrm{ps1}}^{n+1/2} + E_u\Big|_{\mathrm{ps1}}^{n}\right) + \left[\frac{4\tau_\eta - \Psi(\omega, \Delta t)}{4\tau_\eta + \Psi(\omega, \Delta t)}\right] P_{u\eta}\Big|_{\mathrm{ps1}}^{n} + \frac{T_P^{\mathrm{HO}}}{4\tau_\eta + \Psi(\omega, \Delta t)}, \tag{6.24}$$

where T_P^{HO} is again the suitable weighting function. In this way, the expression for computing D_u reads

$$D_u\Big|_{\mathrm{ps1}}^{n+1/2} = \varepsilon_0\, E_u\Big|_{\mathrm{ps1}}^{n+1/2} + \varepsilon_0 \sum_{\eta=1}^{N} P_{u\eta}\Big|_{\mathrm{ps1}}^{n+1/2}. \tag{6.25}$$

Next, analysis substitutes (6.23) and (6.25) into (6.21) to extract the update expression for E_u at $n + \frac{1}{2}$:

$$\begin{aligned}\vartheta_1 E_u\Big|_{i+1/2,j,k}^{n+1/2} - \vartheta_2 E_u\Big|_{i+1/2,j+1,k}^{n+1/2} - \vartheta_3 E_u\Big|_{i+1/2,j-1,k}^{n+1/2} &= \vartheta_4 E_u\Big|_{i+1/2,j,k}^{n} + \vartheta_5 J_u\Big|_{i+1/2,j,k}^{n+1/4} \\ &+ \vartheta_6 \mathcal{S}_v\left[H_w\Big|_{i+1/2,j+1/2,k}^{n}\right] - \vartheta_7 \mathcal{S}_w\left[H_v\Big|_{i+1/2,j,k+1/2}^{n}\right] \\ &- \vartheta_8\left\{\mathcal{S}_u\left[E_v\Big|_{i,j+1/2,k}^{n}\right] + \mathcal{S}_v\left[E_u\Big|_{i,j-1/2,k}^{n}\right]\right\} + \vartheta_9 \sum_{\eta=1}^{N} P_{u\eta}\Big|_{i+1/2,j/2,k}^{n},\end{aligned} \tag{6.26}$$

with parameters $\vartheta_m (m = 1, 2, \ldots, 9)$ defined in terms of the algorithm's inherent traits [spatial increments and functions $\Psi(\omega, \Delta t)$, T^{HO}], the constitutive properties of every material as well as the metrical coefficients, $g(u, v, w)$. Since, ϑ_m have constant values, their calculation – hardly affecting the overall burden – is conducted only once during every time update. Equation (6.26) is repeated for every j along the v-grid direction, where the spatial alternation is performed, and so resulting in a sparse three-band tridiagonal system that can be recursively and rapidly solved by means of several well-known techniques [18]. Upon acquiring E_u component, the prior formulation is identically applied to E_v and E_w, while the respective magnetic and electric polarization quantities are explicitly obtained via their nonstandard FDTD expressions.

The second subiteration for the temporal forwarding of E_u in the interval from $n + \frac{1}{2}$ to $n + 1$ reverses the role of $\mathcal{S}_v[H_w]$ and $\mathcal{S}_w[H_v]$, modifying (6.20) as

$$\begin{aligned}\mathcal{T}\left[D_u\Big|_{\mathrm{ps1}}^{n+1/2}\right] + \tfrac{\sigma}{4}\left(E_u\Big|_{\mathrm{ps1}}^{n+1} + E_u\Big|_{\mathrm{ps1}}^{n+1/2}\right) + \tfrac{1}{2} J_u\Big|_{\mathrm{ps1}}^{n+3/4} \\ = g_{uw}\mathcal{S}_v\left[H_w\Big|_{\mathrm{ps1}}^{n+1/2}\right] - g_{uv}\mathcal{S}_w\left[H_v\Big|_{\mathrm{ps1}}^{n+1}\right].\end{aligned} \tag{6.27}$$

Now, the unknown variables, are H_v and D_u (after the expansion of $\mathcal{T}[.]$) at time-step $n + 1$. Eliminating these terms in a way similar to (6.22)–(6.25) of the first subiteration, one reaches the tridiagonal system:

$$\psi_1 E_u\Big|_{i+1/2,j,k}^{n+1} - \psi_2 E_u\Big|_{i+1/2,j,k+1}^{n+1} - \psi_3 E_u\Big|_{i+1/2,j,k-1}^{n+1} = \psi_4 E_u\Big|_{i+1/2,j,k}^{n+1/2} + \psi_5 J_u\Big|_{i+1/2,j,k}^{n+3/4}$$
$$- \psi_6 S_w\left[H_v\Big|_{i+1/2,j,k+1/2}^{n+1/2}\right] + \psi_7 S_v\left[H_w\Big|_{i+1/2,j+1/2,k}^{n+1/2}\right]$$
$$+ \psi_8\left\{S_v\left[E_w\Big|_{i,j,k+1/2}^{n+1/2}\right] - S_u\left[E_w\Big|_{i,j,k-1/2}^{n+1/2}\right]\right\} - \psi_9\sum_{\eta=1}^{N} P_{u\eta}\Big|_{i+1/2,j/2,k}^{n+1/2}, \quad (6.28)$$

where ψ_m are the counterparts of ϑ_m. Once all electric quantities are computed, then by solving the resulting systems, the remaining fields are explicitly evaluated and the process initiates again for the next time-step.

A completely analogous procedure is followed for media where the frequency-dependent constitutive parameter is magnetic permeability μ. In such a case, Maxwell's equations are accordingly established, while a constitutive relation, similar to (6.17), between η intermediate magnetic polarization vectors and the magnetic flux density $\mathbf{B} = [B_u\ B_v\ B_w]^T$ is derived. Note that the extension of the above method to structures comprising both types of materials is straightforward without any cumbersome constraints [44]. As a matter of fact, every material is rigorously modeled by the appropriate scheme regarding the variation of ε or μ. It is also important to stress that, apart from some fairly limited storage requirements, the algorithm does not considerably increase the total CPU and memory overhead. On the contrary, its ability to accomplish dispersion-optimized simulations of general applications by seriously surpassing the Courant stability condition leads to great computational savings, as will be verified in Chapters 7 and 8.

6.4.3 Stability and Dispersion-Error Analysis

Concerning the stability of the higher order curvilinear ADI-FDTD algorithm, the von Neumann method is applied and the two subiterations are expressed in matrix form as

$$\textit{1st subiteration} \quad n \to n + \tfrac{1}{2}: \quad \mathbf{Z}_1\mathbf{E}^{n+1/2} = \boldsymbol{\Theta}_1\mathbf{E}^{n}, \quad (6.29a)$$
$$\textit{2nd subiteration} \quad n + \tfrac{1}{2} \to n + 1: \quad \mathbf{Z}_2\mathbf{E}^{n+1} = \boldsymbol{\Theta}_2\mathbf{E}^{n+1/2}, \quad (6.29b)$$

where sparse matrices $\mathbf{Z}_l$ and $\boldsymbol{\Theta}_l$ $(l = 1, 2)$ are created by the proper ϑ_m and ψ_m parameters during the creation of the system. If (6.29a) and (6.29b) are combined into a single equation, one acquires

$$\mathbf{E}^{n+1} = \mathbf{Z}_2^{-1}\boldsymbol{\Theta}_2\mathbf{Z}_1^{-1}\boldsymbol{\Theta}_1\mathbf{E}^{n} = \mathbf{M}\mathbf{E}^{n}. \quad (6.30)$$

Observe that $\mathbf{Z}_1^{-1}$ needs to be computed only for the nodes at the left-most area of the lattice, because the electric field at the remaining points can be recursively calculated by the two left neighboring cells. This remark confirms the three-band nature of (6.26) and (6.28) and illustrates a compact way of implementing the ADI technique. The study continues with the solution of the eigenvalue problem, related to (6.30), which gives

$$\lambda_{1,2} = 1 \quad \text{and} \quad \lambda_s = \left[\sqrt{2R_1^2 - 3R_2^2} \pm j(R_2/2 - R_1)\right] / R_1 \quad \text{for } s = 3, \ldots, 6, \tag{6.31}$$

where R_1 and R_2 are functions of $K_\zeta = (\Delta t/\Delta\zeta)\sin(k_\zeta \Delta\zeta/2)$ for $\zeta \in (u, v, w)$, $j = \sqrt{-1}$. Examining the profile of (6.31), the magnitudes of all eigenvalues are found to be less or equal to unity, a fact that reveals the formulation's unconditional stability. This property enhances the convergence of numerical simulation and subdues anisotropy errors independently of mesh resolution opposite to the majority of customary approaches. Hence, the restrictive Courant limit can be *drastically exceeded* with high accuracy levels.

Finally, it would be instructive to examine the improvement of the dispersion relation and verify the large suppression of lattice reflection errors, principally when the temporal increment is selected far beyond the usual FDTD stability criterion. Following a general framework, the dispersion error is defined as

$$\nu_{\text{disp}} = F(u, v, w, t) - F^{\text{num}}(u, v, w, t) = \int \alpha(\omega) e^{j[\mathbf{k}(\omega) - \mathbf{k}^{\text{num}}(\omega)]\mathbf{r}} d\omega, \tag{6.32}$$

in which F is the exact and F^{num} the numerical solution of the problem with their respective wavenumber vectors $\mathbf{k}$ and $\mathbf{k}^{\text{num}}$. Coefficients $\alpha(\omega)$ are the amplitudes of the Fourier transform in (6.32), while $\mathbf{r}$ is a position vector in the (u, v, w) coordinate system. Specifically, F^{num} is assumed to be the superposition of both propagating and evanescent waves, unlike other simplified analyses that employ only the former ones. In this context, the extraction of the dispersion error is based on the estimation of

$$\sum_{\zeta=u,v,w} \left| e^{j[k(\omega) - k^{\text{num}}(\omega)]\zeta} \right| \leq \sum_{\zeta=u,v,w} \left| k(\omega) - k^{\text{num}}(\omega) \right| |\zeta|, \tag{6.33}$$

where $k(\omega) = \omega$ is the free-space dispersion relation and

$$k^{\text{num}}(\omega) \cong \omega \left\{ 1 - \frac{4(\Delta t)^{M+2L}}{873} + O\left[(\Delta t)^{2(M+2L)} \right] \right\} \tag{6.34}$$

the relevant formula of the enhanced ADI-FDTD method. Apparently, (6.34) achieves a substantial improvement, easily adjusted by the order of accuracy M and the complementary degree of freedom L. Indicatively, for $M = 3$ and $L = 2$, selected for most simulations, the second and dominant error term of (6.34) is at least seven orders of magnitude lower than the corresponding term of the common dispersion relation. In the light of these considerations, ν_{disp} becomes

$$\nu_{\text{disp}} \cong \frac{5 + 4(\Delta t)^{M+L}}{873} \omega. \tag{6.35}$$

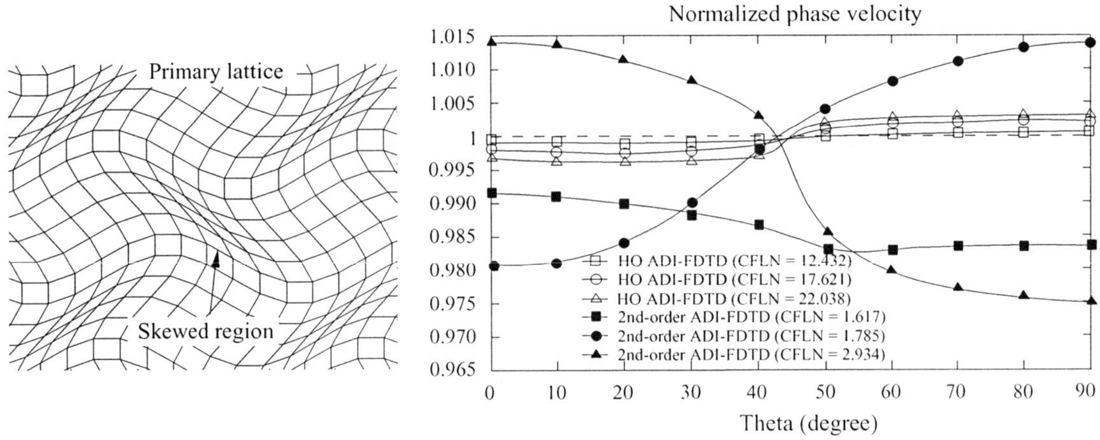

(a) (b)

FIGURE 6.3: (a) A skewed mesh with localized discontinuities and (b) normalized phase velocity versus angle θ for various second-order and higher order (HO) ADI-FDTD implementations

To validate the contribution of the higher order nonstandard concepts to the performance of the ADI-FDTD algorithm, let us consider the skewed mesh of Figure 6.3(a), truncated by an eight-cell PML. Figure 6.3(b) presents the normalized phase velocity versus angle θ for several configurations. Furthermore, Figure 6.4(a) shows the maximum dispersion error versus Courant-Friedrich-Levy number (CFLN) for different lattice resolutions (λ / R_s), while in Figure 6.4(b) the reduction of the accumulating phase discrepancies for various higher order

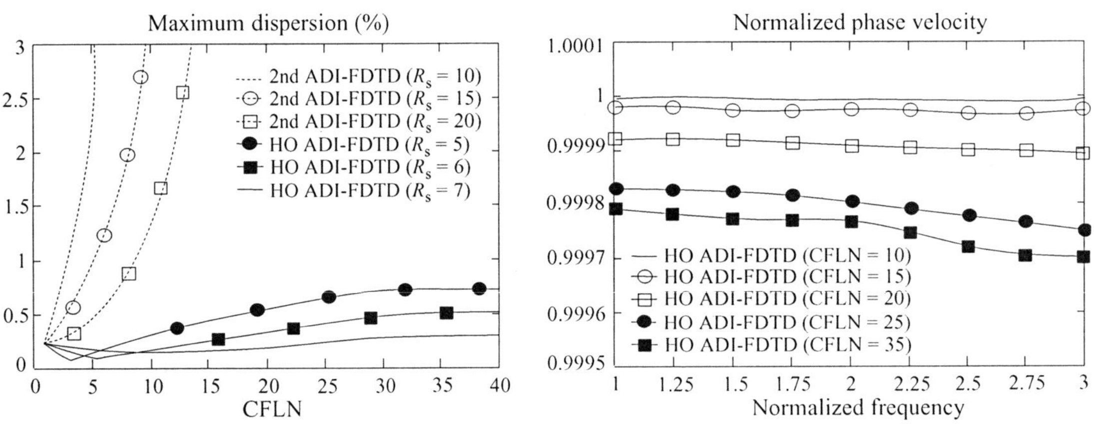

(a) (b)

FIGURE 6.4: (a) Maximum dispersion error as a function of CFLN and lattice resolution and (b) normalized phase velocity versus frequency for diverse ADI-FDTD configurations

ADI-FDTD realizations is displayed. Recall that the CFLN is defined as CFLN $= \Delta t/\Delta t_{\max}$, with $\Delta t_{\max}$ the maximum time-step denoted via the stability criterion of the second-order or higher order nonstandard FDTD method. As can be deduced, the superiority of (6.35) is more prominent when time-steps exceed the Courant limit at a great extent (CFLN > 20). On the other hand, the dispersion error of the improved algorithm is very small even for very coarse grids, whereas its normalized phase velocity is notably close to the free-space one, despite the fact that the Courant limit is greatly surpassed.

6.5 HIGHER ORDER WEIGHTED ESSENTIALLY NONOSCILLATORY SCHEMES IN THE TIME DOMAIN

The purpose of the WENO schemes is to establish a cell-averaged reconstruction framework, which approximates field fluxes at cell boundaries at higher order accuracy levels and alleviates spurious oscillations near discontinuities [45–49]. Hence, instead of evaluating field components employing only one fixed stencil, these schemes incorporate a convex combination of all candidate stencils inside the elementary dual cell. Each of these stencils is assigned a weight that determines its participation in the final approximation of every quantity. The weights can be defined in such a way that in regions of geometric complexity a higher order accuracy is attained, while in smooth areas wave propagation is properly resolved. So, it becomes apparent that their calculation constitutes a significant issue.

To take avail of the WENO advantages and circumvent their original implicit character, a 3-D time-domain (WENO-TD) algorithm in curvilinear coordinates is analyzed in this section. The *fully explicit* method computes the smoothness level of problematic areas and develops a local adaptive finite-difference process that yields the suitable polynomials for spatial differentiation. Moreover, through the selective usage of generally curved higher order FDTD counterparts, the erroneous nonsymmetric gridding of curl operators is ruled out and lattice anisotropy is greatly restrained. In this way, spatial discretization conserves numerical energy, diminishes dispersion errors in a wideband sense, and establishes the correct cell-to-cell continuity conditions for electromagnetic fields across discontinuities. Finally, temporal integration is achieved by a modified leapfrog or Runge–Kutta approach, which annihilates exponentially growing modes.

6.5.1 Development of the Three-Dimensional Algorithm

Let us express Maxwell's hyperbolic time-dependent laws in vector notation:

$$\frac{\partial \mathbf{F}_{\mathrm{A}}(\mathbf{r},t)}{\partial t} + \nabla \times \mathbf{F}_{\mathrm{B}}(\mathbf{r},t) = \mathbf{s}(\mathbf{r},t), \tag{6.36}$$

where $\mathbf{F}_A = [-\mathbf{D} \quad \mathbf{B}]^T$ and $\mathbf{F}_B = [\mathbf{H} \quad \mathbf{E}]^T$ are the electric and magnetic fluxes and intensities in a general coordinate system (u, v, w), $\mathbf{r}$ the position vector, and $\mathbf{s}$ the source term. Regarding spatial derivative approximation in (6.36), the cell average of quantity $f|_{i,j,k}$ is defined as

$$\tilde{f}(t)\big|_{i,j,k} = V_m^{-1} \iiint_{V_m} g(\varepsilon, \mu) f(\mathbf{r}, t)\, dV, \tag{6.37}$$

for a predetermined $(i\Delta u, j\Delta v, k\Delta w)$ discretization, with $V_m = \Delta u \Delta v \Delta w$ the volume of the $m_{i,j,k}$ cell. Given that $g(\varepsilon, \mu)$ are known functions specified by each material in the domain of interest, the control integral formula in (6.37) may be successfully computed via a Gaussian numerical quadrature over the faces and along the edges of the particular element. For illustration if the $m_{i,j,k}$ cell is curvilinear and spans in

$$m_{i,j,k} = \left[u|_{i-1/2}, u|_{i+1/2}\right] \times \left[v|_{j-1/2}, v|_{j+1/2}\right] \times \left[w|_{k-1/2}, w|_{k+1/2}\right], \tag{6.38}$$

the cell average at the $(i + \frac{1}{2}, j, k)$ face is prescribed by

$$\tilde{f}(t)\big|_{i+1/2,j,k} = \frac{1}{\Delta v \Delta w} \sum_{r=1}^{N} \sum_{q=1}^{N} f\left(\mathbf{r}(u|_{i+1/2}, v|_r, w|_q)\right) \Xi_q \Xi_r, \tag{6.39}$$

with the subscripts r and q corresponding to different integration points $v|_r$ and $w|_q$ and grid-oriented coefficients Ξ_r and Ξ_q. Note that (6.39) calls for pointwise values of f, whereas the overall process evolves its volumetric design. In this context, the spatial derivatives of (6.36) can be efficiently obtained through

$$\frac{\partial f(\mathbf{r}, t)}{\partial \zeta} = W_\zeta(\mathbf{r}, t) + O\left[(\Delta\zeta)^{p+1}\right] \quad \text{for } \zeta \in (u, v, w), \tag{6.40}$$

by means of the piecewise polynomials $W_\zeta(\mathbf{r}, t)$. Now, our basic objective is to formulate a weighted convex combination of all functional stencils R_m pertinent to the $m_{i,j,k}$ cell and reconstruct $W_\zeta(\mathbf{r}, t)$. Actually, the proposed $(2p-1)$th-order schemes allocate an *optimal non-linear weight* to each stencil, indicating its contribution to the evaluation of field components. Therefore, the essentially nonoscillatory property is totally exploited in contrast to existing notions with two-point approximations. Assuming u as the direction under study, the resultant stencil classification is

$$\{R_m\}_u^{p-1} = \left\{ f|_{i-s+p/2,j,k}^n, \; f|_{i-s+3p/2,j,k}^n, \ldots, f|_{i+s-p/2,j,k}^n \right\}, \tag{6.41}$$

for $s = 0, 1, \ldots, p-1$. If the nodal positions are appropriately resolved, a set of p interpolating polynomials $Q_{p,u}(\mathbf{r}, t)$ is associated to the stencils of (6.41) with the recursive representation of

$$Q_{p-1,u}^{2p-1}\left(\tilde{f}|_{i-s+1/2,j,k}^n, \ldots, \tilde{f}|_{i+s-1/2,j,k}^n\right)$$
$$= a_p \sum_{l=0}^{p-1} \xi_{l,u}^p Q_{l,u}^p \left(\tilde{f}|_{i-s+p/2,j,k}^n, \ldots, \tilde{f}|_{i+s-p/2,j,k}^n\right), \tag{6.42}$$

TABLE 6.1: Optimal Coefficients $\xi^p_{l,u}$ for Polynomials $Q_{p,u}(\mathbf{r}, t)$

$\xi^p_{l,u}$	$l=0$	$l=1$	$l=2$	$l=3$	$l=4$
$p=2$	$\frac{1}{3}$	$\frac{2}{3}$	0	0	0
$p=3$	$\frac{3}{10}$	$\frac{1}{10}$	$\frac{6}{10}$	0	0
$p=4$	$-\frac{1}{15}$	$\frac{4}{15}$	$\frac{2}{15}$	$\frac{10}{15}$	0
$p=5$	$\frac{7}{24}$	$-\frac{5}{24}$	$\frac{15}{24}$	$\frac{9}{24}$	$-\frac{2}{24}$

at time-step $n\Delta t$ and similarly for directions v and w. Coefficients a_p and $\xi^p_{l,u}$ are acquired via

$$a_p = \frac{1}{2\Delta u}\left|\mathbf{r}(u, v, w) - \mathbf{r}(u - p\Delta u/2, v, w)\right| \quad \text{and} \quad \sum_{l=0}^{p-1} \xi^p_{l,u} = 1, \tag{6.43}$$

with the values of the latter for $p = 2, \ldots, 5$ given in Table 6.1. It is stated that in (6.41) the stencil set is constructed by the original f quantities, unlike (6.42) where cell averages are used. This important aspect adapts all physical mechanisms to the particular algorithm before the numerical integration of (6.39). Therefore, for $p = 3$, the first member of the $Q_{p,u}(\mathbf{r}, t)$ family becomes

$$Q^5_{1,u}(\mathbf{r}, t) = \frac{a_3}{2}\left[\tilde{f}\big|^n_{i,j,k} - \tilde{f}\big|^n_{i-2,j,k} + a_3\left(2\tilde{f}\big|^n_{i-1,j,k} - \tilde{f}\big|^n_{i-2,j,k}\right)\right]. \tag{6.44}$$

Under this perspective, $\mathcal{W}_u(\mathbf{r}, t)$ are derived by the sum of

$$\mathcal{W}_u(\mathbf{r}, t) = \sum_{s=0}^{p-1} \frac{b^u_s}{\sum_{l=0}^{p-1} b^u_l} Q_{s,u}(\mathbf{r}, t) \quad \text{where } b^u_s = \frac{\kappa^u_s}{(\gamma + \mathrm{CL}_u)^p}, \tag{6.45}$$

and γ varies below 10^{-6} to prohibit denominator zeroing. Parameters κ_s are real numbers chosen in consistency with media constitutive parameters. Moreover, CL_u is a smoothness indicator expressed as

$$CL_u = \sum_{\eta=1}^{p-1}\left\{\sum_{\lambda=1}^{\eta}\left(\delta^{p-\eta}\left[\tilde{f}\big|^n_{i-p+\lambda,j,k}\right]\right)^2\right\}\eta^{-1}, \tag{6.46}$$

with $\delta^\kappa[.]$ a 3-D difference operator recursively denoted by

$$\begin{aligned} \delta^1\left[\tilde{f}\big|^n_{i,j,k}\right] &= \tilde{f}\big|^n_{i+1,j,k} - \tilde{f}\big|^n_{i,j,k} \\ \delta^\kappa\left[\tilde{f}\big|^n_{i,j,k}\right] &= \delta^{\kappa-1}\left[\tilde{f}\big|^n_{i+1,j,k}\right] - \delta^{\kappa-1}\left[\tilde{f}\big|^n_{i,j,k}\right]. \end{aligned} \tag{6.47}$$

The above dispersion-optimized single-directional design process repeated along every mesh axis is proven fairly rigorous and convergent, even for coarse lattice resolutions. Hence, spatial derivatives receive an enhanced manipulation without augmenting overall complexity, since (6.42)–(6.46) are computed only once, ahead of the primary simulation phase.

Concerning the explicit update of $\mathbf{F}_{\mathrm{A}}(\mathbf{r}, t)$ in (6.36), the following modified leapfrog scheme is utilized

$$\mathbf{F}_{\mathrm{A}}^{n+1}(\mathbf{r}, t) = \mathbf{F}_{\mathrm{A}}^{n}(\mathbf{r}, t) - \Delta t \left[\nabla \times \mathbf{F}_{\mathrm{B}}^{n}(\mathbf{r}, t) - \mathbf{s}^{n}(\mathbf{r}, t)\right], \tag{6.48}$$

whose key feature is the systematic incorporation of the curvilinear WENO forms in the time integration procedure. Alternatively, to retain higher order temporal accuracy, a three-stage Runge–Kutta process is developed. Denoting the bracketed term on the right-hand side of (6.48) as $\mathbf{U}[.]$, one derives

$$\mathbf{F}_{\mathrm{A}}^{(1)}(\mathbf{r}, t) = \mathbf{F}_{\mathrm{A}}^{n}(\mathbf{r}, t) - \Delta t \mathbf{U}\left[\mathbf{F}_{\mathrm{B}}^{n}(\mathbf{r}, t), \mathbf{s}^{n}(\mathbf{r}, t)\right], \tag{6.49a}$$

$$\mathbf{F}_{\mathrm{A}}^{(2)}(\mathbf{r}, t) = \tfrac{1}{4}\left\{3\mathbf{F}_{\mathrm{A}}^{\mathrm{n}}(\mathbf{r}, t) + \mathbf{F}_{\mathrm{A}}^{(1)}(\mathbf{r}, t) - \Delta t \mathbf{U}\left[\mathbf{F}_{\mathrm{A}}^{(1)}(\mathbf{r}, t)\right]\right\}, \tag{6.49b}$$

$$\mathbf{F}_{\mathrm{A}}^{n+1}(\mathbf{r}, t) = \tfrac{1}{3}\left\{\mathbf{F}_{\mathrm{A}}^{n}(\mathbf{r}, t) + 2\mathbf{F}_{\mathrm{A}}^{(2)}(\mathbf{r}, t) - 2\Delta t \mathbf{U}\left[\mathbf{F}_{\mathrm{A}}^{(2)}(\mathbf{r}, t)\right]\right\}. \tag{6.49c}$$

The stability of (6.48) and (6.49) is mathematically examined by the von Neumann and the energy inequalities method [38]. According to the latter, the inner product on both sides of (6.36) with $\mathbf{F}_{\mathrm{A}}^{n+1} + \mathbf{F}_{\mathrm{A}}^{n}$ gives

$$\left\|\mathbf{F}_{\mathrm{A}}^{n+1}(\mathbf{r}, t)\right\| \leq \left\|\mathbf{F}_{\mathrm{A}}^{n}(\mathbf{r}, t)\right\| + \sum_{\tau=0}^{n} S \left\|\mathbf{U}^{\tau+1}\right\|, \tag{6.50}$$

where S is an amplification factor. Inequality (6.50) reveals that the amplitude of vector $\mathbf{F}_{\mathrm{A}}(\mathbf{r}, t)$ does not grow during its update; i.e., the discrete field energy is fully conserved. This issue yields the criterion of

$$\Delta t \leq T_{\mathrm{C}} \times \min_{ijk}\left(\frac{\Delta u}{c_{ijk}^{n,u}}, \frac{\Delta v}{c_{ijk}^{n,v}}, \frac{\Delta w}{c_{ijk}^{n,w}}\right) \quad \text{for} \quad 0 < T_{\mathrm{C}} \leq 1, \tag{6.51}$$

in which $c_{ijk}^{n,\zeta}$ is the phase velocity of the fastest wave at time-step n traveling along the $\zeta = u, v, w$ axis. From a physical viewpoint, (6.51) serves as a guideline which dictates that the technique's time-step should be smaller than the minimum propagation time in the mesh to attain the correct sampling of any wave interaction.

In order to indicate the advantages of the 3-D WENO-TD method, the radiation capabilities of a 2×1 antenna array comprising circular patches above a dielectric ($\varepsilon_{\mathrm{r}} = 4.4$) substrate are explored. Figure 6.5 presents the evolution of the E_z component between the two elements for various setups. Additionally, Table 6.2 gives the realization aspects along with the

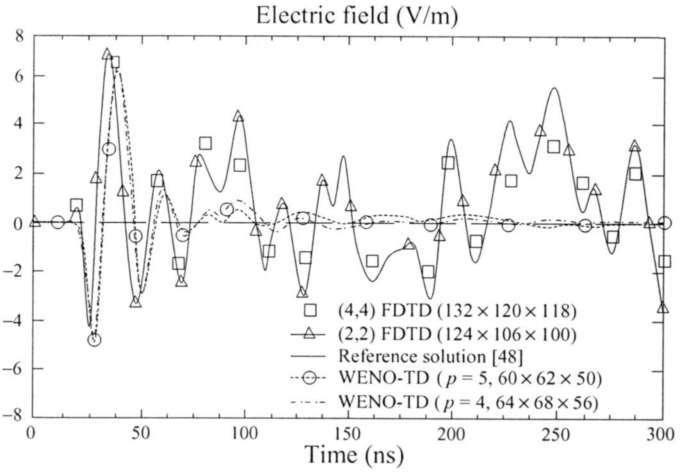

FIGURE 6.5: Evolution of the E_z component for different FDTD and WENO-TD configurations

convergence rate and maximum dispersion error. Results show that the specific algorithm is in very close agreement with the reference solution obtained via [48], unlike the (2, 2) and (4, 4) FDTD techniques, leading, also, to reasonable system demands.

6.5.2 Modeling of Media Discontinuities

The strict standards for high operational attributes in modern structures often lead to complex material discontinuities that restrict the effectiveness of numerical methods. To this problem, the low-cost WENO-TD forms can offer a promising treatment. Assume the interface of Figure 6.6, which crosses the primary lattice at the extra $d_s\,(s = 1, 2, 3, 4)$ points. Defining as h_s the distance of d_s from the relevant surfaces in the medium and focusing on the shaded areas,

TABLE 6.2: Comparison of Implementation Features and Maximum Dispersion

ALGORITHM	LATTICE	CPU TIME (h)	CONVERGENCE	DISPERSION ERROR
(2, 2) FDTD	162 × 156 × 98	28.4	1.7492	8.921×10^{-3}
(4, 4) FDTD	132 × 124 × 82	25.9	3.6571	3.745×10^{-4}
WENO-TD ($p = 3$)	86 × 78 × 54	10.7	2.9853	4.169×10^{-7}
WENO-TD ($p = 4$)	82 × 74 × 50	11.2	3.8217	2.037×10^{-9}
WENO-TD ($p = 5$)	70 × 68 × 46	11.9	4.9502	1.846×10^{-12}

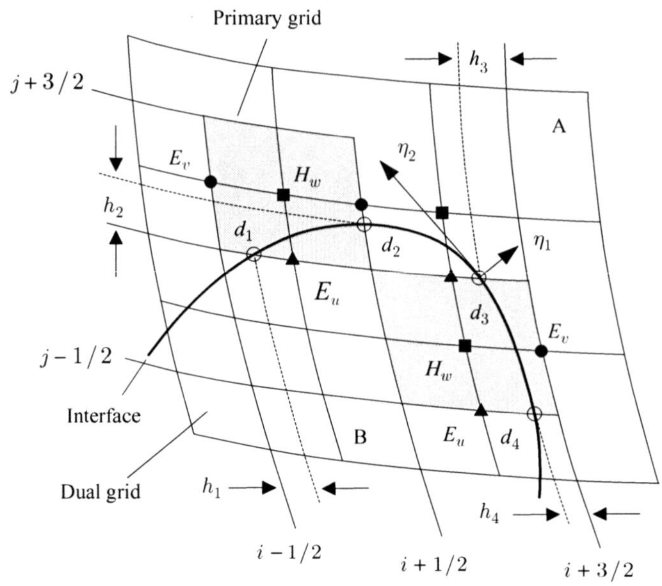

FIGURE 6.6: Transverse cut of a material interface with the auxiliary nodes for the application of the WENO-TD schemes

the expression for $\partial f/\partial u$ spatial derivative at d_2 becomes

$$\left.\frac{\partial f}{\partial u}\right|_{d_2}^{n} = \sum_{\zeta=v,w} \frac{1}{\Delta\zeta}\left[\sum_{s=1,3,4} \beta_{s,\zeta} h_s W_\zeta(\mathbf{r}^{\mathrm{mt}}, t) f\Big|_{d_s}^{n}\right] \quad \text{for mt} = \text{A, B} \tag{6.52}$$

and $\beta_{s,\zeta}$ a smoothness indicator that imposes the proper cell-to-cell continuity conditions. Thus, one has

$$\beta_{s,\zeta} = \sum_{q=1}^{p} \int_m \left(\partial_\zeta^q Q_{q,\zeta}\right)^2 \Delta\zeta^{2q-1}\, d\zeta, \tag{6.53}$$

at the $m_{i,j,k}$ cell. Equation (6.53) is applied near the interface and does not affect the rest of the domain.

For the dispersion relation, the analysis of Section 6.4.3 is employed. Likewise, in the solution both propagating and evanescent waves are superimposed to allow a precise approximation. So, it is derived that

$$k^{\mathrm{num}}(\omega) \approx \omega \left\{1 - \frac{21(\Delta t)^{2p+3}}{1537} + O\left[(\Delta t)^{p+2}\right]\right\}, \tag{6.54}$$

which reveals a serious enhancement adjusted by p. In fact, it is the volumetric pattern of the WENO schemes that circumvents fine meshes and alleviates the contaminating anisotropy discrepancies.

REFERENCES

[1] P. Thoma and T. Weiland, "A consistent subgridding scheme for the finite difference time domain method," *Int. J. Numer. Model.*, vol. 9, no. 5, pp. 359–374, Sep. 1996.

[2] P. Monk, "Sub-gridding FDTD schemes," *Appl. Comput. Electromagn. Soc. J.*, vol. 11, no. 1, pp. 37–46, 1996.

[3] M. W. Chevalier, R. J. Luebbers, and V. P. Cable, "FDTD local grid with material traverse," *IEEE Trans. Antennas Propag.*, vol. 45, no. 3, pp. 411–421, Mar. 1997.

[4] M. Okoniewski, E. Okoniewska, and M. A. Stuchly, "Three-dimensional subgridding algorithm for FDTD," *IEEE Trans. Antennas Propag.*, vol. 45, no. 3, pp. 422–429, Mar. 1997.

[5] K. M. Krishnaiah and C. J. Railton, "A stable subgridding algorithm and its application to eigenvalue problems," *IEEE Trans. Microw. Theory Tech.*, vol. 47, no. 5, pp. 620–628, May 1999.

[6] W. Yu and R. Mittra, "A new subgridding method for the finite-difference time-domain (FDTD) algorithm," *Microw. Opt. Technol. Lett.*, vol. 21, no. 5, pp. 330–333, June 1999.

[7] S. Wang, F. L. Teixeira, R. Lee, and J.-F. Lee, "Optimizing of subgridding schemes for FDTD," *IEEE Microw. Wireless Compon. Lett.*, vol. 12, no. 6, pp. 223–225, June 2002.

[8] F. Mayer, R. Schuhmann, and T. Weiland, "Flexible subgrids in FDTD calculations," in *Proc. IEEE Antennas Propag. Soc. Int. Symp.*, San Antonio, TX, June 2002, vol. 3, pp. 252–255.

[9] L. Kulas and M. Mrozowski, "Low-reflection subgridding," *IEEE Trans. Microw. Theory Tech.*, vol. 53, no. 5, pp. 1587–1592, May 2005.

[10] A. Taflove and S. C. Hagness, *Computational Electrodynamics: The Finite-Difference Time-Domain Method*, 3rd ed., Norwood, MA: Artech House, 2005.

[11] S. V. Georgakopoulos, R. A. Renaut, C. A. Balanis, and C. R. Birtcher, "A hybrid fourth-order FDTD utilizing a second-order FDTD subgrid," *IEEE Microw. Wireless Compon. Lett.*, vol. 11, no. 11, pp. 462–464, Nov. 2001.

[12] S. V. Georgakopoulos, C. R. Birtcher, C. A. Balanis, and R. A. Renaut. "HIRF penetration and PED coupling analysis for scaled fuslage models using a hybrid subgrid FDTD(2,2)/FDTD(2,4) method," *IEEE Trans. Electromagn. Compat.*, vol. 45, no. 2, pp. 293–305, May 2003.

[13] Z. Shao, G. W. Wei, and S. Zhao, "DSC time-domain solution of Maxwell's equations," *J. Comput. Phys.*, vol. 189, no. 2, pp. 427–453, Aug. 2003.

[14] Z. Shao, A. Shen, Q. He, and G. W. Wei, "A generalized higher-order finite-difference time-domain method and its application in guided-wave problems," *IEEE Trans. Microw. Theory Tech.*, vol. 51, no. 3, pp. 856–861, Mar. 2003.

[15] T. Hirono, W. W. Lui, K. Yokoyama, and S. Seki, "Stability and numerical dispersion of symplectic fourth-order time-domain schemes for optical field simulation," *J. Lightw. Technol.*, vol. 16, no. 10, pp. 1915–1920, Oct. 1998.

[16] T. Hirono, W. W. Lui, S. Seki, and Y. Yoshikuni, "A three-dimensional fourth-order finite-difference time-domain scheme using a symplectic integrator," *IEEE Trans. Microw. Theory Tech.*, vol. 49, no. 9, pp. 1640–1648, Sep. 2001.

[17] I. Saitoh and N. Takahashi, "Stability of symplectic finite-difference time-domain methods," *IEEE Trans. Magn.*, vol. 38, no. 2, pp. 665–668, Mar. 2002.

[18] J. C. Strikwerda, *Finite Difference Schemes and Partial Differential Equations*. Pacific Grove, CA: (SIAM Edition) Cole Advanced Books & Software, Wadsworth & Brooks, 2004.

[19] T. Namiki, "A new FDTD algorithm based on ADI method," *IEEE Trans. Microw. Theory Tech.*, vol. 47, no. 10, pp. 2003–2007, Oct. 1999.

[20] F. Zheng, Z. Chen, and J. Zhang, "A finite-difference time-domain method without the Courant stability conditions," *IEEE Microw. Guided Wave Lett.*, vol. 9, no. 11, pp. 441–443, Nov. 1999.

[21] S. Gedney, G. Liu, J. Roden, and A. Zhu, "Perfectly matched layer media with CFS for an unconditionally stable ADI-FDTD method," *IEEE Trans. Antennas Propag.*, vol. 49, no. 11, pp. 1554–1559, Nov. 2001.

[22] B. Z. Wang, Y. Wang, W. Yu, and R. Mittra, "A hybrid ADI-FDTD subgridding scheme for modeling on-chip interconnects," *IEEE Trans. Adv. Packag.*, vol. 24, no. 11, pp. 528–533, Nov. 2001.

[23] J. S. Kole, M. T. Figge, and H. De Raedt, "Unconditionally stable algorithms to solve the time-dependent Maxwell equation," *Phys. Rev. E*, vol. 64, no. 6, pp. 066705(1)–066705(14), Dec. 2001.

[24] S. G. García, T.-W. Lee, and S. C. Hagness, "On the accuracy of the ADI-FDTD method," *IEEE Antennas Wireless Propag. Lett.*, vol. 1, no. 1, pp. 31–34, 2002.

[25] A. P. Zhao, "Analysis of the numerical dispersion of the 2-D alternating-direction implicit FDTD method," *IEEE Trans. Microw. Theory Tech.*, vol. 50, no. 4, pp. 1156–1164, Apr. 2002.

[26] M. Darms, R. Schuhmann, H. Spachmann, and T. Weiland, "Dispersion and asymmetry effects of ADI-FDTD," *IEEE Microw. Wireless Compon. Lett.*, vol. 12, no. 12, pp. 491–493, Dec. 2002.

[27] G. Sun and C. W. Trueman, "Analysis and numerical experiments on the numerical dispersion of two-dimensional ADI-FDTD," *IEEE Antennas Wireless Propag. Lett.*, vol. 2, pp. 78–81, 2003.

[28] Z. Wang, J. Chen, and Y. Chen, "Development of a higher-order ADI-FDTD method," *Microw. Opt. Technol. Lett.*, vol. 37, no. 1, pp. 8–12, Apr. 2003.

[29] S. Staker, C. Holloway, A. Bhobe, and M. Piket-May, "ADI formulation of the FDTD method: Algorithm and material dispersion implementation," *IEEE Trans. Electromagn. Compat.*, vol. 45, pp. 156–166, May 2003.

[30] M. Wang, Z. Wang, and J. Chen, "A parameter optimized ADI-FDTD method," *IEEE Antennas Wireless Propag. Lett.*, vol. 2, pp. 118–121, 2003.

[31] S. G. García, R. G. Rubio, A. R. Bretones, and R. G. Martín, "Extension of the ADI-FDTD method to Debye media," *IEEE Trans. Antennas Propag.*, vol. 51, no. 11, pp. 3183–3186, Nov. 2003.

[32] G. Sun and C. W. Trueman, "Some fundamental characteristics of the one-dimensional alternating-direction-implicit finite-difference-time-domain method," *IEEE Trans. Microw. Theory Tech.*, vol. 52, pp. 46–52, Jan. 2004.

[33] X. T. Dong, N. V. Venkatarayalu, B. Guo, W. Y. Yin, and Y. B. Gan, "General formulation of unconditionally stable ADI-FDTD method in linear dispersive media," *IEEE Trans. Microw. Theory Tech.*, vol. 52, no. 1, pp. 170–174, Jan. 2004.

[34] J. Lee and B. Fornberg, "Some unconditionally stable time stepping methods for the 3D Maxwell's equations," *J. Comput. Appl. Math.*, vol. 166, no. 2, pp. 497–523, Apr. 2004.

[35] A. P. Zhao, "Improvement on the numerical dispersion of 2-D ADI-FDTD with artificial anisotropy," *IEEE Microw. Wireless Compon. Lett.*, vol. 14, no. 6, pp. 292–294, June 2004.

[36] S. Wang, F. L. Teixeira, and J. Chen, "An iterative ADI-FDTD with reduced splitting error," *IEEE Microw. Wireless Compon. Lett.*, vol. 15, no. 2, pp. 92–94, Feb. 2005.

[37] W. Fu and E.-L. Tan, "A compact higher-order ADI-FDTD method," *Microw. Optical Technol. Lett.*, vol. 44, no. 3, pp. 273–275, Feb. 2005.

[38] I. Zagorodnov and T. Weiland, "TE/TM scheme for computation of electromagnetic fields in accelerators," *J. Comput. Phys.*, vol. 207, no. 1, pp. 69–91, July 2005.

[39] B. Donderichi and F. L. Teixeira, "Symmetric source implementation for the ADI-FDTD method," *IEEE Trans. Antennas Propag.*, vol. 53, no. 4, pp. 1562–1565, Apr. 2005.

[40] H.-X. Zheng and K. W. Leung, "An efficient method to reduce the numerical dispersion in the ADI-FDTD," *IEEE Trans. Microw. Theory Tech.*, vol. 53, no. 7, pp. 2295–2301, July 2005.

[41] I. Ahmed and Z. Chen, "Error reduced ADI-FDTD methods," *IEEE Antennas Wireless Propag. Lett.*, vol. 4, pp. 323–325, 2005.

[42] N. V. Kantartzis, T. T. Zygiridis, and T. D. Tsiboukis, "An unconditionally stable higher order ADI-FDTD technique for the dispersionless analysis of generalized 3-D EMC structures," *IEEE Trans. Magn.*, vol. 40, no. 2, pp. 1436–1439, Mar. 2004.

[43] N. V. Kantartzis and T. D. Tsiboukis, "Unconditionally Stable Numerical Modeling and Broadband Optimization of Arbitrarily-Shaped Anechoic and Reverberating EMC

Chambers," in *Proc. 6th Europe Int. Symp. Electromagn. Compat.*, Eindhoven, The Netherlands, Sep. 2004, vol. 1, pp. 48–53.

[44] N. V. Kantartzis and T. D. Tsiboukis, "Wideband numerical modeling and performance optimisation of arbitrarily-shaped anechoic chambers via an unconditionally stable time-domain technique," *Electrical Engr.*, vol. 88, pp. 55–81, Nov. 2005.

[45] X.-D. Liu, S. Osher, and T. Chan, "Weighted essentially non-oscillatory schemes," *J. Comput. Phys.*, vol. 115, no. 1, pp. 200–212, Nov. 1994.

[46] Z. Wang and R. F. Chen, "Optimized weighted essentially nonoscillatory schemes for linear waves with discontinuity," *J. Comput. Phys.*, vol. 174, no. 1, pp. 381–404, Nov. 2001.

[47] Y.-X. Ren, M. Liu, and H. Zhang, "A characteristic-wise hybrid compact-WENO scheme for solving hyperbolic conservation laws," *J. Comput. Phys.*, vol. 192, no. 2, pp. 365–386, Dec. 2003.

[48] V. A. Titarev and E. F. Toro, "Finite-volume WENO schemes for 3-D conservation laws," *J. Compit. Phys.*, vol. 201, no. 1, pp. 238–260, Nov. 2004.

[49] Y. Xing and C.-W. Shu, "High order finite difference WENO schemes with the exact conservation property for the shallow water equations," *J. Comput. Phys.*, vol. 208, no. 1, pp. 206–227, Sep. 2005.

CHAPTER 7

Selected Applications in Waveguide Systems

7.1 INTRODUCTION

The ever increasing needs for more proficient devices in the area of waveguide structures have spurred impressive advances in their technology. The driving motive for this evolution has been a blend of factors. These vary from higher operating frequencies and integration processes on single modules to the demand of decreasing the entire fabrication cycle. As already discussed, the second-order FDTD technique, although it remains the main workhorse in time-domain numerical simulations, suffers from some nontrivial shortcomings. Unfortunately, their impact becomes prominent during the analysis of realistic problems or the design of contemporary arrangements, where the construction of very precise models is deemed crucial. These accuracy levels are required not only to characterize the performance of various waveguide elements, but also to interpret complex interaction phenomena in their interior and predict the attitude of more sophisticated configurations. Toward this direction, the present chapter deals with the application of higher order (HO) FDTD schemes, previously described, to waveguide systems in order to validate their advantages and indicate their algorithmic universality. Therefore, after a brief, yet informative, reference to diverse implementation aspects and the description of a robust modal approach for arbitrary cross sections, a selected set of devices is systematically investigated. Wherever possible, numerical results are compared to exact or measurement data, while in most cases, sufficient evidence for the computational burden is provided.

7.2 EXCITATION SCHEMES AND OPEN-END TRUNCATION

Full-vector electromagnetic analysis of modern waveguide systems via the FDTD method can be a demanding task. Typical structures of engineering interest have arbitrary shapes, apertures, cavities, and embedded passive or active circuit parts that produce complex fields not easily resolved into finite modes. Classifying these difficulties, extensive research has lead to various techniques and mitigation means [1–21]. Despite their different character, all approaches agree that the consistent simulation of the above devices requires *punctual excitations* and *reliable termination* of waveguide open ends. The former issue is of great significance, since it guarantees the separation between incident and reflected waves. Actually, without the correct source model, absorbing boundary conditions (ABCs) cannot be directly imposed on the near-end terminal plane.

It is well acknowledged that the limited dimensions of microwave devices allow the propagation of multimodal fields of diverse spatial distributions. Due to three-dimensional (3-D) curvatures, such distributions in the transverse (relative to the propagation direction) plane may not be numerically implemented when the launch of a single mode is required. Actually, it is the energy content of these modes that controls our selection. Assume, for example, the simulation of a broadband pulse. Should we have utilized the ordinary excitations, it would have been possible to lose a great deal of incident-wave information, as this pulse introduces an energy profile that depends on the wavenumber's features. In contrast, the presence of a discontinuity inside a PEC-terminated waveguide is proven *essential* for the applicability of the FDTD formulation. Thus, if the source-discontinuity distance is relatively small, reflected waves return to the source plane and interact with cutoff modes before the incident field receives its final form. A common (although not always practical) solution to this disadvantage is the use of a sufficiently long uniform section between the source plane and the discontinuity. This treatment, however, produces excessively large computational domains and elongated simulations. The most common excitation techniques are usually divided into the following three categories:

1. According to the first, at the initial time instant, electromagnetic components are defined inside the entire space of the model. This approach is basically used for the analysis of microwave resonators.
2. The second type involves the incorporation of a finite-length pulse. Due to its wideband nature, the transmission properties are determined for a prefixed frequency window spanning in one analysis cycle.
3. The third and most popular approach defines a harmonic oscillation on a boundary or transverse plane of the structure. Note that the specific scheme is applicable only to the problems where no disturbance from inherent discontinuities affects the function of the source plane. Otherwise, S-parameters are erroneously computed and no useful conclusions can be figured out.

Probing on the separation of incident and reflected waves, a concern still remains: two field quantities must be determined on the source plane and updated interactively. For waveguide discontinuities, the regularly implemented scheme of the third case places the source and the near-end terminal plane at the same position, inserts the excitation, and then applies the ABC to the source plane after the pulse has been fully propagated. Nonetheless, before any truncation process is allowed to initiate, DC distortions are induced near the incident waveform by the electric and magnetic boundary conditions. This is another reason why usual techniques cannot be located very close to the discontinuity. An efficient way to alleviate the prior weakness is to shift the source plane several cells away from the terminal one and add the excitation wave as an extra term in the proper update equation [22, 23]. In this manner, the lattice is divided into two distinct areas by means of a fictitious surface, just like the total/scattered-field formulation. Since the evolution of the incident wave is performed along one direction, the region on the left-hand side of the interface will solely accept the reflected part. Therefore, the two fields are fully separated, whereas sufficient annihilation of the latter can be directly attained by the appropriate ABC, now applied simultaneously with the excitation. It becomes obvious that the pulse plane acts as a connecting condition with a triple role: a) precise spatial derivative approximation through neighboring values at the interface, b) successful generation of the desired incident wave, and c) transparent action for all outgoing reflected modes.

On the other hand, the truncation of open waveguide ends must be carefully conducted in order to enable the absorption of evanescent waves. These waves decay so abruptly that cannot be sampled by the usual FDTD mesh, especially at low frequencies [24]. The main consequence is a serious numerical reflection from the perfectly matched layer (PML) interfaces, which is larger below a certain cutoff frequency. This defect is confronted by the HO schemes [25] together with a specific category of PMLs, like the one of [26]. Both algorithms provide fairly sufficient simulations, as will be derived from the experimentations.

7.3 MULTIMODAL HIGHER ORDER FDTD ANALYSIS

In this section, an alternative method for the manipulation of arbitrary waveguide shapes is presented [27]. Based on an exact *multimodal decomposition* and a HO differencing topology, the technique successfully treats complex systems of varying cross section and assures the accurate evaluation of S-parameters or resonance frequencies [28–30]. Let us consider the general waveguide of Figure 7.1(a), including a toroidal-like sector of length θ_t with inner and outer mean radii $R_1(\theta)$ and $R_2(\theta)$, respectively. Actually, the most difficult part is the bent sector that generates nonseparable modes at every frequency, thus not allowing the complete fulfillment of continuity conditions. Comparable observations may be conducted from Figure 7.1(b), where the general three-port Y-junction is depicted. Herein, the discontinuity and the two ports obstruct the separation of Maxwell's equations. The 3-D time-domain methodology decomposes propagating quantities at *prefixed* planes in the bend. Thus, each component f can be written via infinite series as

$$f(r,\theta) = \sum\nolimits_{\kappa} \xi_\kappa(r,\theta) F_\kappa(r,\theta) \quad \text{with } \xi_\kappa(r,\theta) = B_\kappa \cos\left[\kappa\pi(r - R_1(\theta))/R(\theta)\right], \tag{7.1}$$

and

$$B_\kappa = \sqrt{(2-\delta_{\kappa 0})/R(\theta)} \quad \text{for} \int_{R_1}^{R_1+s_l} \xi_\kappa(r,\theta)\xi_\lambda(r,\theta)dr = \delta_{\kappa\lambda}, \tag{7.2}$$

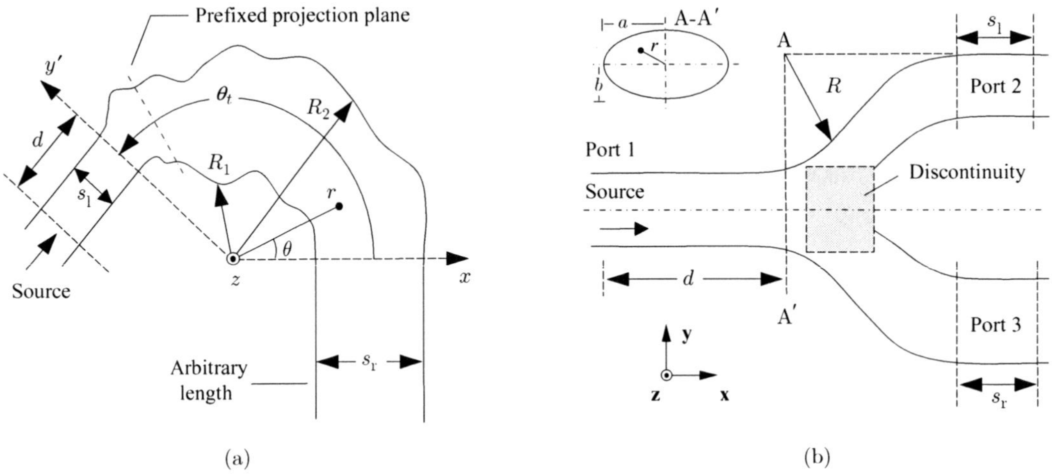

FIGURE 7.1: (a) Geometry of an arbitrarily curved waveguide with a varying cross section. (b) Transverse cut of a three-port Y-junction comprising two ducts of elliptical cross section and different dimensions

where $R(\theta) = R_2(\theta) - R_1(\theta)$, F_κ are scalar coefficients and ξ_κ eigenfunctions complying with the corresponding transverse electromagnetic eigenproblem. Two characteristic functions for $R_1(\theta)$ and $R_2(\theta)$, with $a_\theta = \theta/\theta_t$, $s = s_r - s_1$, $s_r = 1.25 s_1$, and $\tau = \omega/\upsilon_0$, encountered in practical applications are

$$R_1(\theta) = s\tau^2(a_\theta - 1.5) + s_1 - 0.5 s_r \quad \text{and} \quad R_2(\theta) = -s\tau^2(a_\theta - 1.5) + s_1 + 0.5 s_r. \tag{7.3}$$

The key issue in such systems is the initial mode coupling, occurring in two distinct ways: one due to the *curvature* of the waveguide and the other due to its *varying* cross section. The former normally contributes to the generation of HO modes and the latter induces the symmetric ones. It is stressed that existing schemes cannot easily simulate this intricate situation. The specific algorithm overcomes this hindrance by projecting the appropriate governing laws to ξ_κ to extract equivalent ordinary differential equations, which combine electric and magnetic fields. Their solutions lead to robust models and the most substantial, they preserve lattice duality, even when the source is placed quite close to the discontinuity. For illustration, analysis deals with a certain plane in the interior of the bend. Then, Ampère's and Faraday's laws become

$$\begin{aligned} \varepsilon \frac{\partial \mathbf{E}}{\partial t} &= \frac{1}{\tau}\left(\mathbf{W}^{\mathrm{A}} + \mathbf{W}^{\mathrm{B}}\mathbf{W}^{\mathrm{C}}\right)\mathbf{H} - \mathbf{W}^{\mathrm{D}}\mathbf{E} \\ \mu \frac{\partial \mathbf{H}}{\partial t} &= -\frac{1}{\tau}\mathbf{W}^{\mathrm{C}}\mathbf{E} + \left(\mathbf{W}^{\mathrm{E}} - \mathbf{W}^{\mathrm{D}}\right)\mathbf{H}. \end{aligned} \tag{7.4}$$

The elements of matrices $\mathbf{W}^i (i = \mathrm{A}, \ldots, \mathrm{E})$ describe the fundamental curvature details. Hence, after enforcing the boundary constraints, one receives

$$W^{\mathrm{A}}_{\kappa\lambda} = \begin{cases} 0 & \text{for } \kappa = \lambda \\ \Gamma_{\kappa\lambda}\kappa^2/(\kappa^2 - \lambda^2) & \text{for } \kappa \neq \lambda \end{cases} \qquad W^{\mathrm{B}}_{\kappa\lambda} = \left(\tau^2 - \beta_\kappa^2\right)\delta_{\kappa\lambda} \quad \text{for } \beta_\kappa = \kappa\pi/s, \tag{7.5}$$

$$W^{\mathrm{C}}_{\kappa\lambda} = \begin{cases} R_1(\theta) + s/2 & \text{for } \kappa = \lambda \\ \Gamma_{\kappa\lambda}(\kappa^2 + \lambda^2)/(\kappa^2 - \lambda^2)^2 & \text{for } \kappa \neq \lambda \end{cases}$$
$$W^{\mathrm{D}}_{\kappa\lambda} = \begin{cases} \delta_{\kappa 0}\left[R_2(\theta) - R_1(\theta)\right]/R(\theta) & \text{for } \kappa = \lambda \\ W^{\mathrm{E}}_{\kappa\lambda}\kappa^2/\left(\kappa^2 - \lambda^2\right)^2 & \text{for } \kappa \neq \lambda, \end{cases} \tag{7.6}$$

and

$$W^{\mathrm{E}}_{\kappa\lambda} = B_\kappa B_\lambda \left[(-1)^{\kappa+\lambda} R_2(\theta) - R_1(\theta)\right] \quad \text{with } \Gamma_{\kappa\lambda} = B_\kappa B_\lambda \left[(-1)^{\kappa+\lambda} - 1\right]. \tag{7.7}$$

For the sufficient annihilation of vector parasites, equations (7.4) are directly advanced in the time domain via the conventional, (2.14)–(2.20), or the nonstandard, (3.70)–(3.81), HO FDTD schemes. The above procedure provides **E** and **H** values at specific planes inside the bend. These values are next inserted as excitation terms in the FDTD formulas to proceed with the update mechanism in the usual manner.

7.4 APPLICATIONS – NUMERICAL RESULTS

The accuracy and efficiency of HO FDTD schemes in relation to the improved PMLs and the generalized integration schemes of Section 5.5 are verified by means of several 2- and 3-D realistic waveguide problems. These include inclined-slot coupled T-junctions, thin apertures, power-bus printed board circuits (PCBs), and multiconductor microstrip transmission lines. The majority of the discretized models involve consistent grids that are compared to the respective second-order FDTD realizations.

7.4.1 Basic Waveguide Problems

In this section, a set of elementary structures is investigated. These components are essentially embedded in more complex electromagnetic compatibility devices (EMC) devices regarding microwave instrumentations, miniaturized packaging layouts in communication ensembles, power managing tools, and biomedical systems for the estimation of hazardous health effects. Since, their profile may lead to complicated field patterns, it is anticipated that HO FDTD operators will provide the most sufficient results.

The first application is the inclined-slot coupled H-plane T-junction of Figure 7.2 in which the broad dimension of the feed waveguide, forming the T arm, is oriented parallel to the

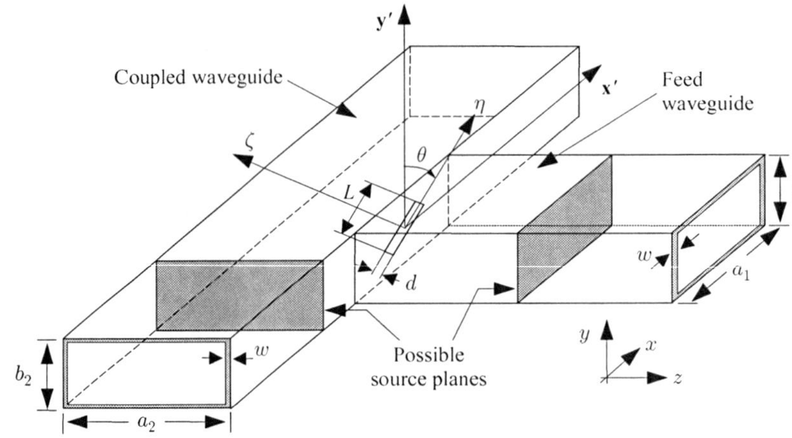

FIGURE 7.2: An inclined-slot coupled H-plane T-junction

axis of the coupled waveguide. Its excitation is performed by a pulsed modulated process, which imposes the source plane several cells away from the ABC's plane to fully separate incident and reflected fields. The dimensions of the T-junction are $a_1 = a_2 = 22.86$ mm, $b_1 = b_2 = 10.16$ mm, $L = 16$ mm, $d = 0.8$ mm, and $w = 1.27$ mm. The domain is discretized into $108 \times 20 \times 72$ cells with $\Delta x = \Delta y = \Delta z = 0.635$ mm and $\Delta t = 1.225$ ps. For the treatment of the widened spatial stencils, a modified version of the one-sided operators of Section 2.4.3 is utilized, while the angle-optimized schemes of Section 2.5.2 attain additional reduction of the dispersion errors. In Figure 7.3(a), the coupling from port 1 to 3, defined as $20 \log |S_{31}|$, is compared with second-order FDTD results (85% finer grid, 70% larger CPU time) and the outcomes of [6]. Their agreement is really very satisfactory, unlike the discrepancies of Yee's technique.

Subsequently, the interesting effect of wall thickness w on the waveguide is examined. While the staircase solutions are not sensitive to such small-scale details, Figure 7.3(b) shows that HO simulations follow the variation of w and react positively to its changes. Note the absence of any oscillations in the frequency spectrum, mainly attributed to the schemes of Chapter 2.

The second example considers the six-port cross junction depicted in Figure 7.4(a). Its dimensions are $a_1 = a_2 = a_3 = 15.799$ mm and $b_1 = b_2 = b_3 = 7.899$ mm. Nonstandard analogs along with the derivative matching fictitious-point technique of Section 2.4.5 are incorporated and the results of [6] are again used as a reference. Moreover, the modified HO (2, 4) FDTD approach of Section 2.5.1 is alternatively employed. Figure 7.4(b) depicts the results that indicate the advanced performance of the nonstandard algorithm for the very coarse $26 \times 22 \times 24$ grid.

Next, analysis proceeds to the iris-coupled elliptical resonator of Figure 7.5(a). Herein, the curvilinear nonstandard operators, (3.70)–(3.81) for $M = 2$ and $L = 2$, are implemented

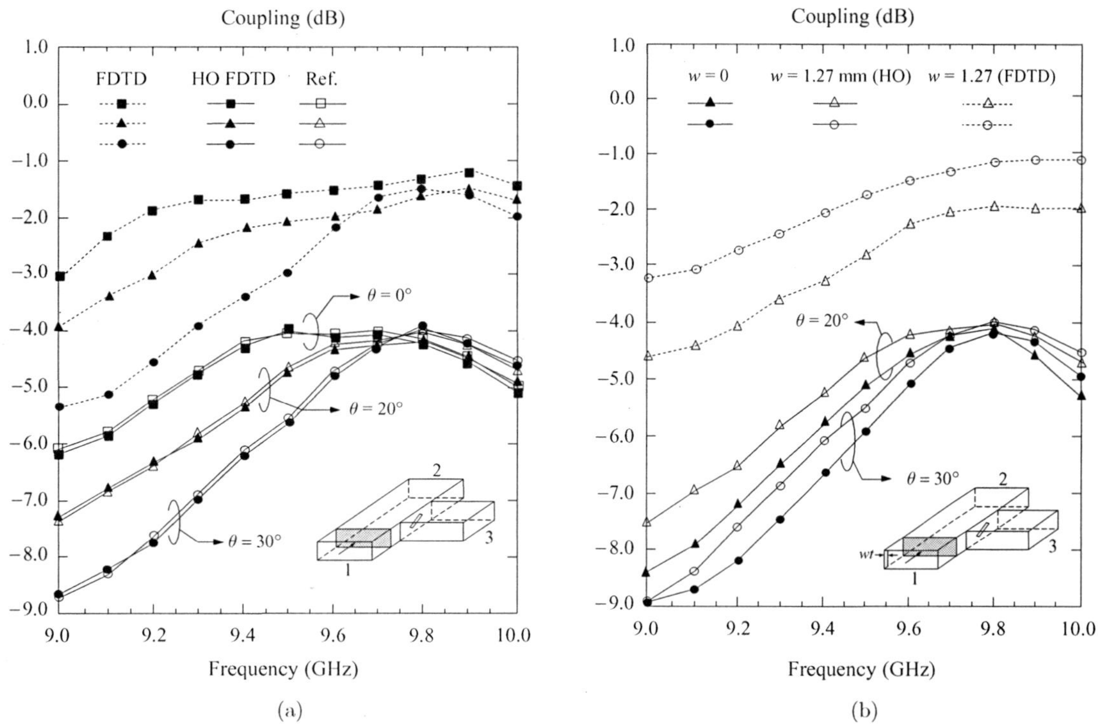

FIGURE 7.3: Coupling coefficient versus frequency of an H-plane T-junction for (a) various angles of slot inclination and (b) a 30° slot inclination considering the effect of wall thickness

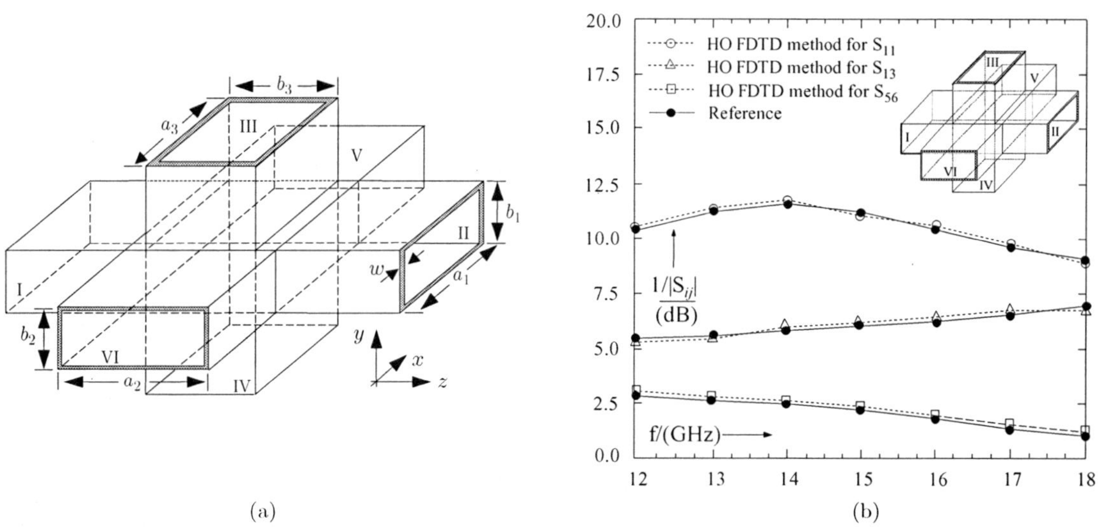

FIGURE 7.4: (a) A six-port cross junction and (b) magnitude of diverse S-parameters computed via the second-order and HO FDTD techniques

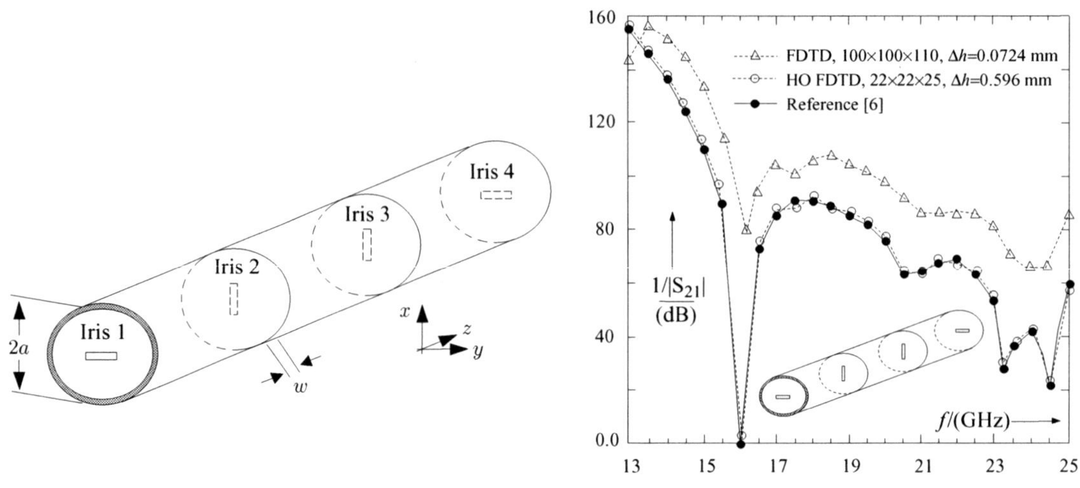

FIGURE 7.5: (a) An iris-slot circular waveguide and (b) the S_{21}-parameter of the structure for a second- and a HO FDTD realization taking into account the effect of wall thickness

in terms of the aspects of Section 2.6. The radius is set at $a = 6.985$ mm, the wall thickness is chosen to be $w = 0.19$ mm, while the irises' dimensions are 2.596×7.409 mm. Spatial stencils are successfully handled via the operators of Section 2.4.1. Figure 7.5(b) gives the variation of S_{21} from which the accuracy and memory savings (almost 80% that of Yee's ones) achieved by the HO FDTD scheme are easily noticeable.

Concerning the absorption of evanescent waves, a 50-mm circular parallel-plate metallic waveguide whose TM_1 mode has a cutoff at 3 GHz is explored. For a wideband attenuation, the PML of [26] is selected. All implementations are performed with operators (3.42) optimized for $k = 1.437$. Various setups are constructed and their reflection coefficient is shown in Figure 7.6. Observe that all absorbers display a promising annihilation of evanescent waves below the cutoff frequency, while HO counterparts attain a considerable decrease of PML depths for specific reflection coefficient values.

7.4.2 Three-Dimensional Realistic Devices

Apart from the aforementioned components, the HO FDTD methodology is applied to various modern waveguide systems whose specialized function requires precise and inexpensive discretizations. Moreover, the need for fast and credible design estimates opts for relatively short simulations without sacrificing the quality of the results. Toward this objective moves the investigation of the present section.

The first problem studies the elliptical cavity of Figure 7.7(a) that has a sidewall narrow inclined slot coupled to a rectangular waveguide. As observed, the cavity comprises an

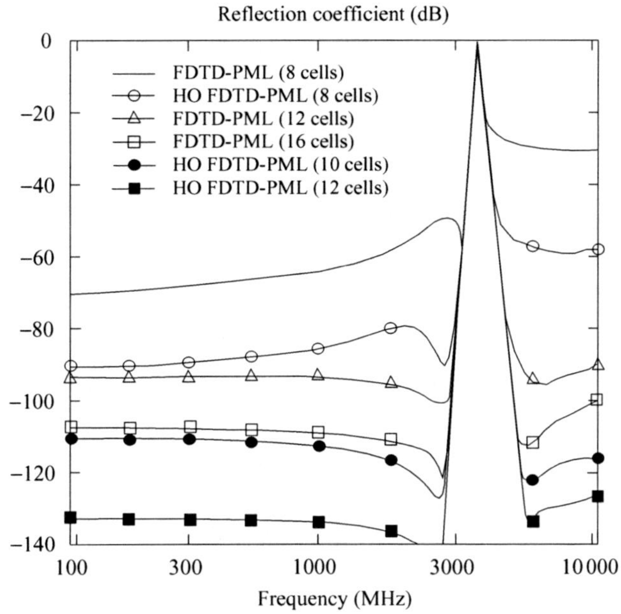

FIGURE 7.6: Reflection coefficients of the TM_1 mode attributed to second-order and HO PMLs

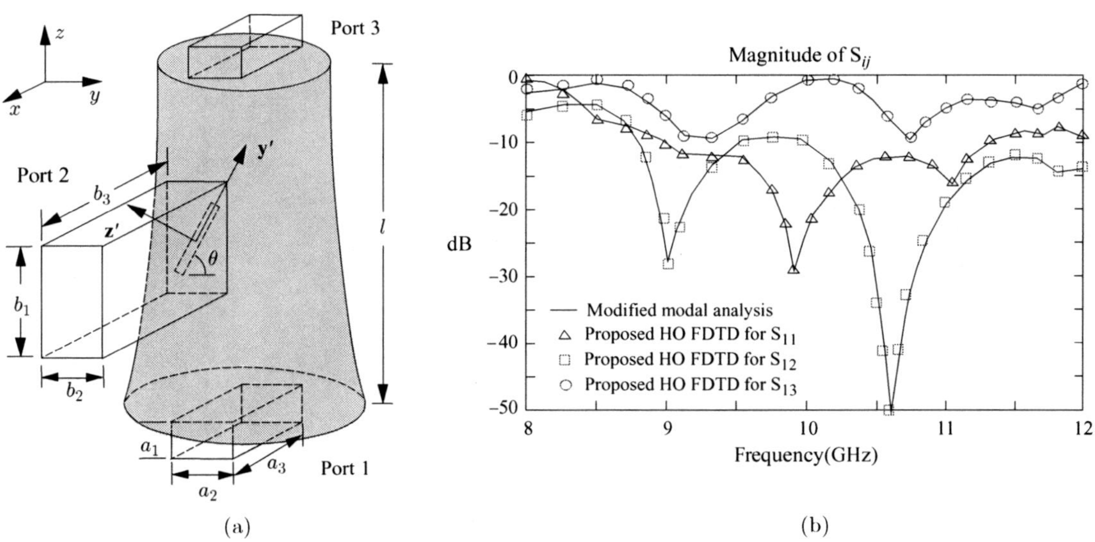

FIGURE 7.7: (a) A sidewall inclined-slot coupled elliptical cavity and (b) magnitude of various S-parameters

elliptical-to-rectangular T-junction as well as two additional ports and is basically used in the design of circular dual-mode filters that are very useful in several EMC devices. The numerical analysis of the above junction is quite demanding due to its curvilinear shape and the arbitrary inclination of the slot. Typical dimensions are $a_1 = 7.39$ mm, $b_2 = 12.16$ mm, $a_3 = 20.44$ mm, $b_1 = 24.98$ mm, $b_2 = 12.63$ mm, and $b_3 = 35.91$ mm. The domain is discretized into $20 \times 35 \times 110$ cells, with $\Delta x = 1.734$, $\Delta y = 1.845$ mm, $\Delta z = 0.931$ mm, and $\Delta t = 1.794$ ps. Also, the slot inclination angle is $\theta = 60°$, while the open ends are terminated by a six-layer PML.

To accomplish a considerable reduction of the undesired dispersion errors, the simulations incorporate the dispersion-relation-preserving (DRP) schemes of Section 2.5.3 along with the operators of Section 2.4.2 that control the larger stencils. Figure 7.7(b) gives the magnitude of different S-parameters (excitation is at port 2) compared with those obtained by the modal method of [11]. As observed, the combined FDTD algorithm is very precise, requiring significantly lower resources (almost 90%) than Yee's scheme. Equivalent deductions for the conventional HO concepts can be drawn from Figures 7.8(a) and 7.8(b), where the magnitude of parameters S_{12} and S_{23} and the shielding efficiency of the structure are computed for $\theta = 75°$.

Similarly, Figure 7.9(b) gives the S-parameters regarding the conducting aperture of Figure 7.9(a) fed by two horn-like waveguides. In this case, the two elliptical slots are discretized through the hybrid (2, 2)/(2, 4) FDTD method of Section 6.2, while for the dissimilar material interfaces the technique of Section 2.4.4 is additionally incorporated. An indicative set of dimensions is $a = 9.27$ mm, $b = 12.78$ mm, $c = 24.53$ mm, $l = 36.54$ mm, $l_1 = 3.92$ mm, $l_2 = 1.56$ mm, and $w = 1.68$ mm. The device is divided into $20 \times 90 \times 52$ cells and Δt is set

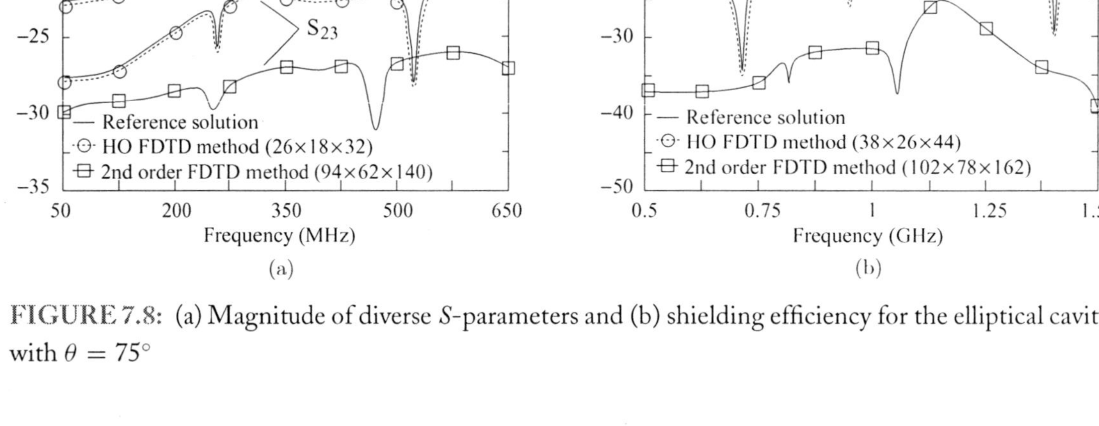

FIGURE 7.8: (a) Magnitude of diverse S-parameters and (b) shielding efficiency for the elliptical cavity with $\theta = 75°$

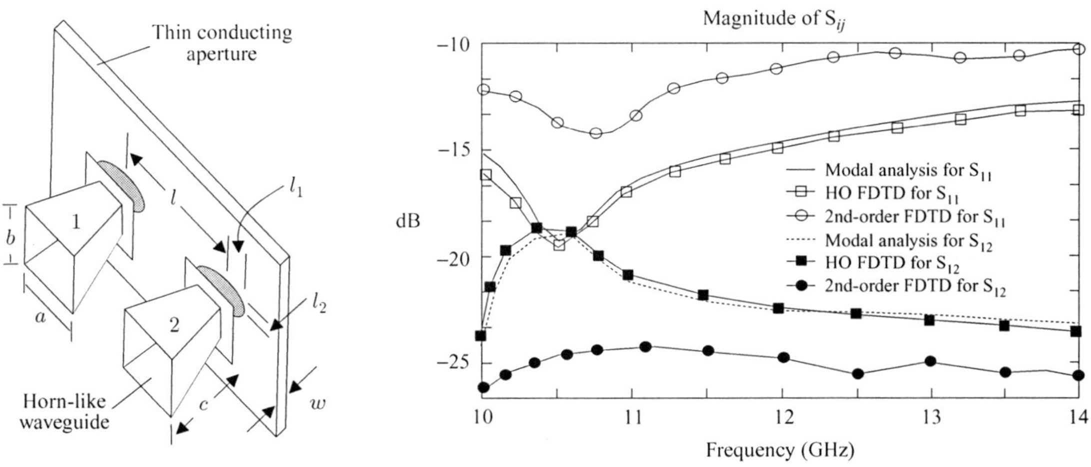

FIGURE 7.9: (a) An aperture with two elliptical slots fed by horn-like waveguides and (b) magnitude of its S-parameters

at 1.013 ps. Note the reasonable agreement between the reference [11] and the HO plots as well as the inability of the second-order FDTD technique to give acceptable results.

In order to validate the technique of Section 7.3, two different structures with varying cross section are considered. The first is a Y-junction, like the one of Figure 7.1(b), whose straight parts have different cross sections and curvatures. It is stressed that a supplementary decrease of the dispersion-error mechanism is achieved by the coefficient-modification algorithm of Section 2.5.5. A usual configuration is $a = 2.97$ mm, $b = 4.56$ mm, $d = 31.84$ mm, $s_l = 20.14$ mm, $s_r = 15.29$ mm, and a PML with a depth of eight cells. The reflection coefficient between ports 1 and 3 is depicted in Figure 7.10. Again, the combined nonstandard methodology overwhelms the classical staircase formulation, an issue also confirmed by the values of Table 7.1 that gives the first five resonance frequencies of the structure.

The reflection coefficient of the second configuration is presented in Figure 7.11. As can be derived from the inlet figure, it is a two-port waveguide of varying cross section. The radii ratio is $R_2/R_1 = 2.69$, $\theta_t = 120°$, and the open ends are truncated by a four-cell PML. Herein, lattice reflection errors are suppressed by means of the optimized finite difference and controllable error estimators of Section 2.5.4 or alternatively via the improved forms of Section 5.4. Results along with the data of Table 7.2 indicate that the HO nonstandard FDTD solutions are in very close agreement with the exact ones [3]. Moreover, it is noteworthy to point out the significant savings in CPU and memory requirements along with the notable reduction of the maximum global error, especially with operators (3.59).

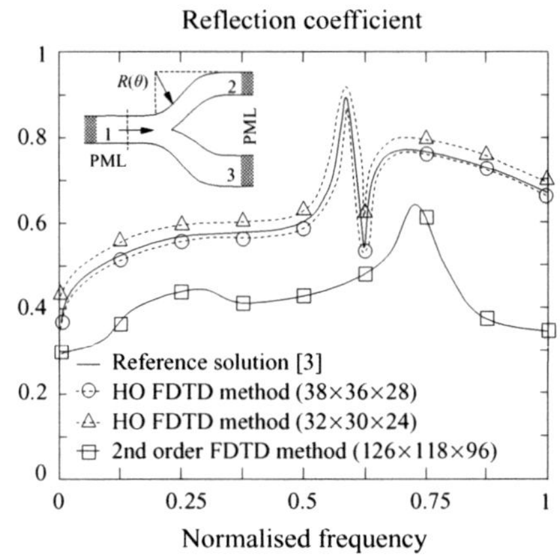

FIGURE 7.10: Reflection coefficient between ports 1 and 3 of the three-port Y-junction for various FDTD simulations

Let us now analyze the more complex shielding application of Figure 7.12(a), which illustrates a rectangular enclosure partitioned by two equal horizontal PEC walls. In the front plane, there is a centered (20×5) cm horizontal aperture. The dimensions are $a = b = 60$ cm, $d = 120$ cm, $l = 70$ cm, $w = 2$ cm, and the excitation is launched by a vertical coaxially fed monopole. Due to the nonstandard operators of (3.43), the domain is discretized into the coarse grid of $30 \times 60 \times 30$ cells with $\Delta x = \Delta y = \Delta z = 2$ cm and $\Delta t = 30.567$ ps. In the area of the aperture, spatial derivatives are computed by the fictitious-point technique of Section 2.4.5, whereas the DRP schemes of Section 2.5.3 are also utilized. Figure 7.12(b) displays the shielding efficiency defined as the ratio of the electric field amplitude evaluated in front of the PEC aperture for a (2, 2) FDTD, a (4, 4) FDTD and a $p = 5$ WENO-TD technique.

TABLE 7.1: Resonance Frequencies for the Three-Port Y-Junction

REFERENCE SOLUTION (GHz) [3]	2ND-ORDER FDTD METHOD SIMULATION $80 \times 60 \times 142$	2ND-ORDER FDTD METHOD RELATIVE ERROR	HO FDTD METHOD SIMULATION $28 \times 24 \times 40$	HO FDTD METHOD RELATIVE ERROR
1.86534	1.82406	2.213 %	1.86511	0.012 %
4.62379	4.46450	3.445 %	4.62203	0.038 %
5.34762	5.01628	6.196 %	5.34467	0.055 %
9.29582	8.49526	8.612 %	9.28754	0.089 %
11.64217	10.45071	10.234 %	11.62948	0.109 %

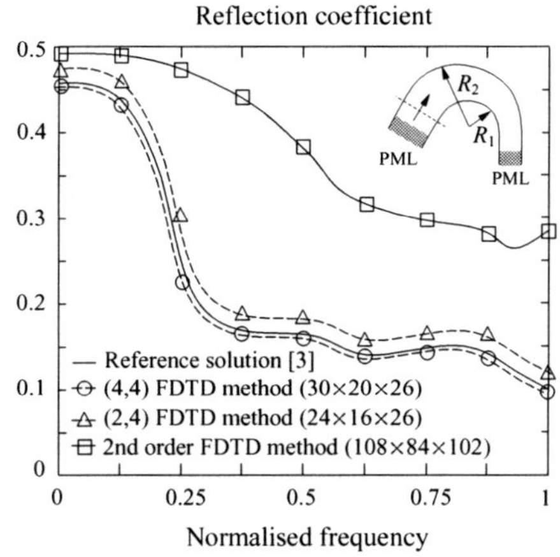

FIGURE 7.11: Reflection coefficient of the two-port system for diverse FDTD implementations

TABLE 7.2: Maximum Global Error and Convergence Rate

FDTD OPERATOR	GLOBAL ERROR	CONVERGENCE RATE
Yee – PML (16 cells)	9.204×10^{-4}	1.792
HO (3.42) – PML (4 cells)	3.829×10^{-3}	2.043
HO (3.42) – PML (6 cells)	5.503×10^{-5}	2.159
HO (3.42) – PML (12 cells)	6.786×10^{-7}	3.984
HO (3.49) – PML (4 cells)	2.214×10^{-3}	2.060
HO (3.49) – PML (6 cells)	8.021×10^{-6}	2.182
HO (3.49) – PML (12 cells)	4.964×10^{-8}	3.957

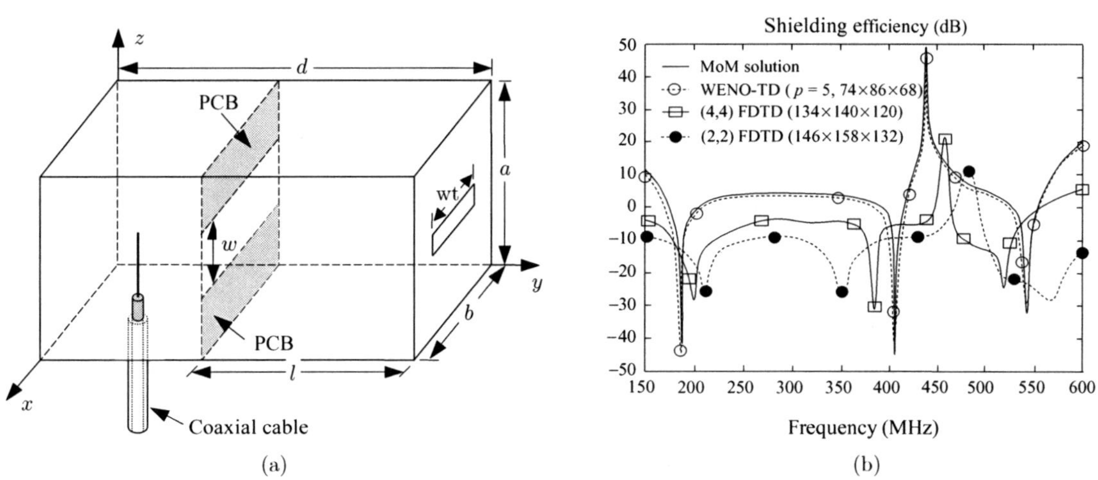

FIGURE 7.12: (a) Geometry of a rectangular enclosure partitioned by two PEC walls. (b) Shielding efficiency for $w = 2$ cm

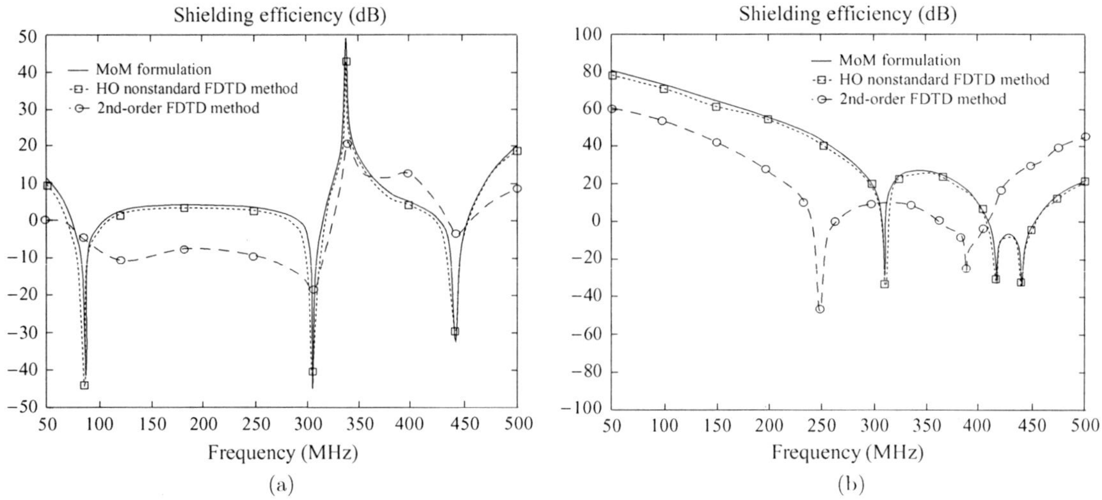

FIGURE 7.13: Shielding efficiency of the rectangular enclosure for (a) $w = 3$ cm and (b) $w = 4.5$ cm

The method-of-moment (MoM) solution provided in [12] is considered as the reference solution. Evidently, the latter algorithm, combined with the prior enhanced formulations, leads to the most accurate results. On the contrary, the simple conventional forms without their contribution cannot sufficiently resolve the first two frequency peaks, even though they employ an 85% finer grid. In analogous lines, promising outcomes are acquired by the nonstandard schemes in Figures 7.13(a) and 7.13(b) for $w = 3$ and $w = 4.5$ cm, respectively.

Next, the effects of wave penetration in an elliptical waveguide with multiple rectangular inclined slots, placed in a conducting plane [Figure 7.14(a)], are investigated. Our setup has the

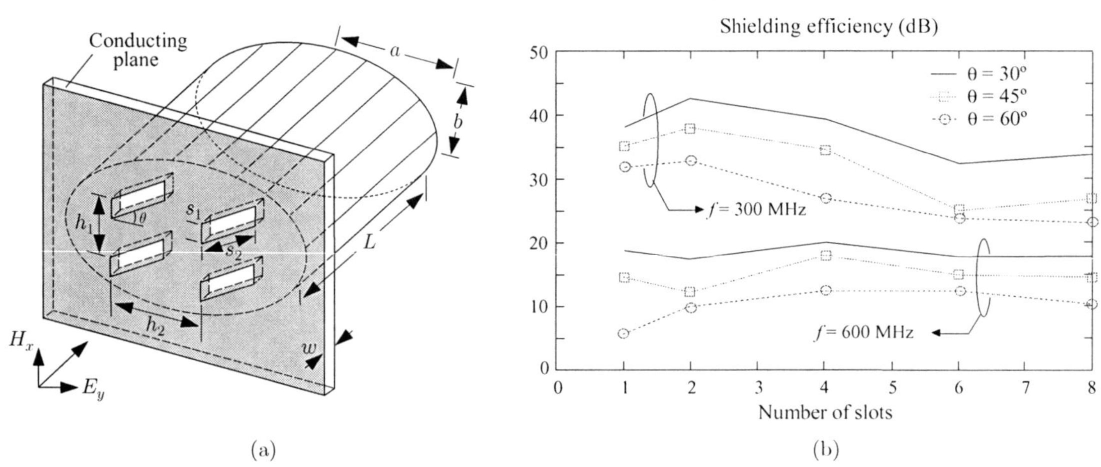

FIGURE 7.14: (a) An elliptical waveguide with multiple inclined slots located in a conducting plane. (b) Shielding efficiency versus the number of inclined slots for various θ angles

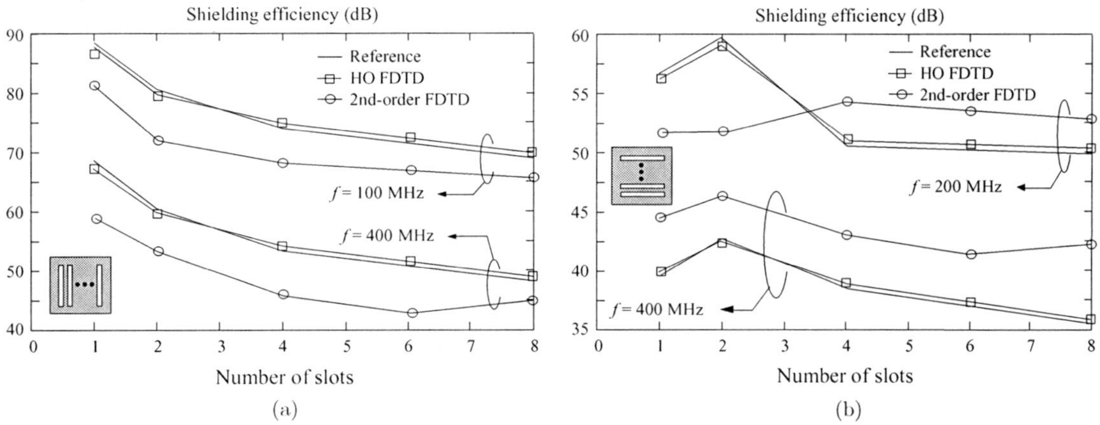

FIGURE 7.15: Shielding efficiency of the elliptical waveguide for (a) vertical and (b) horizontal slots

following dimensions: $a = 10$ cm, $b = 5$ cm, $L = 20$ cm, $s_1 = 1$ cm, $s_2 = 3$ cm, $h_1 = 1.7$ cm, $h_2 = 4.5$ cm, $w = 0.2$ cm with the entire structure receiving a $52 \times 46 \times 94$ discretization. Operators (3.70)–(3.81), for $M = 3$ and $L = 2$, are implemented using an eight-layer PML, whereas the reference solution is a modified version of [8]. Bear in mind that fine geometric details around the slots are modeled by the hybrid method of Section 6.2 and the correction scheme of Section 5.3. Figure 7.14(b) presents the shielding efficiency for three different slot angles, i.e., $\theta = 30^\circ, 45^\circ, 60^\circ$. As expected, HO simulations outperform Yee's ones despite the finer mesh of the latter. Keeping the same arrangement, Figures 7.15(a) and 7.15(b) illustrate the shielding efficiency for vertical and horizontal slots. Also, note that the discrepancy between the two algorithms is around 7–8 dB, which reveals the inadequacy of second-order formulas to discern frequencies in close proximity.

Another significant class of problems, whose waveguiding properties require a systematic study, is the PCBs that constitute the basic parts of many contemporary telecommunication or computer devices. A general PCB configuration is shown in Figure 7.16. Most components

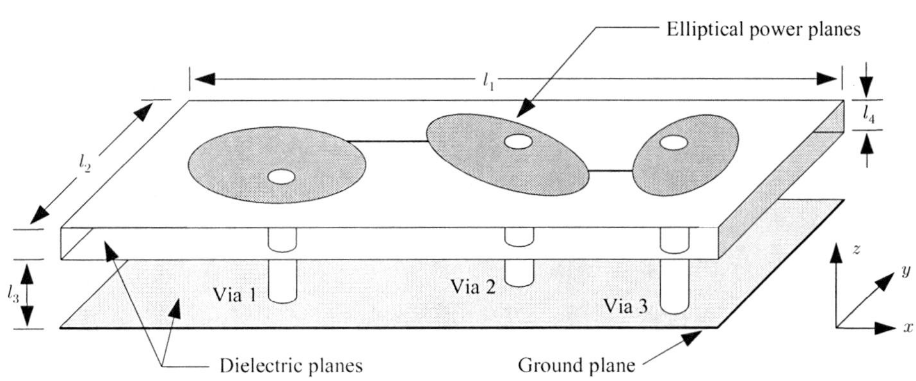

FIGURE 7.16: A power-bus PCB structure with vertical discontinuities

such as integrated circuits and lumped elements are mounted on the first layer, while the power and ground planes are distributed in the inner ones. A typical arrangement has two distinct types of metallization surfaces, namely planar conduction layers (power and ground layers) and surfaces of vertical discontinuities (vias or ports). All dielectric layers are assumed to be infinite, while the power planes can have a completely arbitrary shape. Due to the rapid field changes near these constructions, the geometric peculiarities they may involve and the very precise modeling they require, a HO FDTD analysis will be conducted. Particularly, all media boundaries are manipulated by the technique of Section 2.4.4 and dispersion errors are subdued through the concepts of Section 2.5.2. Moreover, the approach of Section 5.2 incorporates lossy materials in a more physical way.

The basic dimensions are $l_1 = 20$ cm, $l_2 = 15$ cm, $l_3 = 2.5$ cm, and $l_4 = 1.5$ cm. Selecting a material with $\varepsilon_r = 4.7$ and $\sigma = 0.01$ S/m, the infinite space is truncated by an eight-layer PML. The domain is divided into $36 \times 24 \times 8$ cells with $\Delta x = \Delta y = \Delta z = 0.015$ cm, whereas the nonstandard operators of (3.59) are formed by $A_1 = -0.3052042618$ and $A_2 = 0.2078392426$. Figures 7.17(a) and 7.17(b) demonstrate the variation of different S-parameters. The reference solution has been obtained from [10]. As observed, HO results exhibit a very good accuracy without any discrepancies in the frequency spectrum. Conversely, Yee's algorithm is proven to be insufficient in the detection of acute peaks, despite the much finer $82 \times 64 \times 156$ lattice it employs. Hence, a virtually 85% reduction of the CPU and memory requirements is succeeded.

Of equal importance for modern EMC structures are the circularly shielded multiconductor transmission lines and the cylindrical cavity-backed apertures of Figure 7.18. Due to their wideband function range, the former are found to be very useful in the construction of ultra-bandwidth microwave elements, junctions, and couplers, while the latter are commonly involved in the solution of penetration or shielding problems.

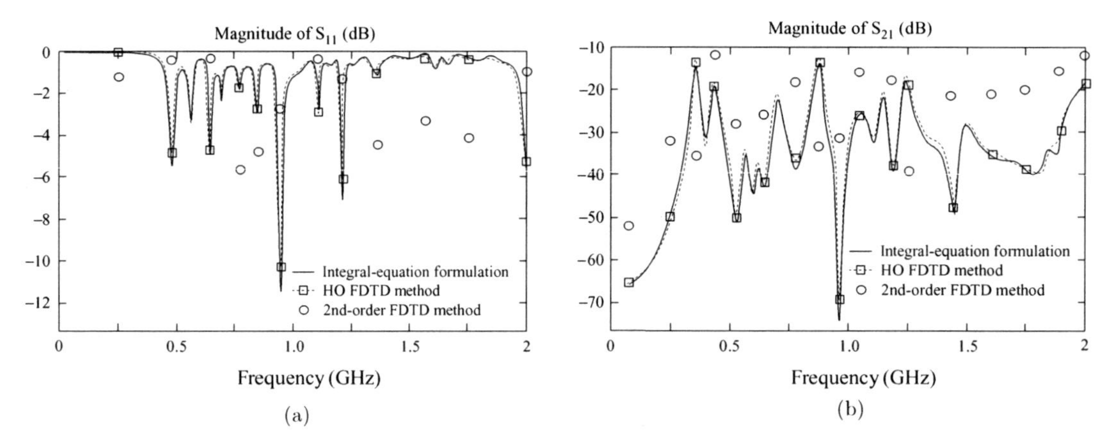

FIGURE 7.17: Magnitude of (a) S_{11}-parameter and (b) S_{21}-parameter for the power-bus PCB device

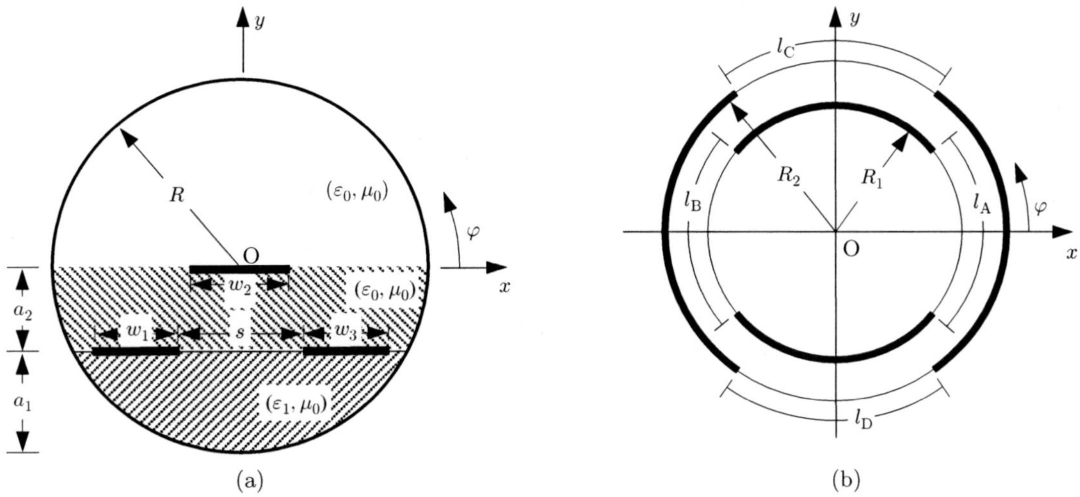

FIGURE 7.18: Cross section of (a) a circularly shielded dual-plane waveguide with three microstrip lines, and (b) a two-layered four-element cylindrical cavity-backed aperture

The modeling of the above structures is performed via the HO ADI-FDTD technique of Section 6.4. In this context, Table 7.3 summarizes the maximum global error and system requirements for three distinct cases regarding a circularly shielded dual-plane structure that has three microstrip lines [Figure 7.18(a)]. Both types of nonstandard operators are employed, namely the ones of (3.59) and (3.70)–(3.81), respectively. The former are designed for $k = 2.721$, $A_1 = -0.1098487022$, $A_2 = 0.0783691643$, and the latter for $M = 3$, $L = 3$. In this context, the curvilinear discretization is conducted according to Section 2.6 and combined with the a 3-D version of the one-sided operators of Section 2.4.3. The main dimensions are $R = 5$ mm, $a_1 = 2.0$ mm, $a_2 = 3.0$ mm, and $s = 2.0$ mm. Moreover, along the z-direction the device is terminated by a six-layer PML. From the results, one can easily discern the superiority of the HO schemes and the notable reduction of the overall simulation time attained by the ADI-FDTD formulation. However, a notable attribute of this method is its minimal dispersion error that allows the selection of time-steps greatly beyond the Courant limit. Thus, the total overhead is drastically reduced. On the other hand, Figures 7.19(a) and 7.19(b) describe the insertion loss of two circularly shielded waveguides with different microstrip locations. Again, all HO ADI-FDTD simulations (time: 112 min) outperform their second-order analogs (time: 5.2 h), even for large Courant-Friedrick-Levy number values.

Finally, concentrating on Figure 7.18(b), two versions of the cylindrical cavity-backed aperture are explored. The lengths l_A, l_B and l_C, l_D of the first case correspond to the angles of 75° and 80°, with radii $R_1 = 10$ mm and $R_2 = 12$ mm. For the second case, both angles are 180°. The whole arrangement is illuminated by a plane wave with diverse contents in evanescent modes. Apart from the curvilinear update procedure, spatial stencils are treated by the modified

TABLE 7.3: Maximum Global Error and Computational Aspects of Different HO ADI-FDTD Simulations

CASE (MM)	FDTD OPERATOR	GRID SIZE	GRID REDUCTION (%)	COMPUTER MEMORY (MB)	TIME-STEPS	CPU TIME (h)	MAXIMUM GLOBAL ERROR
$w_1 = 1.5$	Second-order Yee	$45 \times 67 \times 112$	–	232	4578	15.3	3.561×10^{-4}
$w_2 = 1.0$	HO ADI-FDTD (3.59)	$19 \times 27 \times 48$	92.707	22.9	1432	4.7	4.982×10^{-8}
$w_3 = 1.5$	HO ADI-FDTD (3.70)	$15 \times 22 \times 35$	96.579	10.6	1120	3.9	6.735×10^{-9}
$w_1 = 2.0$	Second-order Yee	$53 \times 82 \times 148$	–	350	5784	22.5	7.034×10^{-3}
$w_2 = 1.5$	HO ADI-FDTD (3.59)	$22 \times 37 \times 61$	92.281	30.6	2140	11.3	2.823×10^{-7}
$w_3 = 2.0$	HO ADI- FDTD (3.70)	$17 \times 26 \times 45$	96.770	15.4	1503	5.1	5.418×10^{-8}
$w_1 = 2.5$	Second-order Yee	$67 \times 93 \times 161$	–	425	6849	31.9	8.926×10^{-2}
$w_2 = 3.0$	HO ADI-FDTD (3.59)	$28 \times 42 \times 72$	91.559	45.2	2647	12.4	4.102×10^{-6}
$w_3 = 2.5$	HO ADI-FDTD (3.70)	$23 \times 37 \times 55$	95.334	29.8	1812	8.3	1.522×10^{-7}

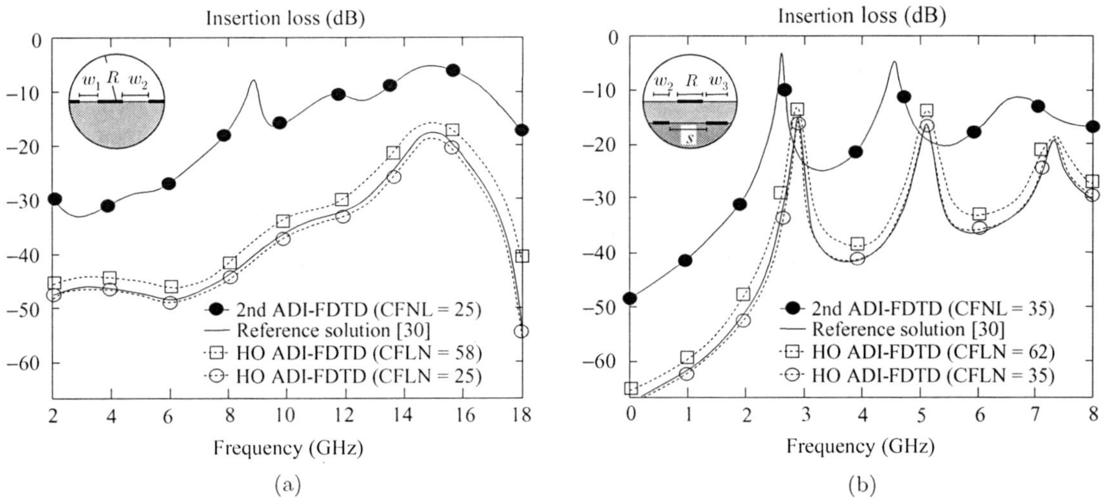

FIGURE 7.19: Insertion loss of two circularly shielded waveguides with three (a) coplanar and (b) dual-plane microstrip lines

operators of Sections 2.4.1 and 2.4.5, while dispersion and anisotropy errors are controlled by the abstractions of Section 2.5.4. Results are compared to a second-order FDTD solution based on a very fine lattice. Figures 7.20(a) and 7.20(b) present the shielding efficiency, where the values computed by the HO ADI-FDTD algorithm are proven to be the most rigorous.

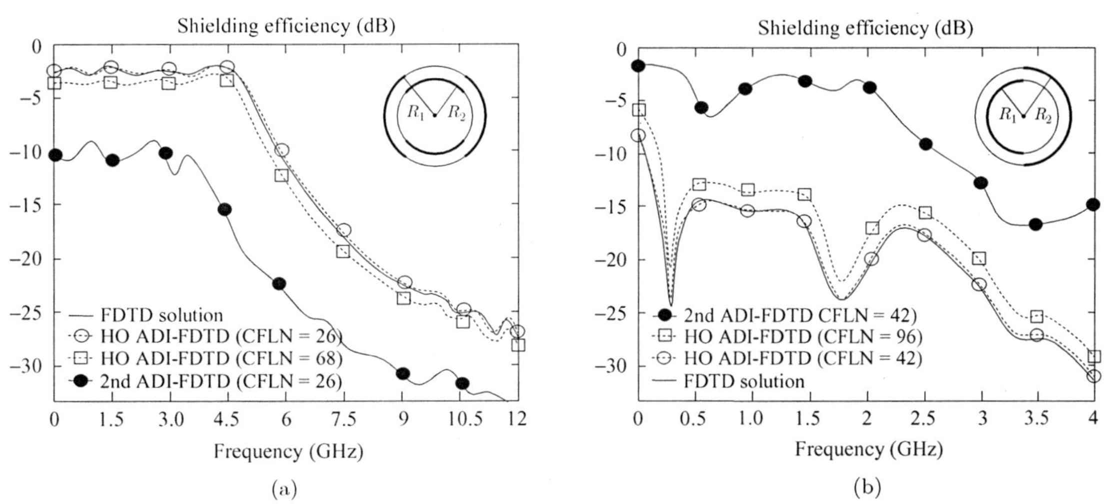

FIGURE 7.20: Shielding efficiency of two cylindrical cavity-backed apertures with (a) four and (b) two cocentric metal elements

REFERENCES

[1] K. L. Shlager and J. B. Schneider, "A survey of the finite-difference time-domain literature" in *Advances in Computational Electrodynamics: The Finite-Difference Time-Domain Method*, A. Taflove, Ed. Norwood, MA: Artech House, 1998, ch. 1, pp. 1–62 (online updated version: http://www.fdtd.org).

[2] T. Itoh and B. Houshmand, Eds., *Time-Domain Methods for Microwave Structures: Analysis and Design*. Piscataway, NJ: IEEE Press, 1998.

[3] A. F. Peterson, S. L. Ray, and R. Mittra, *Computational Methods for Electromagnetics*. Piscataway, NJ: IEEE Press/OUP Series, 1998.

[4] F. Arndt, J. Brandt, V. Catina, J. Ritter, I. Rullhusen, J. Dauelsberg, U. Hilhefort, and W. Wessel, "Fast CAD and optimization of waveguide components and aperture antennas by hybrid MM/FE/MoM/FD methods – State-of-the-art and recent advances," *IEEE Trans. Microw. Theory Tech.*, vol. 52, no. 1, pp. 292–305, Jan. 2004.

[5] J.-F. Lee, R. Palandech, and R. Mittra, "Modeling three-dimensional discontinuities in waveguides using non-orthogonal FDTD algorithm," *IEEE Trans. Microw. Theory Tech.*, vol. 40, no. 2, pp. 346–352, Feb. 1992.

[6] A. J. Sangster and H. Wang, "A generalized analysis for a class of rectangular waveguide coupler employing narrow wall slots," *IEEE Trans. Microw. Theory Tech.*, vol. 44, no. 2, pp. 283–290, Feb. 1996.

[7] A. Rong, H. Yang, X. Chen, and A. Cangellaris, "Efficient FDTD modeling of irises/slots in microwave structures and its application to the design of combline filters," *IEEE Trans. Microw. Theory Tech.*, vol. 49, no. 12, pp 2266–2275, Dec. 2001.

[8] N. V. Kantartzis, T. I. Kosmanis, T. V. Yioultsis, and T. D. Tsiboukis, "A nonorthogonal higher-order wavelet-oriented FDTD technique for 3-D waveguide structures on generalised curvilinear grids," *IEEE Trans. Magn.*, vol. 37, no 5, pp. 3264–3268, Sep. 2001.

[9] J. Fan, J. L. Drewniak, H. Shi, and J. Knighten, "DC power-bus modeling and design with a mixed-potential integral-formulation and circuit extraction," *IEEE Trans. Electromagn. Compat.*, vol. 43, no. 4, pp. 426–436, Nov. 2001.

[10] K.-L. Wu, M. Yu, and A. Sivadas, "A novel modal analysis of a circular-to-rectangular waveguide T-junction and its application to design of circular waveguide dual-mode filters," *IEEE Trans. Microw. Theory Tech.*, vol. 50, no. 2, pp. 465–473, Feb. 2002.

[11] W. Wallyn, D. De Zutter, and H. Rogier, "Prediction of the shielding and resonant behaviour of multisection enclosures based on magnetic current modelling," *IEEE Trans. Electromagn. Compat.*, vol. 44, no. 1, pp. 130–138, Feb. 2002.

[12] M. Feliziani and F. Maradei, "Time-domain prediction of the radiated susceptibility in a shielded cable inside a penetrable shielded box," *Int. J. Numer. Model.*, vol. 15, nos. 5–6, pp. 549–561, Sep.-Dec. 2002.

[13] C. J. Railton and D. L. Paul, "Analysis of circular waveguide filter using enhanced FDTD," *Int. J. Numer. Model.*, vol. 15, nos. 5–6, pp. 535–547, Sep.-Dec. 2002.

[14] C. D. Sarris and L. P. B. Katehi, "An efficient interface between FDTD and Haar MRTD – Formulation and applications," *IEEE Trans. Microw. Theory Tech.*, vol. 51, no. 4, pp. 1146–1156, Apr. 2003.

[15] N. Farahat, W. Yu, and R. Mittra, "A fast near-to-far-field transformation in body of revolution finite-difference time-domain method," *IEEE Trans. Antennas Propag.*, vol. 51, no. 9, pp. 2534–2540, Sep. 2003.

[16] N. V. Kantartzis, T. T. Zygiridis, and T. D. Tsiboukis, "An unconditionally stable higher order ADI-FDTD technique for the dispersionless analysis of generalized 3-D EMC structures," *IEEE Trans. Magn.*, vol. 40, no. 2, pp. 1436–1439, Mar. 2004.

[17] J. Papapolymerou, G. E. Ponchak, E. Dalton, A. Bacon, and M. M. Tentzeris, "Crosstalk between finite ground coplanar waveguides over polyimide layers for 3-D MMICs on Si substrates," *IEEE Trans. Microw. Theory Tech.*, vol. 52, no. 4, pp. 1292–1301, Apr. 2004.

[18] W. Ren, B.-Q. Gao, Z.-H. Xue, W.-M. Li, and B. Liu, "Full-wave analysis of broad wall slot's characteristics in rectangular waveguides," *IEEE Trans. Antennas Propag.*, vol. 52, no. 9, pp. 2436–2444, Sep. 2004.

[19] I. A. Eshrah, A. A. Kishk, A. B. Yakovlev, A. W. Glisson, and C. E. Smith, "Analysis of waveguide slot-based structures using wide-band equivalent-circuit model," *IEEE Trans. Microw. Theory Tech.*, vol. 52, no. 12, pp. 2691–2696, Dec. 2004.

[20] D. Arai, M. Zhang, K. Sakurai, J. Hirokawa, and M. Ando, "Obliquely arranged feed waveguide for alternating-phase fed single-layer slotted waveguide array," *IEEE Trans. Antennas Propag.*, vol. 53, no. 2, pp. 594–600, Feb. 2005.

[21] Y. S. Rickard and N. K. Nikolova, "Off-grid perfect boundary conditions for the FDTD method," *IEEE Trans. Microw. Theory Tech.*, vol. 53, no. 7, pp. 2274–2283, July 2005.

[22] A. P. Zhao and A. V. Räisänen, "Application of a simple and efficient source excitation technique to the FDTD analysis of waveguide and microstrip circuits," *IEEE Trans Microw. Theory Tech.*, vol. 44, no. 9, pp. 1535–1539, Sep. 1996.

[23] S. Wang and F. L. Teixeira, "An equivalent electric field source for wideband FDTD simulations of waveguide discontinuities," *IEEE Microw. Wireless Compon. Lett.*, vol. 13, no. 1, pp. 27–29, Jan. 2003.

[24] E. A. Navarro, T. M. Bordallo, and J. Navasquillo-Miralles, "FDTD characterization of evanescent modes – Multimode analysis of waveguide discontinuities," *IEEE Trans. Microw. Theory Tech.*, vol. 48, no. 4, pp. 606–610, Apr. 2000.

[25] N. V. Kantartzis, "A generalised higher-order FDTD-PML algorithm for the enhanced analysis of 3-D waveguiding EMC structures in curvilinear coordinates," *IEE Proc. Microw. Antennas Propag.*, vol. 150, no. 5, pp. 351–359, Oct. 2003.

[26] J.-P. Bérenger, "An effective PML for the absorption of evanescent waves in waveguides," *IEEE Microw. Guided Wave Lett.*, vol. 8, no. 5, pp. 188–190, May 1998.

[27] N. V. Kantartzis, T. K. Katsibas, C. S. Antonopoulos, and T. D. Tsiboukis, "A 3D multimodal FDTD algorithm for electromagnetic and acoustic propagation in curved waveguides and bent ducts of varying cross-section," *COMPEL*, vol. 23, no. 3, pp. 613–624, 2004.

[28] C. A. Balanis, *Advanced Engineering Electromagnetics.* New York: John Wiley & Sons, 1989.

[29] W. K. Gwarek and M. Celuch-Marcysiak, "Wide-band *S*-parameter extraction from FD-TD simulations for propagating and evanescent modes in inhomogeneous guides," *IEEE Trans. Microw. Theory Tech.*, vol. 51, no. 8, pp. 1920–1928, Aug. 2003.

[30] D. M. Pozar, *Microwave Engineering*, 3rd ed., Hoboken, NJ: John Wiley & Sons, 2005.

CHAPTER 8

Selected Applications in Antenna Structures

8.1 INTRODUCTION

The strong relation between the performance of an antenna and its size relative to the operating frequency is an important topic of research. The growth of telecommunications and in particular the rapid development of mobile telephony and personal communication systems direct scientific efforts to construct multiband and portable devices, which require the analogous antenna structures [1]. These achievements come at a price, however. The complexity of the resulting configurations is constantly increasing, while the incorporation of various hard-to-model materials in their substrates becomes more and more frequent. Obviously, the fulfillment of the preceding specifications entails the systematic analysis of the antenna models. Taking into account the large fabrication cost during the evolution of such systems, precise predictions of their performance should be pursued. Under these circumstances, time-domain numerical methods can be proven a very powerful tool. More specifically, the merits of higher order FDTD schemes are expected to offer reliable solutions to many laborious problems, where traditional techniques lack to provide sufficient treatment.

It is the objective of this chapter to apply these algorithms to several realistic antennas with special properties. Initiating from a short description of excitation schemes, some basic radiating elements are examined in an effort to determine the limits of higher order formulations along with the appropriate perfectly matched layers (PMLs). Next, the class of microstrip, dielectric resonator, and cavity-backed antennas is studied and then our interest is drawn to the considerable category of fractal devices. Finally, the chapter closes with the very promising metamaterial-loaded substrates, whose purpose is to enhance the capabilities of existing arrangements.

8.2 EXCITATION ISSUES AND FEEDING MODELS

The consistent advancement of electromagnetic fields in antenna problems is firmly connected to the incident wave source settings imposed at the beginning of every simulation. As a matter of fact, these settings act like a *connecting link* between continuous and discrete state, since they transfer the past profile (history) of a field and introduce the necessary attributes for the interaction of its components. A carefully selected excitation scheme is the basis for a successful numerical study of an antenna. After all, it can be understood that any irregularity incorporated in the Yee's algorithm via an incident wave is likely to propagate (and not eliminated) in the domain, despite the efficiency of the entire configuration. Not to mention the growth of nonphysical parasitic modes and below-cutoff fields that affects the stability of the discretization procedure.

The appropriate excitation scheme depends upon the particular FDTD setup. For illustration, when the antenna points at which reflections originate are to be found, a *narrow* Gaussian pulse could be the most preferable solution. Thus, for a transmitting antenna, the incident voltage in the feeding line is given by

$$V_{\text{exc}}(t) = V_0 e^{-(t/\tau)^2/2} \xrightarrow[\text{transform}]{\text{Fourier}} V_{\text{exc}}(\omega) = \sqrt{2\pi}\, V_0 \tau e^{-(\omega\tau)^2/2}, \tag{8.1}$$

where V_0 is the amplitude and τ represents a characteristic time constant.

On the other hand, when broadband investigations are required, it is more effective to use a *differentiated* signal and then use the Fourier transform to derive the desired response in the frequency domain. In this case, the corresponding Gaussian pulse becomes

$$V_{\text{exc}}(t) = -\tfrac{V_0 t}{\tau} e^{-\left[(t/\tau)^2-1\right]/2} \xrightarrow[\text{transform}]{\text{Fourier}} V_{\text{exc}}(\omega) = -j\omega\sqrt{2\pi}\, V_0 \tau^2 e^{-\left[(\omega\tau)^2-1\right]/2}. \tag{8.2}$$

For these general considerations to profitably cooperate with the rest of the FDTD aspects, the correct antenna feed model should be constructed. According to its detailed version, the antenna terminals are connected to a transmission line into which a signal is either transmitted or received. In the situation of practical applications, the feed model should be meticulously described, since it launches the basic field data in the entire FDTD process. Hence, inaccurate configurations are likely to spoil the results and lead to completely misleading design data. Normally, for a transmitting antenna, electric components at the gap between the antenna wire and the image plane are associated with a specific voltage, computed through the spatial increment. The required current at the feed point is obtained by applying the integral form of Ampère's law at a prefixed surface with a bounding contour centered on the wire. Such a model, though, may be proven rather abrupt when the time update of specific quantities is our only interest due to the need for repetitive use of the Fourier transforms. This shortcoming can be fixed by means of more advanced and elaborate techniques, such as those presented in [2–7]. It is emphasized that the dimensions of the transmission line and antenna feed region are typically much smaller than the size of the other parts of the structure and therefore are prone to all the inherent FDTD lattice reflection errors. However, if a very fine grid is utilized, the computational demands will be significantly augmented. This is *exactly* where higher order schemes can be implemented and offer considerable improvements. Actually, their efficient stencil management will enhance the accuracy of the simulations and concurrently guarantee their stability and convergence.

8.3 ANALYSIS OF ESSENTIAL STRUCTURES

In this section, the higher order FDTD methods are applied to basic three-dimensional (3-D) radiation and scattering problems that involve simple dipole antennas. As performed in Chapter 7, whenever possible numerical results are compared to reference data, alternative formulations are also taken into account.

Let us examine a radiation problem involving a perfect electric conducting (PEC) sphere of radius r_s centered in a free and lossless space. The structure is excited by an electric dipole source at $(0, 0, z = z_0 > r_s)$ and is polarized along the z-direction. Its temporal profile is given by the smooth function of

$$P(t) = \begin{cases} 10 - 15\cos\omega_1 t + 6\cos\omega_2 t - \cos\omega_3 t, & t \leq T \\ 0, & t > T \end{cases} \tag{8.3}$$

supported in $t \in [0,\ T]$ with $T = 10^{-9}$s and $\omega_m = 2\pi m/T$ for $m = 1, 2, 3$. Due to the symmetry of the problem, only three field components are involved, namely $\mathbf{E} = E_r\hat{\mathbf{r}} + E_\theta\hat{\boldsymbol{\theta}}$ and $\mathbf{H} = H_\varphi\hat{\boldsymbol{\varphi}}$, thus reducing the computational domain to a two-dimensional region with $0 \leq r \leq r_0$ and $0 \leq \theta \leq \pi$. This space is terminated by a spherical PML, as illustrated in Figure 8.1(a). For the solution, we use a 25×120 mesh with $\Delta r = 0.0417$ m, $\Delta\theta = 0.0264$ rad,

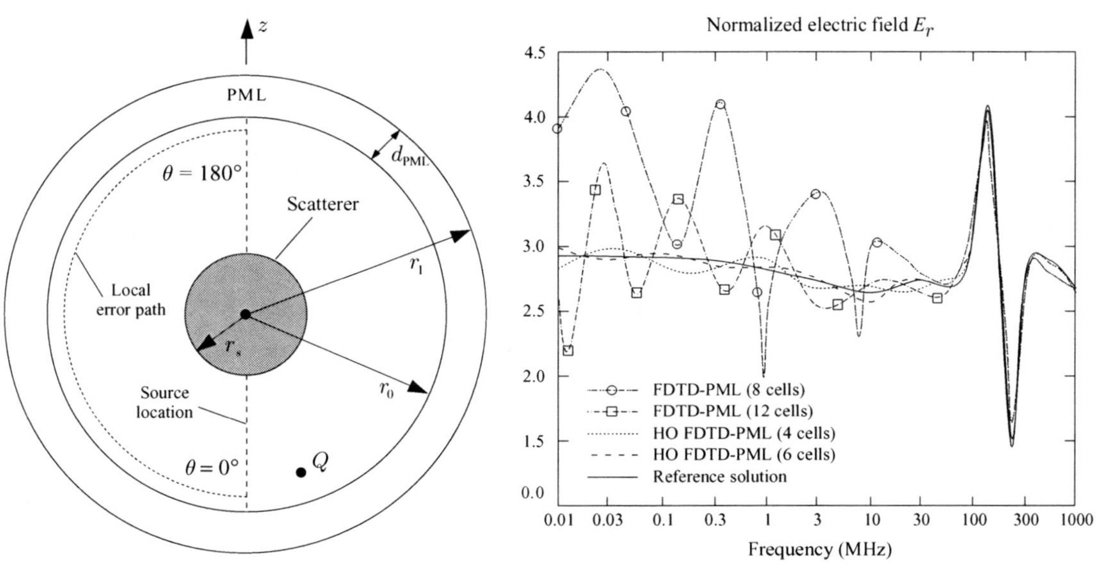

(a) (b)

FIGURE 8.1: (a) Transverse cut of the computational space with the PEC scatterer terminated by a spherical PML. (b) Normalized electric field E_r component of the z-polarized electric dipole

$\Delta t = 7.4367$ps, $r_0 = 1$ m, $r_s = 9\Delta r$, $z_0 = 20\Delta r$ and set the scatterer-PML separation at two cells. By selecting $R = 10^{-5}$, it is derived that $\sigma_r^{\max} = 1.5536 \times 10^{10}$. Figure 8.1(b) gives the normalized electric field E_r for a broad frequency spectrum. The reference solution is obtained in a sufficiently larger domain (150 × 280 cells) and the higher order nonstandard operators of (3.70)–(3.81), for $M = 3$ and $L = 2$, are employed. For the conventional version, dispersion errors are reduced by the modified (2, 4) FDTD algorithm of Section 2.5.1, whereas the vacuum-PML interfaces are treated by the technique given in Section 2.4.4. To reveal the limits of these algorithms we use a depth of four cells. From the results, one can notice that second-order solutions depart from the reference data at a frequency near $f = 35.5$MHz. This is attributed to the reflection of strongly evanescent waves, as discussed in [8], and the dispersion errors of Yee's scheme which for this problem requires a 75% finer grid (88% larger memory). Conversely, the higher order versions exhibit a significant performance, as their solution is very close to the reference one.

The next example explores the behavior of a thin hemispherical shielding positioned between the excitation point and a cocentric dielectric sphere ($\varepsilon = 4.8\varepsilon_0$) centered at the origin with $r_s = 0.5$ m. The shielding is one-cell thick and its radius is 0.625 m. Due to their higher order stencils, operators (3.70)–(3.81), for $M = 3$ and $L = 3$, are combined with the boundary schemes of Section 2.4.1 and the curvilinear formulation of Section 2.6. The domain is divided into 7 × 16 × 32 cells with $\Delta r = 0.125$ m, $\Delta\theta = 0.2094$ rad, $\Delta\varphi = 0.196$ rad, and $\Delta t =$ 0.3494 ns. So, grid resolution reaches the value of $\Delta r/\lambda_{s,\min} = 1/5$, where $\lambda_{s,\min}$ is the minimum wavelength of the Gaussian pulse (8.2) launched at $(0, 0, z_0 = 0.75)$ and supported in $t \in [0, 4t_0]$. Its parameters are selected as $V_0 = 10^{-6}$ and $\tau = 31.446$ ns to avoid the loss of any nontrivial energy content. Note that for the mitigation of dispersion errors the angle-optimized method of Section 2.5.2 is utilized. The respective second-order realization involves a 12-cell PML and a 42 × 68 × 220 domain, stating that the usual FDTD scheme needs a 90% larger overhead. Figure 8.2(a) gives the numerical coefficient in the case of an excitation with a strong traveling/evanescent content. As deduced, the higher order concepts cope very well with this demanding case and most importantly, they enhance the absorption of the PMLs (up to 45dB). Similarly, Figure 8.2(b) presents the global-error convergence versus grid resolution, for different angles of antenna-wave impingement, formed by the wavevector and the normal to the vacuum–PML interface. The scatterer-absorber distance is set at two cells and, for comparison, a second-order 10-cell PML is considered. Outcomes present that, owing to this small distance, the error of the latter absorber does not converge. For instance, when $\Delta r/\lambda_{s,\min} = 1/5$, this PML has an error of 10^0; i.e., it reflects all outgoing waves. In contrast, the higher order counterparts converge very fast, regardless of their depth.

Finally, numerical verification investigates an elliptical structure. Specifically, a 3-D domain in the presence of an elliptical-cylindrical lossy scatterer with a height of 0.75 m and

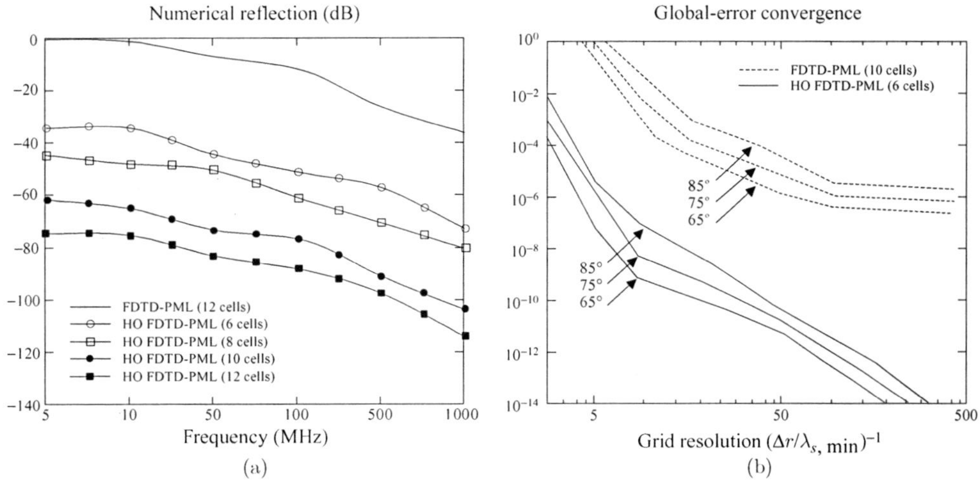

FIGURE 8.2: (a) Numerical reflection coefficient versus frequency for a two-cell scatterer-PML distance. (b) Convergence of global error versus grid resolution

$\sigma = 0.05$ S/m is examined. The primary axis of its bases is 0.3 m, while the secondary one is 0.11 m. Excitation is launched by the electric dipole source of Figure 8.1(a) and the space is discretized in $26 \times 34 \times 62$ cells. Herein, the curvilinear higher order schemes are combined with the lossy formulation of Section 5.2 and the operators of Section 2.4.2. Moreover, reference solution is obtained by means of a fairly larger $78 \times 102 \times 186$ domain. The serious improvement of the maximum global error as a function of distance from the scatterer for a 10-cell PML is illustrated in Figure 8.3. Simulations are conducted via the (2, 4) and the (4, 4) FDTD method as well as the nonstandard operators of (3.42), (3.43), and (3.59). For example, at the frequently used distance of three cells from the scatterer, the error varies between 10^{-7} and 10^{-10} and so enabling a great decrease in the computer burden. This issue is deemed fairly important since structures like the above are encountered in electromagnetic compatibility (EMC) applications. Particularly, when the shielding efficiency or the S-parameters of a device are required, the reduction of the global error is indeed critical in order to assure an acceptable fabrication process.

8.4 CONTEMPORARY ANTENNA CONFIGURATIONS

Having determined the fundamental properties of higher order FDTD algorithms, a set of more complicated problems will be studied in this section. These include the analysis of microstrip, cavity-backed or dielectric resonator antennas with different polarizations, fractal arrays, and metamaterial-loaded structures.

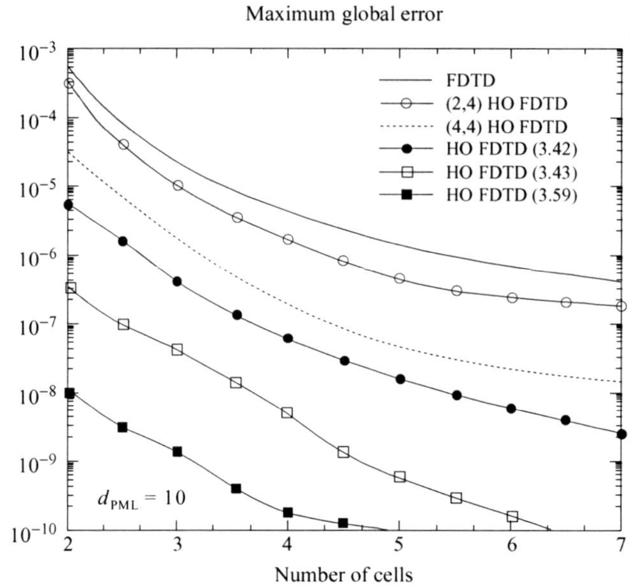

FIGURE 8.3: Maximum global error versus distance from an elliptical-cylindrical structure

8.4.1 Microstrip and Cavity-Backed Antennas

Microstrip patch and cavity-backed antennas constitute an indispensable part of many devices [1, 9–18]. Their notable properties and adjustable features lead to various efficient designs, which comprise the fundamental elements of many up-to-date configurations in telecommunication technology.

The first applications involve two broadband dual-polarized patch antennas. Figure 8.4(a) describes the former structure fed by a capacitatively coupled and an H slot-coupled feed.

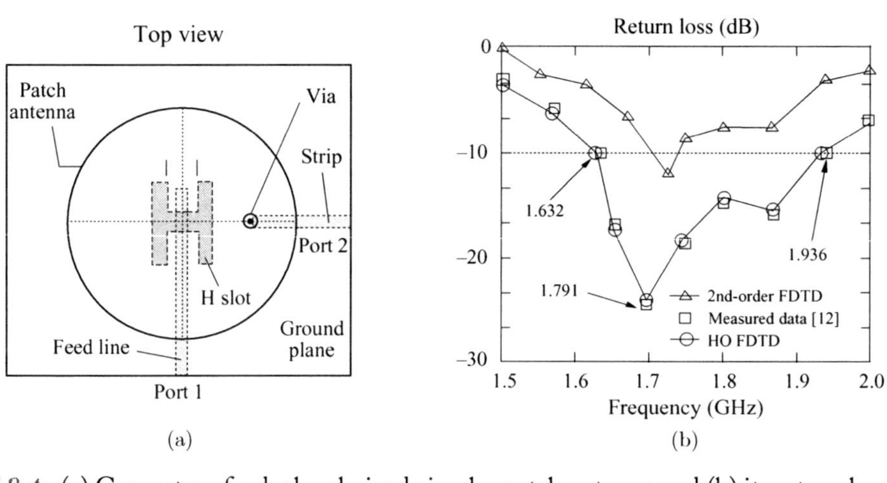

FIGURE 8.4: (a) Geometry of a dual-polarized circular patch antenna and (b) its return loss for port 2

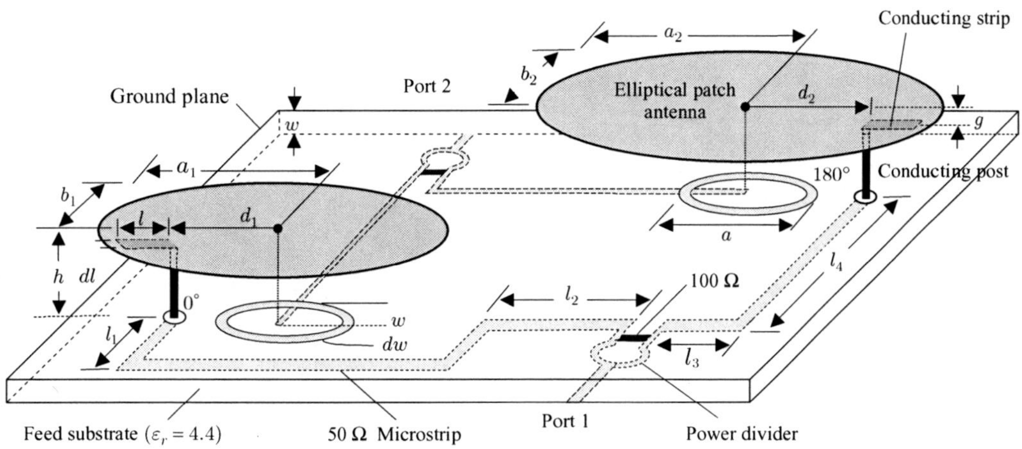

FIGURE 8.5: A two-element dual-polarized elliptical patch antenna

Specifically, port 1 is cut in the ground and centered below the radiating patch. For the discretization of the slot, the hybrid technique of Section 6.2 is combined with the modified operators of Section 5.3. Keeping the dimensions given in [12], a $34 \times 34 \times 40$ grid is constructed and the return loss for port 2 is given in Figure 8.4(b). Termination of the open space is conducted by a six-layer PML with an object-absorber distance set at four cells, while measured data are acquired form [13]. Evidently, the higher order solution is proven very accurate.

Furthermore, Figure 8.5 presents a much more complicated device, i.e., a two-element elliptical patch antenna with high isolation rates [18]. Now, port 1 consists of two capacitatively coupled feeds with a 0° and 180° phase difference provided by a Wilkinson power divider, whereas port 2 is connected to two elliptical slots. The design frequency for the results of Figures 8.6(a) and 8.6(b) is 1.75GHz and 1.8GHz, respectively, and the rest of the parameters are $a_1 = 25$ mm, $b_1 = 10$ mm, $d_1 = 16$ mm, $a_2 = 35$ mm, $b_2 = 14$ mm, $d_2 = 22$ mm, $g = 3.2$ mm, $w = 1.2$ mm, $h = 13.6$ mm, and $l = 12$ mm. These parameters lead to a $120 \times 132 \times 96$ mesh which, despite its size, is almost 80% coarser than the one constructed by the Yee's scheme.

The next example is a wideband 2×1 array consisting of two circular patches with an adjustable air-layer substrate placed over an L-shaped ground plane, as shown in Figure 8.7. The probe feed is oriented in the same plane as the patches, while its circular polarization is realized by two perturbed segments (0.72% of the patch area), cut at the opposite ends of the patch diameter (inclined 40° from the vertical). This combination enforces dominant modes to separate in two orthogonal components with equal amplitude and 90° phase over a broad frequency range. The dimensions are $L_x = 110$ mm, $L_y = 315$ mm, $L_z = 22$ mm, $r = 25$ mm, $d = 35$ mm, $h = 0.19$ mm, $l = 3.56$ mm, $w = 2.51$ mm, and $s = 5.65$ mm. For the simulation, the higher order ADI-FDTD method is combined with the dispersion relation-preserving

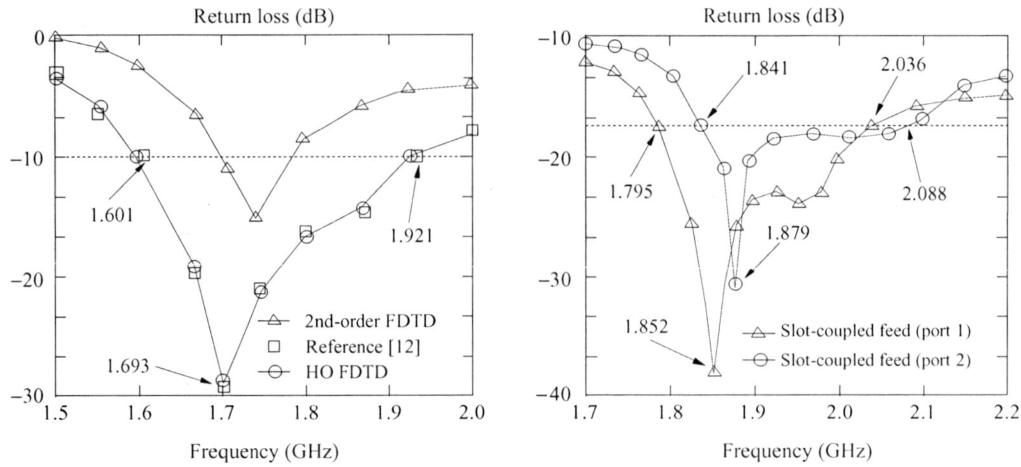

(a) (b)

FIGURE 8.6: Return loss (a) for port 1 at the design frequency of 1.75 GHz and (b) for both ports at the design frequency of 1.8 GHz

schemes of Section 2.5.3. The 3-D domain is divided in $62 \times 114 \times 18$ cells with $\Delta x = 1.775$ mm, $\Delta y = 2.719$ mm, and $\Delta z = 1.223$ mm. So, grid resolution reaches the level of $\lambda/7$ instead of the $\lambda/120$ for the second-order case that corresponds to a $132 \times 258 \times 46$ (91% larger) mesh. Figure 8.7(b) shows the radiation pattern of the antenna and Figure 8.8(a) gives the return loss for the single-element and various 2×1 setups, comparing the results with those

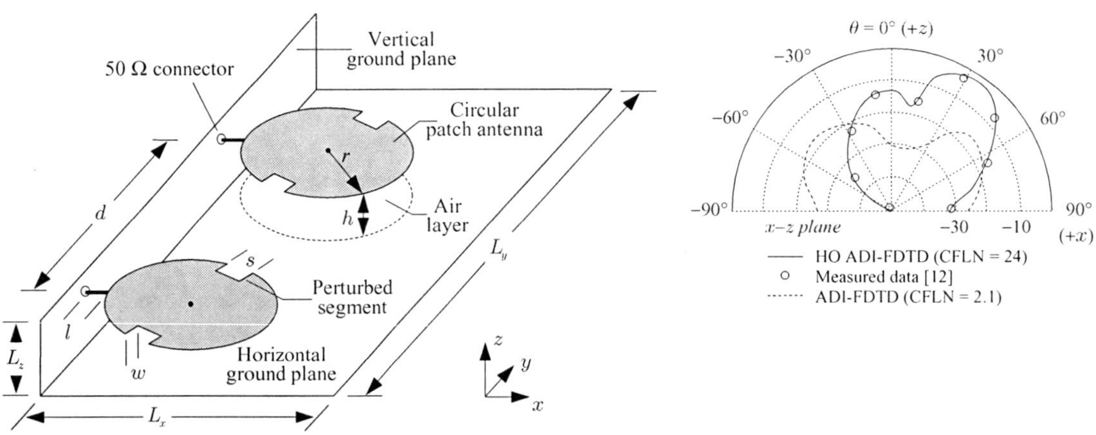

(a) (b)

FIGURE 8.7: (a) A broadband 2×1 array with two circular patches and (b) its radiation pattern in the x–z plane

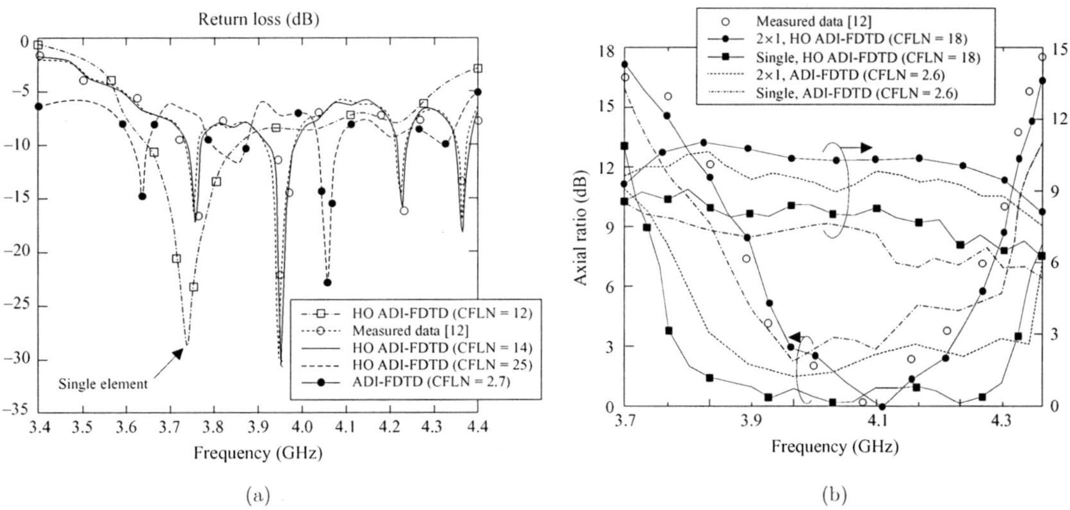

FIGURE 8.8: (a) Return loss and (b) AR and gain versus frequency for the single-element and several 2×1 circular patch array realizations

of [12]. As deduced, the specific technique achieves a very good agreement, despite the large Courant-Friedrich-Levy number (CFLN) values, unlike the common ADI-FDTD approach that cannot compute the resonant frequencies. This behavior is also confirmed in Figure 8.8(b) presenting the axial ratio (AR) and gain of the single-element and 2×1 antenna. Note the greatly widened AR of the optimized array.

Let us now focus on two equilateral-triangular microstrip circularly polarized antennas, placed $b = 1.8$ mm over a $t = 1.6$-mm thick FR_4 ($\varepsilon_r = 4.3$) substrate. The first structure has an H-type slot [Figure 8.9(a)] and dimensions: $a = 42$ mm, $h_1 = 12$ mm, $h_2 = 9.5$ mm, $s_1 = 11$ mm, $s_2 = 5.5$ mm, and $w = 1.2$ mm. The second structure has two θ-inclined narrow slots [Figure 8.9(b)] and $a = 40$ mm, $h = 19$ mm, $s_1 = 2.5$ mm, $s_2 = 1.3$ mm, $s_3 = 6.1$ mm, $s_4 = 4.6$ mm, $g = 2.8$ mm, $w = 1.15$ mm, and $\theta = 30°$. Both antennas are excited by a single feed, positioned at point A or B depending on the polarization. Picking diverse values of M and L for (3.70)–(3.81) and the derivative matching technique of Section 2.4.5 for the interfaces, a $84 \times 82 \times 36$ mesh is generated. Figure 8.10(a) shows the input impedance of the two arrangements, comparing the outcomes with those obtained by the simple FDTD method for a very fine $260 \times 282 \times 74$ grid. Indeed, the higher order FDTD algorithm is fairly accurate. Also to test the method, the optimum return loss frequency f_{RL} and AR frequency f_{AR} versus lengths h and g of the second antenna are provided in Figure 8.10(b). As observed, the usual FDTD approach does not lead to correct results due to its dispersion errors. However, the enhanced schemes prove that such designs are workable, since f_{RL} and f_{AR} are close to each other.

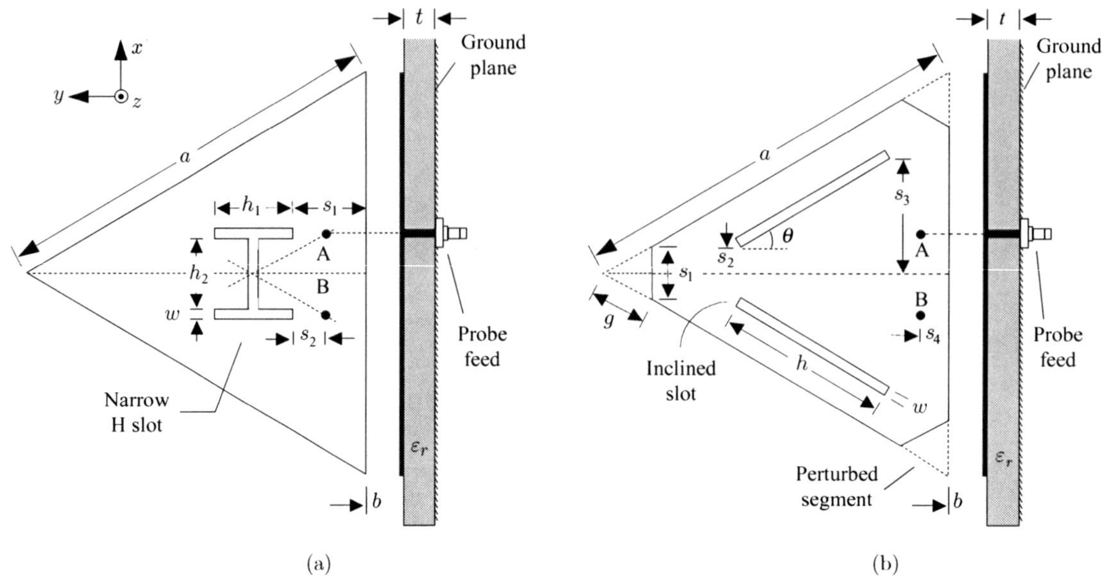

FIGURE 8.9: Geometry of an equilateral-triangular microstrip circularly polarized microstrip antenna with (a) an H-type and (b) two θ-inclined narrow slots

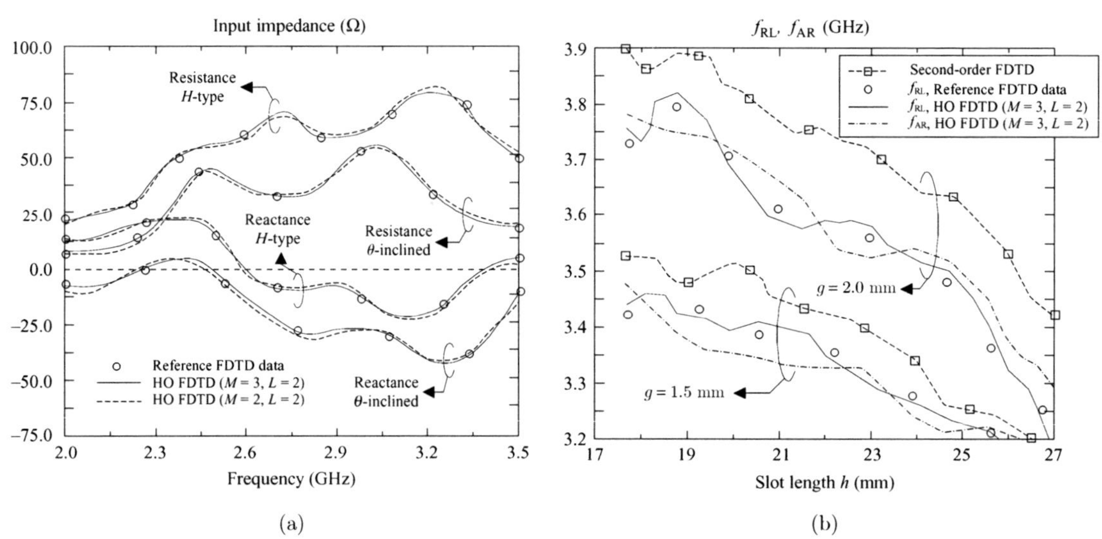

FIGURE 8.10: (a) Input impedance for the two equilateral-triangular microstrip antennas. (b) Optimum return loss frequency f_{RL} and AR frequency f_{AR} for different values of lengths h and g, concerning the θ-inclined antenna

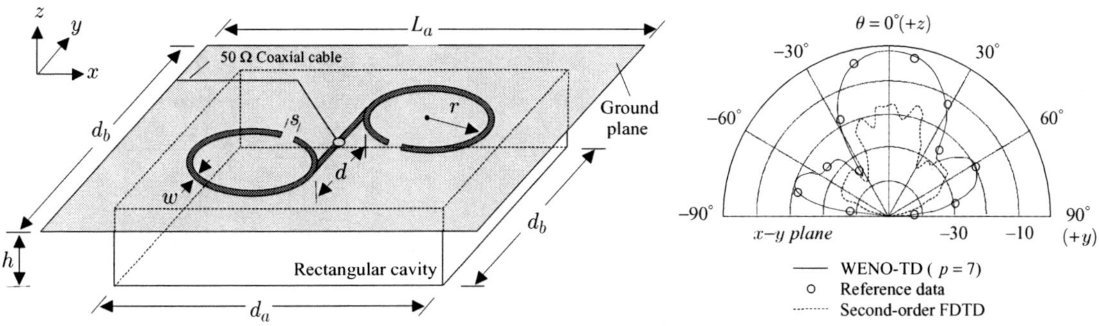

(a) (b)

FIGURE 8.11: (a) A two-element circular loop antenna backed by a rectangular cavity beneath the ground plane. (b) Radiation pattern at the x–y plane

The influence of a rectangular backing cavity beneath a two-element circular loop antenna, described in Figure 8.11(a), is investigated next. Its dimensions are $L_a = 190$ mm, $L_b = 160$ mm, $d_a = 150$ mm, $d_b = 130$ mm, $h = 25$ mm, $r = 26$ mm, $d = 12$ mm, $s = 2.2$ mm, and $w = 1.8$ mm. The structure is excited by a 50 Ω coaxial cable and is discretized in a $86 \times 74 \times 30$ lattice with $\Delta x = 2.209$ mm, $\Delta y = 2.162$ mm, and $\Delta z = 0.834$ mm. The WENO-TD simulations are performed via the controllable operators of Section 2.5.4 combined with the boundary schemes of Section 2.4.1. Figure 8.11(b) gives the antenna radiation pattern and Figure 8.12(a) its return loss. The superiority of the higher order algorithm over the plain FDTD counterparts

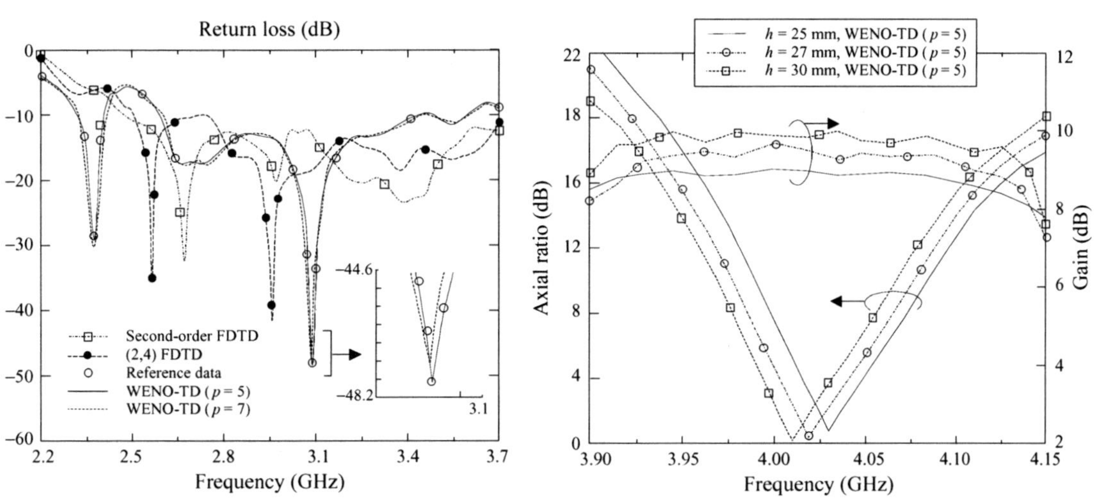

(a) (b)

FIGURE 8.12: (a) Return loss and (b) AR and gain for diverse cavity depths, regarding the two-element circular loop antenna

(88% finer mesh) is evident (reference data are from [15, 16]). Figure 8.12(b), contrarily, displays the role of the backing cavity as a means for fine tuning the AR and gain of a circularly polarized radiator. It is interesting to observe the transition of the AR minimum and gain improvement, even for small changes of h.

8.4.2 Dielectric Resonator Antennas

Dielectric resonator antennas (DRAs) constitute an important family of radiation devices, since they exhibit sufficient directivity and gain, while their small size enables the handling of multipath signals [19–22].

Consider the one-quarter elliptical DRA of Figure 8.13 which irradiates a cylindrical metallic box with a long narrow slot located in its near field region. Apart from the above advantages, the particular antenna produces low return loss and symmetric radiation patterns. Its dimensions are selected as $b_1 = 3.5$ cm, $b_2 = 6$ cm, $d = 1$ cm, $h = 1.7$ cm, while those of the cylindrical box are $a = 3.5$ cm, $H = 8$ cm, $w = 0.4$ cm, $l = 11$ cm with the length of the slot corresponding to $\varphi = 60°$. The antenna is fed by a coaxial cable and the domain is divided into $72 \times 80 \times 64$ cells with $\Delta x = \Delta y = \Delta z = 0.004$ m and $\Delta t = 6.853$.

Since its dielectric constant is rather high ($\varepsilon_r = 12$) and its surface does not align to one of the system axis, the higher order nonstandard schemes of (3.42) and (3.59) are combined with the operators of Section 2.5.5. Due to the interactions between the two parts of the structure,

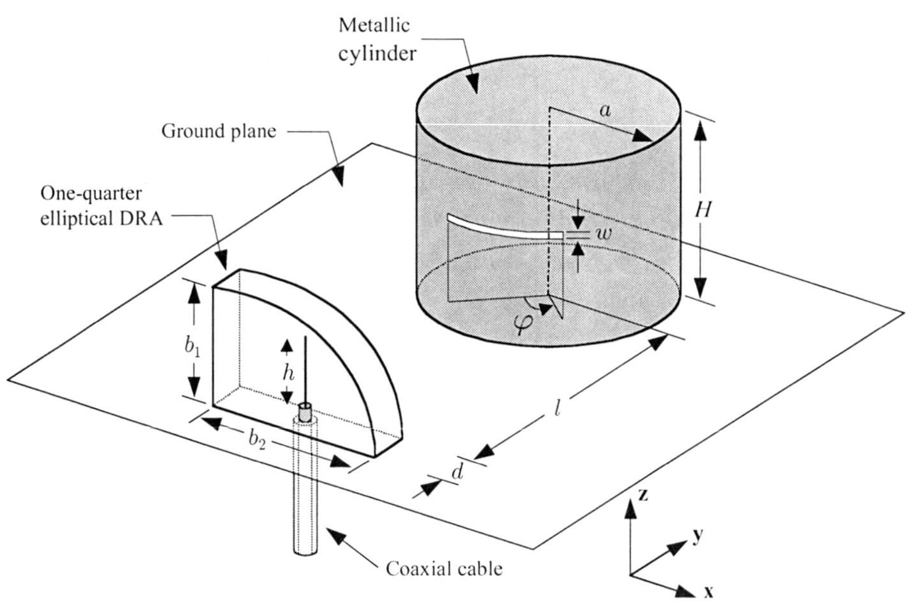

FIGURE 8.13: An one-quarter elliptical DRA irradiating a cylindrical metallic box with a long narrow slot

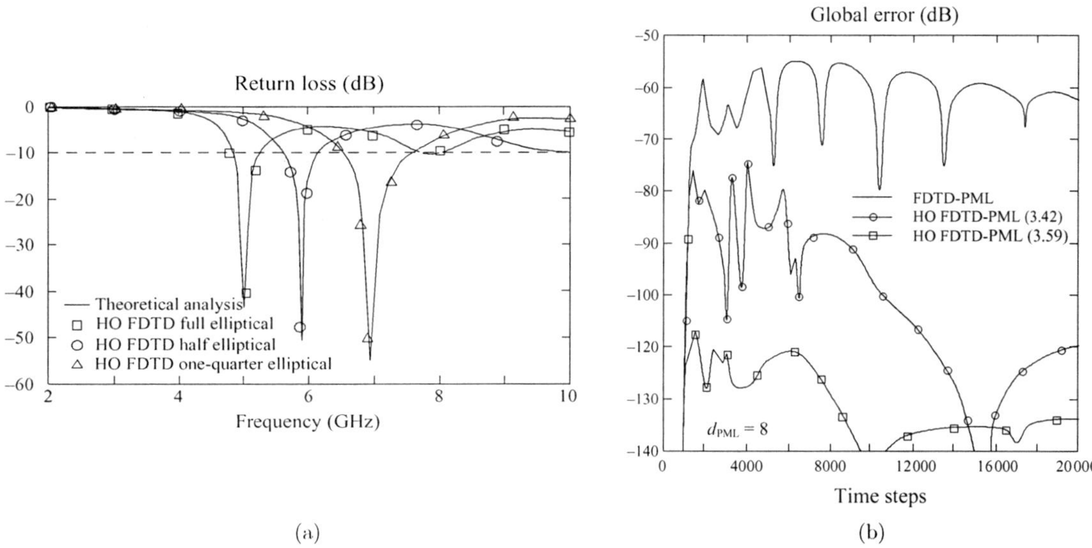

FIGURE 8.14: (a) Return loss of a full, half, and one-quarter elliptical DRA. (b) Global error versus time for three PML implementations

a large number of time-steps are required. This is indeed a pertinent application to validate the stability of the eight-cell PMLs used for external backing. The antenna is theoretically analyzed via [1, 19]. Figure 8.14(a) shows the return loss for the first transverse electric mode of a full, a half, and one-quarter elliptical DRA from which the high accuracy of the higher order algorithm is easily derived. It is stated that the corresponding second-order Yee scheme would have required a 90% larger mesh and 75% CPU time to provide acceptable (but not so precise) results. Also, the global error versus time is computed. The duration is set to 40 000 time-steps to check the convergence of the eight-cell PML. From Figure 8.14(b), it is concluded that the error of the usual absorber still persists. However, this is not the case with the other realizations, since both achieve a significantly smaller reflection error.

The second application is the hemi-ellipsoidal H-slot coupled DRA of Figure 8.15, incorporating two patches as additional degrees of freedom for AR and gain control. The relative permittivity of the cover is $\varepsilon_{rc} = 8.8$ and that of the substrate $\varepsilon_{rm} = 2.45$. A normal set of dimensions is $L = 32.5$ mm, $s_1 = 2.8$ mm, $w_1 = 8.3$ mm, $s_2 = 4.2$ mm, $w_2 = 3.8$ mm, $l_1 = 11.9$ mm, $l_2 = 5.4$ mm, $h_1 = 2.4$ mm, $h_2 = 1.6$ mm, $w = 3.6$ mm, $d = 1.85$ mm, $s = 1$ mm, $\varphi_1 = 30°$, and $\varphi_2 = 290°$.

Constructing a $102 \times 106 \times 48$ grid with a resolution of $\lambda/15$ instead of the 85% denser lattice required by the simple ADI-FDTD technique, the first slot height w_1 is evaluated as a function of the optimum AR frequency, f_{AR}, and other parameters. Reference data are obtained through [19, 22] and Yee's scheme with a $230 \times 240 \times 62$ grid. From Figure 8.16(a), one can

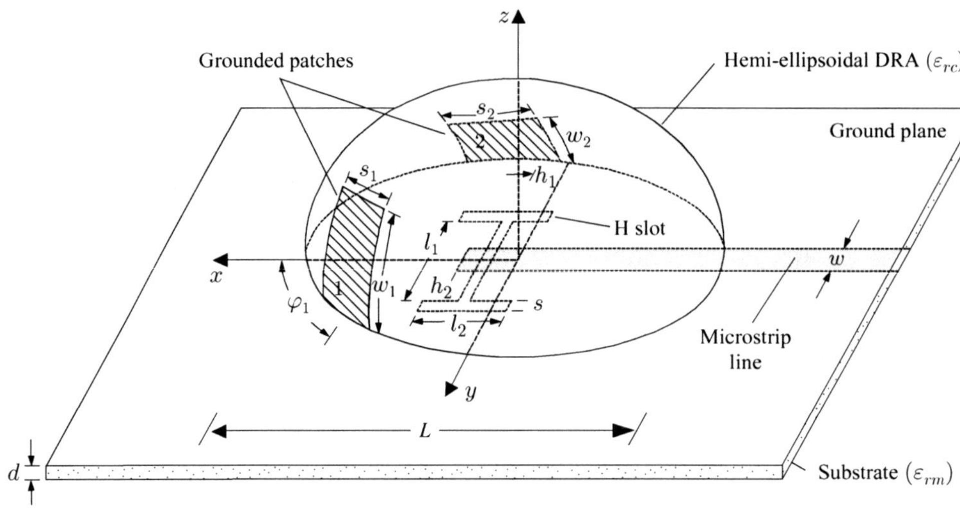

FIGURE 8.15: General description of a hemi-ellipsoidal H-slot coupled DRA with two patches

easily discern the enhanced accuracy of (3.70)–(3.81), for $M = 3$, $L = 2$, and comprehend the role of the two patches in DRA optimization. Likewise, Figure 8.16(b) gives the 3-dB AR bandwidth versus f_{AR} for different dimensions of the H slot. Finally, Table 8.1 provides the f_{AR} along with the maximum dispersion error and diverse implementation issues for a DRA with one patch. Again, the proposed methodology overwhelms the other schemes, even for large CFLN ratios.

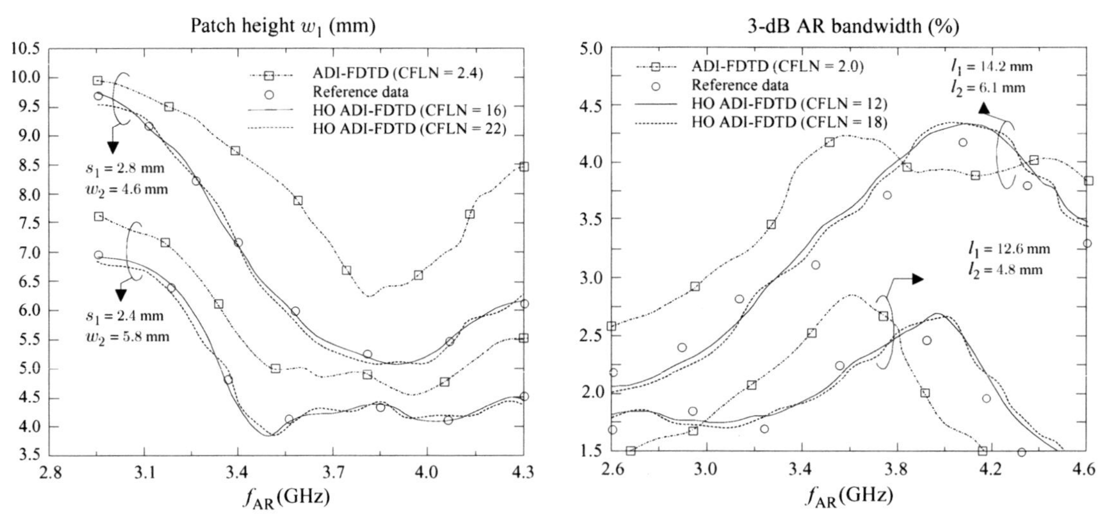

FIGURE 8.16: (a) Patch height w_1 versus optimum AR frequency, f_{AR}, for a hemi-ellipsoidal DRA with two patches. (b) 3-dB AR bandwidth versus optimum AR frequency, f_{AR}, for a hemi-ellipsoidal DRA with one patch

TABLE 8.1: Comparison of ADI-FDTD Computational Issues for the Hemi-Ellipsoidal DRA

REFS [19] (GHz)	w_1 (mm)	ADI-FDTD	CFLN	SOLUTION (GHz)	ERROR (%)	LATTICE	TIME-STEPS	MAXIMUM DISPERSION
3.4672	6.8	Second-order	1.32	3.3069	4.621	154×146×68	58 236	2.362934×10^{-1}
		Higher order	16	3.4667	0.012	76×70×26	3720	4.785917×10^{-9}
3.3865	7.6	Second-order	1.92	3.1903	5.793	172×162×74	64 592	5.632875×10^{-1}
		Higher order	22	3.3854	0.031	82×78×30	3180	3.201059×10^{-8}
3.2196	8.5	Second-order	2.68	2.9671	7.842	190×186×8	72 642	9.059302×10^{-1}
		Higher order	30	3.2177	0.058	88×84×36	2560	7.641055×10^{-8}
3.1428	9.3	Second-order	2.85	2.7541	12.367	198×190×8	84 190	1.823692
		Higher order	36	3.1401	0.084	96×94×42	2416	8.428205×10^{-7}

8.4.3 Fractal Arrays

The majority of approaches for the synthesis of antenna systems have their origins in *Euclidean* geometry. However, there has been a significant amount of recent interest in the possibility of developing new types of antennas that involve *fractal* rather than Euclidean geometric concepts in their design. This rapidly growing research area is usually referred as fractal antenna engineering [23–30]. The main reason for constructing such antennas is twofold. First, fractals have *no* characteristic size. In fact, they have an infinite range of scales, within their structure, comprising an infinite number of clusters that are equal to the entire shape but at smaller size. Hence, this *self-similar* geometric form of fractal shapes, including multiple copies of themselves at several scales, is an attractive candidate for the fabrication of multifrequency antennas. Second, they have a great deal of *irregularities*. It has been long recognized that sharp corners or discontinuities improve the radiation capabilities of electric systems. So, it is sensible to anticipate that these kinds of convoluted shapes should become efficient radiators. Essentially, there are two active areas in fractal engineering, which include the study of fractal-shaped elements as well as the use of fractal ideas in antenna arrays. Apparently, the analysis of the above structures is quite difficult from a numerical perspective. Their constantly repeating patterns require robust techniques, which provide credible results especially during the design procedure. In this section, the higher order FDTD schemes are applied to some characteristic fractal antennas to extend their competence to this challenging class of problems.

Let us study the Sierpiński gasket array of Figure 8.17(a). A subarray employing three dipoles on the corners of an equilateral triangle ($h = 0.5\lambda$ spacing on each side) and an expansion factor of 2 is utilized. Resolution is set to only 8 points/wavelength and three snapshots of the antenna's electric near field are depicted in Figure 8.17(b). The simulations are conducted via the optimized operators of Section 5.4.

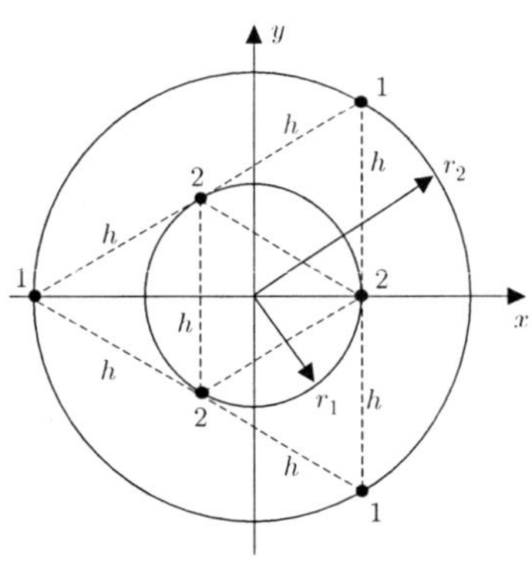

(a)

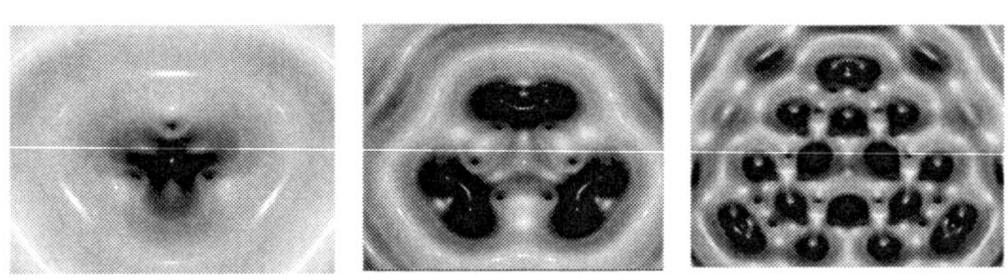

(b)

FIGURE 8.17: (a) The geometry of a Sierpiński gasket and (b) three snapshots of the electric near field

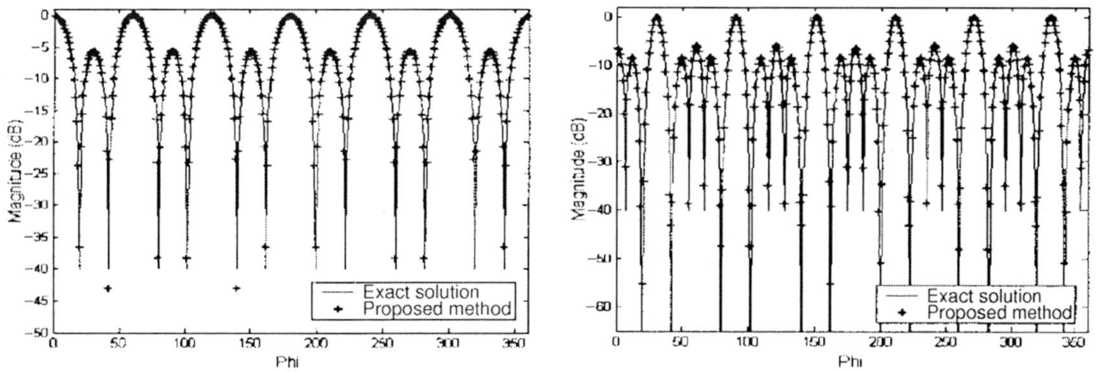

FIGURE 8.18: Radiation patterns of the Sierpiński gasket (third and fourth stage)

The fractal distribution is evident, proving that the iterative array plays a critical role. Similarly, the radiation patterns of a Sierpiński carpet for 8192 time-steps are considered. Herein, analysis involves nonequally spaced and uniformly excited elements with a spacing of 0.5λ. Figure 8.18 shows two stages of the antenna compared with theoretical results. Note the high accuracy, especially at peak points. It is mentioned that the usual FDTD grid would have required 210 × 128 × 64 cells (and yet not attaining this accuracy), while the proposed method needs only a 22 × 14 × 6 grid, i.e., nearly a 90% decrease of the computer memory storage.

To further validate the higher order nonstandard schemes of (3.42) and (3.43), the two-iteration Sierpiński carpet array of Figure 8.19(a) is investigated. As its name reveals, the array requires two stages of construction. The radiating elements are rectangular patches backed by an infinite ground plane a resonant frequency of 2 GHz. Setting the interelement distance to 0.5λ, the termination of the open space is performed by an eight-cell PML [31–33]. The maximum

FIGURE 8.19: (a) A Sierpiński carpet array and (b) the E_z component of the structure

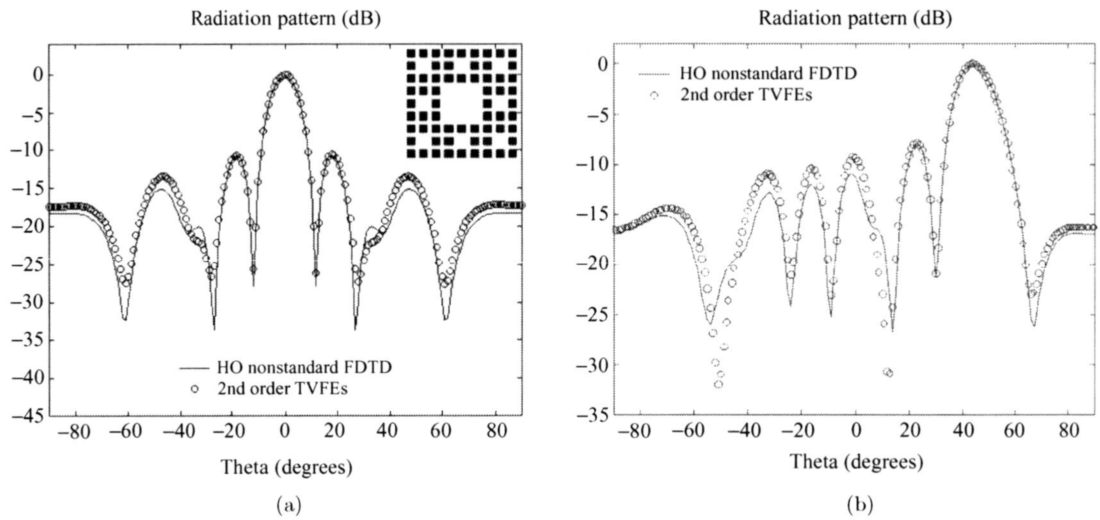

FIGURE 8.20: Radiation patterns of a two-iteration Sierpiński carpet array (a) for the simple case and (b) for 45° scanning

dimension of the array is 4.3λ, while for the analysis a 176 × 176 × 25 mesh is required. Wherever pertinent, the hybrid subgridding method of Section 6.2 is employed. Figure 8.19(b) gives the E_z component, where the smoothness in the evolution is easily noticeable.

Moreover, in Figure 8.20(a), the antenna's radiation pattern is depicted and compared with the outcomes of a higher order vector finite-element technique [34–37] that uses 63 504 tetrahedra and 425 052 degrees of freedom. Evidently, the agreement between the two formulations is very satisfactory. It is also interesting to note that the fractal array produces a narrower main lobe, compared to the uniform planar array, at the expense of the side-lobe levels. On the other hand, Figure 8.20(b) examines the beam steering capabilities of the higher order FDTD algorithm. The example tested is that of a Sierpiński carpet array, where its main lobe is scanned to 45° (for plane $\varphi = 0°$). Again, the two higher order methods are in very good agreement (finite-element solution: 17 424 tetrahedra, resulting in 117 932 degrees of freedom).

As the last example of this category, the 3-D modified Sierpiński-gasket monopole antenna over a ground plane, shown in Figure 8.21(a), is analyzed. Its fractal iterations are carried out only at the lower triangle of each stage (removal of a central triangular piece). For the discretization the higher order nonstandard FDTD formulation ($M = 2$, $L = 2$) is selected. Figure 8.21(b) gives the normalized electric field as a function frequency for various PMLs. Again, both the results and the computational savings are very sufficient, since the grid reduction reaches the value of 87% (usual FDTD: 300 × 190 × 88; higher order method: 36 × 23 × 11).

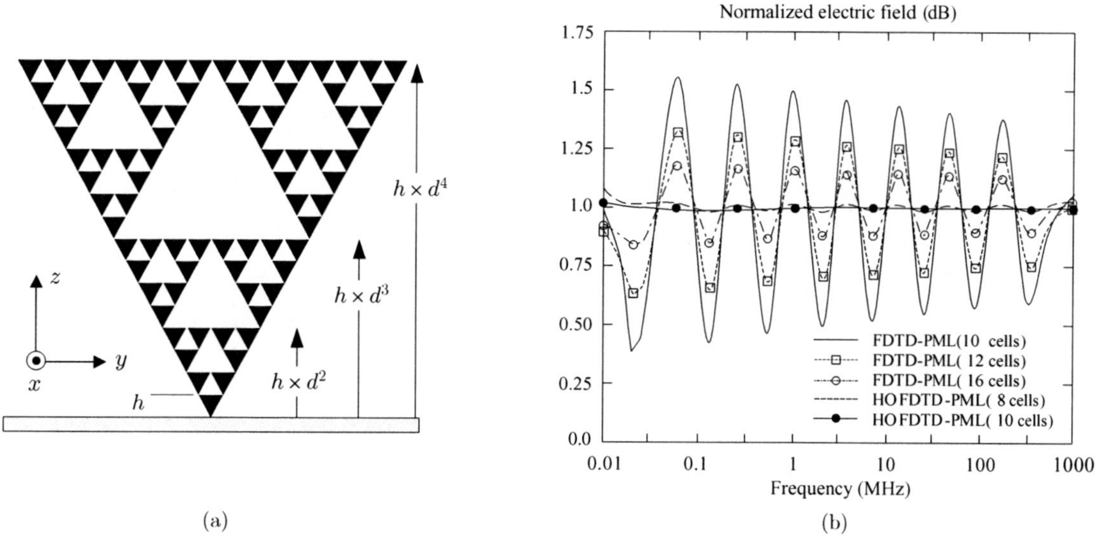

FIGURE 8.21: (a) A modified Sierpiński-gasket monopole antenna and (b) the normalized electric field versus frequency

8.4.4 Metamaterial-Loaded Applications

A major breakthrough in modern electromagnetics is the conception and practical implementation of left-handed metamaterials (LHMs) [38], whose potential applicability has lately gained significant prominence. In an LHM, the concurrent tuning of both effective constitutive parameters to negative values results in a negative refractive index and other noteworthy properties not accessible in nature, such as inverted Snell law, Doppler shift, or Cherenkov radiation. These features have been successfully engineered in the microwave regime [38–46] via thin metallic wires and split-ring resonators (SRRs). Nonetheless, since their electrical size is much smaller than a wavelength, any attempt to numerically explore LHM designs for efficient EMC components entails tools that should avoid prolonged simulations. Therefore, this section explores the possibility of modeling such media by means of the higher order ADI-FDTD method.

A typical LHM is, in principal, synthesized by artificially fabricated classes of inhomogeneities or small inclusions embedded in specific host media with the aim of achieving innovative and physically realizable response functions, not encountered in naturally occurring materials. The most profitable implementation of these composite structures comprises periodic arrays of rectangular or circular SRRs in juxtaposition with metallic rods or strip wires, as shown in Figure 8.22. The SRR grid, which in the millimeter range exhibits a negative effective

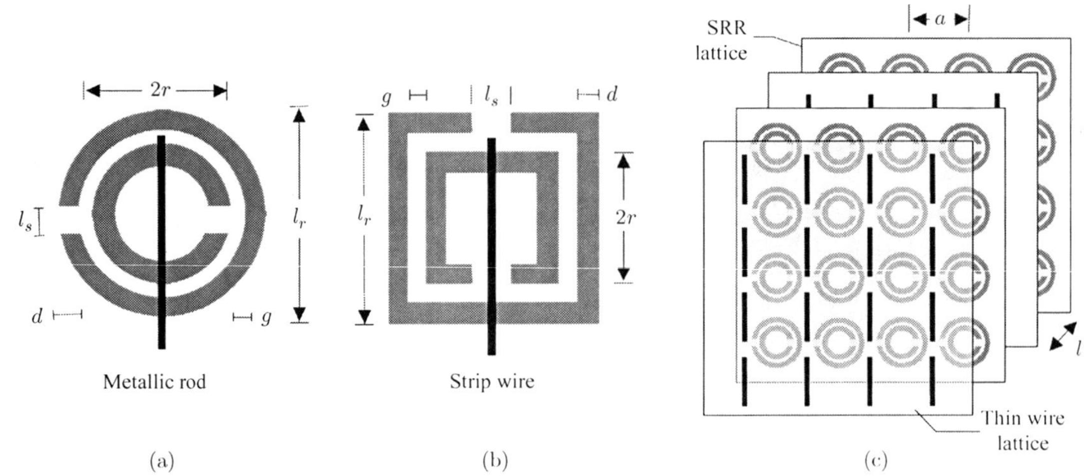

FIGURE 8.22: Layout of (a) a hollow circular SRR paired to a thin metallic rod and (b) a rectangular SRR paired to a strip wire. (c) A periodically arranged composite LHM

magnetic permeability, can be described by the Lorentz (LO) or Drude (DR) lossy model of

$$\mu_{\text{eff}}^{\text{LO}}(\omega) = \mu_0 \left(1 - \frac{F\omega^2}{\omega^2 - \omega_0^2 + j\Gamma_{\text{m}}\omega}\right) \quad \text{and} \quad \mu_{\text{eff}}^{\text{DR}}(\omega) = \mu_0 \left[1 - \frac{F\omega_{\text{pm}}^2}{\omega(\omega - j\Gamma_{\text{m}})}\right], \tag{8.4}$$

where

$$\omega_0^2 = \frac{3l\upsilon_0^2}{\pi r^3 \ln(2d/g)} \quad \text{and} \quad F = \frac{\pi r^2}{b^2},$$

with b the grid constant, r the inner ring radius, and d the width of the SRR. Moreover, ω_{pm} signifies its plasma and $\Gamma_{\text{m}} = 2\rho l/r$ its damping frequency with l the distance of adjacent planes, ρ the perimeter resistance, and υ_0 the light velocity. The double splits and ring gap, g, give an ample capacitance which guarantees that the element's overall resonant wavelength is always larger than the SRR diameter.

On the other hand, the second LHM component, i.e., the network of thin wires acts as a quasi-medium with a negative effective dielectric permittivity and can be, in the microwave spectrum, characterized by

$$\varepsilon_{\text{eff}}^{\text{LO}}(\omega) = \varepsilon_0 \left[1 - \frac{(\varepsilon_s - 1)\omega^2}{\omega^2 - \omega_0^2 + j\Gamma_{\text{e}}\omega}\right] \quad \text{and} \quad \varepsilon_{\text{eff}}^{\text{DR}}(\omega) = \varepsilon_0 \left[1 - \frac{(\varepsilon_s - 1)\omega_{\text{pe}}^2}{\omega(\omega - j\Gamma_{\text{e}})}\right], \tag{8.5}$$

with ω_{pe} and Γ_{e} the analogous plasma and damping frequencies, and ε_s the static dielectric constant. Although both models in (8.4) are theoretically equivalent, the former is proven to be more comprehensive for the detailed analysis of losses, while the latter supports a broader

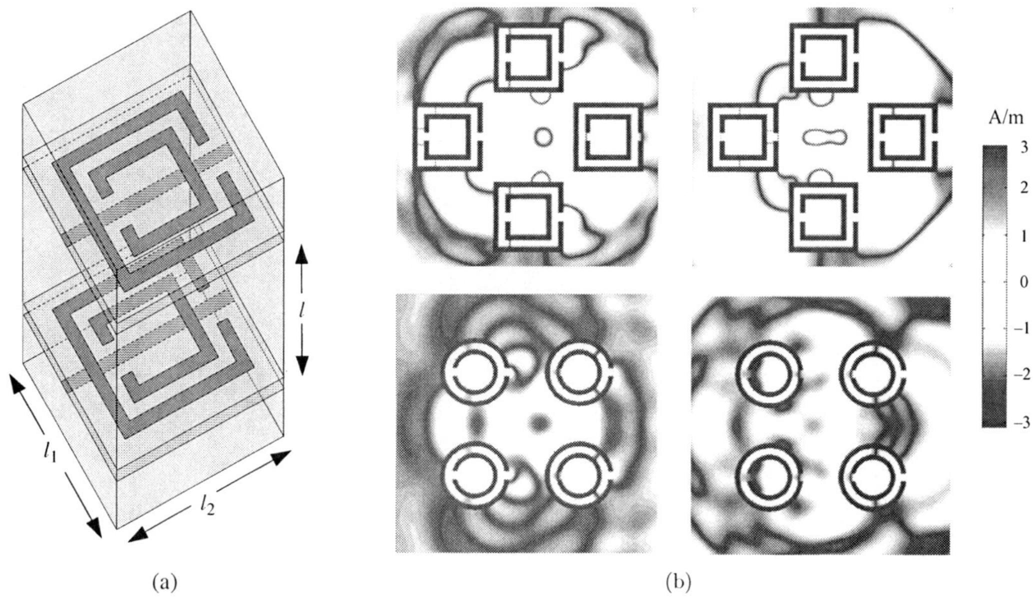

FIGURE 8.23: Part of the LHM substrate of the 5 × 5 planar antenna and (b) H_x variation for two different SRR substrates

bandwidth where constitutive parameters are simultaneously negative. This issue implies that solutions by means of the Drude counterpart will require a shorter time interval to reach steady state. The above dispersive representations furnish the appropriate resonance region for the realistic fabrication of an LHM. Accordingly, when a monochromatic plane wave propagates in such a medium, its Poynitng vector is antiparallel to the direction of its phase velocity.

Consider a 5 × 5 planar patch antenna fixed on the properly tuned LHM substrate of Figure 8.23(a), which is loaded either by rectangular ($l_r = 3.1$ mm, $r = 0.96$ mm, $l_s = 0.35$ mm, $g = 0.31$ mm, $d = 0.28$) or by circular ($l_r = 3.04$ mm, $r = 0.94$ mm, $l_s = 0.32$ mm, $g = 0.3$ mm, $d = 0.28$ mm). The structure is simulated through the angle-optimized schemes of Section 2.5.2 together with a modified version of the one-sided operators of Section 2.4.3. In this context, Figure 8.23(b) shows the variation of the H_x component along a transverse substrate cut, where a very sufficient level of symmetry can be readily seen.

Next, a 16-layer LHM, comprising circular SRRs with $l_r = 3$ mm, $r = 0.8$ mm, $g = 0.36$ mm, $d = 0.34$ mm, constitutes the substrate of a 7 × 7 antenna array. The domain is discretized in 48 × 56 × 32 cells with $\Delta t = 0.8267$ ns rather than the 162 × 174 × 138 cells and the $\Delta t = 0.041$ ns of the typical ADI-FDTD method. The frequency gap in the structure's transmitted power is shown in Figure 8.24(a) (for a CFLN = 22). As observed, a serious improvement is accomplished by the LHM substrate. The last application deals with the behavior

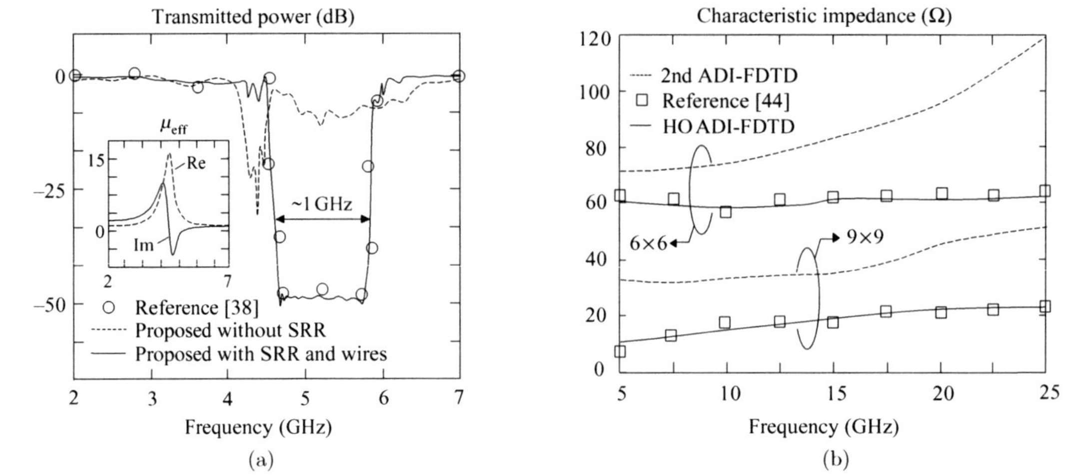

FIGURE 8.24: (a) Transmitted power of a 7×7 antenna array. (b) Characteristic impedance of a high-pass filter for a 6×6 and 9×9 configuration

of a high-pass antenna filter realized by circular SRRs. The higher order nonstandard algorithm of (3.70)–(3.81), for $M = 3$ and $L = 3$, involves a $62 \times 54 \times 30$ lattice that is virtually 90% coarser than the common one and uses a time-step 25 times beyond the Courant limit. Figure 8.24(b) gives the characteristic impedance of the filter for a 6×6 and 9×9 arrangement, indicating that the nonstandard formulation is superior to the second-order ADI-FDTD scheme from an accuracy and convergence point of view.

REFERENCES

[1] C. A. Balanis, *Antenna Theory: Analysis and Design*, 3rd ed., New York: IEEE Press and Wiley Interscience, 2005.

[2] P. A. Tirkas and C. A. Balanis, "Finite-difference time-domain method for antenna radiation," *IEEE Trans. Antennas Propag.*, vol. 40, no. 3, pp. 334–340, Mar. 1992.

[3] J. G. Maloney, K. L. Shlager, and G. S. Smith, "A simple FDTD model for transient excitation of antennas by transmission lines," *IEEE Trans. Antennas Propag.*, vol. 42, no. 2, pp. 289–292, Feb. 1994.

[4] T. Kashiwa, T. Onishi, and I. Fukai, "Analysis of microstrip antennas on a curved surface using the conformal grids FD-TD method," *IEEE Trans. Antennas Propag.*, vol. 42, no. 3, pp. 423–427, Mar. 1994.

[5] J. G. Maloney and G. S. Smith, "Modeling of antennas," in *Advances in Computational Electrodynamics: The Finite-Difference Time-Domain Method*, A. Taflove, Ed. Norwood, MA: Artech House, 1998, ch. 7, pp. 409–460.

[6] J. B. Schneider, "Plane waves in FDTD simulations and a nearly total-field/scattered-field boundary," *IEEE Trans. Antennas Propag.*, vol. 52, no. 12, pp. 3280–3287, Dec. 2004.

[7] A. Taflove and S. C. Hagness, *Computational Electrodynamics: The Finite-Difference Time-Domain Method*, 3rd ed., Norwood, MA: Artech House, 2005.

[8] J.-P. Berenger, "Numerical reflection of evanescent waves by PMLs: Origin and interpretation in the FDTD case. Expected consequences to other finite methods," *Int. J. Numer. Model.*, vol. 13, no. 2–3, pp. 103–114, Mar.–June 2000.

[9] W. Yu, N. Farahat, and R. Mittra, "Application of FDTD method to conformal patch antennas," *IEE Proc. Microw. Antennas Propag.*, vol. 148, no. 3, pp. 218–220, June 2001.

[10] S. V. Georgakopoulos, C. R. Birtcher, and C. A. Balanis, "Coupling modeling and reduction techniques of cavity-backed slot antennas: FDTD versus measurements," *IEEE Trans. Electromagn. Compat.*, vol. 43, no. 3, pp. 261–272, Aug. 2001.

[11] K. L. Wong, *Compact and Broadband Microstrip Antennas.* Piscataway, NJ: Wiley Interscience, 2002.

[12] K.-L. Wong and T.-W. Chiou, "Broad-band dual-polarized patch antennas fed by capacitatively coupled feed and slot-coupled feed," *IEEE Trans. Antennas Propagat.*, vol. 50, pp. 346–351, Mar. 2002.

[13] J. Gómez-Tagle, P. F. Wahid, M. T. Chryssomallis, and C. G. Christodoulou, "FDTD analysis of finite-sized phased array microstrip antennas," *IEEE Trans. Antennas Propag.*, vol. 51, no. 8, pp. 2057–2062, Aug. 2003.

[14] R. L. Li, V. F. Fusco, and H. Nakano, "Circularly polarized open-loop antenna," *IEEE Trans. Antennas Propag.*, vol. 51, no. 9, pp. 2475–2477, Sep. 2003.

[15] M. N. Vouvakis, C. A. Balanis, C. R. Birtcher, and. A. C. Polycarpou, "Multilayer effects on cavity-backed slot antennas," *IEEE Trans. Antennas Propag.*, vol. 52, no. 3, pp. 880–887, Mar. 2004.

[16] K. P. Prokopidis and T. D. Tsiboukis, "FDTD algorithm for microstrip antennas with lossy substrates using higher-order schemes," *Electromagn.*, vol. 24, no. 5, pp. 301–315, July 2004.

[17] N. Bushyager, J. Papapolymerou, and M. M. Tentzeris, "A composite cell-multiresolution time-domain technique for the design of antenna systems including electromagnetic band gap and via-array structures," *IEEE Trans. Antennas Propag.*, vol. 53, no. 8, pp. 2700–2710, Aug. 2005.

[18] N. V. Kantartzis and T. D. Tsiboukis, "A higher order nonstandard FDTD-PML method for the advanced modeling of complex EMC problems in generalized 3-D curvilinear coordinates," *IEEE Trans. Electromagn. Compat.*, vol. 46, no. 1, pp. 2–11, Feb. 2004.

[19] K. M. Luk and K. W. Leung, Eds., *Dielectric Resonator Antennas.* London, UK: Research Press Studies, 2003.

[20] I. Eshrah, A. A. Kishk, A. B. Yakovlev, and A. W. Glisson, "Theory and implementation of dielectric resonator antenna excited by a waveguide slot," *IEEE Trans. Antennas Propag.*, vol. 53, no. 1, pp. 483–494, Jan. 2005.

[21] M. Lapierre, Y. M. M. Antar, A. Ittipiboon, and A. Petosa, "Ultra wideband monopole/dielectric resonator antenna," *IEEE Microw. Wireless Compon. Lett.*, vol. 15, no. 1, pp. 7–9, Jan. 2005.

[22] K. W. Leung and H. K. Ng, "The slot-coupled hemispherical dielectric resonator antenna with a parasitic patch: Applications to the circularly polarized antenna and wide-band antenna," *IEEE Trans. Antennas Propag.*, vol. 53, no. 5, pp. 1762–1769, May 2005.

[23] Y. Kim and D. L. Jaggard, "The fractal random array," *Proc. IEEE*, vol. 74, no. 9, pp. 1278–1280, 1986.

[24] C. Puente-Baliarda and R. Pous, "Fractal design of multiband and low side-lobe arrays," *IEEE Trans. Antennas Propagat.*, vol. 44, no. 5, pp. 730–739, May 1996.

[25] D. L. Jaggard, "Fractal electrodynamics: From super antennas to superlattices," in *Fractals in Engineering*, J. L. Vehel, E. Lutton, and C. Tricot, Eds. New York: Springer-Verlag, 1997, ch. 3, pp. 204–221.

[26] C. Puente, J. Romeu, R. Pous, and A. Cardama, "On the behavior of the Sierpiński multiband fractal antenna," *IEEE Trans. Antennas Propag.*, vol. 46, no. 4, pp. 517–524, Apr. 1998.

[27] D. H. Werner, R. L. Haupt, P. L. Werner, "Fractal antenna engineering: The theory and design of fractal antenna arrays," *IEEE Antennas Propag. Mag.*, vol. 41, no. 5, pp. 37–59, Oct. 1999.

[28] D. H. Werner and R. Mittra, *Frontiers in Electromagnetics*, Piscataway, NJ: IEEE Press, 2000.

[29] J. P. Gianvittorio and Y. Rahmat-Samii, "Fractal antennas: A novel antenna miniaturization technique, and applications," *IEEE Antennas Propag. Mag.*, vol. 44, no. 1, pp. 20–36, Feb. 2002.

[30] D. H. Werner and S. Ganguly, "An overview of fractal antenna engineering research," *IEEE Antennas Propag. Mag.*, vol. 45, no. 1, pp. 38–57, Feb. 2003.

[31] N. V. Kantartzis, T. T. Zygiridis, and T. D. Tsiboukis, "A nonstandard higher-order FDTD algorithm for 3-D arbitrarily and fractal-shaped antenna structures on general curvilinear lattices," *IEEE Trans. Magn.*, vol. 38, no. 2, pp. 737–740, Mar. 2002.

[32] S. R. Best, "A multiband conical monopole antenna derived from a modified Sierpiński gasket," *IEEE Antennas Wireless Propag. Lett.*, vol. 2, no. 1, pp. 205–207, 2003.

[33] T. T. Zygiridis, N. V. Kantartzis, T. V. Yioultsis, and T. D. Tsiboukis, "Higher-order approaches of FDTD and TVFE methods for the accurate analysis of fractal antenna arrays," *IEEE Trans. Magn.*, vol. 39, no. 3, pp. 1230–1233, May 2003.

[34] A. Bossavit, *Computational Electromagnetism: Variational Formulations, Complementarity, Edge Elements*. Massachusetts: Academic Press, 1998.

[35] J. L. Volakis, A. Chatterjee, and L. C. Kempel, *Finite Element Method for Electromagnetics*. New York: IEEE Press, 1998.

[36] J.-M. Jin, *The Finite Element Method for Electromagnetics*, 2nd ed., New York: John Wiley & Sons, 2002.

[37] T. V. Yioultsis and T. D. Tsiboukis, "Convergence optimized, higher order vector finite elements for microwave simulations," *IEEE Microw. Wireless Compon. Lett.*, vol. 11, no. 10, pp. 419–421, Oct. 2001.

[38] J. Pendry, A. Holden, D. Robbins, and J. Stewart, "Magnetism from conductors and enhanced nonlinear phenomena," *IEEE Trans. Microw. Theory Tech.*, vol. 47, no. 11, pp. 2075–2084, Nov. 1999.

[39] T. Weiland, R. Schuhmann, R. Greegor, C. Parazzoli, A. Vetter, D. Smith, D. Vier, and S. Schultz, "Ab initio numerical simulation of left-handed metamaterials: Comparison of calculations and experiments," *J. Appl. Phys.*, vol. 90, pp. 5419–5424, 2001.

[40] R. Marqués, J. Martel, F. Mesa, and F. Medina, "Left-handed media simulation and transmission of EM waves in subwavelength split-ring-resonator-loaded metallic waveguides," *Phys. Rev. Lett.*, vol. 89, paper 183901, 2002.

[41] R. W. Ziolkowski and N. Engheta, Eds., *Special Issue on Metamaterials. IEEE Trans. Antennas Propag.*, vol. 51, no. 10, pp. 2546–2570, Oct. 2003.

[42] R. Ziolkowski, "Design, fabrication, and testing of double negative metamaterials," *IEEE Trans. Antennas Propagat.*, vol. 51, no. 7, pp. 1516–1529, July 2003.

[43] A. Alù and N. Engheta, "Guided modes in a waveguide filled with a pair of single-negative (SNG), double-negative (DNG), and/or double-positive (DPS) layers," *IEEE Trans. Microw. Theory Tech.*, vol. 52, no. 1, pp. 192–210, Jan. 2004.

[44] A. Lai, C. Caloz, and T. Itoh, "Composite R/LH transmission line metamaterials," *IEEE Microw. Mag.*, vol. 5, no. 3, pp. 34–50, Sep. 2004.

[45] T. Itoh and A. A. Oliner, Eds, *Special Issue on Metamaterial Structures, Phenomena, and Applications. IEEE Trans. Microw. Theory Tech.*, vol. 53, no. 4, pp. 1418–1556, Apr. 2005.

[46] T. Kokkinos, C. Sarris, and G. Eleftheriades, "Periodic finite-difference time-domain analysis of loaded transmission-line negative-refractive-index metamaterials," *IEEE Trans. Microw. Theory Tech.*, vol. 53, no. 4, pp. 1488–1495, Apr. 2005.

Author Biographies

Nikolaos V. Kantartzis received the Diploma degree and Ph.D. degree in electrical and computer engineering from the Aristotle University of Thessaloniki (AUTH), Thessaloniki, Greece, in 1994 and 1999, respectively. In 1999, he joined the Applied and Computational Electromagnetic Laboratory, Department of Electrical and Computer Engineering, AUTH, as a Postdoctoral Research Fellow. He has authored or coauthored several refereed journal papers in the area of computational electromagnetics and especially higher order finite-difference time-domain (FDTD) method, perfectly matched layers (PMLs), nonorthogonal discretization algorithms, vector finite elements and alternating-direction implicit (ADI) formulations. His main research interests include time- and frequency-domain electromagnetic and acoustic analysis, EMC modeling, modern microwave circuits, antenna structures and metamaterials.

Theodoros D. Tsiboukis received the Diploma Degree in electrical and mechanical engineering from the National Technical University of Athens, Athens, Greece, in 1971, and the Doctor Engineer Degree from the Aristotle University of Thessaloniki (AUTH), Thessaloniki, Greece, in 1981.

From 1981 to 1982, he was with the Electrical Engineering Department, University of Southampton, U.K., as a Senior Research Fellow. Since 1982, he has been with the Department of Electrical and Computer Engineering (DECE), AUTH, where he is currently a Professor. He has served in many administrative positions, including Director of the Division of Telecommunications at the DECE (1993–1997) and Chairman of the DECE (1997–2001). He is also the Head of the Advanced and Computational Electromagnetics Laboratory at the DECE. His main research interests include electromagnetic field analysis by energy methods, computational electromagnetics (FEM, BEM, vector finite elements, MoM, FDTD method, absorbing boundary conditions), inverse problems, EMC applications and metamaterials. He has authored or coauthored six books, over 120 refereed journal papers and over 100 international conference papers. He was the Guest Editor of a Special Issue of the *International Journal of Theoretical Electrotechnics* (1996).

Dr. Tsiboukis is a member of various societies, associations, chambers and institutions. He was the chairman of the local organizing committee of the 8th International Symposium on Theoretical Electrical Engineering (1995). He has been the recipient of several awards and distinctions.

Breinigsville, PA USA
10 December 2010

250892BV00006BA/1/P